John Quincy Adams

John Quincy Adams

American Visionary

Fred Kaplan

An Imprint of HarperCollinsPublishers

FIRST HARPERLUXE EDITION

HarperLuxe™ is a trademark of HarperCollins Publishers

Library of Congress Cataloging-in-Publication Data is available upon request.

ISBN: 978-0-06-229876-8

14 ID/RRD 10 9 8 7 6 5 4 3 2 1

This book is dedicated
to two much-loved friends and mentors,
Norman Fruman (1924–2012)
and
Carl Woodring (1919–2009).

Contents

Introduction

John Quincy Adams is a president about whom most Americans know very little. In comparative rankings he inhabits the netherworld of the list, not as quickly dismissed as James Buchanan and Franklin Pierce but not even in the ranks of Grant, Madison, Monroe, and Polk, and nowhere near the upper tier of Truman and McKinley, let alone the pantheon of Washington, Jefferson, the two Roosevelts, and Lincoln.

Was Adams indeed a failed one-term president hardly worth deliberating about today? This book tells a far different story. The totality of his life, including his presidency, is a significant tale about accomplishment and failures. It is a story not only about our sixth president's America but about America now. His values, his definition of leadership, and his vision for the nation's future—particularly the difficulty of transforming vision into reality in a country that often appears ungovernable—are as much about twenty-first-century America as about Adams' life and times.

Like Lincoln, Adams, with sharp intellect and scholarly brilliance, was a gifted writer in personal and public prose. He was also a capable poet. Every word that he spoke or published on matters of public policy he wrote himself. "Literature has been the charm of my life," he wrote in 1820, "and could I have carved out my own fortunes, to literature would my whole life have been devoted. I have been a lawyer for bread, and a Statesman at the call of my Country." His poems are often graceful, witty, immediate, and engaging in the eighteenth-century sense; some are lightly occasional, others autobiographical. Gifted with a caustic strain, he also wrote satiric poetry with a bite, and his religious and patriotic poetry engaged his deepest feelings. As he studied languages, he translated Greek, Latin, German, and French writers. He wrote essays on literary as well as political topics, particularly Shakespeare's plays. His book-length history of weights and measurements, on the face of it a dry subject, is an example of a government report turned almost into a masterpiece by its philosophic overview and cogent style. His 1833 "Report of the Minority of the Committee on Manufactures" and his argument in 1841 before the Supreme Court in the *Amistad* case are brilliantly written public documents on public policy and human rights.

From his earliest years, Adams detested slavery: he was one of only two anti-slavery presidents between the founding of the country and the election of Lincoln. Like Lincoln thirty years later, Adams despised the populism, states' rights, and pro-slavery policies of his successor, Andrew Jackson. The Whig Party, which came into existence in 1834 and lasted until 1856, and with which Adams became affiliated, was Lincoln's until 1856. The policies that it advocated and those of the new Republican Party of 1860 were those of Adams' presidency: strong central government, federally supported infrastructure policies, and effective but not expansive executive power. Like Lincoln, Adams believed that the Constitution was a living document; the Western territories belonged to the entire American people; taxing and being taxed were essential to responsible self-government; the country required a modern, national, and regulated banking system; slavery was an abomination that needed to be contained and ultimately eliminated; Native Americans deserved to be respected and brought into the American community; and the federal government had an important role to play regarding the "general welfare" in the creation of educational, scientific, and artistic institutions, such as the Smithsonian Museum, the national parks, the service academies, and land grant universities.

The weakness of Adams' presidency resulted from the determination of his political opponents not to allow him any public policy achievements. With divided government, they had the power to do that. The pro-slavery and land-hungry South and West had good reason to believe that Adams had strong abolitionist sympathies and that he favored a higher price for government land in order to fund federal projects. In 1828, Jackson defeated Adams' bid for re-election. He brought to the office a rush of executive authoritarianism, intemperate demagoguery, the spoils system under the guise of "reform," harsh racism, strong support for slavery, and anti-intellectual ignorance. Without the constitutional provision that for every five slaves, three units be added to the total number of white people counted for representation in a congressional district, the popular and electoral college results of the 1824 and 1828 elections would have had a different appearance: Adams would have been elected in 1824 by popular vote rather than by the House of Representatives, and the 1828 election would have been closer and far from the landslide that it has been misrepresented as having been.

Adams became the only former president to serve in Congress, where he continued for sixteen years, emerging to public notoriety as the most outspoken opponent of slavery and of the Southern stranglehold on the American government in high public office. All his life he feared the dissolution of the Union. Slavery would destroy it, he believed, unless some future president would use the war powers inherent in the Constitution to end slavery. There would be rivers of blood. He both dreaded and hoped for the Civil War.

From his adolescence until his last days, Adams kept a diary: his most varied, brilliantly observed writing achievement, a record of political activities and personalities; theatrical performances and religious reflections; his reading, thoughts, and meditations. That achievement, like all his writing on public subjects, was enlightened by a vision of an America that would move confidently into its destiny, a national union in which individual liberty harmoniously co-existed with federal programs that promoted economic, social, scientific, and cultural progress. It was also deeply personal, a living organism, containing his account of the objective world and his inner life. Much of what went on in his imagination and his heart embodied the dynamic interaction between his reading and what he wrote in his diary. The unpublished original reveals a man of deep feeling. In conjunction with his private letters, it provides us with a moving self-portrait of a loving husband married to an estimable woman; of an engaged father who suffered the misfortune of the deaths of three of his four children; of a son who carried the blessing and burden of his parents' belief in his high destiny; and of a man engaged throughout his life in searching for the right balance between emotion, religious belief, and rational analysis.

In February 1830, Adams wrote a letter that includes a piece of powerful writing that resonates with his own life, with the intensity of his diary, and with the ironies and tragedies of American history.

Every one of the letters of Cicero is a picture of the state of the writer's mind when it was written. It is like an invocation of shades to read them. I see him approach me like the image of a phantasmagoria. He seems opening his lips to speak to me and passes off, but his words as if they had fallen upon my ears are left deeply stamped upon the memory. I watch with his sleepless nights. I share his solitary sighs. I feel the agitation of his pulse, not for himself but for his son . . . for his country. There is sometimes so much in it of painful reality that I close the book. No tragedy was ever half so pathetic. My morning always ends with a hearty execration of Caesar, and with what is perhaps not so right, a sensation of relief at the 23 stabs of the Ides of March. . . . Everything else in the story is afflicting and gloomy.

That is not the case with Adams' story. There is a great deal of pain and gloom. But there is also success and brightness.

Chapter 1
Hooks of Steel
1767–1778

At 5 A.M. on an intensely hot summer day, President John Quincy Adams left the White House by stagecoach for Quincy, Massachusetts. It was July 9, 1826, just two days short of his fifty-ninth birthday. He had been up the night before "in anxiety and apprehension, until near midnight." The heat made him miserable. Candlelight attracted insects. There were no screens. The day before, he had gotten three letters from Quincy with the news that his ninety-one-year-old father, the second president of the United States, was on his deathbed. John Quincy had "flattered" himself that his father "would survive this summer, and even other years." A rider was on his way from Baltimore to tell him he was wrong.

John Adams had died late in the afternoon of July 4. All of Washington already knew that Thomas Jefferson had died a little after 1 P.M. on the same day. Jefferson's final moments came just a few hours after a plea for a cash subscription for the impoverished former president at the capital's Independence Day celebration had met with a meager response. The third president had been an intimate member of the Adams family household in Paris when John Quincy was a young man. In 1800, Jefferson, barely and only by the advantage of counting slaves as voters, had defeated Adams' father's bid for a second term. It was a nasty campaign, mostly on the Jefferson side. In 1808, as a senator from Massachusetts, John Quincy, in an ironic twist, had earned the hatred of many New England Federalists, who later forced him out of his Senate seat, because he had supported some of Jefferson's policies. Adams had thought them in the best interests of the country. Even then he had judged Jefferson's presidency a failure, his personality flawed by self-deception and political hypocrisy. Whether or not Adams would get to Quincy soon enough to see his father one last time was the main thing on his mind. Only some days later did he remark on the obvious coincidence that the second and third presidents of the United States had died on the same day, the fiftieth anniversary of the Declaration of

Independence. It seemed to many a miraculous fact. It has become a cliché of American history.

Like his father, John Quincy agreed with Jefferson on at least one fundamental principle: government derived its legitimacy from the consent of the governed. That meant the rule of the ballot box. The differences were a question of the degree to which government should be structured to reflect the dominance of the popular voice. Neither Jefferson nor John Adams favored democracy in the modern sense, and John Quincy had his doubts. What he had no disagreement about with either former president was that political parties were instruments of bad governance; they were manifestations of individual or group self-interest that would undermine republican government. Good government depended on civic virtue rather than personal greed. Parties were organized lobbies in pursuit of power and money. Every president before John Quincy Adams had believed that presidents should be above party. But an immense, and to Adams appalling, change was well under way. His own presidency was under assault by interests that believed all that mattered was power, that the object of politics was to smash the opposition, and that political parties were the necessary vehicles for obtaining power. Power and financial self-interest were inseparable. In this extreme form, the concept was something new in American politics. And it seemed to Adams both awesome and dangerous.

Intent on reaching Quincy as quickly as possible, he sweated profusely and slept badly as he traveled in ceaseless motion. A little less than medium height, paunchy by genetic inheritance and from occasional overeating, Adams had tight lips and a narrow lower face. His dark dress set off a pale complexion, and the vertical intensity of his lower face moved upward and outward, accompanied by gray sideburns, to a mostly bald head. He traveled in public vehicles, without security or advance arrangements, as was normal practice for presidents. He was accompanied by George Washington Adams, his eldest son and personal secretary. He traveled by water and land, in bone-jarring stagecoaches and dirty steamboats that routed him via Baltimore, Philadelphia, and New Brunswick, New Jersey. There the Raritan River allowed egress to the New Jersey waterways. The ferry to New York ran aground twice. At New York City, he boarded a steamboat to Newport and Providence, Rhode Island. This took almost five days of demanding travel. At best the roads were bad and the ride was tedious; in bad weather the crowded steamboats were inoperative.

But the trip was lightning fast compared with the first time he had traveled even half that route. When his father in 1775 had traveled from Boston to Philadelphia to represent Massachusetts in the Continental Congress, it had taken fifteen days. Twenty-seven-year-old John Quincy had made that same journey in ten days in 1794, when Philadelphia was the temporary capital of the country. The city of Washington did not yet exist. In Adams' later years, the trip from Washington to Boston would take only two days. The railroad had transformed travel. Space and distances had contracted, from horseback, stagecoach, sailing vessel, and canal boat, to the additional speed allowed by improved roads, to the steamboat, and, finally, to the space- and time-shattering railroad.

On July 11, 1826, the day Adams arrived in New York, he noted that it was his fifty-ninth birthday, a fact to record in his diary, not to celebrate. The loss of his father dominated his thoughts. He was returning to the place in which he had been born, now absent the man who had been the most important influence on his life. He had loved his father at least as much as he had loved any other human being. Now that the patriarch of the family was dead, the eldest son would have to assume that role—there was no one else capable. And his mood could not help but be touched by his identification with

his father's fate, the only one-term president of the first
five of the country. John Quincy might possibly be the
second. That, he anticipated, would be a bitter irony.
How would he spend the remainder of his life? And
where would he spend it? On the steamboat *Connecticut*,
heading rapidly toward Providence, he brooded on "the
peculiar circumstances" of his birthday and the question of how long he himself would live. A realist and
a fatalist, he had always tried to accept with resignation the unpredictability of such matters. He could not
know that he would not rest beside his father for another
twenty-two years of intense service to his country.

But whether or not he would have a second term as
president, he did have a strong conviction about where
he would spend much or all of his remaining years. John
Adams' property in Quincy consisted of two modest
eighteenth-century houses on the Plymouth Road,
within fifty feet of one another, in the New England
saltbox style. In one of them John Quincy had been born
and spent the first eleven years of his life. In the other
his father had been born. The property extended for
over a hundred acres toward the Blue Hills. And there
was also a larger house and property a few miles away,
named Peacefield by John Adams, that his parents had
purchased in 1788 when Adams had returned from his
ministerial duties in Europe. It served as his residence

until his death. "It is repugnant to my feelings to abandon this place," John Quincy wrote, "where for nearly forty years [my father] has resided, and where I have passed many of the happiest days of my life. . . . Where else should I go?"

Four days after his birthday, he read his father's will. It was clear that the property could come into his possession only if he saddled himself with heavy financial obligations. There were other heirs who would need to be satisfied. He immediately decided that, at whatever cost, he would preserve for himself and his descendants the two original houses, with the property that his ancestors had farmed, and the attractive larger house with its orchards and gardens. "Though it will bring me heavily in debt," John Quincy wrote to his wife, Louisa, "I cannot endure the thought of the sale of this place. Should I live through my term of service, my purpose is to come and close my days here, to be deposited with my father and mother." To him, the land and the vicinity were a sacred ground of family history and personal memories. "Everything about the house is the same. I was not fully sensible of the change till I entered his bed-chamber. . . . That moment was inexpressibly painful, and struck me as if it had been an arrow to the heart. My father and mother have departed. The charm which has always made this house to me an abode of enchantment is dissolved; and yet my attachment to it, and to the whole region round, is stronger than I ever felt it before."

The attachment was to an idea as well as a place, an idea embodied in John Adams' values and life. No matter how materially his body was entombed, he remained alive in his son's life in a living bequest. Just as John Quincy now inherited the Quincy property, he had also inherited from his father something he valued more than wealth. It was an unbreakable and energizing bond, the living tissue of the ideals that his father, his family, and his extended New England world had pledged themselves to in 1776, and the perpetuation of which the son held to be the only possible definition of himself and his country. It was a pledge to union, a union that, despite the repugnant compromises required to create the Constitution, was based on the values expressed in the Declaration of Independence John Adams had helped to create. He had been designated by the Congress as a member of the committee of three appointed to write the document. In the days preceding its creation, he had been one of the leading, most determined, and most passionate voices arguing for total independence immediately. And it was Adams who thought Jefferson best suited, as a Virginian and a stylist, to compose the initial draft. And though

Jefferson's claim that "all men are created equal" was sincere, any honest application by Jefferson and his Southern associates had, at least in their minds, to involve a significant exclusion. For Adams, there was none. In 1836, on the tenth anniversary of his father's death, when the issue of slavery had begun to dominate his life, John Quincy affirmed the sacred connection in a commemorative poem, "Day of My Father's Birth." From his father's words and life, his own contemporaries could relearn

> that freedom is the prize
> Man still is bound to rescue or maintain;
> That nature's God commands the slave to rise,
> And on the oppressor's head to break his chain.
> Roll, years of promise, rapidly roll round,
> Till not a slave shall on this earth be found!

Arriving a few days too late to see his father alive, John Quincy attended a funeral service at the Quincy Congregational Church. "I have at no time felt more deeply affected by [the death of my father]," he wrote in his diary, "than on entering the meeting-house and taking . . . the seat which he used to occupy, having directly before me the pew at the left of the pulpit, which was his father's, and where the earliest devotions of my childhood were performed. The memory of my father and mother, of their tender and affectionate care, of the times of peril in which we then lived, and of the hopes and fears which left their impressions upon my mind, came over me, till involuntary tears started from my eyes."

In the afternoon he heard a second sermon from a visiting preacher. It was on a text from Revelation: "Be thou faithful unto death, and I will give thee a crown of life." Four days later, he walked to the graveyard and visited the four monuments "to the memory of his ancestors." Five generations now rested there. Afterward he spent hours reading documents, including the accounts of his father's personal estate, his will, and the wills of previous generations going back to 1694. His father had arranged and left them for him. "These papers awaken again an ardent curiosity to know more of these forefathers, who lived and died in obscure and humble life, but every one of whom, from the first settlement of the country, raised numerous families of children, and had something to leave by will. There could indeed be nothing found of them but the 'short and simple annals of the poor.'"

In the twentieth and twenty-first centuries, the Adams family name glitters with semi-mythical power. But it had no such glow in the latter part of the eighteenth century. Five generations of the family had lived in colonial Massachusetts by the time of John Quincy's birth in 1767. No one in the larger world of Massachusetts knew of them. John Quincy's father was an ex-schoolmaster, a hardworking, ambitious, up-and-coming lawyer with strong views about the tension between Great Britain and its American colonies. His ascent to fame and power arose out of revolutionary turmoil, not long-settled inheritance. Until then, the Adams family had no distinction—social, financial, or political. Their local lives defined them. Most of John Adams' American forebears had earned subsistence incomes. They had little to no education, with the exception of a Harvard graduate in 1710 who had a congregation in New Hampshire. When John Adams, also an eldest son, graduated from Harvard in 1755, the college ranked each graduate by social standing. Adams ranked "fourteenth in a class of twenty-four," probably owing to the standing of his mother's family. Twenty-five years later, in 1787, John Quincy Adams was second on the roster of his Harvard class. He was no smarter and had not performed more brilliantly than his father. The list was alphabetical. Times and values had changed.

Like all of his American ancestors, John Quincy, named for a relative whose last name was Quincy and who had died on the day of John Quincy's birth, had been born in rural Braintree, the north portion of which was renamed Quincy in 1792. Along the coast, ten miles south of Boston, it was about three hours away by horseback or coach. By the time John Quincy died, the transportation revolution had changed this country village of small farmers and artisans, whose granite quarries were its main natural resource, into a suburb, now indistinguishable from a much-enlarged Boston. In 1767, though, Braintree was still its own world. Partly arable but not rich, it had a modest market for domestic consumption. Bringing products, whether perishable or not, to distant markets, even to Boston, was difficult. Multiple occupations were commonplace. Wages were low; production was limited; the economy was mostly local. The family that did not harvest vegetables and fruit in the summer was usually hungry in the winter. John Quincy's grandfather was a farmer and shoemaker, a deacon in his church, a member of the fourth generation of Adamses to live in Braintree. His paternal great-grandfather, Joseph, was the youngest of eight sons of the first American Adams. Henry Adams had emigrated from Braintree, in Essex County, England, in about the year 1630, and died in Braintree, Massachusetts, in 1646, "the stock from whom the whole family [is] descended," John Quincy wrote in 1823. Over the years, Adamses served regularly in the governance of Braintree, the New England ritual of town meetings and local elections that made such towns independent of distant authorities. They were all good subjects of the empire, existing happily on the outskirts of its power. For generations the Adams family had nothing to do with Boston and Massachusetts politics. In religion, they were Congregationalists, members of the Protestant denomination that dominated New England, its origins in the first Puritan settlement. But John Adams rejected Calvinism and became a Unitarian. Like his father, John Quincy came to believe that "the Calvinistic doctrines of election, reprobation, and the atonement are so repulsive to human reason that they can never obtain the assent of the mind, but through the medium of the passions; and the master passion of orthodoxy is fear."

John Quincy had no illusions about his paternal family's social standing. What he admired about these ancestors was the decency of their ordinary lives. With the exception of his own father, "the highest worldly distinction which any of them attained was that of Selectman of the Town of Braintree. . . . They were distinguished only as men of industry sobriety . . . at the close of long lives of humble labor, returning their bodies to the soil . . . and their meek and quiet spirits to the god who gave them." He learned nothing that he could not respect about these ancestors. "The absence of information stimulates curiosity which can never be gratified," he lamented in 1823. He never tired of quoting "Elegy in a Country Churchyard." But Thomas Gray's unrecorded "annals of the poor" were mostly on the paternal side. The two Adams men preceding John Quincy made advantageous marriages. They raised the family's social status and provided the network of family connections so important in a kinship society. Early New Englanders took extended family very seriously. The Joseph Adams who was John Quincy's great-grandfather married Hannah Bass, apparently a granddaughter of John and Priscilla Alden, famous names in the history of the Massachusetts Bay colony. The John who was his grandfather married Susanna Boylston, the granddaughter of Thomas Boylston—a well-known surgeon and early émigré from London whose surgeon brother introduced inoculation into North America—and the daughter of a successful Brookline merchant. The Boylston name had presence in Massachusetts.

In 1735, Susanna Boylston became the mother of the future second president. His parents destined him for Harvard and the clergy. That small provincial college was the only game in town, the portal through which passed the young men chosen, mostly by birth and wealth, to be the leading citizens of Massachusetts. It graduated clergymen and a small number of lawyers, a total of about thirty students a year. John Adams' profession turned out to be the law, not the church. When young John Adams observed bitter bickering about Calvinistic doctrine among local Congregational ministers, he found it distasteful and intellectually unacceptable. "In the ordinary intercourse of society, as it existed at that time in New England," John Quincy later wrote, "the effect of a college education was to introduce a youth of the condition of John Adams into a different class of familiar acquaintance from that of his fathers. . . . Another effect of a college education was to disqualify the receiver of it for those occupations and habits of life from which his fathers had derived their support." John Adams' first rise in the world was to the position of schoolmaster in Worcester, which allowed him to be independent of the support of his father, who had done all he could with his modest resources. "Until then his paternal mansion, the house of a laboring farmer in a village of New England, and the walls of Harvard College, had formed the boundaries of his intercourse with the world," John Quincy wrote. John Adams lasted two years in what seemed unpromising drudgery in Worcester. Returning to Braintree and

Boston, he became a law clerk and then, properly credentialed, opened his own practice.

When, in 1764, twenty-nine-year-old John Adams, a man of self-confessed "amorous disposition," married twenty-year-old Abigail Smith, he continued the recent Adams pattern of marrying up. The mutual attraction was strong, the interest in one another intense. But it is doubtful that Adams would have been a successful suitor if he had not been a Harvard graduate. With a sometimes explosive temper and unrestrained frankness, he was a formidable personality. He was also a voracious reader of literature, law, and political philosophy; a capable, hardworking, passionate lawyer with an expanding practice; a witty and entertaining companion; a stubborn, sometimes bold pursuer of what he loved and believed in; and a principled public voice with the possibility of a government position and a political career. A gifted writer, he wrote as readily as he talked. He had begun to publish essays on contemporary subjects in a Boston newspaper, particularly on the marvels of agriculture and the wisdom of ordinary people—he probably had in mind his father and his Adams ancestors.

Both compliant and demanding, Abigail Smith was a presence herself, the second daughter of four children of the Reverend William Smith, a 1725 graduate of Harvard who led the Congregational church at Weymouth, fourteen miles southeast of Boston and a short distance from Braintree. Her mother was Elizabeth Norton Quincy. The Smith and Quincy families had been in Weymouth and Braintree for generations. The extended Smith family was associated with Charlestown on the north side of Boston. In fact, John Adams and Abigail Smith were distant cousins.

Although the Smith family's public presence was modest, its professional and social distinction was noticeably greater than that of the Adams family. Elizabeth Quincy was the daughter of Elizabeth Norton and John Quincy. The Quincy family had arrived in 1633, when Edmund Quincy of Northampton and Lincoln, England, had crossed the Atlantic with two famous clergymen, Joseph Cotton and Thomas Hooker. Active in Boston politics and governance, Edmund had been granted a four-hundred-acre farm, renamed Mount Wollaston, near the Fore River separating Braintree from Weymouth. When Abigail Adams' maternal grandfather, a militia colonel and member of the powerful Governor's Council, died on the same day in 1767 on which his great-grandson was born, the infant was honored with "Quincy" for his middle name. This maternal genealogy also had its romantic side. The Quincy family, and apparently the Nortons too, traced their lineage to the French conquest of Britain under William the Conqueror. The family claimed descent from a Norman baron, the first Earl of Winchester and a signer of the Magna Carta. Abigail's paternal grandfather had an ancient-looking parchment that traced the family back to 1066. And in 1787, the wife of the American minister to Britain, the fiercely antimonarchical Abigail Adams, traveled to Winchester to investigate the claim. With only a slight touch of embarrassment, she wrote to her sister that she found it convincing.

A woman of strong character, Abigail set high moral, religious, and patriotic standards for herself and her children. Although she believed in the values of the patriarchy into which she had been born, she had an independent mind, which she rarely hesitated to express, and became an advocate in her personal circle for a limited version of women's rights. Her values were those of the liberal branch of New England Congregationalism. Like her husband, she had a love of literature and a gift for writing. From an early age, Abigail devoured books, predominantly the canonical writings of the Renaissance and of the eighteenth century, poetry, plays, novels, sermons, and, over the years, pamphlets and newspapers that made her knowledgeable about many things, including politics. Her two sisters, Mary, the eldest, and Elizabeth, six years younger than Abigail, were smart, capable, and pious; the three sisters were close and deeply loyal. Mary and Elizabeth became surrogate parents to two of John Quincy Adams' children. The straight-and-narrow Smith sisters also had one brother, William, an alcoholic who deserted his family and died at the age of forty. His ghost hovers in the background of the moral lessons taught to John Quincy and his siblings. Since John Adams' career of public service kept Abigail apart from her husband and her eldest son for many years, her immersion in literature and her talent as a writer became the texture of a large number of brilliantly written, informative, perceptive, argumentative, moralizing, cajoling, sometimes sensual, and often riveting letters. And when events brought a gathering storm of resentment of British policies toward the colonies to revolutionary explosiveness, she shared her husband's passion for the independence they both became convinced was America's destiny.

The seventh year of John Quincy's childhood came to a close with blasts of cannon from Bunker Hill. He heard them with trepidation on the morning of Saturday, June 17, 1775. From the top of Penn's Hill, close to the Adamses' Braintree farm, he could see flashes from the battlefield and smoke rising from burning Charlestown. The family of four watched with quickly beating hearts. By Sunday, at 3 P.M., the battle had been in progress for twelve hours. His sister, Abigail, was nine years old, and his brothers Charles and Thomas were five and three. They had been told what was at stake, including their possessions and possibly their lives. "Almighty God," Abigail wrote to her husband six hundred miles away in Philadelphia, "be a shield to our Dear Friends. How many have fallen we know not—the constant roar of the cannon is so distressing that we cannot Eat, Drink or Sleep. May we be supported and sustained in the dreadful conflict."

For ten years, tension had been increasing between the colonists and the government, starting with the Stamp Act in 1765 and intensifying in a clash of self-interest disguised as principle to a guerrilla action at Lexington and Concord a decade later. Now there was open war. John Adams, as a lawyer, writer, and representative to the Continental Congress, had become an outspoken (many thought intemperate) advocate for resistance to British infringement on local rule and the freedom from Parliamentary taxation many in the Massachusetts Bay colony believed was their right. They were not used to paying anything but local taxes or being ruled from London in overt ways. That situation had changed. The British government, in a time of revenue shortfalls and overextension, needed and believed it had a right to revenue from its American colonies whose security it had sponsored with blood and money during the Seven Years' War. The colonies benefited from the protection of British arms. They prospered from the fine-tuned machinery of British mercantilism.

Since John Adams also kept a law office in Boston, the family toggled between Boston and Braintree. In 1761, on the death of his father, he had inherited almost one third of his father's Braintree property, consisting of the house in which he had been born, a barn, and forty acres. He was later to purchase the remainder of the estate and additional adjoining properties. Two years before John Quincy's birth, in that same Braintree house, Adams wrote, on behalf of his townsmen, their denunciation of the Stamp Act. Its repeal brought temporary comity and high hopes for the better treatment to which the colonists believed they were entitled. Irrepressibly active as a lawyer, writer, and controversialist, Adams was soon administering, mostly from Boston, the workers on his Braintree farm; traveling the eastern judicial circuit as far as Portland, Maine; writing letters and essays on the rights of the colonists for Boston newspapers; and representing a growing number of clients.

To assert its authority, the crown refused to let the local courts operate. Adams represented Boston in its effort to persuade the British authorities to reopen them. In 1768, he defended John Hancock, by inheritance one of the richest merchants in Boston, in a long trial in admiralty court. In 1769, Adams successfully defended four British sailors who had killed a British naval officer. The next year he also won the freedom of the British soldiers indicted after the "Boston Massacre." The British government, impatient with and affronted by what it considered illegal resistance to lawful acts of Parliament, garrisoned Boston. It became an occupied town. Grievances multiplied. Agitation against occupation and its frictions increased. The Boston population divided into radicals, who began to advocate separation from Great Britain; loyalists, who urged obedience to the crown; and various degrees of separation from either extreme, including many who had little in the way of political views but who prayerfully desired that the whole mess would go away. John and Abigail Adams had begun to favor independence.

Most of John Quincy's earliest memories were of Boston and Braintree. At the age of two, he wandered out onto a busy Boston street in front of the family's rented Brattle Square house. A servant pulled him back, severely wrenching the boy's right hand and dislocating his shoulder, an incident that John Quincy later blamed for lifelong recurring problems with his right arm and hand. A few months later, in February 1770, his fourteen-month-old sister Susanna died. After a number of moves, John Adams purchased a house on Queen Street, which he intended to be his permanent residence in Boston. When John Quincy fractured a finger, Dr. Joseph Warren, an exemplary physician and close family friend, saved it from amputation. After the Bunker Hill battle, when the family got the devastating news that Warren had died on the battlefield, the pain burned itself into John Quincy's memory.

In Braintree, he learned to ride, hunt, and shoot. Warm weather days were refreshed by swims in nearby Black's Creek, and riding and swimming became lifelong recreations. He learned to read, under his mother's tutelage, memorizing the lessons and moral stories in an introductory alphabet book and reader, *The History of Giles Gingerbread, A Little Boy, Who Lived upon Learning*. At six years of age, he wrote to his Aunt Elizabeth, "i have made But veray little proviciancy in reading . . . to much of my time in play [th]ere is a great Deal of room for me to grow better." His grand-parents' home in nearby Weymouth was welcoming, and he later recalled the delight with which he "used to walk over to Grandfather's at Weymouth." He remembered an "old walnut desk" that impressed him, and his grandfather's substantial library in the parsonage. Years later, passing the graveyard where his maternal grandmother was buried, he acknowledged that "she had been a second mother to me." She died in a dysentery epidemic in 1775. "My memory never recurs to her but with a feeling of melancholy affection. . . . She was a guardian angel of my childhood." At the Mount Wollaston mansion the Quincy family provided warm hospitality. John Quincy's paternal grandmother, who had remarried in 1766 and moved into one of the houses on the Braintree farm, added another beneficent grand-motherly presence to the rural scene. Braintree was a restful and calm retreat from Boston's tensions. "Here is Solitude and Retirement," John Adams wrote in his diary in 1769, "still, calm, and serene, cool, tranquil, and peaceful."

The war changed that. Braintree was still a retreat but not a safe one. Abigail, whose revolutionary fervor was as strong as her husband's, prayed with her children for the success of their soldiers and statesmen. The Plymouth Road, which passed only a few feet from the front of the Braintree houses, was crowded with militiamen going toward and away from Boston. John Adams was in Philadelphia. Abigail melted pewter to make bullets and attended to soldiers in need of food, water, and rest. Quincy, accessible by land and water, was of concern to the British, and redcoats sometimes appeared in the vicinity. It was sensible to be frightened. Still, John Quincy was sent to Boston on horseback to see if there were letters from his father. At ten years of age he was the oldest male in the household, and in her regular reading to the children Abigail turned from poets of peace to ones of courage and patriotism. Ironically, and inevitably, they were British writers.

In response to the death of Joseph Warren, Abigail began to read and reread to them two of the best-known eulogies for fallen warriors. William Collins' "How Sleep the Brave" (1746) had been written to honor those who had died maintaining the rule of the British king and his empire against the Stuart-led Scottish rebellion. Collins' "Ode to a Lady on the Death of Colonel Ross" had been written the same year. For Abigail, the power of the poetry transcended the irony of the reversal. In the spring and summer of 1777, as the colonists fought to sustain their rebellion against king and empire, Abigail taught her eldest son to repeat, every morning on rising, the most powerful lines from these poems.

He learned them by heart, and seventy-one years later could still recite from memory, with hardly a slip, many verses, including, "How sleep the Brave, who sink to Rest, / By all their Country's Wishes blest!" and "The warlike Dead of ev'ry Age, / Who fill the fair recording Page, / Shall leave their sainted Rest: / And, half-reclining on his Spear, / Each wond'ring Chief by turns appear, / To hail the blooming Guest." Imagine, John Quincy commented two years before his death, "the impression made upon me by the sentiments inculcated in these beautiful effusions of patriotism and poetry. . . . You may form an estimate by the fact that now . . . I repeat them from memory, without reference to the work. Have they ever shaken my abhorrence of war? Far otherwise. They have riveted it to my soul with hooks of steel."

His abhorrence of war became one of the major themes of his life. The metaphor "hooks of steel" aptly represents the tightly connected elements of family, community, and values that became the essence of John Quincy's sense of himself and his relationship with his country. Beginning with the Battle of Bunker Hill, the war dominated his life. As his family traveled between a town garrisoned by British redcoats and Braintree; as they heard the cannon fire and saw the smoke from Charlestown; as they daily expected British soldiers to come marching over Penn's Hill to ravage the areas south of Boston; as Abigail and her children huddled together and attended to daily sustenance, made difficult by wartime inflation and the absence of men to work the farm, while the father of the family was doing patriotic business in Philadelphia; when they traveled into Boston to be inoculated against smallpox, which was to kill more Americans than the fighting; and when, in 1775, he had his first sight of the newly appointed General Washington, with his ragamuffin army, arriving and setting up positions that encircled the British, and then witnessed the withdrawal of the British forces, John Quincy experienced some of the most formative impressions of his lifetime.

"Those were times of public distress, and terrors, and sufferings," he later reminisced. "I remember the melting of the pewter spoons in our house into bullets immediately after the 19th of April 1775. I remember the smoke and the flames of Charlestown which I saw from the orchard on Penn's Hill. I remember the packing up and the sending away of the books and furniture from the reach of Gage's troops, while we ourselves were hourly exposed for many months to have been butchered by them." He had learned at an early age not to idealize war. "In my early childhood," he wrote to an English correspondent in 1846, "a deep and inexhaustible abhorrence of war was planted in my bosom in the terror of a war then raged by Britain . . . against her own children in my native land." Perhaps because he was a young boy when the rebellion started, he was even more influenced by the emotions that surrounded it and the values that inhered in it than he would have been if he had been either older or younger. And perhaps because he was old enough to feel its momentous impact but not to participate directly, the events he witnessed and lived through in these early years contributed immeasurably to his lifelong detestation of war and his preoccupation with his country's destiny.

In January 1778, John and Abigail Adams made two momentous decisions. Adams had been chosen by the Continental Congress to join Benjamin Franklin and Arthur Lee as part of the joint commission in Paris, representing American interests to the government of the Confederation's only European friend. Congress had recalled Silas Deane, the original third member of the commission, whom Lee had accused of exceeding his authority. Eventually Deane would be exonerated, the victim of political and personal rivalries in a complex situation, and Adams would act in his defense. But at this time he concluded that it was his obligation to accede to the request that he replace Deane. Painful as it was to leave his family, it was his duty as it would be a soldier's to fight. The metaphor was particularly apt since, decades before, Adams had imagined himself as a soldier rather than a lawyer, an inclination that he had not been able to pursue, though he had developed a special concern in his political career for naval matters. John and Abigail's second decision, after much discussion and consultation, was that John Quincy would accompany his father. Europe would be his school.

As the boy walked with his father on the cold, windy morning of February 12, 1778, the few miles from Braintree to the southern extremity of Quincy Bay, he felt the pain of his parting from his mother and siblings. They had all shed tears. Adams had rejected his wife's proposal that she accompany him, unwilling to allow her to be put through the discomfort and danger of the voyage. Father and son were now to be the first Adamses to cross the Atlantic since 1630, going, as they knew, in the wrong direction. There was the possibility that they would not survive the voyage; winter crossings were especially dangerous. Once they were in Europe, under what circumstances they might return would be determined by events beyond their control. What Adams most feared was capture by British warships. He could expect to be executed for treason—and what then would happen to John Quincy?

His young son's inclusion was partly an attempt to alleviate the pain that John Adams felt at the thought of going entirely by himself. But it was mostly justified as an opportunity for the boy to learn more about the world than he could at home. His education was much on his parents' minds. A young lawyer, Abigail's second cousin John Thaxter, clerking in Adams' law office, had been tutoring John Quincy under the close oversight of his parents. The reading lists and instructions that Adams never tired of preparing made clear that his parents had in mind a grand career for their oldest son. "I wish to turn your thoughts early," Adams wrote from Philadelphia, "to such studies as will afford you the most solid instruction and improvement for the part which may be allotted you to act on the stage of life," since "the future circumstances of your country, may require other wars as well as councils and negotiations, similar to those which are now in agitation." Thucydides and Hobbes were the short but meaty list of the moment, and the education of a future statesman or successful lawyer was under way. John and Abigail Adams were willing to take the risk of this voyage to further their vision of his future.

Concern for the secrecy of Adams' mission had determined a clandestine departure from an unlikely place. Few lanterns lit the shoreline in the late afternoon darkness for fear of British troops or boats in the area. That morning Adams, close to the point of departure, had sent a note to his wife: "Johnny sends his Duty to his Mamma and his love to his Sister and Brothers. He behaves like a Man." In a rough sea, with a high wind blowing, they stepped into the waiting barge. Heavy coats, provided by the captain, and dry hay at their feet kept them warm. They were rowed, accompanied by Adams' servant Joseph Stevens, to the 24-gun frigate *Boston*. Friendly and patriotic hands took them aboard. "Master Johnny was very happy in his papa's consent to accompany him," his mother wrote to John Thaxter. She was deeply worried about the dangers of the voyage, especially that her young son would be tempted by the loose customs of European decadence. He might learn ways that would make him more European than American, a transformation that would have serious implications for his earthly career and his heavenly salvation. It was a theme that would preoccupy her for over a decade. What John Quincy felt as he stepped aboard the *Boston* can only be assumed to be excitement. With a high sea and then a snowstorm, they were obliged to put in to Marblehead. The next day the weather turned fair, the wind propitious. "Poor Johnny is gone," Thaxter responded. "I think he is now laying the foundations of a great man." For Abigail, amid the pain of parting, ambition for her son narrowed to concern for his well-being and what mattered most: character. "Injoin it upon him," she wrote to her husband, "never to disgrace his mother, and to behave worthy of his father." She consoled herself "with the hopes of his reaping advantages under the careful eye of a tender parent which it was not in my power to bestow upon him."

The atmosphere aboard ship, with its 172 men, had less to do with morals than with survival. Part of the threat was boredom. Adams compared life aboard a ship on a trans-Atlantic crossing to life in prison. The didactic father soon had John Quincy studying French with the help of books and the tutelage of two French passengers. A few days out, the boy had his first taste of seasickness. Almost everyone was ill, including Silas Deane's son, Jesse, slightly older than John Quincy, whom Adams had undertaken to deliver to his father. But they soon had excitement enough. The sails of three frigates were sighted, probably British. One gave unsuccessful chase for an entire day. Then the weather turned brutally bad. "To describe the ocean, the waves, the winds, the ship, her motions . . . the sailors, their countenances, language and behavior, is impossible," Adams wrote in his diary. "No man could keep upon his legs." Father and son braced themselves with their feet against the timbers in their cabin. Suddenly they heard "a tremendous report." A thunderbolt had struck four men, one of whom died three days later, and split the mainmast. To the captain's surprise, it held together. If repairs over the next three days had not been successful, the voyage would have been either over or long delayed. "I often regretted that I had brought my son," Adams confessed to his diary, as they sailed eastward. "I was not so clear that it was my duty to expose him, as myself, but I had been led to it by the child's inclination and by the advice of all my friends." It was Adams' inclination also. Ironically, this first was to be the worst trans-Atlantic crossing that either of them ever experienced. Despite that, throughout these trials, "Mr. Johnny's behavior gave me a satisfaction that I cannot express— fully sensible of the danger, he was constantly endeavoring to bear it with a manly patience."

By mid-March, they were in especially dangerous waters. The captain initiated preparedness drills, commanded by the experienced first lieutenant, William Barron, a Virginian, "an excellent officer, very diligent and attentive to his duty." As a naval vessel, the *Boston* had the obligation to engage the enemy if it had the advantage, but otherwise to evade combat, always taking into account the safety of its important passengers. When it encountered an armed vessel that seemed about to attack, there was a brief engagement.

Disobeying the captain's request that he stay below, Adams felt a cannonball swish directly over his head and into the sails. When the *Boston* swung broadside to attack, the British ship *Martha,* realizing that it was heavily outgunned, struck its colors. It turned out to be a private but armed vessel, a privateer carrying valuable cargo with a letter of authorization from the British to attack enemy ships at will. All governments engaged in this practice, which encouraged a legal form of piracy in which profiteering was the main motive. The *Boston* now had its own prize. It soon also had a disaster. A few days later, it chased another possible prize. Unable to determine its provenance, Lieutenant Barron had a cannon fired to signal the ship that it was required to identify itself. "The gun burst, and tore the right leg of this excellent officer, in pieces." An immediate amputation followed during which the lieutenant told Adams that he had a helpless family and begged him to make sure his children were provided for. After two weeks of suffering, he died. John Quincy could never forget the event. "I have a distinct recollection," he wrote fifty-four years later, "of being present at the performance of the funeral service over his remains when they were committed to the deep . . . one of the most painfully impressive events which I have witnessed in the course of my life."

Numbers of ships came into sight, among which were British cruisers. "But here We are, at Liberty, as yet," John Adams wrote in his diary. High winds and strong currents drove the enemy away and opened a clear path to Bordeaux. The coast of Spain came into sight, but when an adverse wind prevented the *Boston* from entering a harbor it headed for open sea. Two days later, six weeks almost to the day after leaving Quincy, they entered the stream of marine traffic into and out of Bordeaux. Soon they could see its lighthouse. The pilot boat that came alongside navigated the *Boston* down the Gironde into the city. The countryside seemed almost inexpressibly beautiful to the relieved and exhausted voyagers. And the Americans soon learned heart-lifting news. On February 6, 1778, France had formally recognized the United States. A treaty of alliance and commerce had been signed at Versailles. "It gives me a pleasing melancholy to see this country," Adams wrote. "Europe thou great theater of arts, sciences, commerce, war, am I at last permitted to visit thy territories. May the design of my voyage be answered"—and also his design for his son.

Chapter 2
A European Education
1778–1783

A week later, they were in Paris. To their surprise, it seemed at first as if there was not a single hotel vacancy available. Advance reservations were almost impossible in a world in which the uncertainties of travel and the slowness of communication made such matters catch-as-catch-can. And what they were able to catch was unsatisfactory. So they accepted the invitation of Benjamin Franklin, whom Adams knew and respected from their service together in Congress, to lodge in his spacious villa in suburban Passy. Franklin's generous invitation was at least partly self-interested. He had no intention of forfeiting to any of his colleagues his position as the primary American in Paris, a role that he performed to perfection. He dressed and behaved the part of an American amusement, as if he were a frontier or homespun exotic. The French, particularly the aristocracy and the government, adored him. So too did the ladies, who found his role as elderly flirt irresistible. His strategy was purposeful—the best way, he believed, to further American interests. Adams found in Franklin an acceptance of aristocratic moral standards that embodied just what Abigail had warned her son against. Franklin's life was "a Scene of continual dissipation," he wrote to her, by which he meant lazy and luxurious. But that did not overly distress him or unbalance his judgment. "That [Franklin] was a great genius, a great wit, a great humorist and a great satirist, and a great politician is certain," he later commented. "That he was a great philosopher, a great moralist and a great statesman is more questionable." John Quincy would come to share his father's assessment.

Within days of arriving in Paris in early April 1778, the ten-year-old boy was immersed in his own activities. In the background, there were grand events and notable personalities. But he seems to have taken no more notice of the diplomatic corps than he did of the American businessmen, some of them honest, many of them intriguers, who hung around the mission hoping to obtain information that could be turned to profit. Paris and the mission at Passy bulged with Americans eager for refuge, company, and news from home, especially since London was no longer comfortable for those who sided with the revolution. Adams quickly immersed himself in the "extensive correspondence" with Congress, with the French court, "with our frigates, our agents, and with prisoners, and a thousand others" that his job required. The administration of the commission, for which Franklin was responsible, appeared lax, almost irresponsible. Expenditures seemed excessive. Appalled by the jealousies and dissension among the commissioners and the Americans associated with it, Adams considered Franklin too tolerant of factional behavior that damaged American interests. Soon he discovered that Franklin was a primary player in the bitter factionalism; Franklin and Arthur Lee, the third member of the commission, were enemies. Whatever one favored, the other opposed. That made doing even the simplest official business difficult. Lee and Silas Deane, who remained in Paris for the time being though Adams had replaced him, also hated one another, in part because Franklin had sided with Deane against Lee. Ralph Izard, a leader of the American community in Paris, and Franklin were likewise passionate enemies. Each tried to prejudice Adams against the other. It was a hornet's nest.

———————

The young boy's eyes were on the attractions of the city, his new friends, and the school he began attending in the middle of April 1778. "Life was new, everything was surprising, everything carried with it a deep interest," he noted sixty years later. "It is almost surprising to me now that I escaped from the fascination of Europe's attractions." He immediately wrote to his mother and brother Thomas about the delights of Paris, particularly impressed by the "many fine public walks" and the "fine rows of trees in the gardens," so different from Braintree and Boston. The trees that impressed him might have started what was to become years later a passion for planting trees at Quincy to beautify the landscape and replace the forests that Americans were clearing with self-interested abandon. And he immediately had a taste of the city's cultural attractions, "very fine music and singing." Whether or not he was homesick, he had been trained to know his duty, which was to "not let slip one opportunity in writing to so kind and tender a mamma as you have been to me for which I believe I shall never be able to repay you." Abigail missed him terribly, as if a limb had been torn away from her body. She would not learn of his safe arrival until after the middle of June, through a newspaper report that got to Boston more quickly than any of her son's or husband's letters. The first letter she received took three months to cross the Atlantic. She worried that they had been captured or were dead. "I hope I shall never forget the goodness of God in preserving us through all the dangers," her son wrote to her, one of the first of hundreds of letters to his mother in the next decades, "and that by his almighty power we have arrived safe in France after a troublesome voyage." He knew the formulas for assurance and sent them honestly, in sincerity and love.

Before the war, his days had been partly a pastoral idyll. Now the urbanity of Paris provided new vistas, an actual school, and city attractions. In Braintree he had had home schooling, effective but limited, and had become an avid reader of the books the household provided, particularly the authors who were touted by his parents and from whom his mother read to her children—not children's books, but Shakespeare, Pope, and Milton. He had begun to read them on his own, vaulting successfully into Shakespeare's plays, the start of a lifelong passion. Reading lists from his father were a regular lesson in pedagogic overreaching, mainly ancient and modern history. His Latin studies were under way, overseen in Braintree by John Thaxter. Soon John Quincy was imitating his parents' instructions, sending moral and educational advice to his two brothers, setting the example of his own high standards, urging them to study French, and providing them in a long letter, mostly in French, with a list of the books that they should study to learn French, including an analysis of their relative merits. But why, he wrote, do I send you this long list? "I answer that you may have the means in your possession of furnishing yourself . . . of a complete collection of books for learning the French tongue." His brothers had no such opportunity. But his own engagement was a necessary passion, a given of why he was there. "You have entered early in life upon the great Theater of the world," his mother wrote to him. That theater spoke French.

Soon there was a slightly comic dance under way, father and son attempting to move to the same French rhythms. John Quincy's age was in his favor. His father's was not. Adams stumbled, studying French as if he were a schoolboy, attempting to learn it on his own, aware of the disadvantage of not speaking the language but unwilling to take time away from his duties for lessons. He struggled, frustrated, perplexed, almost comic in his failures. "Papa laments very much his having neglected this study in his youth, in terms so pathetical," John Quincy wrote to his brother Charles, "as to have made a deep impression upon my mind."

John Adams' French improved over time, but the fluency he sought escaped him. John Quincy took to it quickly, just as he would in the next decade to learning Dutch, German, and some Spanish, as well as becoming a masterly reader of Latin and an adequate student of ancient Greek. By late July, after four months in France, John Quincy began "to read and speak French, pretty well," Adams wrote to his wife. He had tutoring, he had a French school, and he had Paris, which meant urban encounters and the theater.

Two weeks after arriving in Paris, John Quincy was attending a boarding school in Passy, not far from the Franklin establishment. Franklin's grandson, Benjamin Bache, who was one of John Quincy's friends at the villa, already attended. Two decades later, as a journalist and printer who subordinated truth to ideology, Bache would be imprisoned for seditious libel with the blessing of John Adams' administration. Young Jesse Deane was also enrolled. John Quincy "packed up his things and went to school, much pleased with his prospect because he understood that rewards were given to the best scholars, which he said was an encouragement. Dancing, fencing, music, and drawing, are taught at this school, as well as French and Latin." The students spent their days in a strictly timed regimen. At the end of his first week, John Quincy wrote to his mother

that he liked the school "very Well." His French profi-
ciency improved rapidly. "Johnny . . . reads and chat-
ters french like a french Boy," Adams told Abigail. "If
I do not make myself master of French," John Quincy
asserted, "it will not be for want of opportunity or of
books but that this talent . . . may be improved to the
best advantage it is necessary to be a good husband of
my time." No more frivolous amusements, he wrote to
Charles. "We are sent into this world for some end. It is
our duty to discover by close study what this end is and
when we once discover it to pursue it with unconquerable perseverance." At the age of eleven, he already
had the perseverance. He had a mother and father who
preached and practiced the rigorous management of
time. His New England culture gave high value to productive work. And that culture posited some end, some
goal for each individual on earth that paralleled God's
providential plan for the human race.

Already John Quincy seemed to himself, with his
father's approval, always to have a pen in hand, writing letters to his family. His father advised that he keep
a letter book in which there would be a copy of every
letter he wrote, since originals were often lost in transit or simply irretrievable. John Quincy implied to his
mother that he was about to do that. His father, who
gave him a blank book for the copies, also urged him to

keep a diary, "yet I have not patience, & perseverance, enough to do it so Constantly as I ought." The copybook and diary got off to a faltering start, if started at all, though "Pappa . . . takes a great deal of Pains to put me in the right way." Since there was almost no age too young to start, the Adamses believed, performance sometimes needed to catch up to intention. It would mortify him, he confessed, speculating about letters he had not yet written, "to read a great deal of my Childish nonsense" a few years hence. But he also would have "the pleasure, and advantage, of remarking the several steps, by which I shall have advanced, in taste, judgment, and knowledge." Still, "a letter book of a lad of eleven years old," he told his mother, as if standing outside himself and making an objective observation, "cannot be expected to contain much of science, literature, arts, wisdom, or wit." Already, he lamented, he had missed the chance to record the interesting sights he had seen, "which if I had written down in a diary, or a Letter Book, would give me at this time much Pleasure to revise, & would enable me hereafter to Entertain my Friends, but I have neglected it & therefore, can now only resolve to be more thoughtful, & industrious, for the Future." In the years to come, he would see a fulfillment of this dedication to writing— for self-exploration, for the historical record, and for the embodiment of the writer's compulsion to record and make the world—that even a precocious eleven-year-old could not have imagined.

If he had kept a diary, he would have given primacy to the impressions made on him by Parisian entertainments, the "surfeit" of diversions that the city provided. He listed for his mother the sights that every visitor needed to see, the "scenes of Magnificence" from "the Palace and Gardens of Versailles" to "the Church of Notre Dame," many of which he visited in his father's company. He also roamed by himself or with friends. At Versailles, he had the distinction of being his father's son, an introduction that accompanied him in most of his activities and that gave his father the pleasure of seeing his eleven-year-old performing with a dignity and courtesy that evoked expressions of praise from their French hosts. He is "much esteemed here," John Adams wrote to his wife, "which gives me constant pleasure." Some praise was likely to have been merely polite. But some of it was undoubtedly sincere and well observed, the noticeable phenomenon of this American youngster carrying on conversations like a man or at least a precocious boy, listening and speaking with attentive and knowledgeable modesty. And in very good French.

To the young boy, Parisian theater was fascinating, different from any experience available in Boston, where performing plays was forbidden. Contrary to the Puritan tradition, John Adams believed the stage was a place of art and culture. And the theater in Paris was a schoolhouse for the improvement of their French, so much so that Adams brought with him the text of the classical plays he and John Quincy attended in order to reinforce the spoken word with the words on the page. Parisian theater ranged from low comedy, vaudeville, spectacle, and circus to opera and the classical drama of the French Renaissance, the plays of Corneille, Racine, and Molière. John Quincy loved it all. His father trusted in the self-censorship that Adams family moral training had instilled. "He lets me go now and then especially if there happens to be a tragedy of Corneille, Racine or Voltaire. I have been twice to the operas and several times to the Italian Comedy but I have been oftenest to the French comedy. . . . The language, the wit, the passions, the sentiments, the oratory, the poetry, the manners, and morals are at the French Comedy."

Theater was becoming one of the passions of his life. It enticed another passion, his first love. "The first woman I ever loved was an actress, but I never spoke to her, and I think I never saw her off the stage. She belonged to a company of children who performed at the Bois de Boulogne near Passy, when I lived there with Dr. Franklin and my father. She remains upon my memory as the most lovely and delightful actress that I ever saw. . . . Of all the ungratified longings that I ever suffered that of being acquainted with her, merely to tell her how much I adored her, was the most intense. I was tortured with desire for nearly two years, but never had the wit to compass it. I used to dream of her for at least seven years."

In Paris in 1778, father and son began a number of years in which they were essentially a family of two. The bond between them deepened. His school schedule and his father's work required separation. But weekends and holidays they reserved for themselves. John Quincy became more his father's son than ever could be the case for the two brothers who had been kept at home. A careful observer, he found in his father an admirable and desirable model. Whatever Adams' flaws, his son embraced him in his entirety, as an expression of duty but also of respect and love. There were indeed particular traits of character to be taken into account. In his family life, John Adams was a self-willed, domineering, but loving taskmaster. With his children, he believed he was rarely wrong. He expected them to honor their parents and their parents' values. He believed he knew what was best for them. As a father, Adams was a middling version of what was typical of his New England colonial generation: there were some who were less assertive, others more. But both John and Abigail believed that children were brought into the world to be taught moral rectitude. Parents owed their children admonition and discipline, not expressive sentiment.

Although strict about values and conduct, Adams still embraced the interests of this world. He loved literature, art, and history. Work and pleasure were inseparable. He was politically ambitious, sometimes self-righteous, and quick to suspect that people and groups were in league against him. And he never, for himself or anyone else, denied that the flesh had its natural needs. Such matters, though, were to be informed and guided by the highest moral standards, and by reason. What was to be feared was excess and corruption. He happily drank wine and beer but stopped far short of the alcoholism to which two of his sons were to succumb. Of his three sons, one was to die prematurely, impoverished and emotionally ill; another had a frustrated career and struggled with alcohol. Only his heir apparent, John Quincy, managed a stable personal life. Still, the children deeply loved their parents—anything else was inconceivable. They all faced the most difficult of challenges: nature and society demanded that they love parents who loved their children in ways that sometimes made their lives difficult. This was to be less true for John Quincy than for his brothers, though there were tests to come. But he had an advantage as the eldest son who was now his father's only intimate companion.

Both felt in their Paris life how much they had a taste for its attractions. At the same time, their attachment to family and country, the people they missed and their evocations of the familiar scenes from which they were absent, kept them aware of the price to be paid. Within two months, Adams was "wearied to death with gazing wherever I go at a profusion of unmeaning wealth and magnificence." He missed his wife and children. John Quincy's presence was all the more precious. He is "the Joy of my Heart." Although he liked his situation "pretty well," John Quincy confessed to his mother that he would "much rather be amongst the rugged rocks of my own native town than in the gay city of Paris." He was sincere, even if exaggerating for home consumption, when he wrote, "I had rather be in America than in any part of France." No doubt father and son were homesick, and Adams worried about the welfare and even safety of his family, struggling to maintain a semblance of a normal life in a country at war and partly occupied. As some of Abigail's letters began to arrive, the exigencies of inflation, of having to make do, and of danger to family and country were much on Adams' mind. John Quincy had the personality and discipline to mitigate homesickness with the pleasures of his Paris life. But his father had to contend with a different scenario. Congress had sent him to Paris. All his satisfactions depended on the fulfillment of this mission. Gradually, in the fall and winter of 1778, he became aware that he had no mission to fulfill.

Voices from across the Atlantic began to be heard in Paris. There was a power struggle in progress, with personal and financial interests at stake. The Silas Deane–Arthur Lee accusations and enmity inevitably permeated Congress. Since there was no executive branch of government, Congress handled all aspects of foreign affairs. The rumor reached Adams that he would be recalled, which he hoped would be the case. His presence in Paris, he had concluded, was futile and irrelevant. After all, he was to the French and the English "a man of whom nobody had ever heard before, a perfect cipher, a man who did not understand a word of French,—awkward in his figure—no abilities—a perfect bigot—and fanatic." From the time of his arrival, his role had been marginalized. The commission was paralyzed, three heads with one body. And Franklin and Adams, though on cordial terms, had incompatible virtues. The senior diplomat worked by anecdote, indirection, charm, and tactical compliance. Adams favored directness, sound argument, and moral earnestness. That Franklin would remain as a commission of one seemed to Adams better than the current situation. By February 12, 1779, news of Congress' inaction on the Deane matter and the election of its new president, John Jay, a Deane partisan, reached Paris. Adams was appalled. But he felt only relief at the news that Franklin had been made sole minister to Versailles. "On Dr. F. the Eyes of all Europe are fixed." He himself was considered not to be "of much Consequence." There were good reasons for this, Adams wrote. But Franklin's "age and real character render it impossible" for him to do the job properly.

Another rumor circulated: that Adams would be sent to Vienna. Or maybe Holland. Recall seemed more desirable. "My evils here arise altogether from Americans," he wrote to Abigail. His attempt to remain impartial had been undermined by an open letter that Deane had published, "one of the most wicked and abominable productions." It slandered Lee, whom Adams thought a good though intemperate man, and damaged American interests. Deane would be required to testify before Congress. Adams let Congress and the French minister of foreign affairs know his views. But Congress itself was split by similar divisions about power, money, ideology, personalities, sectional influence, the proper relationship with France, and the terms of a peace treaty with Britain. The French minister in Philadelphia exerted strong influence, since French aid was the lifeline for American survival. The government in Paris pulled the strings as if America were its puppet, and accusations were leveled in Paris and Philadelphia by self-interested parties. The strongly pro-French faction in Philadelphia favored subordinating American policy to French interests. Another faction desired France as a supportive ally, not a controlling superior. And amid all the estranging bitterness, now dipped in blood, between the United States and Great Britain, there was a relationship of language, culture, and history that needed to be taken into account. In 1779, there was good theoretical reason for an American who believed in the claims of the Declaration of Independence to approve more of the constitutional monarchy of Britain, despite its rigid elitism, than of the monarchical government of France, still almost feudal in its political and economic arrangements. An American colonist under British rule had considerably more civil rights and personal liberty than did the vast majority of the subjects of Louis XVI.

When, in the middle of June 1779, in fair weather, John Quincy and his father sailed westward to Boston

on the French frigate *La Sensible*, with no intention of returning to Europe ever again, neither could
have had any idea of how closely the rest of their lives
would be engaged with the complicated and contentious relationships, at home and abroad, among
France, Great Britain, and the United States. All
the issues existed in embryo in the years of this first
American mission to France, and in the next decade
the revolution in France and then the rule of Napoléon
would add an ideological dimension that would bitterly
divide American politics between those who identified
more with France and those who identified more with
Britain. As they sailed homeward, this was beyond
both their horizons.

John Adams' mind was mostly on his much-longed-for
reunion with his family. Abigail had been tugging at
his heartstrings with complaints that he was not writing to her as much anymore. John Quincy, as concili-
ator, explained how busy his father was, that many
letters had been written by both of them which must
have been lost. After ten years of daily intimacy since
their marriage, John and Abigail had been apart much
of his three years in Congress and now almost an additional year and a half. As soon as Adams had become
certain of his irrelevance, he desired to be recalled as
quickly as possible. When Congress kept silent, he

took matters into his own hands, despite the possibility
that such action would meet with disapproval. He did
not want to wait for instructions that might send him
to another country. Father and son left Paris in early
March but, to their disappointment, their attempts to
sail were frustrated. For two months they made do
with sightseeing and visiting in Nantes, where they
met the American tobacco merchant Joshua Johnson,
who had set up residence with his family, for tea and
conversation. The French government finally came to
the rescue. As a diplomatic courtesy, the Adamses were
given passage on the 24-gun frigate *Sensible*, whose
mission was to transport to America the new French
minister, the Chevalier de La Luzerne, and his staff of
almost two dozen.

As they waited for departure, his son was much on
John Adams' mind. "My dear fellow traveller is the
comfort of my life," he wrote to his wife. "He has
enjoyed a great opportunity to see this country, but this
has unavoidably retarded his education in some other
things." His father now assisted him in making translations from Latin and French, particularly Cicero's first
oration against Catiline, an early moment in what was
to become a lifelong identification with and passion for
Cicero the writer and Cicero the man. John Quincy has
"enjoyed perfect health from first to last," Adams wrote

to Abigail, "and is respected wherever he goes for his vigor and vivacity both of mind and body, for his constant good humor and for his rapid progress in French, as well as his general knowledge which for his age is uncommon." Once on board, John Quincy agreed to teach English to his two prestigious fellow voyagers, La Luzerne and his secretary, François de Barbé-Marbois. They are "in raptures with my son. . . . I found this morning the ambassador" seated "on the cushion in our state room, Mr. Marbois in his cot at his left hand and my son stretched out in his at his right—the Ambassador reading out loud . . . and my son correcting the pronunciation of every word and syllable and letter. The Ambassador said he was astonished at my son's knowledge. That he was a master of his own language like a professor."

This was to be, both father and son believed, a final homecoming, a return to much-loved people and familiar places. They sailed into American waters at the beginning of August 1779 after a smooth voyage of a little over six weeks. The French mission went into Boston for a formal welcome. John Quincy and John Adams, who had disembarked earlier, were rowed to the same beach from which they had departed a year and a half before. The boy of almost twelve had no reason to think that he would even return to Europe, let alone that in less than four months he would be crossing the Atlantic again, not to return for six years.

When the *Sensible*, with a new, mostly American crew, sailed from Boston for France on November 15, 1779, John Quincy and his father were aboard. The twelve-year-old boy did not want to go. After all, he had just returned from eighteen months away from his family and home. He had barely had time to re-acquaint himself with his sister and two brothers. His mother's pleasure at embracing him again had to have filled the boy with the happy expectation that he would remain for some time in the arms of those he loved and that he would now have an American edu-cation, directing him, in the footsteps of his father, to Harvard. The secretary to the French mission, though, had urged Adams to take "the young gentle-man your son" with him back to France so that he would learn to be "one day useful to his country" and an advocate of French-American friendship. "My little son, sir, is very sensible of the honor you have done him," his father replied, "but I believe it will be my duty to leave him at home, that his education may be where his life is to be spent."

But what education other than home tutoring was feasible in war-torn America? The decision was almost last minute, not John Quincy's but once again his parents'. His mother took him aside privately and convinced him, which was not really different in the end from issuing a command. It was in his, his father's, and his country's interest that he return to Europe. Patriotism required that they all make this sacrifice. His nine-year-old brother Charles would go as well. "My dear sons I cannot think of them without a tear, little do they know the feelings of a mother's heart," Abigail wrote. "May they be good and useful as their father." Summer was over. No more playing in the marshes. No more swimming in Black's Creek. No more roaming with his brothers.

Since their return, John Adams had been busy. He arranged the evidence of expenditures to submit to Congress for reimbursement of his outlays in France. Almost as soon as he touched Massachusetts soil, his Braintree neighbors elected him to represent them at a convention to draw up a new state constitution, which Congress had encouraged every state to do. He spent much of his time at meetings in Cambridge and at his desk at home, drafting what became, almost verbatim, the state constitution, adopted the next year by popular referendum. One of his colleagues was his contemporary, Theophilus Parsons, a Newburyport lawyer who was gaining a statewide reputation. Just as his son's education was on Adams' mind, so too was that of the citizens of his state. The constitution provided government support of learning, literature, and the arts. With its initial sponsorship of Harvard, it was to make Massachusetts the national leader in education for centuries to come. In conjunction with these constitutional provisions, Adams proposed the creation of an American Academy of Arts and Sciences. Years later, John Quincy would advocate similar values in his own political career.

In September 1779, John Adams had received the surprising news that Congress had unanimously appointed him minister plenipotentiary to negotiate a treaty of peace and commerce with Great Britain. British councils were divided, with policy disagreements and shifting views among both the Tory and Whig parties. Gradually, the likelihood of a negotiated settlement was becoming fixed in many minds on both sides of the Atlantic. British trade was suffering. The cost of the war was aggravating an already substantial public debt. The main issues in any peace settlement would be boundaries and trade, though who would have the upper hand would be determined by future military action. Congress was more divided about people than policies. It had, though, begun to realize that if the military situation continued to be a draw, eventually America's short supply lines, with the help of the French Navy, would defeat Britain's longer ones. It made sense to send a well-instructed peace commissioner to Paris, where the first stage of negotiations would take place. The French minister in Philadelphia, Conrad Gérard, had run Congress behind the scenes. His replacement, La Luzerne, did the same. The French minister of foreign affairs in Paris, the Comte de Vergennes, ran everything else.

After its bitter airing of the Deane-Lee affair, criticism of Franklin's role, and months of paralysis, Congress concluded that Adams was the best available choice to negotiate a peace treaty and a commerce treaty with Britain. He had just returned from France. He knew the players and the situation but was not besmirched by corrupt activities or the mudslinging between the factions. His adequate French was an asset. Franklin would remain in Paris, the minister to the French government. The duties of the two positions would be distinct. If events made it desirable, additional commissioners could be appointed. Vergennes was soon told by his ambassador in Philadelphia that Franklin was out of favor but that Adams had been instructed to defer to France in all matters. Adams' instructions contained no such directive. But the French believed they did, and Adams had only partial knowledge of these crosscurrents. If he had known more, it is likely that he would not have accepted the appointment. "Let me entreat you," he wrote to Abigail, "to keep up your spirits and throw off cares as much as possible. . . . We shall yet be happy, I hope and pray, and I don't doubt it. I shall have vexations enough, as usual. You will have anxiety and tenderness enough as usual. Pray strive not to have too much." He was especially right about vexations.

The day before John Quincy boarded the *Sensible*, he started to keep a diary. With a youthful sense of his own undisguised subjectivity, he titled it on the first page, "A Journal By Me JQA VOL: 1st," later to be revised to "A Journal by JQA From America to Spain VOL. 1st. begun Friday 12th of November 1779." He was soon keeping it daily, with omissions, a practice he continued for the rest of his life. He could have no idea that it would become an immense record and analysis of himself and his times, the most valuable firsthand account of an American life and events from the last decades of the eighteenth century to the threshold of the Civil War. It was to become, over time, the companion with which he spent the most hours, a history that he consulted to corroborate the past and that he wrote each day to affirm the reality of the present. In it he was to write history, poetry, narrative, literary analysis, travel accounts, religious musings, prayer, meditation, self-analysis, and autobiography. At the start, it was exclusively a daily record. As the Adams party sailed out of Boston Harbor, in daylight and in expectation of a safe passage, he had already made his first entry. It was the simple statement that he had taken leave "of my Mamma, my Sister, and Brother Tommy, and went to Boston." He said nothing about how he or they felt. The emotions of the moment were reserved for the spaces between the lines. His father had urged him to keep it as a record of fact. Eventually, like his father's diary, it was to become a record of fact and *feeling*. Now, as he put pen to its first page, he began his own distinctive literary career.

The crossing turned out to be a frightening adventure. There were 350 crew and passengers, including Adams' friend Francis Dana, who had been appointed secretary to the peace commission and chargé d'affaires, though he knew no French, and John Thaxter, tutor to Adams' sons and his private secretary. Soon the ship sprang a leak. "The passengers are all called to the pump four times a day," John Quincy noted in his diary; two pumps were constantly going. The captain decided to make emergency landfall in northwestern Spain. Ashore, Adams had a decision to make: whether to resume travel by boat or go instead by land across the Pyrenees. He chose the land route.

Although the Spanish authorities treated them as dignitaries, the trip was a nightmare. Spanish roads were primitive, accommodations nonexistent or filthy, without beds, pillows, sheets, or provisions. They slept close to farm animals. People did not wash themselves or clean their houses, which, without chimneys, were soot-covered. Their caravan slowly trudged and rose through mud and mountains, frequently stopping to repair carts and carriages. John Quincy, noting places and facts of interest, made no complaints to his diary. Father and son began to learn Spanish. The landscape, stripped of vegetation by an impoverished people, "looks like a man's face that is newly shaved," John Adams noted. The pervasive poverty was depressing; the heavy Catholic ambience and clerical domination were repellent. Only the clergy looked well fed and prosperous. "I thank Almighty God that I was born in a country where anybody may make a good living if they please," John Quincy wrote in his diary. They all caught bad colds. A miserable and exhausted John Adams remarked that in twenty-five years of traveling he had "often undergone severe Tryals, great Hardships, cold, wet, heat, fatigue, bad rest, want of sleep, bad nourishment. . . . But I never experienced any Thing like this Journey."

As father and sons entered Paris in February 1780, after almost three months en route, none of them could anticipate how long they would remain in Europe. Certainly, given how long it took to travel across the Atlantic and Adams' official instructions, they knew that it would not be less than a year. That it would stretch to over five for John Quincy and his father was beyond anticipation, and it would have astounded if not frightened them if there had been reason to anticipate at the start, or even as each year went by, that it would be that long. Adams might have brought his family with him if he had known, though he soon wrote to Abigail that the misery of travel through Spain made him sigh with relief that he had insisted that she, Nabby, and Thomas stay at home. The day after their arrival, John Quincy and Charles were enrolled in a school in Passy that attracted children of the American community, including Silas Deane's son. John Quincy felt quickly at ease, absorbed by the challenge of his studies. "I am . . . very content with my situation." Charles was soon achingly homesick.

Adams' diplomacy, except for formal rituals such as his presentation at Versailles, did not go well. Differences of temperament and policy between Vergennes and Adams surfaced quickly. Neither liked the other. To Vergennes, the New Englander was too much the plain-talking and presumptuous American. The French minister of foreign affairs seemed to

Adams the embodiment of the amoral European courtier. More important, Adams was soon confirmed in his belief that, though French and American interests sometimes overlapped, they were often considerably different. When he argued that his mission should be announced to the British, Vergennes forbade it, angering Adams. Vergennes concluded that Adams was chafing against Congress' instructions to defer to France. Adams rejected Vergennes' criticism. Speaking out, he offended the minister, who did not believe that French interests would be advanced by peace negotiations at this time. In any event, he preferred the more amenable Franklin.

For John Quincy and Charles, boarding at the school in Passy while their father stayed in Paris, parental guidance came by letter from Paris and from Braintree. John Quincy's focus was mostly on his studies, particularly Latin, Greek, French, geometry, arithmetic, writing, and drawing, the standard curriculum for a boy whose parents had the means to promote a liberal education and a path to a profession. His father had asked the headmaster to eliminate fencing and dancing, activities that would not be germane to John Quincy's life in America. Cicero and Erasmus headed the long list called "My Work for a day" that he sent to his father with the note that "as a young boy cannot apply himself to all those things and keep a remembrance of them all I should desire that you would let me know what of those I must begin upon first. I am your dutiful son." Latin and Greek, his father responded, were of the highest importance. The other subjects should be secondary, the writing and drawing only amusement. "I hope soon to hear that you are in Virgil and Tully's [Cicero's] orations, or Ovid or Horace or all of them. I am, my dear child, your affectionate father. P.S. The next time you write to me, I hope you will take more care to write well. Can't you keep a steadier hand?" His father approved of his son's next, more disciplined effort, complimenting him on the improvement.

Moral lessons and exhortations came from a mother who worried about her son. She most feared that he would wander from the straight and narrow. God was to be thanked and worshipped, she counseled him. He has created man "after his own Image and Breathed into him an immortal Spirit capable of happiness beyond the Grave, to the attainment of which he is bound to the performance of certain duties which all tend to the happiness and welfare of Society and are comprised in one short sentence expressive of universal Benevolence, 'Thou shalt Love thy Neighbor as thyself.'" This is "elegantly defined by Mr. Pope in his *Essay on Man*," Abigail reminded her son.

"Justice, humanity and Benevolence are the duties you owe to society in general. To your Country the same duties are incumbent upon you with the additional obligation of sacrificing ease, pleasure, wealth and life itself for its defence and security. To your parents you owe Love, reverence and obedience to all just and Equitable commands." What he owed most to himself was to know himself. That would allow him to avoid self-love and self-deception. "You my dear son are formed with a constitution feelingly alive, your passions are strong and impetuous. . . . Few persons are so subjected to passion but that they can command themselves." But the passions, especially anger, can be controlled by reason and love. "The due Government of the passions has been considered in all ages as a most valuable acquisition." And self-knowledge is the key. "I will not over burden your mind at this time. I mean to pursue the Subject of Self-knowledge in some future Letter, and give you my Sentiments upon your future conduct in life when I feel disposed to resume my pen." That was soon and often.

Abigail gradually introduced into her many letters to her son two other subjects of great interest to both of them, politics and family matters. Politics, people, literature, moral conduct, human nature, love, and concern about family and friends: the letters attest to how brilliant the Adamses were as a family of writers. And while John Quincy, as an adult, sometimes had theological doubts and came to believe that religion should be a matter of personal faith, not coercion or persuasion, he became an admirable example of fidelity to the ethics and spirit of his parents' and New England's religious spirit.

In late July 1780, John Quincy was again on the road, this time from Paris to Brussels and then to Amsterdam. Although the thirteen-year-old schoolboy had been content at Passy, his father had come to the conclusion by late spring that his own diplomatic efforts in Paris had not made, and at least for the time being, would not make progress. On the contrary, he had offended the French foreign minister who would hardly deign to see him. He had made Franklin's situation awkward and more difficult. He had been expressively himself in his interactions with French officialdom. The result he blamed on the French, who had to sense that beneath Adams' outspoken contentiousness was a pervasive contempt for the pomp and corruption inherent in the aristocratic regime. Unlike Franklin, Adams was too much the blunt American who hated his country's being subordinate and resented its dependence on France. He had little to do in

Paris but write informational letters and send printed matter to Congress and to the busy foreign minister, whose hands were full with complicated domestic politics and relations with European governments. He even gave Vergennes advice about how best to use French naval power against the British.

Rather than stay unprofitably in Paris, Adams, on his own initiative and with his two sons, left for Amsterdam, which he chose because he glimpsed the opportunity of persuading Dutch bankers to lend money to the impoverished United States. Unlike Franklin, he believed that America needed to be less deferential in its dealings with all the European powers, including France. Adams wanted to make the argument that it was in the Netherlands' self-interest to assist the new country. Any opportunity to make the United States less dependent on France, whose tone of noblesse oblige irked Adams, seemed to him worth pursuing, no matter the likelihood of rejection. The possibility of such a diplomatic venture had been in the air for a number of years, with Congress making tentative but unrealized efforts in that direction. Before returning home the previous year, Adams had considered going to the Netherlands and, by letter from Braintree, had discussed with John Jay, president of Congress, the possibility of an alliance. The Netherlands and Great Britain were commercial rivals, and Amsterdam was the financial capital of the Continent. In October 1779, Congress had appointed Henry Laurens, a South Carolina businessman, slave trader, and former president of the Continental Congress, minister to the Netherlands. He had been tasked with pursuing political recognition and financial aid but was captured by the British and imprisoned, the fate that Adams most feared in his trans-Atlantic voyages. Since Adams felt he was useless in Paris, Amsterdam seemed worth trying.

At the end of August 1780, the two Adams boys were enrolled in the well-regarded Latin School in central Amsterdam. John Quincy was initially enthusiastic. But since he knew no Dutch, the rector assigned him to classes in Latin and Greek far below his level, claiming that he needed to know Dutch before he proceeded to anything more advanced. John Quincy objected to what seemed to him a demotion to the company of younger boys and childish lessons. First he complained to his father, who had recently written to Abigail that he himself had been forced to study politics and war so "that my sons may have liberty to study mathematics and philosophy . . . geography, natural history, naval architecture, navigation, commerce and agriculture, in order to give" their children the "right to study painting, poetry, music, architecture." Now he had an unhappy son who did not have the liberty to study any of the subjects he and his father respected. Adams confessed to his wife that he feared that his affection "got the better of my judgment in bringing my boys." The issue intensified in the fall. "Your son . . . acts badly," the rector wrote to Adams. "Please remove him before his punishment here will become a public scandal." Surprised and pained, he immediately withdrew both boys. "I should not wish to have children," he later told Abigail, "where a littleness of soul is notorious. The masters are mean spirited wretches, pinching, kicking, and boxing the children, upon every turn." The Dutch valued pennies more than they valued ideas and independent spirit, Adams believed. America needed their pennies. But never at the cost of self-respect.

Within a few weeks an alternative developed. A young American doctor from Newport, Rhode Island, named Benjamin Waterhouse, who had left America in 1775 to study medicine in Edinburgh and London, had been studying at the University of Leiden. He recommended Leiden to Adams, whom he had met in Paris. Smart, ambitious, generous, and an admirer of Adams' outspoken patriotism, Waterhouse was eager to be of help. Leiden had a surfeit of well-qualified tutors, he told Adams. John Quincy was old enough to benefit from public lectures. Young boys were regularly admitted to Harvard and to European universities, and no one was prohibited from attending public lectures. "If the gentlemen should come, I can insure them an agreeable society and a genteel circle of acquaintance." There would be no discrimination against Americans, as Adams thought there might have been in Amsterdam. Expenses would be less, which would be an advantage since Adams was feeling pinched by the high cost of travel and life in France and the Netherlands, as salary and reimbursement were slow in coming. What had come from Congress, when it learned of Laurens' capture, was a commission to procure Dutch loans for the United States. His instinct to do this without authorization had been validated.

Accompanied by Thaxter, John Quincy and Charles arrived in Leiden, twenty-five miles southwest of Amsterdam, a little after the middle of December 1780. They were soon set up in lodgings in the same building as Waterhouse, a situation that proved the start of a lifelong friendship. Guided by Waterhouse, the new arrivals attended some lectures at the school of medicine. Within the week, John Quincy was, to his joy, immersed in Latin and Greek studies with a capable teacher. From Amsterdam, John Adams sent instructions and exhortations, even to the minor importance of diversions, such as ice skating, for exercise. But even such entertainments ought to be mastered. "Everything in life should be done with reflection, and judgment, even the most insignificant amusements. They should all be arranged in subordination, to the great plan of happiness, and utility." John Quincy's days soon became happily crowded with Greek and Latin lessons; with reading Shakespeare and Pope, a set of which he requested from his father; and with lectures in subjects from law to natural science to religion. He was immersed in the studious life he so much valued. Waterhouse and Thaxter provided sensible oversight, and through Waterhouse he had an adequate recreational and social life. Adams often visited Leiden or was nearby in The Hague.

In Amsterdam, John Adams had sublet a well-located house that served as his residence and as the unofficial American embassy. In Leiden, he stayed at his sons' lodgings, where he spent weeks writing a memorandum to the Dutch States General, the assembly of delegates representing the United Provinces of the Netherlands, advocating recognition of American independence. Without such recognition there would be no financial aid. To his disappointment, he soon discovered that there had been a change in his instructions from Congress, now dominated by Francophiles under the influence of La Luzerne and others who distrusted Adams' judgment. James Madison believed Adams' temperament made him an undesirable negotiator who would unduly favor New England interests in trade negotiations. Franklin was in the anti-Adams and Francophile camp, hostile to Massachusetts and New England in general, particularly to its insistence that a peace treaty include protection for its long-standing practice of fishing and curing in Canadian waters and on Canadian beaches. The French were indifferent to this issue, but their congressional supporters were eager to trade these privileges for British concessions in the West and Mississippi Valley. Congress now explicitly ordered Adams to defer to Vergennes in all aspects of the peace negotiations, including the French minister's ongoing refusal to allow Adams to negotiate directly with the British. At the same time, his commission to negotiate a treaty of commerce with the British was revoked. To add insult to stupidity, three additional members had been appointed to the peace commission: Benjamin Franklin, John Jay, and Henry Laurens. No longer the exclusive peace commissioner, Adams was hurt and angry.

John Quincy's mind was on happier matters: Greek, Latin, law, history, and English literature, without the restrictions of a formal school or the heavy hand of schoolmasters. This idyll ended in June 1781. Charles, who had been ill during the winter and spring, was homesick, to the extent that he seemed not to be benefiting from his European experiences. He was not particularly studious, and he desperately missed his mother and siblings. His father's company was sporadic and not fully satisfying. John Adams decided to send him home, accompanied by Waterhouse, who had completed his medical studies and had decided to return to New England, soon to be appointed the first professor of the principles and practices of medicine in the newly founded Harvard Medical School.

Before long, John Quincy was traveling in the other direction. On his fourteenth birthday, July 11, 1781, he was in western Prussia on his way, with Francis Dana, to the court of Catherine the Great in St. Petersburg. It was a last-minute arrangement, and he had only one week to prepare for departure. He and Dana scurried around Amsterdam stores to outfit themselves. On his son's birthday, John Adams, "distracted with more cares than ever," wrote to Abigail—who had not heard from her husband and sons for almost a year—the astounding news that John Quincy "is gone, a long journey with Mr. Dana." He was to be Dana's private secretary. Dana had accepted the appointment of minister to Russia, though a parsimonious Congress had not provided a penny for secretarial assistance. He was to be in effect a one-man legation, at a lower salary than that of other American ministers. Dana spoke hardly any French, the language of the Russian court. He asked Adams for the services of John Quincy, who spoke excellent French and wrote a legible hand. "I was at first very averse to the proposition," Adams later wrote, "but from regard to Mr. Dana, at last consented." He assured Abigail that it would not cost any more than if John Quincy had stayed at Leiden, for eventually Dana would repay what Adams laid out for his son's expenses. And a likely benefit was that, since John Quincy would be "satiated with travel in his Childhood," he will "care nothing about it, I hope, in his riper Years." It is unclear whether Adams paid due attention to the likelihood that he would not see his son again for at least a year or two.

John Quincy had been, off and on, in the thirty-six-year-old Francis Dana's company since they had sailed together on the *Sensible* in November 1779. He felt comfortable with Dana and confident in his stewardship. Dana's appointment as secretary to the peace commission had been a title without duties. A Harvard graduate and successful Boston lawyer, he had served in the Massachusetts revolutionary government and the Continental Congress. With Adams' backing, he had now been appointed to present himself to the court at St. Petersburg to persuade Catherine the Great "to favor and support the sovereignty and independence of these United States and to lay a foundation for good understanding and friendly intercourse . . . to the mutual advantage of both nations." Catherine had instituted a league of neutral nations whose purpose was to be a counterweight to British power and provide protection for neutral commerce. Dana's mission was to open diplomatic relations with Russia and negotiate American membership in the league. Whether he would be recognized by the St. Petersburg court was an open question. French influence was immense. Vergennes' policy was to limit freedom of American diplomatic action. The more independent America's relations with other European nations became, the less dependent it would be on France. As long as France was in a state of declared or undeclared war with Britain, its own interests would not be served by peace between Britain and America. American membership in the league of neutral nations was a long shot.

To fourteen-year-old John Quincy, the fifty-five-day, 2,500-mile journey, first to Utrecht, where Dana purchased a carriage, then to Cologne, Frankfurt, Leipzig, Berlin, Danzig, and on to St. Petersburg, was more interesting than exhausting, though it was both. There was no public transportation. Long-distance travelers used their own vehicles, handled by a servant and professional guide. Horses and drivers were hired, if available, at postings at established distances, usually at extortionist rates unless limited by government regulations. The variety of currencies created confusion and added expense. At the Rhine, John Quincy and Dana crossed by a rope ferry. They crossed again at Cologne, opposite which John Quincy noticed "a village, inhabited by Jews; a nasty, dirty, place indeed, and fit only for Jews to live in." He was puzzled that the Protestants seemed to suffer more restrictions in this Catholic area than did Jews, who "are tolerated and have their synagogue." In Frankfurt, he had his first sight of a ghetto. "There are 600 Jew families here who live all in one street which is shut up every night, and all day on Sundays, when the gates are shut they can only come out upon occasions of necessity, but the Jews can keep their shops in any part of the city." He seemed mostly puzzled, and would clearly have had no sense that he was being anti-Semitic or even of what anti-Semitism was. Ignorant of the complicated history or even the basic facts of the relationships among Catholics, Protestants, and Jews in the Rhineland or anyplace else in Europe, and the comparatively tolerant policies of Frederick the Great, John Quincy was years later to know much more and have different views.

At Leipzig, Dana accelerated the pace by traveling night and day. The first night they narrowly avoided injury when the carriage overturned. In late July they spent nine days in Berlin, where Dana bought a replacement carriage. Berlin was "the handsomest and the most regular city I ever saw . . . a very pretty town, much more so than Paris," John Quincy noted. It was a city undergoing forced urban renewal under the direction of Frederick the Great, who was both an enlightened ruler and an authoritarian despot. His subjects, John Quincy noted, "have a great reason to complain of him, for he certainly treats them like slaves." He was appalled that Prussia was a military dictatorship, and, even worse, in the area of Poland they passed through on the journey to Riga, mostly under Russian control, "all the farmers are in the most abject slavery, they are bought and sold like so many beasts, and are sometimes even changed for dogs or horses. Their masters have even the right of life and death over them, and if they kill one of them they are only obliged to pay a trifling fine; they may buy themselves but their masters in general take care not to let them grow rich enough for that." In Russia, John Quincy noted, the serfs were slightly better off than in Poland, "though the common people are all slaves."

Slavery was a reality in the Boston and Braintree world into which John Quincy had been born. But it was minimal: barely in sight, often disapproved of, and gradually disappearing. Some Boston, Plymouth, and Providence merchants dealt in slaves, mostly indirectly, and fortunes had been made. South of New England, especially south of Philadelphia, slavery was an inescapable part of American life. At its worst, it was no different from what John Quincy was seeing now. At its average, it was modestly better. And America was developing a small cadre of free blacks, an advance impossible for serfs in Poland and Russia. But what John Quincy saw in Russia had no equivalent in his personal experience, and in the abstract the notion of slavery was as abhorrent to him as to his mother and father, who paid much more than lip service to the affirmation that "all men are created equal." It was a statement that, as he grew into intellectual manhood and a public career, was to become a guiding principle of John Quincy's life. In Russia, now, he clearly deplored what he saw, though his record of his surroundings was observational, not emotional. It seems likely, though, that the slavery he saw in Poland and Russia formed part of the subtle matrix of awareness taking shape in his mind over the following decades that American slavery was a great evil.

Their carriage rolled into St. Petersburg in September 1781. The city had been transformed from wretched urban primitiveness to European modernity at the beginning of the eighteenth century by the vision of Russia's autocratic ruler, Peter the Great, and by the work of serfs. Its splendors soon became the cityscape against which John Quincy's and Francis Dana's expectations were to be disappointed. At first, all was optimism, based on innocence and ignorance. The weather immediately turned cold. As the freezing, snow-filled winter set in, it did not take long for John Quincy to realize that having been sent to St. Petersburg was a mistake which had nothing to do with climate. His European venture was, from the start, for the sake of his education. Adams and Dana had assumed that the Russian capital, like other European cities, had an educational establishment that provided tutors and schools for young men. But St. Petersburg had no formal or informal educational institutions. The Russian elite were educated in Paris. The rest of the population was mostly illiterate.

For a young man eager to pursue studies for entrance to an American university, St. Petersburg offered nothing. Dana did some limited tutoring, and a German tutor was eventually found. But, compared with Leiden, this was an educational desert. Both Dana and John Quincy soon realized that little could be done other than home tutoring and self-education, but neither provided what John Quincy needed. He worked at his Greek and Latin, especially Cicero. He was able to borrow or buy canonical works of English literature, particularly poetry, drama, and history. He read Dryden's poems, Molière's plays, and Samuel Richardson's novel *Clarissa*. After setting up a schedule, he worked hard on his own despite the difficult conditions. Double windows offered some insulation, and fireplaces burned constantly, but their lodgings were still always cold. There were some entertainments and company, mainly provided by the diplomatic community, and walks in the city whenever the weather allowed. Yet even if he had wanted to learn Russian, there was no one to teach it to him. By March 1782 it was clear to all the principals that this Russian venture had been a mistake. Letters from Dana and his son had opened John Adams' eyes to the situation. John Quincy was by now desperately homesick. He envied John Thaxter when he learned that Thaxter was going home. Braintree seemed like paradise compared with this.

Dana's own mission also went poorly. The French minister, under instructions from Paris, played a double game. He openly advocated that the Russian court accept Dana's credentials. Behind the scenes he discouraged Russia from recognition of, let alone alliance with, the United States. Catherine also had games to play. She needed to sustain her alliance with France, to keep Prussia at bay, and to maintain a balance against British naval power, which was damaging neutral commerce. This was not a propitious time for the initiation of relations between Russia and America. Since none of this was personal, the French minister and the residents of the small diplomatic community in St. Petersburg were modestly hospitable to the frustrated American minister and his personal secretary. But with the court closed to them, they could not attend its entertainments or partake of its social activities. Public spectacles provided some pleasure and instruction, particularly military parades and religious holidays. There were occasional concerts and theater performances. In better weather, they sailed out a distance into the Gulf of Finland. Both Americans, though, looked longingly westward.

By late winter, Dana and John Quincy had requested permission for John Quincy to return to Amsterdam as soon as possible. If his return was to be delayed beyond June or July, exit by ship would not be possible. By land, he would be delayed at various places en route as the weather turned colder. Intensely busy in the Netherlands, John Adams had by spring of 1782 realized that the Russian assignment was not benefiting his son enough. A letter written on May 13 reached John Quincy's hands on the 6th of September. "I want you with me," his father had written, "I want you to pursue your studies too at Leyden. Upon the whole, I wish you would embark in a neutral vessel and come to me." This was easier to propose than arrange, and the season was already "too far advanced to think of going by water." John Quincy could not get to Holland before the end of January. "I might very possibly be obliged to pass the whole winter in Norway or at Copenhagen." And Dana had not received permission to leave.

A carriage taking a group of travelers to Sweden provided a safe means for John Quincy to travel the hardest stage of the route. Sweden's representative at St. Petersburg entrusted him with letters to deliver to Stockholm, and when the carriage lumbered out of St. Petersburg on October 30, 1782, John Quincy carried with him letters of introduction from the Swedish ambassador and other diplomats. Since there were no taverns or accommodations on the road, the passengers slept in the moving coach. Many towns along the way were walled and minimally hospitable. After a week, they crossed the Gulf of Bothnia into Finland. They were soon in Helsinki. Water crossings, many of which had frozen over, were difficult. But the weather improved, and at the end of the third week of November they arrived in Stockholm, eight hundred miles from St. Petersburg. Forced by the weather to stay there for five weeks, John Quincy discovered that there were businessmen and government officials happy to be hospitable to the son of a well-known American diplomat. His French allowed him to converse with every educated Swede in a country in which "strangers are treated with a great deal of politeness and civility." Sweden's monarchy seemed relatively benign, especially compared with Prussia and Russia. Some businessmen asked him to tell his father that they were interested in opening trade with America. "Sweden is the country in Europe which pleases me the most," he noted.

On New Year's Day 1783, he and his fellow travelers were on the road again, through heavy snowfalls from Stockholm to Göteborg. His name and letters of introduction gave him additional oversight and hospitality along the route from Göteborg to Helsingborg, where they crossed the Øresund Strait to Elsinore. In the middle of February, he arrived in Copenhagen. Europeans with American connections or interests provided company and guidance. The French ambassador welcomed him. He scurried around to the museums and went to the theater. Northern European winter storms and freezing cold kept him in Copenhagen, but in Amsterdam John Adams, whose last letter from his son was dated February 1, was deeply worried. "You cannot imagine, the Anxiety I have felt on your Account."

He transformed his anxiety into an all-points bulletin to the French and Dutch consular services to search for any sign of his son and report back immediately. Meanwhile, John Quincy waited for favorable weather to travel to Kiel by boat. It never came. The wind continued to blow in the wrong direction, and the harbor stayed frozen. After a fruitless three-week wait, he took the land route to Hamburg. He stayed a month, as if the closer he got to Amsterdam the less compelled he felt to travel swiftly, or at least the more inclined he became to take advantage of where he was. His father's son, he took stock of the commercial and manufacturing assets of the city, a fifteen-year-old boy directed toward serious adult matters. Hamburg, he wrote to his mother, "will I dare say, carry on hereafter a great deal of trade with America." He finally reached The Hague at the end of the third week of April 1783. It had taken him almost six months to arrive at his father's house, the newly purchased U.S. embassy in the Netherlands. It was not the same as being home, but it was as close as he had been for a long time. He would have been distressed to know it was as close as he was to be for over another two years.

Much had changed since he had left for St. Petersburg, including John Quincy himself. "He is grown up a Man, and his Steadiness and Sobriety, with all his Spirits are much to his honour," John Adams reported to his wife. Slightly taller than his father, he was a little over five feet seven. Round in the face in a boyish way, he was of medium build, with dark eyes touched with hazel and plain features that made him handsome in the non-distinct Adams way. A miniature portrait painted by a Dutch artist in the summer of 1783 shows the pursed lips and small smile of both his parents, and a nose full enough to be noticeable but with a touch of the aquiline. He wears a wig, probably put on to fulfill the formality of having his portrait painted, and the sedate clothes of the personal secretary of a minister plenipotentiary. Something of restrained pride is in the body language. No doubt the Russian venture had given him a sense of himself and more than a glimpse of adult independence, though he remained still very much, by personality and sense of duty, his parents' child. All of his movements were dependent on them, except for those of a young man's body and imagination. In Sweden, he had been aroused by "the beauties of the women," not a word of which enters his diaries and letters of the time. "I have not forgotten the palpitations of the heart which some of them cost me," he wrote twenty years later, "and of which they never knew."

At the end of 1781, while John Quincy was in St. Petersburg, Lord Cornwallis had surrendered. American independence had become a certainty. In the spring of 1782, John Adams had achieved, with the help of fast-moving events, the most important accomplishment of his diplomatic career. Impressed by the American victory at Yorktown, the States General made the momentous decision to accept Adams' credentials. The principals signed a treaty of amity and commerce. Adams then persuaded Dutch bankers to extend a substantial loan, which the United States desperately needed. For the first time, it could pay its pressing bills at home and abroad. That the bankers thought the United States creditworthy had a multiplying effect, which soon led to four long-term loans. Between 1782 and 1787, this was the main source by which America remained solvent. It was to the credit of the Netherlands and its bankers that recognition and the loans came despite adverse pressure from the British, the British faction in the Netherlands, and France and the French faction. France still wished to keep American interests subordinated. It had good reason to fear a British-American alliance, and America feared to be allied with either Britain or France. In London, the leadership changed and changed again as peace faction, war faction, and delaying faction tussled with difficult issues. But the pragmatists dominated. Events had out-paced French restraints on American action. In the late fall of 1782, Adams, Franklin, Jay, and Laurens signed a preliminary peace treaty with Great Britain. A year later, John Quincy and his father were for the first time walking the streets of London.

Chapter 3
Slow Voyage Home
1783–1787

At Dover, in October 1783, father and son stepped onto English soil. The two latest generations of American Adamses were in the homeland of their ancestors and their language. John Adams was thrilled to find himself on the corner in London where John Street and Adams Street met. But they were also on soil where, only a year before, they could not have walked without arrest, subject to the penalties of treason. And yet, despite seven years of war, they felt, as did most Americans, the close connection between British and American culture.

Dashing around London as fast as good horses and a carriage could take them from one famous sight to another, they visited St. Paul's, Westminster Abbey,

Buckingham Palace, the British Museum, and the Royal Academy, then Windsor, Bath, Oxford, and Cambridge, where the new astronomical observatory made a great impression on John Quincy. They attended art galleries and theaters. At Westminster Abbey, John Quincy was overwhelmed with "Awe and Veneration" at the monuments to the great poets, especially the inscriptions, the quotation from Shakespeare's *The Tempest*, and the invocation "O rare Ben Jonson." At Drury Lane and Covent Garden, he reveled in every sort of play, from Shakespeare to *Tom Thumb*. His theater of the mind became a theater of the stage. The monuments to great warriors struck a different chord, for how much to love and how much to hate England was both a personal and a political negotiation. With his father, he attended the opening of Parliament. The king "made his most gracious speech from the Throne: All the Peers were in their Robes which are scarlet and white: the King's and the Prince of Wales's were of purple velvet." His father years later published an account of their reception on entering the lobby of the House of Lords. The usher appeared "in the room with his long staff, and roared out with a very loud voice, '*Where is Mr. Adams, Lord Mansfield's friend!*' I frankly avowed myself Lord Mansfield's friend, and was politely conducted to my place." That distinguished jurist had not too long before told "that same house of lords, 'My Lords, if you do not kill him, he will kill you.' " It was great political theater, and a lesson for John Quincy about the conduct and courtesies of international relations: an enemy today can become a friend tomorrow.

At Dover, as if empowered by finding himself on the soil of his native language, John Quincy wrote his earliest extant poem, without rhyme and in irregular iambic pentameter:

> There is a cliff whose high and bending head
> Looks fearfully on the confined deep—
> How dizzy 'tis to cast one's eyes so low!
> The crows and choughs that wing the midway air,
> Seem scarce so gross as beetles. . . .
> The fishermen that walk upon the beach
> Appear like mice, and yon tall anchoring bark
> Diminished to her cock; her cock a buoy,
> Almost too small for sight. . . . I'll look no more
> Lest my brain turn, and the deficient sight
> Topple down headlong.

With an exuberant eye that was both proud and self-critical, he disparaged his own poem to Peter Jay Munro, John Jay's sixteen-year-old nephew, who had accompanied his famous uncle to Paris. In December, the Adamses visited "*Twickenham*, formerly the Residence of *ALEXANDER POPE*," John Quincy's favorite author next to Shakespeare. At the theater and with text in hand, he thrilled to *Hamlet*, *The Merchant of Venice*, *King Henry VIII*, and *Measure for Measure*. He particularly noticed the difference between the styles of acting in Paris and London. That English audiences reacted so emotionally to the famous Mrs. Siddons' performance of Isabella in *Measure for Measure* surprised him. "A young lady, in the next box to where we were," almost fainted. She "was carried out. I am told that every night Mrs. Siddons performs this happens to some persons. I never heard of anything like it in France." His love of Shakespeare, though, did not prevent him from being critical; this was the start of his practice of making even genius subject to analysis. His standards were rational and moral. From the start, he had a strong sense of language as the place at which all literary criticism begins. Noble as *Hamlet* is, he wrote to Peter Munro, "I am told that every night Mrs. Siddons performs, foolish things in it which can come out of the head of man. Only think of the following line, O woe is me! *I have seen what I have seen, seeing what I see!* . . . for a person in deep distress, is it not most pitiful, and it is full of puns and Quibbles. . . . But here I am plaguing you with a criticism of a play;

which I suppose you think I might as well leave; for if you wanted a criticism you would make it yourself." Munro disagreed but acknowledged that he could provide his own insights. "I am not a severe critic," John Quincy responded. "As to my *proving* by a Line in *Julius Caesar*, the meaning of others in *Othello*, it is not more, nor (I may say) so much, as what all the able Commentators my predecessors have done, and therefore I have great Reason in so doing: I explain a Passage in an Author, by another in the same Author." With youthful enthusiasm, they argued about the meaning and effectiveness of phrases in *Hamlet*, *The Merchant of Venice*, *Othello*, and *Julius Caesar*.

Although Hamlet's exile to the land of madness was an ironic theatrical joke, the young American recognized that the reality was a different matter. Father and son visited Bedlam, the hospital for the insane. They saw "a great number of fools and madmen, but as that is no more than what we see every day in the Street, and in Society," he observed, "there is no Necessity of my giving you a detailed account of the poor wretches I saw there." Was England mad; were its leaders self-destructive, its people imprisoned in poverty, many of the poor literally driven crazy? To American visitors, it seemed possible. France's support of the United States made it undesirable for visiting Americans to take notice of conditions there. Few Americans, even longtime residents, foresaw the revolution that was on the near horizon. Joshua Johnson, who had fled from England to the safety of France and now had returned to London, where the Adamses visited him at his Tower Hill home, typified the American businessman, equally comfortable in both countries. Francophiles saw in Great Britain what they were conveniently blind to in France. Even Americans who embraced their English origins viewed Britain as a nation divided between a rich, all-powerful elite and an impoverished population groaning under the burden of an unequal distribution of wealth and the taxes that wars of empire created. Civil liberties were repressed. Beggars lined London streets. In the countryside, hovels predominated. To the Adamses, revolution in Britain seemed a distinct possibility, a new order to be formed on the American style of republican government. Until then, they felt it was likely that the rulers of Britain would find new occasions and means of making war on the United States.

Still, on this first visit to London, John Quincy's spirits were so buoyant that he copied for Munro's amusement a comic drinking song about the heavy-boozing "Old Toby," the clay of whose body had been transformed into a brown jug that was "Now sacred to friendship, to mirth and mild ale, / So here's to my lovely sweet Kate of the vale." He slyly explained that he had his own "sweet Kate." Why did he go to the theater so often? he rhetorically asked Munro. "I go pretty often to the play here, because, if there was no other enticement than this, that you are sure to find a number of fine women there, it would be enough for me. For a long time, every evening I went, I was in love with a new object." He did acknowledge that there was a particular object of his infatuation. *"Mr. Joshua Johnson, Great Tower Hill; London.* That's all I have to say. . . . P.S. Alas! . . . I am in a desperate Situation: I sometimes think of hanging, shooting, or drowning myself for—I won't tell you what for." The Johnsons entertained handsomely; Joshua was now the American consul in London, a nonpaying position of financial advantage to an American businessman, and the Adamses used his Cooper's Row home as their mailing address. John Quincy did not say who at Tower Hill he was in a passion about. Johnson's eldest daughter, Nancy, was ten; his next oldest, Louisa Catherine, eight. Perhaps Johnson's beautiful twenty-six-year-old wife, Catherine, was the object of momentary infatuation. Maybe it was at least partly a tease and a joke to entertain Munro. "Alas! Alas! I have left her. Heaven knows when I shall see her again," he wrote in January 1784 from The Hague. In another letter, he confessed that it was "entirely a joke, and to have something to say; for there is not a female in England, that would give me a half hour's pain, if I never were to see her again."

But an untitled poem that he wrote in early 1784 and sent to Munro suggests that, amid the joking, posturing, and exaggeration, there was indeed someone in London to whom he felt enough attraction to evoke poetic language to express his feelings. The myth of the Judgment of Paris provides part of the conventional frame of a ten-stanza poem about an impassioned lover of a beautiful woman who does not return his feelings. The object of his desire is called Chloe, one of the names of Demeter, a Greek goddess of fertility, and a popular name in eighteenth-century English poetry.

1. *Oh love, thou tyrant of the breast,*
   Thou hast deprived me of my rest,
   Oh thou hast changed me quite,
   I lay me down upon my bed
   Chloe comes straight into my head
   And keeps me 'wake all night.

2. *Or, when sleep comes to soothe my cares*
   Chloe again to me appears
   How charming does she seem!
   And then I think my Chloe's kind;

But soon I wake and straight I find
That all was but a dream.

By 1783 it had been decided that Abigail and Nabby would sail to England. Adams desperately wanted to have his wife with him. He missed his only daughter, who was now of an age to come out fully into the world. That issue had arisen with an uncomfortable twist. Nabby had been pursued by and fallen in love with Royall Tyler, a young Boston lawyer from a successful family. Adams, who learned about it at the beginning of 1783, had reason to be suspicious of the man's character. Tyler's reputation was that of a carouser and ladies' man. "He is but a prodigal son," he wrote to his wife, "and though a penitent, has no right to your daughter, who deserves a character without a spot." There were indications that Abigail had been encouraging the courtship. "I am so uneasy about this subject, that I would come instantly home, if I could with decency." But since he could not, then his family must come to him, at least his wife and daughter. They would then all return to the United States together the next year, unless Congress appointed him to another European position, which he hoped would not happen. Thirteen-year-old Charles and eleven-year-old Thomas would stay with Abigail's sister Elizabeth and her husband, John Shaw, who would tutor them. "Will you come to me this fall and go home with me in the spring? If you will," Adams wrote from London, "come with my dear Nabby." Whatever the cost and trouble, "I am so unhappy without you that I wish you would come. . . . I am determined to be with you in America or have you with me in Europe, as soon as it can be accomplished consistent with private prudence and the public good." If his encouragement to her to sail would come too late for them to leave that summer, travel in the fall, he urged.

Finally, in June 1784, mother and daughter sailed from Boston, accompanied by two servants. Assuming that they had left months before, Adams did not write. By July, he was almost distraught and wrote to Abigail as if she were still at Braintree. "I have been in constant and anxious expectation of hearing of your arrival in London. Your letters encouraged me to hope and expect it, otherwise I should have been with you at Braintree before now. . . . My own opinion is that you had better stay. I will come home . . . and leave politics to those who understand them better and delight in them more."

John Quincy stayed at The Hague through the spring of 1784, plunging back into his Latin and Greek studies. "I am still here in my Solitude," he wrote to Peter

Munro, "and have got quite accustomed to it. . . . I don't regret the amusements of the great cities. Was it not for the desire I have of seeing you, and enjoying your company, I should not have the least inclination of returning to Paris." He was not "in the merriest mood." If he did not have the distraction of an excellent library, he would "almost despond." The Hague "is one of the prettiest Places, in the world. . . . Yet I had rather be almost anywhere else . . . there is no such thing as Sociability." Money is all everyone seemed to care about. "I scarcely know a person of my age in the whole place." In response to Munro's criticism, he defended his poem about Chloe, word by word, line by line, and now assured him "for certain [Chloe] is a real living person." Books and ideas helped the days to pass quickly. As he read the Roman historians, he had no doubt about their lessons for American life. He and Munro "differed in opinion upon the subject of Julius Caesar. . . . You thought him a great and a good man; carried away by his ambition. I regarded him and I do still, as a tyrant, and as a bold audacious villain, whose determination was to enslave his country, no matter by what means. I find that his whole life was a continuing encroachment upon the liberties of his country."

When Munro disagreed with John Quincy's claim that *Paradise Lost* was as great a poem as the *Aeneid*, the young critic vigorously defended his view, arguing the importance of a balance between original thought and respect for the opinion of acknowledged authorities. But an independent mind came first. Munro seemed to John Quincy to have little of it. "I will give you the opinion I have formed myself of the two Poems. . . . I am pretty well acquainted with both; for I have read *Paradise Lost* attentively; and within the last four months I have translated that whole Poem into writing with my own hand, and therefore I hope you will allow me to have a Sentiment of my own upon that matter, and I declare I think *Paradise Lost* . . . very near if not quite equal to the *Aeneid*." He advised Munro "never to decide a thing in your own mind upon hearsay alone but to examine things yourself and judge for yourself. . . . For why in the name of Heaven, should not men be capable of as great things now as they were two thousand years ago?"

Great things were on his mind but so was discretion, which he valued. As a sixteen-year-old much in the company of adults, he had early on developed a standard of discretion often at odds with his temperament. On the one hand, spontaneity and frankness were natural to him. On the other, he was aware that, though frankness could be an asset, masking one's views and feelings was sometimes a strategic necessity. He realized that there were always people ready to use his own words against him, and that the mask of convention could protect him from attack. He did not want to be different. He wanted to be better, to be an embodiment of the highest fulfillment of the New England values into which he had been born. And those values, as they had developed, emphasized free speech and original thought under the guidance of discretion and a due respect for other people. Originality of mind was consistent with conservatism. With Munro he could be uninhibitedly himself. But he reminded Munro twice "not to let my Opinions when they are contrary to the common ones be known to anyone but yourself."

Well before his mother was at sea, John Quincy was on his way to London to meet her and Nabby on the assumption that they had sailed at least a month earlier. His father had important business in Amsterdam and could not come. John Quincy's instructions were to meet his mother and sister in England, make travel arrangements, and hasten them to Holland for the long-desired reunion. But the slowness of delivery and the possible loss of letters made it partly a comedy of the unknown. Delighted to leave The Hague, he arrived in London a little after the middle of May 1794. While waiting for their arrival, he had the pleasure of hearing the twenty-four-year-old arch-Tory William Pitt, the youngest prime minister ever, and his opponent, the radical, pro-American Whig James Fox, in Parliament. He soon heard William Burke addressing the House for over two hours, criticizing the king for having dissolved the previous Parliament, an attack on arbitrary royal authority that any American patriot would have admired. When an American captain brought news that there had been talk of a wedding "in our family," John Quincy thought it likely, he wrote to his father, that "we shall not have the pleasure of seeing my sister here." It seemed reasonable to conclude that both ladies were still in Braintree.

A disappointed John Quincy returned to the Netherlands and resumed his studies, only to learn toward the end of July that his mother and sister had landed in England three days earlier. Nabby and Tyler had decided on a trial year's separation; they would be married on her return. John Quincy was immediately dispatched to London. "I send you a son who is the greatest traveler of his age," John Adams wrote to Abigail, "and . . . I think as promising and manly a youth as is in the world." On July 30, Abigail's servant came running into their hotel room. "'Young Mr. Adams is come.' 'Where where is he,' we all cried out?" "I drew back not really believing my eyes—till

he cried out, 'Oh my mamma! and my dear sister.'"
Nothing but the eyes at first sight appeared as the boy
he once was. "His appearance is that of a man, and in
his countenance the most perfect good humor." Adams
was overjoyed with the news of their safe arrival. "I am
twenty years younger than I was yesterday." Traveling
as quickly as he could, he was in Abigail's arms on
August 7, after four and a half years of separation. He
learned from her that Congress had appointed him,
with Benjamin Franklin and Thomas Jefferson, a commissioner to negotiate treaties of commerce with all
the major European states. Jefferson was in London
as well. It was John Quincy's first meeting with the
forty-one-year-old Virginian.

The beauty of the late summer countryside
impressed the Adamses as they drove from Calais to
Paris, where Adams and Jefferson, Franklin's replacement, were to begin negotiating treaties with various
European states. Nabby and her brother renewed their
friendship, big sister now with a little brother who had
become an impressive young man. In the middle of
August 1785, they settled into a grand house in suburban Auteuil, four miles west of Paris. Abigail was
overwhelmed by its size and luxury, less so by Paris.
She had come from the small, stoic houses of New
England and the stringency of wartime Massachusetts.

London she had thought beautiful and elegant—not semifeudal Paris, "a horrid dirty city. . . . But where my treasure is here shall my heart go," she wrote to her sister Elizabeth, and her treasure had to be in Paris for the time being. She soon partly adjusted to the attractions of French life for people of rank and means, which included dinners (some of which the Adamses hosted); court occasions at Versailles; and the theater, which they went to frequently, sometimes to Abigail's discomfort at the immorality of the French stage. Even the comparatively latitudinarian John Quincy noted in his diary that the plays in Paris "almost universally are very indecent." The public taste "seems to be entirely corrupted." Fascinated by a balloon ascension he had seen the previous year, he saw another well-publicized launching. "I heartily wish they would bring [the] balloon to such a perfection, as that I might go to N. York, Philadelphia, or Boston in five days time." Pomp and ceremony got on his nerves. Religious superstition seemed absurd. Standing in a shop, he heard the tinkling of a bell that signified that a priest was on his way to administer last rites. All those present but he fell on their knees and "began to mutter prayers and cross themselves."

The Adamses' social circle in Paris was large and delightful, especially for Nabby, who was introduced to a social life she had not experienced before. John Quincy had one old friend from his Passy days, Franklin's grandson Benjamin Bache. But he envied the good fortune of his closest Paris friend, Peter Jay Munro, who had returned to America, and looked forward to his own return, though no date had been set. All the Adamses were on easy terms with the diplomatic corps, with Franklin's American and French circle at Passy, with wealthy and accomplished Americans in Paris, with the French friends of America, particularly the Marquis de Lafayette and his wife, and especially with Jefferson. They were uneasy, though, with the proliferation and hierarchy of servants, the code of legitimated theft by servants and shopkeepers, the lack of a serious work ethic, and the frivolous use of time, energy, and money that characterized the French upper class. At Carnival time, John Quincy noticed that the government encouraged diversions and entertainment in the Paris streets, even hiring people to wear masks and run about. "Thus does this government take every measure imaginable to keep the eyes of the people shut upon their own situation: and they really do it very effectually." An actor at a play he was attending made an indirect reference to the king that was menacingly hostile. "I shall never forget the effect of this incident upon my reflections at the time." That spring, John Quincy made note of

Beaumarchais' criticism of the king in *The Marriage of Figaro* and his consequent imprisonment. He had to be thinking that something potentially explosive was in the air.

The Jefferson and Adams families were on familiar, come-any-time terms. Without a son, Jefferson took a special interest in John Quincy. They talked about books, philosophy, history, and the Latin and Greek authors the young man was studying. In carriage rides, on walks, at one another's homes, they were regularly together. John Adams was delighted to have Jefferson as a colleague and friend. On political matters, he seemed to John Quincy "a man of great judgment." When Jefferson discoursed about his native Virginia, he told his eager young listener that tobacco was a soil-destroying crop, which should be replaced by wheat, and that "the blacks . . . are very well treated . . . [and] increase in population more in proportion than the whites." Lanky, freckled, with reddish-blond hair, Jefferson was a combination of Virginia courteousness and an ideological iron fist. Shy and uncomfortable in large groups, he was charming and relaxed in personal conversation. Adams and Jefferson had been colleagues in Philadelphia, with hardly a difference between them on the issue of independence. Like the Adamses, Jefferson had an unshakable American patriotism.

He and Adams, both well educated and widely read, having worked closely and well before, were to work closely and well for much of the next year. Jefferson would remain in France, replacing Franklin early the next year as the U.S. minister there. But John Adams' days in Paris were numbered, as he learned in April 1785. In early February, Congress, after divisive rivalries and split votes, had appointed him the first American minister to the Court of St. James's. Abigail was to return to London, the city to which she had taken an immediate liking. John Quincy, if all went well, would soon be on his way home.

The lonely eighteen-year-old felt every bump of the bad road from Paris to the coast. He had not been home for five and a half years. He had been away for seven of his eighteen years and had become, oddly enough, in his father's resonant phrase, "the greatest Traveller, of his age," having spent long periods in Amsterdam, The Hague, St. Petersburg, and Paris; crossed the Atlantic four times; and traversed by mule, horse, carriage, and sled the abysmal roads of Spain, Germany, Russia, and Sweden, and the better roads of France and the Low Countries. He had traveled at least fifteen thousand miles. He wanted very much to go home. He was leaving because he and his parents believed that if he did not leave, he would not be able to identify himself fully enough or be identified by others as an American. It was a given that every Adams male had to earn his own living. If it were to be done as an American, it had to be preceded by an American education. It was also a firm Adams family belief that John Adams had gone into public service at considerable financial sacrifice. The family thought it likely that John Quincy would follow in his father's footsteps. But whether it was to be the law or government, they agreed that it was essential he return to America, despite the pain of separation. Fifty years later he wrote in his diary that "my return home in 1785 from Auteuil, leaving my father when he was going on his mission to England, decided the fate and fortunes of my after-life. It was my own choice, and the most judicious choice that I ever made." It was to some great degree, he believed, a testimony to the power of attachment to one's native place, of a primal and emotional patriotism, beyond rational explanation. At Harvard, he would meet young men of his own age "and form connections in early life amongst those with whom he is to pass his days," as Abigail wrote to her friend Mercy Warren. "My son has made a wise choice."

Aboard the *Courier de l'Amérique* bound from Lorient to New York City, he had the amusing though irksome charge of supervising the care of seven greyhounds, a gift from Lafayette to George Washington. On the evening of May 25, 1795, they sailed into the night sky. When John Quincy awakened, "we had nothing but the sea, and the azure vault bespangled with stars, within our sight." He was seasick for four days as the ship rocked in calm weather, without any breeze at all. He feared that the voyage would be a long one. One of the ship's officers seemed barely competent, the result, he believed, of the French practice of making government appointments strictly by birth and favoritism. By day, the rocking of the ship kept him from much reading or writing. He had pledged that he would keep a daily diary, which he would afterward share with Nabby. He focused on the weather and on character sketches of his fellow passengers. At night, fear of fire mandated that the ship sail in darkness. With only five passengers, social life was limited. The personality of a Dutch merchant who had traveled the world impressed on John Quincy that "every nation seems to have a peculiar characteristic, which nothing can efface: whether it is owing to education, or to the nature of the different climates, I cannot tell. I rather think to both." He stopped short of self-analysis, but he noted that the sailors "prefer being mistaken, to being right by the information of another." They sailed into tropical summer weather, then northward. He had wanted to be on American soil by July 4, "the greatest day in the year, for every true American," and wrote from memory into his diary lines from a favorite poem by James Thomson that, in John Quincy's mind, identified America as the place where liberty will survive despite its loss everyplace else. On July 4, they were still at sea. Seven weeks after sailing from Lorient, they finally sighted land. Guns were fired off Sandy Hook for a pilot to come aboard as the *Courier de l'Amérique* waited for the tide to turn. John Quincy was soon walking on Manhattan streets.

The young man who spent three midsummer weeks in New York had a distant resemblance to the eleven-year-old boy who had sailed to Europe six years before. Introduced as someone of importance, he was indeed a young man with a pedigree and credentials, the son of the first American minister to the Court of St. James's, with powerful friends like Jay and Jefferson and Franklin. In the small world of American politics, that meant a great deal. John Quincy's share in this patronage society was small and fragile. But it made his weeks in New York a pleasurable whirl of dinners, at one of which he met Tom Paine. He had letters to deliver, especially to the newly appointed secretary for foreign affairs, his father's good friend, John Jay. With Jay, he visited the Dutch minister, and soon he was introduced to Elbridge Gerry and Rufus King, two Massachusetts delegates; James Monroe of the Virginia delegation; Henry Knox, the secretary of war; and George Clinton, the governor of New York. He accepted the invitation of the president of Congress, Richard Henry Lee, to board at his home. His days were filled with parties, walks, and excursions. "I have been introduced . . . to almost all the members of Congress," he wrote to his father, "and to a great number" of New Yorkers. He knew that attentions were paid to him for his "father's sake."

But he was also of special interest for what he had to say about Europe, as if he were an authority. Members of Congress were eager for news from London about what steps the British were taking to implement the peace treaty. Its terms required that they evacuate their forts on the northwestern frontier, but they apparently were being reinforced. The British had decided not to evacuate without good faith evidence that American debtors were repaying British merchants for debts incurred before the war, which the treaty also required. A crisis seemed at hand. Various states had passed laws freeing American debtors from any legal obligation to pay. In London, John Adams quickly became aware that no progress would be made on settling outstanding issues until the anti-repayment laws were repealed, but state legislatures and American businessmen opposed repeal. Foreign policy conflicted with domestic pressures. "The politicians here wait with great impatience to hear from you," John Quincy wrote to his father, summarizing the political situation as if he were writing a diplomatic dispatch.

Having overstayed his time in New York, he was already too late to be home in time for the Harvard commencement, an entertaining holiday for those in the Boston area with connections to the college. "You will perhaps think I had better be at my studies," John Quincy wrote to his father, "and give you an account of their progress, than say so much upon politics. But while I am in this place I hear of nothing but politics." Accompanied by a recent French-American acquaintance, he made his way northward and eastward on poor roads, having been persuaded that he would have a better chance to see the country by road than from the packet from New York to Providence. At New Haven, he delivered a letter to Ezra Stiles, the president of Yale College, whom Jefferson had described as "an uncommon instance of the deepest learning without a spark of genius." Since the country did not have a developed postal system, John Quincy was sometimes given let-ters to be delivered along his route. In exchange, he had convenient introductions to notable people who provided hospitality, which slowed his pace. "I am very impatient to get home to Boston," he wrote to his sister. At Hartford, he delivered a letter from his father to the poet and lawyer John Trumbull, who had been Adams' law student. Crossing into Massachusetts, he was pleasantly surprised when "the mistress of the tavern where we dined told me my name." She said "she knew me from my resemblance to my father who had passed several times this way."

On August 25, 1785, having been on the road for ten days, he finally reached State Street in Boston. At Cambridge, he embraced his brother Charles, a member of the new freshman class, and other family members. "I shall not attempt to describe the differ-ent sensations I experienced in meeting after so long an absence the friends of my childhood, and a number of my nearest and dearest relations. This day will be for-ever . . . deeply rooted in my memory. . . . It has been one of the happiest I ever knew."

Everything at first seemed vivid in the light of the present moment, enriched by comparison between six years before and now. "No person who has not experi-enced it," he noted, "can conceive how much pleasure there is in returning to our country . . . when it was left at the time of life that I did. . . . The most trifling objects now appear interesting." His aunt Mary noticed that he was "quite a stranger in his own country," an exotic bird to himself and others, a young American who had spent so much of his boyhood in Europe. But he had little difficulty relearning his American self. Fifty years later, he commented in his diary, "There is a character of romantic wildness about the memory of my travels in Europe from 1778 to 1785, which gives to it a tinge as if it were the recollection of something in another world." Over the next weeks he visited aunts and uncles, his extended family, and the Adamses' social world. He was deeply moved when he entered the family home at Braintree in which his father had been born, now occupied by Aunt Mary and Uncle Richard Cranch. He was accompanied by Royall Tyler, who was boarding with the Cranches and studying law. "It reminded me of the days of my childhood, most of which were passed in it, but it looked so lonely and melancholy without its inhabitants." He "went to the library, and looked over the books, which are in good condition; only somewhat musty and dusty, which shows that their owner is not with them." Like his father, he had become a lover and collector of books, gratified even by their feel and smell.

His father's books without his father made him miss his father all the more. But John Quincy floated with pleasure in the ambience of the family that he had not seen for so long, including his brothers and especially his paternal grandmother. She asked him over and over again, "When will they return? . . . I could only answer with a sigh." As he sat at meeting, listening to Reverend Anthony Wibird, it all seemed so familiar that it was as if he "had heard him every week" since he had left Braintree. Looking at the congregation, he recognized only some of the faces.

The president of Harvard, who questioned him about what Greek and Latin texts he had studied, advised him "to wait till next spring before I offer, and then enter for three months in the junior Sophister class." He had not read, let alone mastered, all the texts required for admission. Soon he was on the road to Haverhill to submit himself to the tutelage of his uncle, Reverend John Shaw, at whose home he would live. His thirteen-year-old brother Tom, also preparing to qualify for Harvard, was already Shaw's pupil. John Thaxter, who had opened a law practice in Haverhill, introduced John Quincy to various local families, especially those in which there were young ladies. Seventeen-year-old Nancy Hazen, an orphan adopted by a former Haverhill resident, lived with the Shaws. Slender, of medium height, with dark hair and good features, she had blue eyes that sparkled "with natural Wit, sweet sensibility, and the most perfect good humour," Elizabeth Shaw had written to her sister Abigail the previous year, assuring her that Charles was too young to be in danger. John Quincy's cousin Elizabeth Cranch lived close by, at the home of Leonard White, who was also preparing for Harvard. Elizabeth's brother William divided his time among Harvard, Braintree, and Haverhill. Another attraction was Royall Tyler. Soon Tyler and John Quincy were having long conversations. But at the same time, a letter breaking off Nabby's engagement was on its way across the Atlantic, something Tyler may have sensed coming.

Between May 1785, when John Quincy had sailed from Lorient, and August, Nabby felt she had increasing reason to end her engagement. Through the spring and summer she was disappointed, bewildered, and then pained that not one letter from her fiancé arrived. Word came from the Cranches implying that Tyler was behaving erratically and unreliably. He had boasted that he had not written to Nabby. He failed to deliver letters that she had sent him to pass along to others, and did not mail letters that people had asked him to mail to her. If his judgment in small matters seemed poor, what could be expected of him in large ones? Was he being irrationally possessive or manipulative? Or was this his way of expressing ambivalence? It was all worrying and suspicious. Had she made a mistake? Nabby asked herself. In late May, soon after John Quincy's departure, she met the newly arrived secretary to the legation, William Stephens Smith, a thirty-year-old Princeton graduate and former Revolutionary War officer. Within weeks he approached Abigail about his interest in Nabby. In consultation with her parents, she decided to break her engagement to Tyler. That she had a new suitor in the wings was an advantage but not a precipitating factor, though she saw Smith daily if only because he worked in the building that Adams had bought to serve as the first American legation and the minister's residence. In December, the month in which Smith formally asked for permission to marry Nabby, Tyler announced that he was going to London to explain everything and reestablish the engagement.

Meanwhile, John Quincy had ample opportunity to fall in love with Nancy Hazen. The infatuation kept him in minor turmoil for some months. In New York, he had enjoyed meeting lovely women and made note in his diary of their attractions, especially expressing interest in why he found one type more attractive than another. Did blondness and fair skin appeal to him more than dark hair and complexion? He came to no conclusion. But he strongly preferred women who were modest, engaging, and unblemished by vanity. It is "rather a dangerous situation," his cousin Elizabeth wrote to Abigail. "He tells *me* that his heart is wonderfully *susceptible*, that he falls in love one moment and is over the next." He felt passionate but ambivalent about Nancy. She already had a number of suitors, of which he was not one, and he had neither means nor desire for a formal relationship. "She either treats her admirers too well or too ill," he noted. He continued to rate the qualities of many of the local ladies, though "Miss Nancy" was most frequently in his company and on his mind. Much of the rest of his mind and time was devoted to Latin and Greek studies. Harvard was more important than Nancy, though he struggled with his feelings, his reason at variance with his heart. In London, Nabby met a beautiful Swedish lady whom she had reason to think, she wrote teasingly, had been someone who had made her brother's heart flutter. "I have heretofore more than once," he confessed to his diary, "been obliged to exert all my resolution to keep myself free from a passion which I could not indulge, and which would have made me miserable had I not overcome it. I have escaped till now more perhaps owing to my good fortune than to my own firmness, and now again I am put to trial. . . . I never was in greater danger."

In mid-November 1795, he felt enough endangered to hope that Nancy would indefinitely extend a week away, for when "the passions are high and the blood is warm, it is impossible to make a choice with prudence necessary upon such an occasion," and imprudence could lead to fatal error. "May it be my lot, at least for ten years to come, never to have my heart exclusively possessed by any individual of the other sex." Affairs were out of the question; parental and societal disapproval would be restrictive and severe. And Nancy, even if she had had a special interest in John Quincy, who had neither a profession nor means, would not have considered him suitable. Unintentionally, she helped him overcome his passion. He soon felt "in much better spirits" than "for a considerable time. . . . I am now fully satisfied that I have nothing to fear. . . . I never saw Nancy coquet it quite so much," he noted. "She seemed really determined to outstrip herself." Infatuation faded; friendship commenced, and he began to worry about Nancy's imprudence and her welfare. He comforted himself with a poem, "To Delia," written in rhyming iambic pentameter couplets of the sort he so much admired in Pope's poetry, with the same moral as *The Rape of the Lock.* He had, he told his sister, a rage for rhyming.

> *Let poets boast in smooth and labored strains*
> *Of unfelt passions and pretended pains.*
> *To my rude numbers, Delia now attend,*
> *Nor view me, as a lover, but a friend. . . .*
> *I, whom neither love nor passion blind,*
> *Seek the unfading beauties of the mind.*

The conventional lesson was neatly expressed:

> *For all the gifts that nature can impart*
> *Are vain without the virtues of the heart.*

And his own heart was at issue:

> *The flames of passion seek not to excite*
> *Unless you wish that passion to requite.*

Bad things happened, he warned, to cruel and reckless women who raised passions without regard to the consequences. Did he give Nancy the poem to read? Did she respond?

He began three months of redoubled effort at his studies. "His candle goeth out not by night," Aunt Elizabeth

wrote to Nabby. "I really fear he will ruin his eyes."
He focused on Virgil, Horace, Homer, Xenophon,
Greek grammar, the Greek New Testament, geography, a logic textbook, and Locke's *Essay on Human
Understanding*. That Harvard required Locke's *Essay*,
which denied the existence of innate ideas, marks how
far it had already distanced itself from Calvinist orthodoxy. "I am very much inclined to think him right. It
has been said that his argument to prove that the existence of a God is not an innate idea may be injurious,
but they make no alteration in the reality." God can be
seen in his creation, in nature, so the argument that
each human being is born with the knowledge that God
exists seemed unnecessary to John Quincy. What he
found conclusive and indisputable was that justice and
virtue are not innate in human beings. They needed to
be taught, as he had been by his parents, society, and
his own efforts. And, he concluded, the multiplicity of
religious sects in New England was a good thing. The
more there were, the less likely that any one of them
could dictate to others. They all seemed to him more
or less equally good and bad. And the idea of a divinity who condemned people to everlasting hell "for what
they could not in any measure help or prevent" was
abhorrent. It was a view he never changed. The purpose of religion was to help and comfort people, not to

condemn them. Its teachings should be predominantly ethical and practical, not theological. And "whatever a man's religious principles," it was "impolite and improper for him to ridicule the general opinion."

On New Year's Eve 1785, he consoled himself with the thought that, whatever his errors of the past year, at least he did not have to reproach himself "with vice, which it has always been my principle to dread, and my endeavor to shun." In February, Nancy moved to another welcoming home in the neighborhood. "Her going away has given me pleasure, with respect to myself; as she was the cause of many disagreeable circumstances to me. There was a time when I was sensible of being more attached to her than I should wish to be to any young lady." Still, he liked her, except for her vanity. They parted as friends, though in mid-February, when she and John Thaxter came for dinner, he recognized that he could never quite make up his mind about her. That night she seemed a less admirable person than he had sometimes thought and always wanted her to be. None of that mattered now.

In mid-March 1786, in clear mild weather, he mounted his horse, crossed the river by flatboat on a path cut through the ice, and rode to Cambridge. The next day he was examined, orally at first, before the president, four tutors, three professors, and the librarian, almost the entire Harvard faculty. He may not have noticed that it was the Ides of March. There were some questions he could not answer at all. He was put briefly to the test of Locke and Isaac Watts' *Logic*, for which he was prepared, and of some geography questions for which he was not; and was asked to construe three stanzas of Latin poetry and some lines of Homer. Then, in another room, he was asked to translate in writing a paragraph from English into Latin. When he was done, the president took it and returned fifteen minutes later, "marching as the heroes on the French stage do," and said, "You are admitted, Adams," as an upper junior, with a waiver of fees out of respect to his father's service to his country.

On March 22, 1786, the night sky over rural Cambridge was emblazoned with "the most extraordinary northern lights" John Quincy had ever seen, so bright that in a night without moonlight he could read ordinary print outdoors. In a world in which the alternatives for reading were daylight or candlelight, it was a moment of pleasurable wonder. It also illumined a college world that he found both enlightened and provincial. He sized up President Joseph Willard quickly: a man of great pomposity and vanity who insists that "there are no misters among the undergraduates. . . . He calls them Sir" and declines to speak to the two undergraduates who live in his house. John Quincy was addressed as "Sir Adams." The incompetence of most of the tutors, recent graduates, made them targets of John Quincy's sharp tongue. "Your brother is exceedingly severe upon the foibles of mankind," Elizabeth Shaw wrote to Nabby. "And if anyone says to him, Mr. Adams you are too satirical— Not more severe than just," he replies.

His attraction to satire and irony had its dangers. On the one hand, he could not help making fun of pretension and vanity. On the other, those he satirized had virtues. His passion for fairness required that he provide an objective assessment. Willard, the butt of ridicule for being stiff and pedantic, was also "esteemed and respected for his learning." John Quincy also needed to take into account that a sharp tongue could be counterproductive. It could be witty, intelligent, and literary as practiced by some of the writers he most admired, from the Roman satirists to Pope and Swift. But it also had an abrasive edge, a critical deflation that could give its targets pain. For a young man intent on making his way in a highly socialized world, a sharp tongue would make enemies. And it would make rough the path to either the bar or government service. "I have already come to the resolution of showing all the respect and deference to every member of the government of the College that they can possibly claim," he wrote to Nabby, "but to you I can venture to give my real sentiments, such as arise spontaneously in my mind, and that I cannot restrain." And he could also give them to his diary. He was, though, then and much more so later, to think of his brief time at Harvard as a turning point in his life. "My short discipline of fifteen months . . . was the introduction to all the prosperity that has ever befallen me, and perhaps saved me from early ruin," he wrote fifty years later.

That he was argumentative required finding a balance between deference to elders and self-assertion. He gave high value to reason and evidence. But with more experience of the world beyond Boston than most of his elders, and already widely read, he had a high regard for and stubborn adherence to his own views. In private, he was often vigorously outspoken. "He had imbibed some curious notions," Aunt Elizabeth wrote to his mother, "and was rather peculiar in some of his opinions, and a little too decisive and tenacious of them." But, she assured Abigail, "in company Mr JQA was always agreeable, pleasing, modest, and polite, and it was only in private conversations that those imperfections of youth were perceivable." Sensible Aunt Elizabeth, who loved and cherished him, knew him well. She was, though, at best an unreliable prophet. He might change some particular tactics but not his personality. Over time, he learned enough control of his argumentative stubbornness to use it as an effective tool of negotiation. And he learned to use his satiric skills masterfully in political debate, especially in his congressional career. But he often struggled not to use them disadvantageously. His aunt's prophetic eye was not totally partisan when she observed that in John Quincy, "I see the wise politician, the good statesman, and the patriot in embryo." And for those with whom he did not put on a public face, he had a warmth and openness that caused Aunt Elizabeth to shed tears of regret at his departure from her home in March 1786. "I wish Mr JQA had never left Europe," she wrote to her sister, "that he had never come into our family. Then we should not have known him. Then we should not have been so grieved. Then we should not have this occasion of sorrow." There may have been more than a touch of drama in this letter for the pleasure of her sister. But Elizabeth was sincere.

What Adams most valued about Harvard was that it gave him the opportunity to reenter American life. He was, though, soon aware of its limits. More so than he had anticipated, it was provincial. The small faculty was less talented than he had expected, more learned than communicative, and almost always dictatorial. Most faculty members were Christian clergymen; the tilt was in the direction of liberal Congregationalism. The system was based on lectures; students were required to parrot back what they had read and recite their translations from Latin and Greek. A good memory and diligent preparation were great assets. The schedule required early morning communal prayer. At Haverhill, John Quincy had gotten used to studying late into the night. Now it made more sense to rise early, the start of a lifelong pattern. He soon got a taste of the full regimen, lectures and recitals in Latin, Greek, Euclid, Locke, Hebrew, religion, and science. He was comfortable with numbers and mechanical devices, with barometers and record keeping. The sky above and the weather below fascinated him. More than any other subject, astronomy excited him. The night sky always compelled his attention. "Mathematics and natural philosophy are studies so agreeable," he wrote in his diary, "that the time I devote to them seems a time of relaxation." Professor Williams closed the science course with lines from Pope's *Essay on Man*, "All are but parts of one stupendous whole, / Whose Body, Nature is, and God, the Soul," marrying liberal Christianity with the eighteenth-century scientific worldview. What John Quincy equally valued was the atmosphere of collegial community, of which he had had almost none in his European years.

Sometimes the community functioned in unattractive ways. "It seems almost . . . a maxim among the governors of the College," Adams observed, "to treat the students pretty much like brute beasts." Morale suffered. Resentments built. "If anything . . . can teach me humility, it will be to see myself subjected to the commands of a person that I must despise." Rivalries among classes and students encouraged practical jokes and fistfights. High spirits, rowdiness, and mob judgment led to smashed windows and broken furniture. Soon after his arrival, the sophomores "assembled . . . some of them got drunk . . . then . . . broke a number of windows . . . and after this sublime maneuver staggered to their chambers. Such are the great achievements of many of the sons of Harvard." His parents' standards kept him to the straight and narrow. "Drunkenness is the mother of every vice," he noted in his diary, alcohol the fuel that drove asocial acts. Wine, which appealed to John Quincy who had learned in Europe the attraction of moderate amounts, sometimes sent him tipsy to bed. But his internal monitor kept that to a minimum, especially during his year of hard study in 1786–1787. Warnings and advice came from Abigail. He was well out of Europe because the temptation to vice would be less in America, but young men always needed to be advised against drink and sex. The first would lead to the second. Then came gambling, which seemed to Abigail a national preoccupation in England, along with lying and scandal. At Harvard, they were companionable sports. Abigail particularly worried that Charles would be tempted. Would John Quincy stand guard over his brother? What the Adamses most feared was personal dissoluteness. For their nation, they feared that prosperity would sap the new country's moral rectitude, its dedication to the principles of the revolution.

Vacations were spent mostly at Haverhill and Braintree, except for the Christmas holiday of 1786. Since there was an acute shortage of wood, Harvard started the vacation early. John Quincy got permission to stay on campus, where, despite the chill, he read and studied. The quiet and solitude appealed to him. Crowded conditions at Braintree kept him from studying or sleeping. The boys "make such a noise in the morning as would make you laugh," Mary Cranch wrote to her sister. "In Charles," Elizabeth Shaw wrote, "I behold those qualities that form the engaging, the well accomplished gentleman, the friend of science, the favorite of the *Muses*, and the *Graces*, as well as of the ladies." Abigail might have seen in this assessment, disguised as a compliment, things to be concerned about. "In Thomas . . . I discern a more martial, and intrepid spirit . . . a love of business . . . indefatigable in everything." In both cases, Elizabeth Shaw proved a poor prophet.

From the start, it seems as if Abigail and John Adams put more intensity into and raised higher hopes for their eldest son than for his brothers. Charles was already at Harvard when John Quincy arrived, and Thomas enrolled the next year. Neither was particularly studious. A friend reported to Abigail that at Haverhill John Quincy was "as studious as a hermit." The brothers hunted and fished at Haverhill and Braintree. At the Shaws', they slept together in one large bed, companionable siblings in a world in which beds were often shared. In July 1786, Thomas entered as a freshman. Their cousin and friend Billy Cranch studied shoulder to shoulder with John Quincy. "If you [were] to see them all together it would give you great pleasure," Mary Cranch wrote to her sister. "Four more promising youths are seldom seen, may nothing happen to blast our hopes!"

No matter its flaws, Harvard was, in John Quincy's view, "upon a much better plan" than any university he had seen in Europe. It took seriously its mission to teach morals and civic responsibility; its commitment was to the values and welfare of the new republic.

After all, the purpose of education was to train men to be effective stewards of their own lives and civic leaders in religion, law, and government. The good life for the individual and nation depended on the inseparability of moral conduct and character; on honorable conduct; on respect for piety, learning, and the law; on discipline and hard work; on unshakable patriotism; and on everything else that John Adams had taught his son to associate with Christian and republican virtue. The test was conduct, not catechism, and to the extent that many of his fellow students failed it, John Quincy hoped that some of them would eventually do better. Such was human nature, not to be condemned but to be educated and encouraged into improvement. Parentage counted, but training was crucial, and the opportunities Harvard provided for betterment became the grounds of his gratitude. Despite the rote nature of the curriculum and the inadequacy of the tutors, he flourished in his refusal to make the institution's flaws an obstacle to his own progress. And he embraced its main strengths: the science, mathematics, and philosophy courses, and the system of essay preparations on set topics for oral presentation and debate, usually about moral choices and civic responsibilities. He could already write with expressive facility and a high level of precision and energy. His diary kept the synergy flowing between his mind and pen. He now needed to combine these talents into cohesive and persuasive structures, and to learn how to give them effective oral expression.

In April 1786, he heard his first forensic disputation, a set assignment on "the question of whether a democratical form of government was the best." The entire class participated, seriatim, one arguing that it was best, the next the contrary. "This is one of the excellent institutions of this University," John Quincy wrote to his sister. The question put to him in his first disputation was "whether the immortality of the human soul is probable from natural reason." He had been assigned the affirmative. "But it so happens that whatever the question may be, I must support it." In an essay on the widespread belief that death liberates the human soul from its body, he invoked the example of "the enslaved African, bending under the weight of oppression and scourged by the rod of tyranny," who "sighs for the day when death shall put a period to his woes, and his soul again return to be happy in his native country."

There was an unleashed energy about John Quincy's studiousness startling even to his family and more so to his Harvard contemporaries. He aimed to study at least six hours a day in addition to classes. This was partly a rebound from his desultory European years, partly an effort to emulate his father, for whom he had unbounded admiration, even to the extent, his aunt Elizabeth remarked, of imitating his posture and walk. And he had inherited his father's tendency to gain weight, though Abigail took some credit for this feature also. Exercise, she urged, to keep the fat off. He had been trained from early on to believe not only that he had a great future but that success depended upon his unremitting efforts. "Near as we are to Boston," he told his father, "I have been there only once. . . . A person who wishes to make any figure as a scholar at this University must not spend much time, either in visiting or in being visited." There was another price to be paid. Some of his classmates resented him. That he was the son of the famous John Adams did not help. He wanted to be liked, but he wanted more to be successful. He also worried that his ambition had the potential to be self-destructive. If he reached too high, he might too readily fall. Ambition was neither a Christian nor a republican virtue, and merit should be its own reward. It should not need to tout itself. Unfulfilled ambition might gnaw at the spirit and body. Fulfilled ambition, however, might attract resentment. Honors and rewards should search out deserving men, not be sought after. It was a view that father and son shared, always in principle if not in performance. Having been bred to be ambitious, John Quincy tried to make his ambition as little visible as possible.

But his work ethic and his talents made him unusually noticeable. Attracted to music, he learned to play the flute. There were two elite societies at Harvard, each with ten or so members. Both held regular meetings, requiring the preparation of written essays on set subjects for oral presentation and debate. John Quincy was elected to the small Harvard Phi Beta Kappa chapter in June 1786. The A.B. Club, a local creation, also met regularly in student rooms, for collegiality and intellectual stimulation. He rarely missed meetings of either, part of a select company that included his closest college friends, his cousin Billy Cranch, Leonard White from Haverhill, and James Bridge from Maine. At Phi Beta Kappa, he had his competitor for class honors, Nathaniel Freeman; Moses Little from Newbury, bright and engaging, became a friend; James Forbes was youthful and unfocused but charming; and Henry Ware was the tutor with whom he roomed during his first term. One of the prices he paid for keeping his nose to the grindstone was self-sacrifice. "It is against the law for me to look at a young lady until [graduation]," he joked to his sister, "and then I suppose it will be too late." But the law was observed in the breach. Young ladies were particularly plentiful in Boston, where he dined occasionally at the home of Francis Dana.

In the spring of 1786 he wrote for the A.B. Club an argument in favor of education and civilization, a topic that elicited from him the remark that "ideas of happiness appear always to be local, and always adapted to the situation of men." For Phi Beta Kappa he wrote an essay on "whether civil discord is advantageous to Society." It was, he proposed, an advantage in a republic to have an organized opposition to the governing party. "Which so ever of the Party is at the head of the government is sensible that the other will take advantage of every error, every mistake, and even every ill success that may attend the administration; and will consequently make more exertions to preserve and increase the favor of the people in general than if it was perfectly secure in power." There is no indication that he had been assigned the affirmative. He apparently assumed that both parties would be honest and honorable in placing the country's welfare before their own. Before an audience of almost four hundred, he and Billy Cranch debated "whether inequality among the citizens is necessary for the preservation of the liberty of the whole." Mary Cranch reported to her sister that "they did not either of them speak loud enough. . . . Other ways they performed well." Over the next year, John Quincy addressed more than a dozen such subjects, from the power of music and poetry to whether Christianity had been a force for good.

His heart and mind seemed especially engaged with the nature of the love between Shakespeare's Desdemona and Othello on the topic "whether love or fortune ought to be the chief inducement to marriage." He had strong feelings about the issue. Although *Othello* was in every other way a perfect play, making a black man the object of Iago's jealousy and of Desdemona's love seemed to him an unrealistic and undesirable transgression of racial boundaries. And to present Desdemona's betrayal of her father's authority in positive terms seemed reprehensible. It was a view of the play he was to return to and develop at length later in his life. It perplexed him that his revered Shakespeare could confound what he and his contemporaries believed the normative understanding of race and marriage. Blacks deserved the freedom that all humans deserved, and slavery was detestable. But physical attraction between whites and blacks, let alone marriage, broke sacred codes of nature. It rose to the level of the repellent. And, he argued, nature had not intended men to marry entirely for love. Youthful passion over time inevitably diminished. It alone could not sustain a marriage. Neither wealth nor passion should be the chief pursuit in selecting a wife. The sustaining bond ought to be "mutual esteem." And "the only difference between mutual esteem and love is that the one is founded only upon reason, to which the other is diametrically opposed." He charmingly confessed, though, that he was not "obstinately attached" to these views, "and should any arguments be produced sufficient to convince me that they are erroneous, I shall retract them without hesitation."

He was also learning how to write. He had no need to give thought to the importance of that—it was a given. Good writing and good character were connected. Anyone with the ambition to contribute to knowledge, society, and country needed to add skill to talent, and to learn how to write effectively in a variety of genres. That required discipline, training, and persistent effort. At Harvard, he learned to write essays. By the time he graduated, the level was high: effective sentences, substantial paragraphs, precise word choice, a range of literary and other sources to draw on, a sense of overall structure, and a firm grasp of logical progression in a prose that was increasingly fluent, firm, and persuasive. At the same time, he could not keep away from poetry, some of it emotionally expressive, some humorous or satiric. Good prose style had no genre limitations, as he had been educated to see, especially under the influence of his parents, talented and skillful letter writers. By the late 1780s, John Quincy had learned to be a masterful letter writer too. He had some of his mother's spontaneous fluency and expressiveness, and he learned from his parents and his reading the varieties of personal and formal tones that the letter as a genre allowed. His diary, though, often embodied his writing at its best. There, in privacy, he could express himself spontaneously. It was a record of daily life, but it was also an analysis of self and society, of what he thought, of who he was, and of what other people seemed to him to be. It was a work in progress, extendable to the limits of his life, to become a repository of poems, prayers, pen portraits, self-analysis, descriptions of places and travel, accounts of political events and ideas, notes on family life, and thoughts and speculations about religion and philosophy.

The pen portraits of some of the classmates he liked are precise, expressive, finely balanced, and mostly laudatory. "*William Cranch* of Braintree was 17 the 17th of last July. The ties of blood, strengthened by those of the sincerest friendship, unite me to him. . . . Our sentiments on most subjects are so perfectly similar that I could not praise his without being conscious of expressing a tacit applause of my own." Those of classmates he disliked are even more riveting examples of his skill. "Solomon Vose of Milton Suffolk C, was 20 the 22d of February; a vain, envious, malicious, noisy, stupid fellow, as ever disgraced God's creation; without a virtue to compensate for his vices, and without a spark of genius to justify his arrogance; possessing all the scurrility of a cynic with all the baseness of a coward." Joseph Jackson "was 19 the 27th of last October. His countenance is of a brown inexpressive cast, and his face is as perfect a blank as his mind. His eyes are black, and always in an unmeaning stare. He is extremely dull of apprehension, and possesses no other talent that that of pouring forth with profusion the language of Billingsgate. If I was called to point out the smallest genius in the class, I should show him: if the most indolent and negligent student, he would be the man; but at the same time I must do him the justice to say he is not vicious; and when all the faults which the man has may be attributed to nature, perhaps we ought not to find fault with him. Died. August 1790." That John Quincy went back to this diary entry some years later to provide the finality of the date of Jackson's early death emphasizes just how self-consciously literary his relationship to his diary was.

In early July 1787, twenty-year-old John Quincy wandered alone in the churchyard in Braintree, rambling through high grasses waving in the breeze.

He attempted to read "the inscriptions which love and friendship have written on the simple monuments." Suddenly he was "startled by a rustling noise." Looking around, he noticed "a large snake" slithering through the bending grass. "I pursued him but he soon found his hole into which he slipped and escaped my pursuit. Was it the genius of the place? . . . If it were *a gentle spirit,* some more amiable shape than that of a serpent might have been assumed; some shape which might engage the affections, and call forth the soft and pleasing passions."

There were ungentle spirits in public places and in John Quincy's thoughts. For almost a year, the nation had been confronting the rebellious actions of farmers in western Massachusetts. They believed themselves more heavily taxed and oppressively ruled by elite American legislators in faraway cities than they had been by the British colonial government. Money was tight: taxes were high, personal and war debts pinched, government costs considerable. Led by Daniel Shays, they disobeyed laws, pilloried tax collectors, prevented judgments against debtors, and defied state and federal law, borrowers in rebellion against lenders. The snake of "Don't Tread on Me" had been reborn as a rebellion against the downside of self-rule, the American objection to paying the cost of state and national governance.

In the fight between those who had money and made the laws and those who had debts, the debtors felt powerless except to respond with violence. Massachusetts was the battleground.

In September 1786, when riots broke out in the Berkshires, the governor urged the government to quell them immediately. In Boston, which feared it was the ultimate target, the militia vowed to defend the city, the state government to crush the rebels. A rumor spread that thousands were coming to attack the courts and the legislature. "Where this will end time alone can disclose," Adams wrote in his diary. "I fear it will not be before some blood is shed . . . a civil war, with all its horrors." Like his family and his class, he had no doubt of who or what was at fault. "The people complain of grievances . . . the court . . . the Senate, the salaries of public officers, the taxes in general, are all grievances because they are expensive: these may serve as pretences, but the malcontents must look to themselves, to their idleness, their dissipation and extravagance. . . . These have led them to contract debts, and at the same time have rendered them incapable of paying them." At Harvard, students, reviving a Revolutionary War Latin tag that meant "as much for Mars as for Mercury," drilled and paraded in the college yard. Although John Quincy did not join, he shared the view that the rebellion had to be crushed. By December, government troops made various arrests. In January 1787, reinforcements marched westward to repulse an attack on the arsenal at Springfield. The rebels were killed, captured, or dispersed. Shays fled to New Hampshire.

The rebellion, though successfully repressed, left a strong impression on the minds of Adams and his contemporaries. Could the interests of different groups in society be reconciled peacefully? Would disparities in wealth create unsustainable tensions? Would the nation be rational and prosperous enough to create a system of sound money? Would a standing army be necessary to ensure obedience to the law and respect for the country's institutions? Could the thirteen colonies be sustained as a single nation? And could the work ethic, the abhorrence of debt, and the commitment to progress through self-improvement, which Adams believed to be the essence of the New England character, also prove to be the character of the country as a whole? If not, the snake that slithered through the grass would cast dark, even poisonous shadows on the nation's future.

As commencement day approached, what would come next was much on John Quincy's mind. Harvard had "engaged" his affections. It had provided for a year and a half a "guardian spirit" that would accompany him wherever he should find himself in the future. It had provided a protected space for study "without the avocations of business or the hurry of life." The coming separation from four or five of his classmates "saddens very much the anticipation of commencement, when we must part, perhaps forever. . . . We shall never meet again, all together." Whatever he would do next would not be, he believed, as congenial as what he was leaving. "Here void of every care, enjoying every advantage for which my heart could wish, I have passed my time without the perplexities with which life is surrounded." Already he was anticipating a future in which, with mingled pleasure and pain, he would remember his college days. He projected himself into a future in which he would relive the depression that he was feeling in the present. In his bleak mood, he noted, "these disagreeable reflections haunt me continually and embitter the last days of my college life."

In London, his parents believed that his commitment should be to the law. That was the road to self-sufficiency. John Quincy accepted that as a given, but not with enthusiasm. The clergy was out of the question. He had no interest in business or medicine. The notion of being his father's law student was attractive and would save fees. But, though John and Abigail yearned to return to America as soon as possible, Adams' recall had been delayed for at least another year. And whether he would be asked to continue in government service or be allowed to return to private life remained an open question. Adams made inquiries for his son, and John Quincy looked around. A three-year clerkship in Francis Dana's law office in Boston had seemed attractive, though the young man worried that Boston's attractions might undercut his concentration. When Dana, in March 1787, suffered a stroke from which he would never fully recover, that avenue was closed, though Dana proved well enough over the next decades to continue his law practice and to serve as chief justice of the Massachusetts Supreme Judicial Court. "To me, he has been a second father," John Quincy wrote of his traveling companion on the journey to Russia. He spent some days and nights at the Dana home, attempting to be of help. "I was shocked at seeing him; pale, emaciated and feeble, he scarcely looks the same man he was three weeks ago." Theophilus Parsons, the thirty-seven-year-old Newburyport lawyer who had participated with John Adams in the Massachusetts constitutional convention in 1779 and was soon to have a distinguished judicial career, was another possibility. He had been mentioned to John Quincy the previous year. "I should be very glad to study with him," he wrote to his father, who approved of Parsons. They met at a dinner at Dana's in June, a month before graduation, and Parsons seemed "a man of great wit, as well as of sound judgment and deep learning." Cotton Tufts, Abigail's Boston cousin who handled the Adamses' financial affairs, negotiated the details. By the end of June, John Quincy's apprenticeship had been arranged. He was to spend much of the next three years in Newburyport.

Chapter 4
Most Beautiful and Beloved
1787–1794

The stagecoach from Boston to Newburyport left at 3 A.M. It was a summer morning in September 1787. Twenty-year-old John Quincy had been John or Johnny to his family. Occasionally he had been JQA. To the world, he was now Adams, a youthful, slimmer version of his father, almost trim enough to look fit, with a full head of hair, expressive eyes, and a fair complexion, not handsome but attractive, a face molded in the Adams character. The twelve-hour journey gave him time to review the events of the previous month. Commencement had been held in mid-July. Parts for the ceremony had been distributed a month before. Ranking second, he had no reason to be ashamed of his class standing, though Nathaniel Freeman had bested him in the competition for valedictorian. Not that he felt envy or resentment—they were equals, he believed. He regretted not being even more successful, but he admired Freeman, especially his writing and speaking skills. What Adams most regretted is that he had not left Europe for Harvard a year and a half earlier.

His class ranking entitled him to the reward of "An English Oration." The subject assigned was "The importance and necessity of public faith to the well-being of a Community," a subject he found appealing. For the first time, he had a public forum, an audience to address what he and others thought a momentous subject: the difficulties facing the new country as it tried to deal with its transformation to successful self-rule. "I am led unawares into political ground," he wrote to his father, "and now I am there I must indulge myself." The issue was the national debt. It had three levels: the debts owed to British merchants stipulated by the peace treaty, those incurred by individual states to pay their war costs, and the debts the federal government owed to Holland and France for loans that had kept the government afloat. Without repayment, no one would value the nation's credit. The crux of the matter, as Adams saw it, was national honor.

Nervous, hesitant, sharper in intellect and writing than in speaking skills, he made an appeal to the "spirit of patriotism." An adequate performance, it was at its best when most direct, less effective when his rhetoric soared. It was addressed to a friendly audience, the more sophisticated of whom knew that the grandiloquent appeal, with references to Greece, Rome, and Great Britain, was in essence a political speech. Between the lines, it advocated that the federal government be given taxing power to allow it to repay its war debts. The debate on a new constitution was under way. There were sharp differences of opinion about whether to redistribute, and to what degree, power between the states and the federal government. And, unsurprisingly, the Boston newspaper accounts of the commencement discussed John Quincy's address in the context of his father's well-known views, some of them widely shared: a nation's word in a peace treaty was its bond; sound money, otherwise known as hard currency, was mandatory; paper currency was thievery; personal and public debts needed to be repaid over time and on a stipulated and sacred schedule; and a strong federal executive, with veto power over the acts of a popularly elected legislature, was essential to the security and prosperity of the country. John Quincy ended his oration, as he had begun, on a personal note, his "consciousness" of his "own insufficiency. . . . Warmed by that friendship," he concluded, directing his voice to his classmates, "which will ever be the pride and comfort of my life, I can . . . therefore only exhort you, when you shall be advanced upon the theater of the world; when your country shall call upon you to assist in her councils, or to defend her with your fortunes and your lives . . . to retain those severe republican virtues . . . which alone can effectually support the glorious cause of Freedom and Virtue. . . . And may national honor and integrity distinguish the American commonwealths, till the last trump shall announce the dissolution of the world, and the whole frame of nature shall be consumed in one universal conflagration."

Glowing in the praise of classmates and family, John Quincy was gratified and slightly frightened when, after the commencement ceremony, Jeremy Belknap, a well-known historian, asked to publish his oration in the *Columbian Magazine*. It would be published in Philadelphia, where the convention deliberating on a new federal constitution was meeting. Many in his audience would have been Federalists, ready to support increasing the power of the national government, not that the proposals of the Constitutional Convention were known in July 1787. Its work was not to be made public until mid-September, the start of a vigorous yearlong public debate. But though the delegates were pledged to secrecy, the major issues were widely acknowledged, and John Quincy himself would only gradually come around to supporting the Constitution as a whole. He was in a quandary about whether to give Belknap his assent. Not willing to take the step himself, he left it to Belknap whether or not to publish, which was tantamount to consent. John Adams in London read it with unmitigated pleasure. In 1822, John Quincy reread it "with humiliation; to think how proud of it I was then, and how much I must blush for it now."

There was no blushing for the total commitment to a life in which love of country was inseparable from his private and public voice. And "country" was always in the air at Harvard during his brief time there. Over fifty years later, he recalled a striking moment that took its power from a quotation from Cicero. It was the end of his senior year. The military company, formed to respond to Shays' rebellion, had assembled for the retirement ceremony of the graduating officers and the swearing in of their successors. William Cranch was the adjutant. Solomon Vose was the captain. Adams was there as a spectator. On June 2, he had written a scathing pen portrait of Vose that verged on the nasty. In 1839, that was not in his memory. Vose was "a fine, handsome fellow, six foot tall" who "delivered . . . a very handsome valedictory as they were drawn up in the yard fronting Harvard College." Adams recalled that he was "much struck" with Vose's introduction of a passage from Cicero into his speech. In 1787, even poor Harvard students knew and expressed themselves in Latin: "But when with a rational spirit you have surveyed the whole field, there is no social relation among them all more close, none more dear, than that which links each one of us with our country. Parents are dear; dear are children, relatives, friends; but one's native land embraces all our loves; and who that is true would hesitate to give his life for her, if by his death he could render her a service?" That passage, he wrote, "has never gone out of my memory since that day."

As he entered Newburyport, John Quincy felt that three years in this provincial seaport town, forty miles northeast of Boston, would be a challenge. "He knew no one," he tearfully told his aunt Mary, "and nobody cared for him." She consoled him by emphasizing that his aunt Elizabeth, "who loved him like a parent," lived only fourteen miles away. He had chosen Newburyport partly because he feared that Boston's social attractions would interfere with keeping his nose to the legal grindstone. Now he worried that it would be a social desert. He would be isolated, with nothing to do but study, which would undermine his spirits and health. In April, at Harvard, he had spoken in defense of the profession in a debate on "the comparative utility of *Law, Physic, and Divinity.*" Attacks on the legal profession were pervasive, Adams granted, fueled by those who had lost disputes and who blamed their lawyers, by the popular conviction that lawyers prospered from conflict, by widespread antagonism to the courts, and by the common conviction that lawyers dominated an exploitative governing structure. But, he argued, the law and a court system were necessary to ensure the rights and well-being of the individual. And "the intimate connection between" the law and government "must be obvious to everyone." In a free government, lawyers like Demosthenes and Cicero become the embodiment of the society's commitment to justice and patriotism.

Anyway, there was no practical alternative to the law, and that meant assiduous study. Legal education was essentially an intensive reading course, starting with William Blackstone's *Commentaries on the Laws of England* and proceeding through various eighteenth-century British texts on common law, statute law, pleadings, admiralty law, and so on. Law clerks spent a good deal of time copying documents, since there were no devices for mechanical duplication. Before leaving Boston, John Quincy had spent a day as an observer at court, where registered law students could see the law in practice. He soon did the same in

Newburyport, where he watched Parsons in action. He quickly discovered that he would have little direct contact with his mentor; the popular lawyer was busy with clients and cases, often in distant places. It was an education, at least in the first two years, mostly by reading. And John Quincy had to put aside books like Henry Fielding's *Tom Jones*, "one of the best novels in the language," for Blackstone. *Tom Jones* "cannot lead a person to form too favorable an opinion of human nature," he noted, "but neither will it give a false one." Adams well knew that the same could be said of the law. Although for almost three years he concentrated on legal texts, he still preferred books like Jean-Jacques Rousseau's *Confessions*, "the most extraordinary book I ever read in my life." He had hoped to have more time to read "books of entertainment. . . . But [not] after passing eight hours a day in the office, and in spending four hours more in writing minutes, and forms at home." He did, though, find time to read poetry. "In Thomson's *Castle of Indolence* I first read 'I care not, Fortune, what you me deny: / You cannot rob me of free nature's grace. / You cannot shut the windows of the sky / Through which aurora shows her brightening face.'" But as an attempt, even forty years later, to argue that he had found ways to overcome low spirits, it seems more willful than convincing.

During much of the winter and spring of 1787–1788, he worried that his health was suffering from late night studying. Most nights he took notes, summarizing his reading until his hand hurt. Often his eyes ached. "I nearly destroyed myself by late evening study," he wrote years later, "by sitting up till one and two o'clock in the morning reading and writing. . . . I lost nearly the whole of the succeeding winter by sleepless nights followed by days of languor, dejection and debility, which disqualified me for all close study or intense application of mind." Actually, he was robust in the essentials, though his eyes sometimes became inflamed and runny, which reading and writing in weak light may have exacerbated. "My eyes and health begin to fail," he complained at the time, "and I do not feel that ardor for application which I should have." He was partly driven by desperation, which contributed to low spirits, his life heavily confined by hard study, which he often did not enjoy, and his realization that the legal profession was already overcrowded. Far from home and family, he was depressed. "My prospects appear darker to me every day." He turned to his diary, to the act of writing, for solace. Feelings and thoughts rather than daily events dominated his entries. He speculated about his methods and qualities as a diary writer. Was it not simply too difficult to make up arrears when he missed a day or more? And was the diary's most useful function not as a place in which to whine and complain? Could he always count on its privacy being secure and, if not, how might that influence the entries? He made speculative remarks about affection and love, especially in the relationship between parents and children. By the end of the year he felt better. Was it worth keeping this diary, he asked himself, considering how much time it demanded? On New Year's Day 1788, he resolved to keep it for another year. "When I look back at these volumes . . . I am at least able to say, at the close of the day, that day I did something."

No matter how busy or low-spirited he was, there was one unavoidable topic: the debate about the Constitution. Through the fall and winter of 1787–1788, divisive discussion preoccupied the country. At the statehouse in Boston voices on both sides argued passionately, though it seemed likely that Massachusetts would vote for approval. Nationwide, the division was close. To some, an empowered federal government under a strong executive seemed essential. The proposed Constitution provided rules for commerce between the states, established uniform national import taxes, prevented states from whirling off in their own directions, curbed the excesses of state legislatures, arranged for orderly development and protection of

Western lands, spoke with one voice in foreign affairs, and created a defense system to safeguard the nation's sovereignty. To others, it offered both too little and too much. That it did not contain a Bill of Rights frightened those who feared the reestablishment of a British-like tyranny. To those who preferred a loose confederacy in which one's primary loyalty was to a state, it was a horror. The least government possible was the best protection for slavery. Local independence protected constituencies from national values and majority rule. And while most Northern states had equitable tax policies, minimal corruption in government, and an eighteenth-century version of democratic representation, Virginia and most Southern states were governed by dictatorships of the elite.

During the autumn, John Quincy and Billy Cranch carried on a miniature debate of their own. Was it really desirable, John Quincy asked, to have three branches of government? Would that not create confusion and competition? How would the president defend himself from encroachment on his authority? Worse, the almost unlimited power granted to Congress might result in congressional despotism. By early February 1788, when it received the approval of the Massachusetts convention by a majority of only nineteen votes, he had been converted though he was not convinced. He worried that the Constitution was neither democratic enough nor cohesive enough. The Union would be too weak. "I am still of opinion that if this constitution is adopted we shall go the way of all the world: we shall in a short time slide into an aspiring aristocracy and finally tumble into an absolute monarchy or else split into twenty separate distinct nations perpetually at war with one another; which god forbid!" But "I think it my duty to submit without murmuring." Representative government, he agreed, requires assent to majority rule, and this Constitution was on its way to becoming the law of the land. It seemed to him better than the alternatives. "I must freely acknowledge that I still lament the want of an adequate representation of the people; and of rotation in the offices of government: the blending of the legislative and executive powers in the Senate; and the indefinite powers granted to the administrators: but I am convinced that opposition now would be attended with immediate evils without being productive of any good effects and you may now consider me as a strong *federalist*." It was this Constitution or a worse one. Over time, he was to become its outspoken defender and an advocate of its broad interpretation.

Reading law had its pleasures and pains. Blackstone was a delight. Edward Coke was turgid and abominable. While Adams shared Parsons' pro-federalism, his mentor's boasts about the tricks he had pulled during the debates in Boston were off-putting, the kind of chicanery that revealed poor character and bad government. At the office, when his clerks protested that Parsons' loud conversations kept them from concentrating, Parsons responded that it was necessary to learn to read despite distractions. He preferred talking to his students about literature and politics rather than law. Sitting in a rocking chair, cutting his chewing tobacco with a penknife, he quoted Francis Bacon: "Reading makes a full man, conversation a ready man and writing a correct man. Now young gentlemen I would have all full, ready, and correct." John Quincy always remembered it as good advice. In midwinter he went by sleigh with friends to Exeter, New Hampshire, to hear the state's delegates debate whether or not to ratify the federal Constitution. Their level of argument seemed so low that it shed no light other than on the poor quality of the delegation.

In Newburyport, he attended church on Sundays, sometimes twice, his religious feelings rational and sincere. And, after all, what was there in the way of entertainment on Sundays in midwinter in eighteenth-century Massachusetts? Sunday sermons provided an intellectual and a writing focus. He made evaluative entries in his diary about the rhetorical and intellectual abilities of the ministers, the start of a lifelong practice of devoting his Sunday entries to the effectiveness of that day's sermons. He noted that a visiting preacher was "common-place: his ideas were trifling, his language was inelegant and his manner was an unsuccessful attempt at the florid. . . . I had quite enough in hearing him once, and therefore in the afternoon I went to hear Mr. Spring, who entertained me much better, though I am not a great admirer of his doctrine."

Newburyport social life also had its pleasures. When he first arrived, he knew almost no one. That soon changed. He visited Catherine Jones, a young lady he had met in Cambridge, and others to whom he had introductions, the daughters of successful Newburyport merchants. His classmate and friend Moses Little, studying medicine with a local doctor, and Samuel Putnam, another classmate, clerking elsewhere because there had been no room for him at Parsons', became daily companions. At the beginning of the new year, Putnam moved into the same lodgings. The three shared walks and discussions, joined by Horatio Townsend, a Harvard friend who also clerked in Parsons' cramped space. Soon they were joined by James Bridge, John Quincy's closest Harvard friend, forming an intellectual club with some other young men, eight in all, meeting once a week. There was occasional late night carousing. And they all participated in teas, walks, and dances in mixed company. Unable to focus on his studies, by mid-February 1788 John Quincy had been "somewhat in the dumps" for six weeks, he complained to William Cranch. "I have in great measure . . . gone into a course of dissipation. . . . Balls sleighing parties and visiting occupy the whole of my time."

Female company often challenged his values and self-image. His attitude was partly conventional: women were, in general, vain and heartless. There were exceptions, pure and generous, beautiful in body and soul. At the moment, though, there was no such lady in sight, and he was not an eligible suitor even if there had been. Still, he had to work at restraining desire. In this regard, he had the advantage of being self-conscious about his awkwardness in the company of women. That was sufficiently deflating to cause him pain and anger, and his criticism, often scathing, of women he met was mostly self-defensive. But he rarely declined to go to social gatherings. There he found music, singing, and dancing. Kissing games were played, courtships in progress: "I trust my Reason will for at least seven years to come, preserve my heart as free, as it has been here," he had written to his mother in 1787. The "here" was Cambridge, where campus life encouraged ascetic self-regulation. At Newburyport, there was less pressure for self-restriction. He was annoyed when told by teasing mischief makers that he was in love, "which you know is as good as to be caught with a sheep under my arm," he joked to Billy. But he left open the inference that he was stealing the sheep or making other use of it. "When my sullens leave me, I will give you some female characters, from among my acquaintance: you may depend upon it, the theme will be rich whether for satire or for panegyric." The satiric tradition was readily at hand. His much-admired Pope provided the cultural touchstone for the struggle between beauty and virtue and vanity and good sense in the widely read *The Rape of the Lock.*

John Quincy began a longer version of what he had started the year before, a satiric exposure of the foibles of women he knew, embedding the few lines he had previously written into a substantial poem called "A Vision." Working on it for the next year and a half, he finished it in June 1790. It satirizes seven women, modeled on Newburyport young ladies, each an embodiment of a particular female flaw. In a dream vision, the narrator pleads with Cupid not to pierce him with an arrow except in the presence of the person of his own choice, for he is someone "whom *nature* formed without an art / To win the soul or captivate the heart." Assenting to the narrator's request, Cupid gives him the opportunity to select one from a parade of women. "For in the search we possibly may find / Some who may possess the beauties of the mind." Lucinda, Belinda, Narcissa, Vanessa, Corinna, Nerea, and Almira pass in sequence before him. Each is flawed. One lacks a "*feeling heart,*" another is "a compound strange of *vanity* and *pride,*" another snobbish; one is nasty, one is a prattler, another stupid, another . . . Then an eighth woman appears:

> *With graceful steps the lovely CLARA moved,*
> *I saw, I gazed, I listened, and I loved. . . .*
> *The partial gods presiding at her birth*
> *Gave Clara beauty when they gave her worth.*
> *Kind Nature formed of purest white her skin,*
> *An emblem of her innocence within. . . .*
> *While Venus added, to complete the fair,*
> *The eyes blue languish and the golden hair.*
> *But far superior charms exalt her mind,*
> *Adorned by nature and by art refined;*
> *Hers are the last beauties of the heart.*
> *The charms which nature only can impart;*
> *The generous purpose and the soul sincere,*
> *Meek sorrow's sigh and gentle pity's tear.*
> *Ah, lovely Clara! can a heart like thine,*
> *Accept the tribute of a muse like mine?*

Clara almost certainly is Mary Frazier, whom John Quincy met in early 1788; she was the younger of the two daughters of a Newburyport merchant, Moses Frazier. Born in 1774, with blue eyes, blond hair, and a fair complexion, Mary was "to me," Adams recalled fifty years later, "the most beautiful and beloved of her sex." Probably most of the poem was written in 1789, the last section perhaps late in that year or early in the next. By May 1788, Adams and his friends were frequent visitors at the Frazier home. In mid-August, he joined a group of young friends who spent many evenings there. "The young Misses have assumed an importance rather above their years. . . . I receive not much satisfaction in their company, and as they are handsome, I had rather look at them for five minutes than be with them five hours." An acquaintance, Samuel Breck, to whom Adams gave a copy of "The Vision," years later recalled that "there was an exception to [the poem's] widespread vituperation" of women, Mary Frazier. "She was so exquisitely beautiful, so faultless in feature, complexion, expression. . . . I knew Miss Frazier well, and can testify to the mildness of her blue eyes, the plumpness of her cherry lips and her carmine cheeks, but above all the fascinating charm of her eyes." Breck recalled that John Quincy "was exceedingly in love with her. . . . His was a passion of unusual violence."

Mary was to be sixteen in 1790, a marriageable age, and John Quincy twenty-two. He was with her on her sixteenth birthday. From November 1789 to August 1790, he saw her frequently. They took private walks together, suggesting that Mary's parents as well as John Quincy's Newburyport circle knew that this was a serious relationship. "In the 15th year of her age," he wrote in his diary fifty years later, "she gave me then 22 the assurance of her affection and the pledge of her faith." What were the conditions under which a mutual pledge could be made and sustained? He had only the distant prospect of being able to support a wife and family. Although he was close to the end of his legal clerkship, an independent career beyond simple quali- fication was charged with uncertainty. Where would he practice law? And in a crowded profession, would it not take him years to attract enough clients to support himself, let alone a family? In the meantime, he would be dependent on his father, a condition he found barely tolerable even now. He could anticipate his parents' response to the announcement that he was in love. If they should discover this from a source other than him- self, it would be even worse. His primary pledge, to his parents and to himself, was to avoid all impediments to establishing a career, which meant not marrying until a career was well under way. If he did indeed pledge

himself to Mary, they would have to agree to postpone the marriage indefinitely, which she or her family
might find unacceptable. "*You* may know (though it
is known to very few)," he wrote to his friend James
Bridge, who quoted back to him his own words, "that
all my hopes of future happiness in this life, centre, in
the possession of that girl." But a passionate love was
confronting an unmovable obstacle.

Nabby and William Stephens Smith had married in
London in June 1786. Abigail had extolled Smith's virtues, eager to erase the potential stain on her daughter's reputation from the abrupt termination of her
engagement to Tyler and the short time that had
elapsed before her commitment to Smith. Prickly,
proud, and confident, Smith believed he deserved the
rewards of his country as a man of proven service and
high character. A connection with the Adams family
confirmed his sense of himself. John Adams hoped
that Smith would study law, which could provide an
honorable livelihood for a man of independent character. His son-in-law probably thought that view came
too easily to a man who continued to be the benefi-
ciary of government appointments.

In January 1787, Adams informed Congress that
he would resign his office within the year and return

home. In March 1788, Nabby left for New York with her husband and infant son, and in April John and Abigail sailed for Boston, eager to be reunited with their sons—Charles and Thomas at Harvard, John Quincy in Newburyport. The younger sons, who had participated in a student riot, were a worry. The official account does not make clear the extent to which they were culpable, but Charles was relieved of his job as a dining hall waiter for his refusal to provide eyewitness testimony, and Thomas was fined for participation. Before sailing, John and Abigail received both condemnatory and exculpatory accounts. John Quincy, who had reason to believe the worst, lectured his brothers, both of whom denied his right to play that role.

When the Adamses arrived in Braintree in mid-June 1788, after a celebratory welcome by the Boston elite, they were astounded to realize that they had returned to a different country from the one they had left. It seemed bitterly divided, the unity of the Revolutionary War days gone. The self-sacrifice that had characterized the struggle for independence had apparently been replaced by slothful and materialistic self-indulgence. And what exactly would John Adams do for the rest of his professional life and to support his family? The thrifty Adamses had found ways to save. Abigail had used her profits from the sale of dry goods sent to her from Europe to purchase government bonds, which had increased in value. And she made regular additional purchases in a depreciated market. Whereas Abigail preferred paper assets, John Adams had only one material passion. Whenever he had cash available, he instructed his business manager, Cotton Tufts, to keep his eye out for Braintree real estate, especially anything adjacent to the four properties he already owned. And that he was returning to Braintree meant that he and Abigail needed a residence there that suited their needs. His birthplace and the adjacent house were not suitable or appropriate for expansion. He and Abigail did not need a mansion. But they were delighted to be able to buy a fifth Braintree property, a modest, attractive, and expandable early-eighteenth-century house with eighty-three acres known as the Vassall-Borland estate. John Quincy came from Newburyport to help get it in shape for his parents' return.

It was the home in which John and Abigail were to live the rest of their lives. But it was not to be their year-round residence for over another decade. On the last day of April 1789, John Adams was sworn in as the first vice president of the United States. He had received thirty-four of sixty-nine possible electoral votes; the rest were divided among ten other candidates. The divided vote for vice president expressed what was soon to be dramatically evident: the new country had deep geographical, political, and philosophical divisions, and Federalists and Antifederalists (soon to call themselves Republicans) were on the verge of mutual hatred. His new position meant residence in New York City, the temporary capital, then Philadelphia. The pay of $5,000 a year without any moving or expense money seemed painfully low. "I must be pinched and Straitened till I die, and you must have to toil and drudge as I have done," he wrote to John Quincy. Nabby, whose residence in New York had helped to make the city tolerable for Abigail, returned to London in 1792 with her husband, who went abroad twice in pursuit of a diplomatic or military post with a European government. Charles skipped his graduation in 1789 to join his parents in New York, where arrangements had been made for him to clerk in the legal office of Alexander Hamilton, though a few months later, when Hamilton was named secretary of the treasury, Charles found another mentor. Thomas briefly visited New York after his graduation in July 1790, then—to study law—continued on to Philadelphia, as did his parents and the federal government.

When John Quincy left Newburyport in the summer of 1790 to open a practice in Boston, he was pessimistic about his future. The day before he had walked "in the evening with M.F.," he noted in his diary. He "*had troubles of the heart, deep and distressing.*" That Mary Frazier would one day be his bride seemed increasingly unlikely, and the road to a successful legal career long and daunting. He was admitted to the Massachusetts bar in mid-July. Ten days later, he went back to Newburyport to see Mary. In August, he set himself up an office in the front room of his father's house on Court Street in Boston, across from the courthouse, and put out the late-eighteenth-century equivalent of a shingle. The location was good, but business was not. He returned to Newburyport for a few days in September 1790 to visit at the Frazier home and saw Mary at least once the next month in Medford, where she visited relatives. But her family required a binding promise, a public acknowledgment of his irrevocable commitment. Gradually, over the next year, his parents learned about the relationship. They and his siblings weighed in on the matter, and Nabby advised against an imprudent commitment. Actually, he had already given it up, even before his parents knew of its existence. He would not commit himself without their consent. He did not have the present means of supporting even himself. And he had no prospects good enough to allow him to think that he could support a wife and family in the near future.

Still in love with Mary, he retained some hope that he might preserve the relationship. "It was a consuming flame," he remarked years later. "Love such as I felt for that lady" was "a distressing malady: it made me restless, sick, unhappy; indeed, I may say wretched." But the lovers accepted the reality. "Dearly!—how dearly did the sacrifice of her cost me, voluntary as it was," he wrote in 1838, "for the separation was occasioned by my declining to contract an unqualified engagement. . . . Four years of exquisite wretchedness followed." He felt strongly that to make a commitment in the face of these conditions would be dishonorable. His only alternative was to suffer pain and loss. The pain "*gradually wore away.*" Very gradually, and never entirely.

When, in October 1790, the twenty-three-year-old novice lawyer stood before the Court of Common Pleas to argue his first case, his voice trembled. He had had only three hours in which to prepare—presumably his employment was last minute. "I was too much agitated to be possessed of proper presence of mind. You may judge of the figure I made," he told his mother at the tag end of a letter of bitter lamentation about his loneliness, "three hundred miles distant from every member of the family; alone in the world, without a soul to share the few joys I have, or to participate in my anxieties and suspense, which are neither few nor small." Having lost the case, easily outperformed by Harrison Gray Otis, one of Boston's young legal stars with whom he was to have a long relationship, he felt deeply his failure, especially how paralyzed he had been. His father provided perspective. "Your agitation, your confusion . . . are not at all surprising. Had you been calm and cool, unaffected and unmoved, it would have been astonishing," and why should he expect to have much business? That would take time, perhaps years. After all, people would naturally be reluctant to entrust their interests to an inexperienced lawyer. But he would, John Quincy promised himself, become better at it. And, over time, he did, though his strength was to be in the substance more than the delivery.

Settling in Boston was a choice of last resort. Newburyport was a possibility, but it had more than enough lawyers. Braintree, where his father had practiced, already had a lawyer, William Cranch. There was hardly enough business for him, and John Quincy would have no place to board other than with his aunt Mary, Cranch's mother. At Boston he had advantages: his Welsh family cousins, free office space, the primary courts, friends passing through or in residence, the entrée that his name provided, and a front-row seat from which to observe the drama of state politics. But he felt deeply the want of friends in the legal profession. Gradually, clients started to come to him, very few, then a few more. Over the next year, he began to have regular business and to handle it, with long hours of diligent preparation, competently. Still, it was tough slogging, with small remuneration. He had some friends, and a few classmates lived close by. Others, like James Bridge, were sometimes in town. John Hall, John Gardner, Daniel Sargent, and Nathan Frazier, Mary's cousin, became regular companions. He and Billy Cranch continued their intimacy, both treading the well-worn path between Boston and Braintree, where John Quincy hunted and fished. As always, he found Parson Wibird's Sunday sermons boring. Some evenings were spent at a loosely formed Wednesday Evening Club, others at social gatherings. The parties seemed mostly dull and awkward. A capable dancer, he went to assemblies, the formal dances organized by Boston society. More often than not, though, he found his dance partners as lacking as himself. In his diary, he strove for accurate pen portraits, warts and all. Restless, especially in hot weather, he took nighttime walks on the Commons. There were titillations of the eye and mind with streetwalkers. But they were temptations he exposed himself to for the purpose of resisting them.

In the background of his every moment lingered his sense of loss. For that, he believed, the only antidote was to put his energy into becoming self-sufficient and making a name for himself.

Before leaving Newburyport, he had been suffering from insomnia and lassitude. By September 1789, he felt better enough to travel to New York to visit his family. The president and Congress were setting up the new government. Crucial issues were being debated. Drawing passage and expense money, as he did his quarterly allowance, from Cotton Tufts, he took the stage to Providence, where he visited John Brown's College, still called the College of Rhode Island. Brown's house was the "most magnificent and elegant private mansion, that I have ever seen on this continent." But the Rhode Island legislature, which he visited twice in one day, seemed a parody of republican good sense. At the state supreme court, "they were doing nothing: and the appearance of the Judges was a perfect burlesque upon justice." Newport, attractive from the river, up close "exhibited a melancholy picture of declining commerce and population," its former prosperity "chiefly owing to its extensive employment in the African slave trade, of which some remnants still continue to support it." Houses were in disrepair, the streets dull. The sailing packet took six days to get to

New York. Passing through Hells Gate, he enjoyed the late summer vistas and attractive country homes on both side of the East River. From the downtown wharf, he walked to his parents' home at Richmond Hill, now in Greenwich Village, on the Hudson. President Washington and his family were dinner guests. A fatigued John Quincy ate alone and went to bed.

But he had a number of other occasions in his three weeks in New York to be in the president's company. The dignified Washington was at the pinnacle of his influence, acknowledged to be the human pillar on which the new republic was rising and the most popular man in America. Like his parents, John Quincy thought Washington could do no wrong. His wisdom, prudence, and dignity made him inimitable. As the first Congress debated the implementation of the new Constitution and attempted to deal with the unfinished business of the war, the Adams family agreed entirely with the policies of his administration. With the House and Senate in session, John Quincy attended the debates and made visits, including one to Fisher Ames, an important Federalist leader from Massachusetts; had dinner with the secretary of war, General Henry Knox; and sat in at the courthouse, where he witnessed a murder trial. At the House he listened to "the debates . . . upon the judiciary bill," starring Elbridge Gerry from

Massachusetts and James Madison from Virginia. "But I confess, I did not perceive any extraordinary powers of oratory displayed by any of these gentlemen. The subject had been already so much discussed that . . . the eloquence had all been exhausted, but the spirit of contention still remained." He immediately had a sense of "the difficulty of adjusting the opposing sentiments which direct the conduct of men living in different climates and used to very different modes of living." For the time being, President Washington was the adhesive that kept together Federalists and Antifederalists, North and South, the elite and the general population.

In 1791–1792, no matter how hard John Quincy worked at the law, no matter how many court sessions he attended, it was tediously slow going. And his energy was at a low ebb. But with so much time on his hands, why not use it to read, particularly Livy, Tacitus, and Cicero, which would enable him to improve his Latin? He also renewed his enthusiasm for Shakespeare, Milton, Pope, Swift, Voltaire, Hume, Samuel Johnson, and Laurence Sterne. He read Edmund Burke, whose speeches and essays on the American and French revolutions dazzled him. And he could generate energy for writing, not only in his diary but about political ideas. His father discouraged him from speaking out in the public forum: it was best inhabited by those who did not need public approval or government employment. In public service, John Adams warned him, he would not have a more comfortable life than as a lawyer. The public was fickle, the pay modest. But if you enter the public arena, "do it, my dear son without murmuring. . . . Independence, my boy, and freedom from humiliating obligations are greater sources of happiness than riches." If he had a successful career in the law, he would be quite comfortable. If he pursued a political career, he would perhaps be both a drudge, with comparatively low income, and the target of envy and criticism. Still, service to one's country and the fulfillment of some high civic purpose were an obligation that both John and Abigail had taught their son. Could he possibly avoid it?

Ambitious to fulfill this civic vision that he and others had for him, John Quincy struggled with his own and the widely held view that ambition could be dishonorable. But his heart was "not conscious of an unworthy ambition." And he struggled with the conviction that he was not working hard enough. He was not using his time well, a constant theme, no matter how hard he worked or how much he accomplished. "The consideration is equally painful and humiliating . . . that the ambition is constant, and unceasing, while the exertions to acquire the talents which ought alone to secure the reward of ambition, are feeble, indolent, frequently interrupted, and never pursued with an ardor equivalent to its purposes." How to express that ambition honorably, effectively, and successfully? How to avoid the humiliation of talents and ambition unfulfilled? Like his father before him, he put pen to paper, a process that was becoming as natural to him as breathing.

In April 1791, with a keen nor'wester blowing, John Quincy walked up to the highest point on Beacon Hill to observe a partial solar eclipse. He was soon driven down by the severity of the wind. Whatever his earthly problems, he was always fascinated by astronomical phenomena. "Hurt my eyes much by observing this . . . eclipse without a glass." With susceptibility to eye irritation and infection, and with considerable reading by candlelight, he had made matters worse. He had days in which he was "almost blind." But the affliction was intermittent. Most days he could see quite well enough to write, which he started to do intensively and quickly. Although complaining about low spirits, ill health, too much time on his hands, lack of focus, very few clients, and his dislike of "these sandy deserts of legal study," he had a sudden surge of energy and ideas. It kept him writing effectively for seven weeks. He now found himself for the first time composing a sustained narrative, his inaugural as a writer on issues of public policy. His father's and his own cautionary voice had urged that he stay out of politics. But he found that he could not restrain himself.

Between early June and late July 1791, he published a long essay in the form of eleven letters in the Boston *Columbian Centinel,* a weekly newspaper friendly to moderate Federalist views. The title was *Letters of Publicola,* "the people's friend." He had just read Tom Paine's *The Rights of Man,* a response to Burke's *Reflections on the French Revolution.* Jefferson had sponsored its American publication, which had as its introduction his letter praising Paine's views and criticizing "political heresies." Savvy American readers had no doubt that the phrase referred to John Adams' recently published criticism of the French Revolution. Adams had argued that a successful government needed to contain and balance all the prominent elements in a society. In principle, the division of the British government was an exemplary model. The moral corruption of its leaders, not the structure of its government, was at fault. Since a strong executive was an asset to effective government, there was nothing necessarily adverse in having a constitutionally limited monarch in a system of checks and balances. Jefferson counterattacked, charging that the vice president was a monarchist. He would put a king on an American throne.

John Quincy's response was a bold initiative. No one had invited him to write these letters; he took that on himself. Although it was a pseudonymous publication, there was the possibility that eventually its authorship would be revealed. A defense of his father's views, it was also an act of self-assertion. And, once he had started, nothing held him back, partly because he was able to believe that writing for the public was not a public career. He distinguished between addressing the public and working for the public, though he had to be aware that one might lead to the other. For a young man noted for caution, it was an unexpected and risky venture. If he failed, would he not be failing his father? And, after all, who was he to take on the famous author of *Common Sense*?

Praising the French Revolution and its National Assembly, Paine believed that the best government was an unmediated and direct democracy. The next best was a unitary legislature that could respond quickly to the voice of the people. Radical change should be unimpeded by checks and balances. The French had created a single legislative body without an executive. The British should do the same, Paine argued. In Paris, the first stage of the revolution had resulted in the creation of the National Assembly as the sole governing body. A republic was in place, though the royal executions were two years off. So too were the preventive wars against the republic's European enemies. In 1791, the American pamphlet press was abuzz with the oracular words of those who thought the new order the gateway to heaven and those who thought it the portals of hell. Many Americans identified with Paine and France. After all, the United States had recently completed its own successful revolution. Britain had been the enemy. And a pro-French foreign policy seemed reasonable repayment for French support during the American Revolution. Also, whatever the excesses in France, they were in support of a republican government. If the mainstream Federalists, like John Adams, were pro-British, it was in the limited sense of not wanting to throw the baby out with the bathwater. Like Burke, the Federalists favored checks and balances and the stability that evolutionary change promoted. But, John Quincy argued, both Burke's extreme attack on the French Revolution and Paine's extreme defense were to be avoided. The revolution still had promise. But it would never be realized if Paine's views became the political reality. Paine "has hunted for epigrams where he ought to have sought arguments." His practice is to accommodate "the facts of history to his political purposes." As for Jefferson? "I am somewhat at a loss to determine what this very respectable gentleman means by *political heresies*. Does he consider this pamphlet of Mr. Paine's as the canonical book of political scripture?" Paine's central claim, John Quincy believed, was false. "That a whole nation has a right to do whatever it pleases, cannot in any sense whatever be admitted as true." The eternal and immutable laws of justice and morality are paramount to all human legislation. If those in the majority are "bound by no law human or divine, and have no other rule but their sovereign will and pleasure to direct them, what possible security can any citizen of the nation have for the protection of his unalienable rights?"

Since the purpose of governments is to protect the rights of citizens, the best bulwark against infringement, Adams argued, is a written constitution. It defines and limits executive and legislative action. It is the law, created by the people themselves, to which all legislative law must conform. The next best is an unwritten constitution, such as Great Britain's. The English constitution is imperfect. The people suffer under unfair representation. But "I venture to assert that the people of England have no right to destroy their government, unless in its operation the rights of the people are really oppressed, and unless they have attempted in vain every constitutional mode of obtaining redress." So, "when Mr. Paine invited the people of England to destroy their present Government and form another Constitution, he should have given them sober reasoning and not flippant witticisms." He "should have proved what great advantages they would reap as a nation from such a revolution, without disguising the great dangers and formidable difficulties with which it must be attended." A revolution that did not embody in its counsels and passions "the eternal and immutable laws of justice and of morality" would most likely create more misery than the regime it replaced. That, Adams feared, was what was happening in France. The worst, he thought, was yet to come. Cataclysms, exacting a heavy cost in life and liberty, would provide the object lesson to demonstrate the superiority of the American Revolution and the American Constitution, which "appears to me to unite all the advantages, both of the French and of the English, while it has avoided the evils of both. Divide your power so that every part of it may at all times be used for your advantage, but in such a manner, that your rights may never depend upon the will of any one man or body of men."

Why, he asked, had Americans been able to accomplish a constructive revolution, an example of how it can best be done if it must be done? Partly because Great Britain was only moderately repressive but also, he argued, because the revolt against Britain had been undertaken only as a last resort, after every effort at redress of grievances had been tried. The United States took as its model a variant of the British constitution, and, most important, Americans and their elected representatives believed and acted in conformity to "the immutable laws of justice and morality." These ex-British colonial Americans knew what those laws were because of their immersion in the Judeo-Christian tradition. Americans respected the biblical command-ment to be a people of justice and mercy in a tradition that mandated righteousness and right action. Although the United States was a collage containing orthodoxy, heterodoxy, deism, freethinkers, sectarian fervor, and sectarian indifference, the "distinction between *power* and *right*" was widely shared. It was both religious and secular.

And the implementation of that distinction came from a prevailing belief in fundamental rights that pre-ceded governments and constitutions. "Happy," Adams concluded, "thrice happy the people of America! whose gentleness of manners and habits of virtue are still suf-ficient to reconcile the enjoyment of their natural rights, with the peace and tranquillity of their country; whose principles of religious liberty did not result from an indiscriminate contempt of all religion whatever, and whose equal representation in their legislative councils was founded upon an equality really existing among them, and not upon the metaphysical speculations of fanciful politicians, vainly contending against the unalterable course of events, and the established order of nature."

Letters from Publicola anticipates some of the main themes of Adams' career. It marks his coming-of-age as a writer. Writing with fluency and speed, he combined analytic and intellectual nuance with a prose style notable for conciseness and grace. By modern standards, his sentences have a touch of eighteenth-century periodicity: they have no difficulty carrying parenthetical clauses, are often long by the standards of modern prose, and are comfortable with rhetorical devices drawn from Latin. Cicero was his model, and, like some of the best eighteenth- and early-nineteenth-century English prose, it sometimes feels to the modern reader as if it were the work of a brilliant translator from an unidentified Latin source who had subordinated Latin origins to English expressiveness. Adams' prose shares that quality with the best of his educated contemporaries. For his time, Adams' sentences are finely drawn; they have a rhythmic sense of variety in length and structure that compel the eye and create powerful effects. He invariably chooses the right word; he never overdresses his phrases with repetitions and unnecessary adjectives. The reader feels in the presence of a precise mind and a muscular prose.

Having gotten started writing for the public, John Quincy looked for an occasion to do it again. *Letters of Publicola* had been a great success, reprinted in New York and Philadelphia, where it was widely quoted in Federalist circles; and in London and Edinburgh. There was chastisement from Francophiles, especially as soon as its authorship became known, though John Adams was at first thought to be the writer. Father and son were condemned as monarchists, the all-purpose term used by the emerging Republican Party to refer to those who showed insufficient enthusiasm for France. Some of the vice president's most ruthlessly partisan opponents believed that he literally desired America to have a king. Publicola, a Boston critic wrote, "seems to have some talents, but perverted as they are, they are worse than thrown away. Like Burke he has attempted to raise a structure upon a rotten foundation. . . . It is a circumstance highly honorable to the political character of our country that an *host* of enlightened writers have arisen, in every part of the United States, to oppose the abominable heresies of Publicola." Praised by friends and ideological sympathizers, John Quincy readily tolerated the criticism. His father's enemies were inevitably his own. In *Letters from Publicola*, John Quincy had not hesitated to commit himself to Federalist principles. There was not the slightest distance between family loyalty and shared ideology.

In early 1791, some residents of Braintree's north precinct petitioned the legislature for incorporation as a separate town, to be named Quincy, after John Quincy's namesake. He and William Cranch, strongly identifying with the north precinct, argued the case before the legislative committee, with John Quincy as the lead attorney. In February 1792, Quincy came into existence. It was an additional tie between the person and the place, between the Adams family and their town of residence. John Quincy also had a voice in two factional dramas that were playing out in Boston in 1792. An unruly city, it had enough theft, arson, street crime, and mobs to make it desirable to expand its police force. Many favored new rules of governance that would create a reformed police department. To his surprise, at a meeting to which he had come as a spectator, Adams was elected to the committee to report on the issue. The reform effort turned into an ineffectual

mess. Eloquence and reason were on the side of the supporters of reform. But every change had to be agreed to
by a majority vote of "the whole body of the people,"
who, when the opposition packed the hall with "seven
hundred men," looked to Adams "as if they had been
collected from all the jails on the continent. . . . From
the whole event I have derived some instruction, and
above all a confirmation of my abhorrence and contempt of simple democracy as a Government. . . . The
whole affair has given me some additional knowledge
of human nature."

When a traveling theater company came to Boston
in the fall of 1792, John Quincy was delighted. He
would once again have the opportunity to see drama
on the stage. His experience as a theatergoer had been
one of the pleasures of his European years. He also
believed in theater as an educational and moral force.
But he had not seen a play performed since he had left
Paris. Theatrical performances of any kind had been
made illegal by the Massachusetts legislature. Even
attending a performance was against the law. A young
Swiss immigrant, Albert Gallatin, noted in 1780 that
"Boston is highly boring. There are no public amusements and many superstitions of the sort that prevent
one from singing, playing the violin, playing cards,
bowling, etc. on Sunday. . . . They are concerned about

nothing more than their probity." In 1792, the newly created Boston Tontine Association, a civic-minded experiment in capitalism and real estate, petitioned the legislature to repeal the prohibition. It seemed at variance with the Massachusetts Constitution. "If your legislators have any regard for the morality of the people," Harrison Gray Otis' grandfather responded from London, "they will not give the least countenance to the stage, which . . . is called the *Devil's Chapel* . . . for what's there to be met with but lewd laughing, but smut, railing and buffoonery." To force the issue, the Tontine Association built the Board Alley Theatre. In late November, John Quincy responded, probably both amused and determined, to the announcement that "this evening will be delivered a Moral lecture, in five parts, called Hamlet, Prince of Denmark. . . . End of the lecture, a Hornpipe. The whole will conclude with an entertaining lecture called Love a-la-Mode." David Garrick's *The Lying Valet* was advertised as a "Comic Lecture." By attending the theater, John Quincy was breaking the law.

When Governor John Hancock referred in passing to the existence of the theater, the legislature demanded that the law be enforced. Hancock instructed the attorney general, James Sullivan, to have the performances stopped. After the sheriff, stepping to the stage, arrested an actor, the audience rioted. At the same time, an angry anti-theater crowd headed toward the governor's house and then the theater, which they intended to tear down. They were dispersed by the threat of the Riot Act. Although the defendant was dismissed, the theater company was advised by its supporters against more performances. At a mass pro-theater meeting in Faneuil Hall, John Quincy was appointed to a committee to request the governor's assistance in enacting repeal. Sullivan, self-righteous and defensive, argued in the Boston *Independent Chronicle* under the heading "A Friend to Peace" that the remedy for an unconstitutional law was repeal, not criminal defiance. Adams responded in the *Columbian Centinel*, under the pseudonym Menander, a fourth-century B.C. Greek writer of comedies: "In a free government the minority can never be under an obligation to sacrifice *their rights* to the will of the majority, however expressed. The constitution of this State is expressly paramount to the laws of the legislature." It was perfectly legitimate to force the legislature to reconsider an allegedly unconstitutional law by acts of civil disobedience. "As to the *violent measures* which the *Friend to Peace* mentions as having been resorted to, *they* have all been on the part of the government." Adams had no sense of being on thin ice in his claim that each citizen has the right to act according to his conscience if he disagrees with a law, providing that the disagreement is expressed by civil—not violent—disobedience, and that the purpose of the infraction is to provide the most favorable circumstances to test the law. It was the first expression of a principle he was to hold throughout his life. The Massachusetts legislature repealed the anti-theater law in 1793.

John Quincy now had a public voice. On the last day of April 1793, the selectmen of Boston came to his office to tell him he had been appointed to deliver the town's official July 4 oration. The honor was not to be refused. The family rejoiced. The success of *Publicola* had made him a public figure of minor distinction. With his usual restraint, he made note of the honor in his diary, without further comment. He kept himself focused on his reading, his law practice, and social events, all of which seemed to him mixed blessings, as if daily life took its emotional tone not from his accomplishments but from his anxieties. He was afraid of a number of things, especially of showing emotion in public. He still worried that he would be unable to earn his own living. He feared that the successes he had already had would breed jealousy among his peers. And his father's experiences and views heightened that anxiety, as if the world were out to get any Adams but especially the vice president and his son. In fact, only part of the world was—mostly for political and, in a few instances, personal reasons. But John Quincy had learned to be cautious, suspicious, and self-protective. He needed to restrain his spontaneous feelings and satiric tongue. But the attraction was irresistible. At the same time that he was invited to deliver the Independence Day oration, he began publication in the *Columbian Centinel* of a three-part article under the pseudonym Marcellus—a celebrated third-century B.C. Roman general credited with preserving the independence of his nation—on a political issue that was, in various forms, to be part of the American political drama for the next twenty-five years. This literary Marcellus defended the wisdom of America's Marcellus, President Washington, who on April 22, 1793, had issued a "Proclamation of Neutrality."

Great Britain and France were at war again, and Washington declared that neutrality was the best policy for the United States. The pro-French faction, including Jefferson, who by the end of the year was to resign as secretary of state, was furious; Jefferson would spend the next six years planning a coup d'état. "I am really astonished at the blind Spirit of Party," John Adams wrote to Abigail, "which has Seized on the whole soul of this *Jefferson*: There is not a Jacobin in France more devoted to Faction." John Quincy strongly supported Washington's neutrality policy. Otherwise the United States, without an army or navy, without the capability or the will to pay for a military, would be torn apart by the British lion or destroyed by the machinations of the French. America's interest would be sacrificed to the interests of the belligerents. The claim that it had a treaty obligation to ally itself with France against Britain was spurious. "No stipulation contained in a treaty can ever oblige one nation to adopt or support the folly or injustice of another," John Quincy maintained. If the United States did not prevent those Americans who wanted to enrich themselves as privateers from illegally capturing British merchant ships on the high seas, how could it claim, he asked, "freedom of neutral shipping" for our own ships? "As a nation whose happiness consists in a real independence, disconnected from all European interests and European politics, it is our duty to remain the peaceable and silent though sorrowful spectators of the sanguinary scene." After all, "we have a sea-coast of twelve hundred miles everywhere open to invasion, and where is the power to protect it? We have a flourishing commerce, expanding to every part of the globe, and where will it turn when excluded from every mark of the earth? . . . We are in a great measure destitute of the defensive apparatus of war, and who will provide us with the arms and ammunition that will be indispensable? . . . To advise us to engage voluntarily in the war is to aim a dagger at the heart of the country. . . . War is murder," Adams told his readers.

But he also had no hesitation in telling the crowd that gathered at the Old South Meeting House in Boston on Independence Day that there was an important exception. The war for American independence had not been murder. It had been liberation. On July 1, in blisteringly hot weather, he rehearsed the speech before two friends, then rehearsed twice the next day and twice again the following. He recognized how anxious he was. "Delivered the Oration at the request of the Town; the performance well received—for which I feel grateful as I ought," he reported in his diary. The national narrative that Adams embraced avoided any sober analysis of the British point of view, as if the British were simply irrational tyrants or stupid bureaucrats, and it erased from history the brutal civil war between pro-British and pro-independence Americans, recasting the revolution as a simple conflict between tyrannical oppression and the partisans of liberty. Parliamentary tyranny was entirely at fault. Revolution was justified. He made no mention of the fact that the colonists had had no obligation to pay any direct taxes to Great Britain, despite the benefits received, or their conviction that the Indian lands, which Britain preserved for the native inhabitants, were rightfully theirs for the taking. In the creation of the new nation, patriotism determined reality.

Adams set the American bar for justifying the violent overthrow of an existing regime much lower than he had set it, in *Publicola*, for overthrowing the French or British governments. Indeed, the heroic Americans of the revolutionary generation "did not hesitate . . . when your coasts were infested by a formidable fleet, when your territories were invaded by a numerous and veteran army, to pronounce the sentence of eternal separation from Britain, and to throw the gauntlet at a power whose recent triumphs were almost co-extensive with the earth." The challenge, he told his audience, was to keep faith with the vision of the founders, to be ever ready to sacrifice to keep it alive and well. "Should the voice of our country's calamity ever call us to her relief, we swear by the precious memory of the sages who toiled and the heroes who bled in her defence, that we will prove ourselves not unworthy of the prize . . . of republican virtue." Published at the middle of the month in a brief pamphlet, it had nothing controversial about it, unlike *Publicola*. But, Adams wrote thirty-five years later, it was received "with a warmth and animation which gave me cheering encouragement at the threshold of life. It has dwelt like a charm upon my memory ever since." The plaudits did not cheer him for long. Since to advocate neutrality was to be pro-British, pro-French Antifederalists viewed John Quincy as a minor officer in the army of the enemy. On August 1, his name was on a list "posted on the mast of the French frigate la Concorde," in Boston Harbor, "as an aristocrat. . . . Defamed; proscribed;—what next?"

That summer, still boarding with the Welsh family, John Quincy alternated hard work with listless days. He attended the usual social and cultural events with little sense of pleasure, and wandered Boston streets for exercise and the occasional excitement of street encounters. What they were and with whom he never says, even to the privacy of his diary. His eyes continued to bother him intermittently. As usual, he was restless. In September, he began writing again. It was something to do that he could do well. His aim was to defend the Washington administration. The newly appointed French minister, Edmond Genet, had arrived in April 1793, within days of the issuance of the Neutrality Proclamation. In Europe, where he had the reputation of being hotheaded, he had been expelled from the Russian court. Jefferson, still secretary of state, welcomed him. Celebrated by huge pro-French crowds, Genet challenged the administration to prevent his engaging in partisan activities, intending to negate the neutrality policy or even, he hoped, overthrow the government. In defiance of the Neutrality Proclamation, he openly sponsored anti-British activities, including commissioning privateers and organizing volunteers to fight in Florida. To the Federalists, his conduct insulted American sovereignty. Through the summer and fall, the administration felt under siege by an enemy within the gates. The cabinet weighed the consequences of demanding that France recall Genet: such a demand would further antagonize pro-French Americans, who would claim that only Congress, not the president, had that power; there was no precedent for expelling a foreign minister; and it also might cause a rupture with the French government. The pro-French faction called for a declaration of war against Britain, which routinely captured American ships doing business with France. John Quincy had no doubt that the French, when it was in their interest, would no more respect the freedom of neutral shipping than the British. But he believed a war with either of these powers would be a disastrous mistake.

What he had begun writing in September 1793 soon coalesced into a manuscript in progress. After a pause, he commenced again. "But for what purpose," he asked in his diary. "The cause of my Country?"

His eyes hurt, especially when he was reading and writing. He had, though, the energy of a man outraged that American sovereignty and President Washington were being humiliated. Genet is a "hare-brained Hotspur of an Envoy," he wrote in the last of three letters published in the *Columbian Centinel* under the pseudonym Columbus, "the most implacable and dangerous enemy to the peace and happiness of my country. . . . Where then was the commission, where were the credentials, which authorized him to treat with the people of America, through any other medium than that of their government?" Worse, Genet connected himself "and his interest with a particular party of American citizens . . . a party professing republican sanctity beyond the rest of their fellow-citizens" but not disguising "sentiments hostile to the national government of the country." Not, Adams argued, that parties themselves are necessarily disadvantageous. "In a state of civil and political liberty, parties are to the public body what the passions are to the individual. And as the passions are said to be the elements of life, so the animating and vivifying spirit of party seems to be essential to the existence of genuine freedom." But, "like the passions, too, it is a prolific source of misery, as well as of enjoyment." Parties needed to be restrained within the boundaries of the Constitution. The alliance between the emerging Democratic-Republican Party and Genet was extraconstitutional. As to the president's power to expel an offending diplomat there could be no doubt. "There is in this country, as in all sovereign states, a power competent to dismiss the agent of a foreign power."

By late fall of 1793, the Washington administration had had enough—even Jefferson had. Genet's arrogance and presumption had become a political liability. In late December, the secretary of state conveyed to the French government the demand that Genet be recalled. Within days, Jefferson himself resigned. "I have so long been in an habit of thinking well of his Abilities and general good dispositions," John Adams wrote to his wife of Jefferson's resignation, "that I cannot but feel some regret at this event: but his want of candour, his obstinate prejudices both of aversion and attachment: his real partiality in spite of all his pretensions and his low notions about many things have so nearly reconciled me to it, that I will not weep. . . . I know . . . his soul is poisoned with ambition . . . his temper embittered against the Constitution and Administration."

What next for John Quincy? His law practice stumbled on, with enough cases to pay his expenses, which had been his goal. He was approaching independence, though of a bare cupboard sort. The domestic happiness he had hoped for was far out of reach. Boston friends and his Quincy family were dear to him. But anxieties and unhappiness still kept the register of his days. "Life in this State is no blessing," he confided to his diary. "I wish it restored with innocence to the giver," a wish for oblivion that was rhetorical, not real. The law, even when he won cases, gave him little pleasure. His appearance as Columbus caused its own anxieties. Its reception among some of his friends startled and grieved him. There seemed a coolness that implied resentment at his success. Would such success always come at a heavy price?

Returning to his lodgings in the evening on June 3, 1794, he picked up a letter at the post office. The new secretary of state, Edmund Randolph, had called on the vice president with startling news. The president had nominated John Quincy Adams "to go to The Hague as Resident Minister from the United States." Suddenly, his life took a new and tumultuous turn. He had no expectation of any appointment, he later claimed. "I wrote those papers without expecting that they would ever be seen by President Washington. . . . I wrote them with the wish rather than the hope that they would contribute to cool down . . . public feeling . . . to the benefit of the country; and that all the selfishness that I was conscious of . . . was a hope that it would

gratify my father and mother and give me some reputation among my townsmen." Perhaps the thought of
some reward had crossed his mind. Despite his reservations about public employment, it was impossible to
sidestep such an invitation. Service to the country was
an Adams obligation. He had, after all, been trained
by his parents to believe that this was his destiny. He
would be following in his father's footsteps, returning
to a country he already knew something of, including its language, though his excellent French was even
more to the point. And it would provide a new outlet
for his restlessness and ambition. He would have the
company of his brother Thomas, who would receive
a salary as his private secretary. If John Quincy was
to have the distinguished public career that had been
forecast for him, this was a first step toward it. And it
came from the mind and hand of the man in public life
he most revered other than his father.

His close friends Nathan Frazier and Daniel Sargent
accompanied him on board ship as far as the lighthouse
in the outer harbor. Nathan was Mary Frazier's cousin;
she was still in John Quincy's heart. None of them
could have guessed that Daniel Sargent would become
her husband. "The pain of separation from my friends
and country was felt as poignantly by me at the moment
when these two young men left the ship as it ever has

been at any period of my life. It was like the severing the last string from the Heart. I looked back at their boat as long as it could be seen, and when it had got out of sight, I did not, but I could have turned my eye and wept." He had no doubt that this farewell to family and friends would be for a long time, perhaps even for the last time. On the afternoon of September 17, 1794, as the *Alfred*, bound for London, picked up a gale that took it out to sea, he saw "a beautiful rainbow in the East a pledge of fair weather for the morrow." He was once again sailing to Europe.

Chapter 5
Remember Your Characters
1794–1797

As they approached London Bridge at twilight in mid-October 1794, a sleepy John Quincy, who had been dozing, heard "a sound as of a trunk falling from the carriage." He and Thomas had left Dover at 3 A.M. Having been instructed to travel to The Hague via London in order to deliver important dispatches to John Jay, who was negotiating with the British in the hope of preventing another war, John Quincy had hired a private carriage rather than having their luggage shipped separately. At customs, he had made certain that the two trunks containing the documents were secured by leather straps to the outside of the coach. As it rattled over cobblestones through noisy London streets, the unexpected sound startled him. Thomas jumped out before the horses came to a stop. Both trunks were gone. They found one under the carriage, the other about thirty feet behind, about to be crushed by an oncoming vehicle. The straps had been cut.

It was "a hair-breadth escape" by sheer luck. Strictly speaking, the loss would not have been his fault. It seemed likely that the driver who rode the lead horse had been in league with thieves. Probably the straps had been cut by a child with a razor who had slipped under the carriage when it stopped to pay the turn-pike toll. But if they had been lost, there would have been gossip and criticism, the national interest would have been damaged, and the question would have been asked, Was he responsible? Enemies would have said yes, without a fair examination of the evidence. That would always be so in a public career. Making enemies was in the nature of the occupation. It was a danger he needed always to keep in mind. He had narrowly escaped a damaging incident beyond his control. As long as he was in the public eye, he would have his share of such contingencies. "I once more wish you a prosperous Voyage an honorable Conduct and a happy Life" were the last words his father had written to his sons before they sailed from Boston. "Remember your Characters as Men of Business as well as Men of Virtue." Luck, as John Adams had learned, determined parentage, place and time of birth, opportunity, and quotidian contingency. But the quality of individuals and nations depended on character, which meant reputation. It was a reputation for honesty, integrity, and competence that he desired for his sons.

At sea the *Alfred* proved true to the worst John Quincy had heard about it. He had been advised to wait for a better ship, but by early September his patience had been exhausted. In Philadelphia in July, he had taken a few weeks to study the relevant papers at the State Department. Then he had to wait until early August for a meeting with Alexander Hamilton, the secretary of the treasury, since his mission required protecting America's credit with the Dutch bankers. Then he decided to sail from Boston to say good-bye to family and friends. But since fewer boats sailed from Boston than New York, that created further delay. With no other vessel available, in mid-September he and Thomas boarded the *Alfred*. She was "old, crazy and leaky," like sailing in "an egg-shell," with a captain who navigated only by experience. Fortunately, the voyage was blessed with good weather, producing the speediest crossing John Quincy had ever experienced, twenty-eight days from Boston to European landfall.

In London, he delivered the precious trunks to Jay. He also had letters for other Americans, including the

American minister, Thomas Pinckney. Jay shocked Adams with the news that Robespierre was dead. "I repeated with utter astonishment, 'Robespierre dead,' more times than was perfectly decent; and could scarcely believe I had heard right." The French Revolution delivered startling news, much of it gruesome, with such rapidity that it made American heads spin. The next day he could provide, in return for Jay's startling news, only an account of the comparatively dull but highly partisan goings-on in America. Jay was of course interested to hear what was being said about his peace mission and about his friends and colleagues with whom John Quincy had dined and socialized. In New York, Adams had been at a dinner at which he had met Talleyrand, former president of the Constituent Assembly, who was "reserved and distant." Excluded from France because of politics and from England because of his role in the revolution, Talleyrand had taken temporary asylum in America. John Quincy agreed with his father's comment, when Talleyrand asked the ex-president what he "thought of the new Constitution of France. I answered that the Shortest Expression I could give him of my Opinion of it, would be by comparing the King to Daniel in the Lyons Den, and his Ministers to Shadrach, Meshach and Abednego in the fiery Furnace . . . if either of them ever got out alive it must be by a miraculous Interposition of Divine Power and not by the Efficacy of any known natural Causes." The rapid deposition of French leaders, John Quincy thought, did not speak well of French democracy.

American politics, including relations with the Indian tribes, had its own twists. In Philadelphia, where he had marked his twenty-eighth birthday, John Quincy attended, at the invitation of the president, a smoking ceremony between Washington and a group of Chickasaw Indian chiefs. "The President began, and after two or three whiffs passed the tube . . . all round. . . . These Indians appeared to be quite unused to it; and from their manner of going through it looked as if they were submitting to a process in compliance with *our* custom. Some of them I thought smiled with such an expression of countenance as denoted a sense of *novelty* and of *frivolity*, too; as if the ceremony struck them not only as new, but also as ridiculous."

In London, Adams was flattered when he was included in a series of working meetings with Jay and Pinckney, who consulted him about the proposed treaty article by article. A realist, Jay knew that the Americans were to the British as the Chickasaw were to the Americans. It was a peace pipe they would have to smoke. And the provision that each country indemnify the losses incurred by the citizens of the other for each country's noncompliance with the peace treaty of 1783 seemed fair. The treaty, Adams responded to Jay, who asked his opinion, was likely to be the best one the United States, an impoverished country with almost no military, could get. "I told him that as to the whole project, I felt myself inadequate . . . but assent to the idea in which he and Mr. Pinckney concurred, *that it was better than war.* . . . My observations were made with the diffidence which naturally arose from my situation, and were treated with all the attention that I would expect or desire." They knew that at home the treaty would provoke nasty controversy.

In England, John Quincy compared what he saw with what he had seen almost ten years before. The countryside looked more attractive and wealthy; London itself seemed larger; the distances between people's residences were so great that he could make delivery each day of only two or three of the letters he had brought with him. In comparison, Boston, New York, and Philadelphia were small towns. "Since the peace with America," he wrote to his mother, "this Country has been prosperous in its Commerce beyond all conception, and at this moment its opulence is incredible." Theater in London compelled his interest.

He saw *Henry the Eighth* at Drury Lane, which had undergone a complete renovation since he had last been there; it was now more elegant but also too large and the acoustics poor. The celebrated actress "Mrs. Siddons appeared in the character of Queen Catherine. She is as much as ever and as deservedly the favorite of the public, but the enthusiasm of novelty is past and her appearance alone no longer crowds the houses. . . . She performed the part . . . to great perfection." At Covent Garden, he saw *Romeo and Juliet*. During the intermission, the orchestra began to play *God Save the King*. There was a "thunder-clap of loud applause" from every section of the house.

Everywhere he heard and saw an exaggerated patriotism that struck him as hollow, servile, and fearful. Monarchy worship was so extreme, he noted, that anyone who did not know better would think Great Britain an absolute monarchy. "Disaffection is silent," he wrote to Thomas Welsh. "It strikes the eye and vanishes like a flash of lightening in the darkness of night. The Administration is strong, but the War is unsuccessful. The people ardently wish for Peace, but their Government are ashamed to ask it." And while he did not fully agree with the view that the British "from the sovereign to the beggar have a most inveterate hatred against America," he had reason to think

REMEMBER YOUR CHARACTERS · 209

that many still did not accept the United States as an independent nation. In fact, "neither the Government, nor the people of this Country, have any real friendship for America."

He received current reports about the Netherlands from Jay and Joshua Johnson. French troops were already or would soon be on Dutch soil. Johnson, who continued as the nonsalaried American consul in London, had interests to protect in Holland. Deeply invested in American tobacco warehoused in Amsterdam, he feared for its security. As American minister, Adams had an obligation to help facilitate American business interests. On broader matters he would find himself, Jay told him, in an "extremely delicate" situation. The House of Orange had allied itself with Britain. The opposition party, known as the Patriots, had allied itself with France. To the royalists or Orangists, the invading army was a criminal enemy. To the Patriots, it was a liberator. The Stadholder and his royalist regime would be suspicious of the intentions of America and its minister. The Patriots would assume that the United States would side with them. And the French, whose armies everyone believed would easily conquer Holland, would continue to resent American neutrality whether in the Netherlands or anyplace else. Finally, though the Dutch bankers strove to be as apolitical as possible, political turmoil might affect their actions and damage American interests. Adams would have to watch out for all these things. The Patriots would try to draw him to their side. The royalists would watch him closely.

Unlike James Monroe, the new American minister to France, Adams had already made clear that his enthusiasm for the revolution was qualified. By 1794, its anarchic brutality had become abhorrent to him. In Paris, Monroe was, in Adams' view, committing the cardinal diplomatic sin. He was publicly siding with France, despite his government's neutrality policy. "Be always upon your Guard," John Adams would urge his son. "No Character in human Life requires more Discretion, Caution, and Reserve, than that of a Public Minister in a foreign Country." John Quincy had puzzled over exactly what his charge was to be, beyond upholding America's neutrality policy, even before leaving America. After all, he had told his father, if it was only to manage the American loans and protect American credit, then he would certainly not do that for more than three years. But the French invasion changed the situation. Now there were serious tensions and crosscurrents that needed to be taken into account. And, as his father had emphasized, whatever the situation in Holland, The Hague was a listening post for everything going on in Europe. His primary task was to be the eyes and ears of the Washington administration.

Soon after he arrived at The Hague, the coldest winter in memory set in, with the canals and rivers frozen until April; it was a land of "frightful foggs," as Thomas reported home. Adams attended to Johnson's tobacco, arranging for safe protection and facilitating its sale in Paris. At first, there was a bustle of official activity. His credentials were accepted, which included a meeting with the Stadholder, Prince William of Orange, soon to be deposed. He "is well disposed," Adams noted. "With a good heart and a feeble mind, he is the man of his councils and not of his own energy. The princess detested almost universally. Haughty, domineering, incapable of submitting to misfortune with dignity." The prince and princess were soon crossing the channel in a small fishing boat, blaming an ungrateful world, so report said, not themselves, for their forced exile. Apparently the princess "broke out in transports of rage, until she was totally exhausted and sunk into a state of sullen apathy." Adams was astonished to find "everything so perfectly quiet. . . . You would imagine yourself to be in a land blessed with a profound peace." Early in 1795, the French troops occupied The Hague and then Amsterdam. With no desire to fight on Dutch soil for the sake of the House of Orange, Britain had withdrawn its forces.

Suddenly Adams heard the "Marseillaise" everywhere. He had been instructed to stay in the Netherlands only if the French occupiers were accommodating, which they were. With a pro-French American minister in Paris, the French, though still confiscating American ships, expected that a gradual application of honey would bring the United States to their side. Adams accepted the French hospitality. He thought well of General Charles Pichegru, who seemed to know that the French general of today might find himself in exile or guillotined tomorrow. Meanwhile, Paris was deliberating about the future of Holland: would it be allowed independence as a puppet state or absorbed into greater France? "Everything," Adams wrote, "has the appearance of profound tranquility, and an absolute and total political revolution, has hitherto left *everything* else, exactly as it was before." The French officers and men "live in perfect harmony, with the people of this Country upon whom they are quartered, and to whom however unwelcome, they are far from being uncivil guests." By midspring, there was a Franco-Dutch treaty of friendship in place. The Netherlands had become, in essence, a French tributary, in existence for the welfare and glory of France.

To Adams' relief, the diplomatic community still functioned normally. With the Portuguese minister, Antonio de Araujo de Azevedo, he discussed Rousseau. He enjoyed the hospitality and goodwill of the Swedish ambassador, Count Carl Löwenhielm. The Prussian minister, Baron von Bielfeld, became a friend with whom he took regular walks. They amused themselves at the theater, with a trip to the fair at Rotterdam, and with conversation about political matters, especially "the rights of Man, the origin and foundation of human Society and the proper Principles of Government. He says that in his opinion no consideration whatever can in any case justify a violation of truth. I told him that such a sentiment was rather extraordinary coming from a diplomatic man." On overall worldview there seemed general agreement. On the rights of women, he found Löwenhielm and Bielfeld agreed with him that "it is in vain to labor . . . against the prescriptions of Nature. Political subserviency and domestic influence must be the lot of women, and those who have departed the most from their natural sphere are not those who have shown the sex in their most amiable light." The intellects of women like his mother should shine behind the scenes, their influence felt in private conversation. The three European ministers puzzled about and worked to find ways to live in a changing world. Löwenhielm, like

Adams, represented a country that struggled to maintain neutrality. De Araujo's Portugal resisted being pulled into the French sphere of influence. Bielfeld's Prussia had recently become neutral. What they all saw ahead were endless European wars, though Adams observed that "the course of events is always so different from everything foreseen that one would think it is one of the professional employments of the fates to baffle all human penetration." As readers of Rousseau and Voltaire, they discussed natural rights, the origin of governments, the relationship between religious belief and government, and the ferment, creative and destructive, that the French Revolution had made. His colleagues would have agreed that "it is remarkable that in France with a government founded upon the supremacy of the people and the rights of man," the government consists of people "chosen by a small revolutionary committee, and in whose appointment the people had not any agency other than that of acquiescence." Why was this kind of tyranny better than a monarchy?

Outside of these discussions Adams was often bored. The local theater was turgid. There were excursions to take, such as one to Leiden, where he had been a student. Thomas' presence was comforting but not stimulating. What irked John Quincy the most was that he had so little work to do. He shopped for books and read a great deal. The pay was ample, $4,500 in salary with an equal amount to be drawn for expenses. With Thomas' salary of $1,500, it was enough to live comfortably, even to save money, which he sent to Charles, who had touted good investment prospects in New York. "An American Minister at the Hague," Adams wrote to the secretary of state, "is one of the most useless beings in creation. . . . At present I am liberally paid for no service at all," except for reporting European news that he got from newspapers. Letters from America took as much as six months to reach him. It was "something like drudgery to be obliged to write perpetually without receiving any letters in return," though most of his lengthy, fact-filled, and sharply analytical letters to the State Department got through. Among their other virtues, they conveyed a comprehensive overview enlivened by enough specific detail to give vividness to abstractions. His stylishly written dispatches were read and valued by the cabinet. John Adams passed along the letters he personally received to the president, who read them all. "Mr. J Q Adams Your Son," Washington wrote to Abigail, "must not think of retiring from the work he is now in. His prospects if he continues in it are fair and I shall be much mistaken if in as short a period as can be expected, he is not found at the head of the Diplomatique Corps."

It was praise enough to turn anyone's head. But John Quincy still had difficulty not feeling useless: he wanted the security of a reliable profession. Government service was not that. He missed his family, and he wanted a wife and family of his own. His father responded, "I want your society as well as that of my other Children, but I must submit to your and their Arrangements in Life. I wish you to come home and be married after two Years." But, when you do, "you must return with the Spirit of a Stoic—a determined Spirit to bear any neglect . . . from your countrymen without resentment, to go obstinately to the Bar . . . and attend patiently in your office for business."

This was hardly cheering advice. "We are indeed once more scattered about the world," John Quincy wrote to his sister, "and our destiny from our childhood, has been that of wanderers, beyond the common lot of men." He was not reconciled to this. And it had none of the romantic glow with which his father looked back and recollected "the adventures of myself and my wife and daughter and sons. . . . I see a kind of romance, which, a little embellished with fiction or exaggeration or only poetical ornament, would equal anything in the days of chivalry or knight errantry." John Adams could afford, late in life, this literary self-dramatization. But he had not at all felt that way at the time. Separation from family and country had been painful, though he had had the advantage of a wife and children with whom to reunite. Like his father, John Quincy was an avid reader of *Don Quixote*, and he too had a predilection, even a preference, for tilting at windmills. But there had not been, so far, the sense of a demanding mission. And while he had the company of Thomas, he was still separated from home and without a companion for his bed and heart.

In late October and early November 1795, Adams was stuck in a tiny Dutch port on the English Channel. Instructions had come from Philadelphia that he was to go immediately to London to exchange the ratifications of the Jay Treaty. He had ambivalent feelings about his assignment. He believed, as did Washington, Hamilton, and Jay, that the treaty served American interests, though the prediction that it would be a bitter pill for America to swallow had proved accurate. And there was no doubt that the French would feel they had been stabbed in the back. In Philadelphia, the treaty had barely survived the onslaught from the Republicans. Many Americans thought it an insult to national dignity. It reaffirmed that debts to British merchants incurred before the war had to be repaid, with compensation awarded to those whose property had been unjustly confiscated. In return, the British would evacuate their forts in the Northwest and American commercial vessels would be allowed limited access to the West Indies trade. The British declined to negotiate about compensation for slaves the United States claimed had been abducted. Britain would not remove its warships from the Great Lakes, renounce arming its Indian allies in the event of war, or recognize that American ships had neutral rights on the high seas.

As a gale pummeled the Dutch coastline, the captain of the pilot boat John Quincy had engaged to take him to England refused to leave the harbor. Stuck on the wrong side of the Channel, he felt himself in both a timeless vacuum and a time-pressured situation. And he could not liberate himself from anxiety about his career. No matter how much his father kept urging the advantages of the law, John Quincy emphasized his distaste for it. "I shall therefore always consider the bar as a resource, but I shall certainly consider it as the last." He might have to resort to "the last," but he hoped not. Being stuck in the town of Hellevoetsluis brought strongly to mind other long-standing concerns, particularly his loss of Mary Frazier. Perhaps no one had empathized more with his heartache than Aunt Elizabeth, who had had a ringside seat and to whom he had confided. After brooding about this for over a year, Elizabeth wrote to him that "perhaps no one knew better than myself, the strong emotions which tore, and agitated your Mind. . . . I longed to lighten your Heart—and to have you pour out all your Grief into my feeling—faithful Bosom." But like all broken hearts, his would mend. He would find a woman worthy of him. "Amiable minds are said to be the most susceptive of love. Your heart has felt his power. . . . Yet let me hope, that whenever you may *wish* to pay your vows at the altar of Hymen, you may find a daughter who excels them all—in real worth, as in beauty. Who is deserving of your highest esteem, and tenderest love." There would be "some new candidate for the nuptial state" who would be the right one for him and at the right time.

That the time had not come was much on his mind when he learned that Charles and William Stephens Smith's sister had been married in New York. "He is at last safe landed," Nabby wrote to John Quincy, "after all the hair breadth escapes and imminent dangers he has run." Unlike the eldest brother, the less prudent Charles had defied his parents. "Ay! Charles has got the start of me, to be sure. I was not so restive under correction, and submitted to sacrifice my chance. . . . If you should follow his example," he told Thomas, "and get into a snug corner of matrimonial enjoyment, while I continue to be buffeted about the world in solitary celibacy, you may be assured of having the same cordial good wishes of your affectionate brother." Thomas responded to his mother, the fount of caution and admonition who advised him to stay celibate and single, that "five and twenty years ago you would perhaps have spared a young man so impracticable, not to say *unfeeling*, a task." Nevertheless, "filial respect and affection still reign supreme in the heart of . . . Thomas B. Adams." It also reigned in John Quincy's. But he was angry and resentful. From Hellevoetsluis, he carefully and at length wrote to his mother a statement of what he had sacrificed and a proclamation of liberation. It was, he implied, his mother who had most stood between him and an engagement to Mary Frazier. Was he now, he asked, being advised by his parents to come home and find a bride? And sustain that bride and family by, if necessary, a career at the bar? He had torn his heart out of his breast "by voluntary violence. . . . If after such wounds have been healed; after all the impressions once so dearly cherished have been effaced," then maybe he would be ready to consider marriage. "But to sacrifice the choice of the Heart is all that prudence or duty can require. . . . It must henceforth pursue its own course; if it can choose again, its election must be spontaneous. . . . The deliberate sacrifice of a strong passion to prudential and family considerations is indeed so widely distant from the orthodox doctrine of Romance, that there is not I believe a novel writer of the age, who can get rid of such an incident without the help of a pistol." He did not, he assured them, have a pistol in hand. "But the real lessons of life are seldom to be found in novels."

Aware that whether or not one finds the lessons of life in novels depends on what novels one reads, he had time at Hellevoetsluis to read a novel that startled him, Choderlos de Laclos' *Les liaisons dangereuses*. Expecting to be in London shortly, he had brought only a few books. With his usual intensity, he threw himself into a close reading on the two levels that mattered most to him, the literary and the moral. His own liaison with Mary Frazier had been dangerous in its clash between the heart and the head, romantic spontaneity and prudential caution. Laclos' novel dramatized the thrill of seduction, exploitation, and revenge in a morally vacuous world. It "is very well written," he wrote, "but it is a portrait of human depravity which one would . . . hope is exaggerated." A New England and an American hope, it was one the liberal Adams family maintained, though they regularly saw examples of its breach: gambling, heavy drinking, sexual adventures.

Yet the characters of *Les liaisons dangereuses* existed in a world of privilege, boredom, and sexual licentiousness totally unlike eighteenth-century Boston. "The woman, who is represented . . . as the most prominent and most detestable character . . . is very defective in point of moral effect. Her calamities have no sort of connection with her crimes; they are such as must be supposed to have happened equally in case she had been entirely innocent. . . . Poetical Justice consists not merely in representing the wicked as miserable; but so miserable because they are wicked. To load them therefore with calamities to which all men are equally liable cannot produce a good moral effect. . . . In truth the rain falls and the sun shines equally upon the just and the unjust. But it is also true that guilt is followed generally by punishment, and that the misery inflicted upon vice, can be just only as it proceeds from it." The best literature was wisdom literature, and *Les liaisons dangereuses* taught the wrong lessons.

By the time he arrived in London in mid-November 1795, he was too late to witness the signing of the treaty. The chargé d'affaires had stood in for him. John Quincy had reason to expect that his stay in London would be brief, a few months at most. Thomas Pinckney would return soon from Spain with, it was expected, a treaty of amity and commerce in hand and resume his role as the senior American diplomat in London. In the meantime, Adams would watch the store. To his annoyance, Undersecretary of State George Hammond kept announcing that Adams was minister plenipotentiary to the Court of St. James's. Finally, after much insistence to the contrary, he was able to pressure Hammond into acknowledging that Adams was the minister to The Hague on special assignment to London. The misrepresentation seemed purposeful, perhaps to irritate and test him, perhaps to fabricate an embarrassment to the American government. "I most sincerely hope," he wrote to his father, "that I may never again be called to a station, which in the eyes of truth or even of Envy, may appear to have been the gift of favor rather than the reward of merit." In December, he sent a long dispatch describing his meeting with William, Lord Grenville, the minister for foreign affairs, at which they reviewed the major issues and expectations of both governments about the implementation of the treaty. "I have been accustomed all my life to plain dealing and candour, and am not sufficiently versed in the art of political swindling to be prepared for negotiating with an European Minister of State." He felt fortunate that there was little he had authority or instructions to do other than keep the chair warm. By early 1796 he was eager to be back at The Hague.

When, in mid-January, Pinckney returned, John Quincy was happy to be relieved of responsibility, though he had to wait for his instructions to return to the Netherlands. An alleged discovery of a cache of manuscripts in Shakespeare's handwriting, including an alternative version of *King Lear*, a new tragedy called *Vortigern and Rowena*, and a lock of hair with a love letter to Anne Hathaway, was the talk of the town. Richard Brinsley Sheridan paid £500 for the rights to put *Vortigern* on at Drury Lane. The owner of the manuscripts gave Adams and others the opportunity to examine them, with the London cultural world divided between believers and deniers. "The internal evidence," he wrote to his mother, "is indeed so very strong in their favor that it becomes a little suspicious from its minuteness. For instance, at the end of the Lear, is written, 'The end of my Tragedy of King Lear. William Shakespeare.' Does not such a minute look a little as if it was made on purpose to answer a question very natural now, but which the author probably never foresaw?"

At first London did not provide the pleasure it had on previous visits. And he felt the absence of anyone who cared about him personally. At The Hague, he had Thomas. In London, he had the art galleries and theater, but he could not overcome his sense of having seen it all. At least, he told his mother, "I have not yet . . . lost my attachment to poetical beauty, and still recognize with delight the flashes of original genius. Shakespear therefore retains almost unimpaired his empire over my mind." Still, a sense of dissipation and solitude dominated. It was not created by a lack of company. Three touring Boston friends, John Gardner, Nathan Frazier, and Thomas Crafts, kept him company on walks and at dinners. John Hall soon joined them, and an excursion Adams took with Gardner and Crafts to Cambridge in February lifted his spirits. It allowed him to affirm that his health was better, though there had been nothing especially wrong with it. But he often felt low-spirited and without energy as the wait for instructions began to seem endless. He and others speculated that the delay was caused by what might be a change in assignment. Maybe he would not be returning to The Hague. He did his best to put that possibility out of mind.

He did, though, have news from his family and from newspapers that the Jay Treaty continued to be the focus of partisan divisiveness. Although the Senate had approved it, a faction in the House created a constitutional crisis by attempting to deny funds for implementation. "It is a Mortifying Consideration," John Adams wrote to Abigail, "that five Months have been wasted upon a Question whether National Faith is binding on a Nation." Finally, the funds were allocated and the precedent established that the Constitution gave responsibility for treaties exclusively to the Senate. Once that body ratified a treaty and the president signed it, the House had no alternative but to fund its provisions. By 1795, Washington himself had become the target of those who hated the Jay Treaty, thought the Federalists crypto-monarchists, and believed the House should have more power than the president and Senate. In early 1796, Washington let his inner circle know that he had had enough and would retire at the end of his term. It was momentous news for the Adams family and the nation. In London, the British ministry and the American community well knew that John Quincy's father was likely to be the next president.

Meanwhile, two expatriate families warmly entertained John Quincy for his own sake. He had frequent dinners at the home of John Singleton Copley. A friend of the Adams family, the Boston-born artist, who had been living in London since 1775, had painted an iconic portrait of John Adams in 1783 and attended Nabby's wedding in 1786. The Copleys had two attractive daughters, one of whom John Quincy thought "handsome, if not beautiful. . . . There is something so fascinating in the women I meet with in this Country, that it is well for me, I am obliged immediately to leave it." Whether Englishwomen found him fascinating is undetermined. But if they had seen only the portrait, intended as a surprise gift to Abigail, that Copley began to paint in February 1796, he need not have worried about his own attractiveness: lightly powdered hair, worn puffed in the style of the day; lips slightly fashioned into a restrained smile; a narrow but firm chin, aquiline nose, and expressive brown eyes; a young man of medium build, in a dark black coat, with a white cravat wound around his neck and tied into a small knot. The background of the upper two thirds of the portrait is a deep maroon velvet curtain. The lower right of the curtain has been drawn back, revealing a pastoral landscape with the dawn breaking, the light from behind. John Quincy's face and cravat are illumined from the front, as if by a light that comes from within. It is Copley's vision of John Quincy at his handsomest, a portrait that emphasizes youth, intelligence, and sincerity. But an authenticity shines through the idealization. What Copley saw and painted represents a version of John Quincy's character and the expectation of achievement that the Adams family believed would be his. After his final sitting, John Quincy sent it to his mother. "He has made a good picture of it," he wrote in his diary.

The home of his other favorite American expatriate family, the Johnsons, had become his as much as his rooms at the Adelphi Hotel. Soon after arriving in London, he brought Johnson letters to be sent to America, a courtesy the consul had been providing since John Quincy had been in Holland. By the start of the new year, Adams had become an intimate guest of the Johnsons. He had also done what he had deluded himself into thinking he might never do again—fall in love. Having first met the family in Nantes in 1779, John Quincy probably had only dim if any memories of the Johnson children, seven girls and one son. Sixteen-year-old Thomas Baker Johnson was now at Harvard. John Quincy gained a sense of the daughters when the eldest and most musical one, Nancy, who played the harp and piano, entertained the company. Her sister Louisa Catherine, the second daughter, sang. In late January, he danced until three in the morning at a ball in honor of Louisa Catherine's twenty-first birthday. The family assumed that he was interested in Nancy, but they were mistaken. It was Louisa. Slim, with brown eyes, light brown hair, and pale complexion, at about five feet six inches nearly as tall as John Quincy, she "rattled on" in his company about whatever came to mind. Soon he was visiting almost every day and taking afternoon excursions with the ladies, particularly Louisa, Nancy, and Mrs. Johnson.

"I wish he would come home," John wrote to Abigail, and return to his law practice "after marrying his girl if she is still disengaged." But John Quincy apparently never considered resuming his relationship with Mary Frazier. They had last met in the fall of 1791. During his three years in Boston they had not seen or written to one another, though he confided to friends how much pain he still felt at having lost her. When he left for Holland in 1794, he had a position and a salary, and it would still have been possible to attempt to sound her out. But he did not. Not a word appears in his diary or his letters that even implies that, once the primary obstacle had been overcome, he considered trying to reclaim the woman he had adored. Their separation had not been, he told his mother, "the result of necessity or of an angry moment. It was a mutual dissolution of affection. . . . The flame was not covered with ashes. It was extinguished with cold water." He and others, though, had represented the separation and its aftermath quite differently in 1791. And other evidence suggests that the passion continued for years, the regret even longer. Still, he explained at too great length to his parents, the only promise he had made "to her . . . was that I never would marry a woman whose character or conduct should make *her* regret that she too had once made the same choice; or in other words that her successor in my affections should not be unworthy of a place which *she* had held. She made me the same promise."

By February 1796 he had fallen in love with Louisa Catherine. For some weeks, soon after becoming a regular visitor at the Johnsons' home on Cooper's Row, he teased her about the songs she sang to her sister's accompaniment, as if he objected to them. When he responded to Louisa's challenge that he write a better song, she began reading it aloud in company. The children's governess interrupted her, snatching it from her hand. Louisa blushed, aware for the first time that there might be some truth to the governess' guess that it was she to whom John Quincy was attracted. Late in the month, he went shopping, he wrote to his mother, with Mrs. Johnson and her second daughter, "a very amiable young lady." Early in March, he confided to his diary, "Ring from Louisa's finger. Tricks played. A little music and dancing. Placed in a very difficult dilemma. Know not how I shall escape from it." Louisa's family was startled. The Johnsons had assumed his interest was in the outgoing, sparkling, and flirtatious Nancy.

In late March, conveying the news that the Copleys were well, he gave a broad hint to his parents: "Perhaps you will hear of another family that has been still more attractive to me; but of this I may write more on a future occasion." Abigail guessed which family it was.

"Whom you call yours shall be mine also. . . . I would hope for the love I bear my country, that the siren" is at least half American. When John Adams wrote to his son that he wished that the "attachment had been made in America," John Quincy replied that he too would have preferred that she be in all ways "*wholly* American," but "the destiny which is said to preside over these things" did not cooperate. "I can only hope that neither my friends nor myself will ever have any other reason to regret my choice." Abigail feared that Louisa's birth and upbringing would make her life in America difficult, and that John Quincy's career, if it was to be on the public stage, would be damaged by the taint of a foreign wife, as if an American wife were not good enough for him.

Louisa's fifty-two-year-old father, Joshua Johnson, who, Louisa later wrote, "always had a prejudice towards the *Yankees* and insisted they never made good husbands," had come to London in 1771 from his native Maryland, a partner in an Annapolis firm specializing in exporting tobacco and importing dry goods. Its transportation facilities, insurance policies, credit arrangements, and hedged investments facilitated the sale of tobacco from American planters to European merchants and of dry goods from English manufacturers to American wholesale and retail customers.

The firm handled these items either on commission or consignment. Profits came partly from its charge on the commission sale of merchandise. The consignment arrangement offered the opportunity for higher profits but carried greater risk. Johnson managed the firm's London business well enough to help keep it and its successor, Wallace, Muir, and Johnson, profitable for twenty-five years in a volatile tobacco market. Large sums were owed to British firms that supplied goods on credit. Large sums were owed to Wallace, Muir, and Johnson by Maryland and Virginia planters. It was a highly leveraged business with little room for misjudgment.

Johnson had the advantage of an important family connection and an engaging personality. His grandfather had practiced law in London before emigrating to Maryland in the late seventeenth century. His older brother, a successful lawyer, represented Maryland in the Continental Congress, drafted the declaration of rights for Maryland's constitution, served as an associate justice of the Supreme Court, and turned down the opportunity to serve as Washington's secretary of state. Johnson was Anglican in England and Episcopalian in America. Like his brother, he was an ardent patriot who supported independence. During the Revolutionary War, he moved to France, with three children, one of whom was three-year-old Louisa Catherine, and his wife, twenty-one-year-old Catherine Nuth Johnson, though they may not have married until 1785. Thirteen years younger than her husband, Catherine was a Londoner by birth and by all accounts a beautiful woman. It was a marriage of love and mutual dependency: "She was his pride, his joy, his love, and in her and his children was concentrated all that made life desirable," Louisa later wrote. "The greatest fault he had was believing every one as good, as correct, as worthy as himself." During flush times, Johnson maintained large and comfortable residences, up to eleven servants, and schools and tutors for the children. When business faltered, there were contractions and economies. After the war, he prospered enough to lease a country home. Each of his daughters, he told them, would have a dowry of £5,000. At Great Tower Hill in London, the family welcomed visiting Americans, especially when Johnson was appointed American consul by Secretary of State Thomas Jefferson, a friend of Joshua's brother. When Nabby Smith and her husband returned to England in 1791, they and the Johnson family became close friends.

In France, Louisa Catherine attended a Catholic school taught by Ursuline nuns specializing in the education of girls. French became her first language. An ecumenical family, the Johnsons attended a Catholic church in Nantes. Deeply impressed by its rituals and solemnity, the little girl became a fervent believer. When, in 1783, the family returned to London, eight-year-old Louisa found herself immersed in English, a secular school, and an Anglican world. She soon became an ecumenical Anglican, with strong beliefs and a religious sensibility. Shy, sensitive, occasionally ill, she was given to fainting. Pampered by her parents, especially her father, Louisa thought herself superior to her schoolmates, by whom she felt respected but not loved. At home all was joy. "All the scenes of my infancy," she later wrote, "float upon my fancy like visions which never could have had any reality yet like visions of delight in which all was joy and peace and love."

An avid reader from an early age, Louisa loved books, as did her mother, and to have other people read aloud, and music. With an attractive voice, she sang well and learned to read music by sight. Poetry and the theater particularly appealed to her. Reading the classics of French and English literature with equal facility, she absorbed Milton, Shakespeare, and the eighteenth-century canonical writers. "Much, much depends," she later observed, "upon the reading of our early life. Children appear to read carelessly, and to note little of the subjects which are placed before them.

But the impressions which are made on them are durable, and though apparently lost for years; frequently stamp the taste and mark the character of the mind at a later period of their lives." None of her youthful writing exercises survive, but that she wrote a great deal of poetry later makes it likely that she wrote some in her youth. Her letters and diary reveal her gift for writing, combining fluency, spontaneity, and a sharp mind, intuitive rather than logical.

She also had an inclination to worry, to be attuned to others, and to place her trust in higher forces, from her father to a providential God. Thought of as more steady than her elder sister, she became her father's household assistant, especially since her mother was often ill. Louisa also seemed fragile and prone to sickness. When the Johnsons brought the three oldest girls out into society, Louisa, her nerves often on edge, preferred to stay at home. To Louisa, her father "was the handsomest man I ever beheld," and her life at Great Tower Hill was filled with "domestic felicity." She always remembered it as the golden period of her life. As Nancy came of age in the mid-1790s, there were suitors, young men, British and American, coming to the house. Apparently there were also young gentlemen for Louisa, who hardly took notice, as if it were impossible that anyone would be interested in her. "If I ever had any admirers no woman I can assure you," she wrote to her sons in 1825, "ever had fewer lovers than your Mother." When she became aware that John Quincy was paying special attention to her, she was startled into eligibility and womanhood.

At a little before eleven in the morning on July 26, 1797, Louisa Catherine and John Quincy were married in the Church of All Hallows Barking. A close friend of the family, who had married Louisa's parents, performed the ceremony. Joshua and Catherine Johnson, Thomas Boylston Adams, and John Quincy's Boston friend John Hall attended. No one in America of either family would have thought it reasonable to cross the Atlantic to be there. Louisa's sisters witnessed the first of them to take the sacred vow. The groom had written the previous summer to his mother that "the matrimonial union you know is always professedly for better [and] for worse. . . . Prudence is a sorry matchmaker." He was to find over the next year that the parameters of "for better and for worse" were to be tested by events beyond his and Louisa's control. They were about to travel a long road together.

But the year and a half that had elapsed between their engagement and the wedding resulted from John Quincy's relapse into prudence. In April 1796, he and her parents differed about a marriage date. The Johnsons preferred it be soon, perhaps that summer. So did Louisa. They had assumed the wedding would take place before he returned to The Hague or, if not then, sometime later that year, perhaps in the autumn. But Adams was adamant. When it would be, he could not say. He believed that he could not support a family and his professional expenses on his current salary. So the marriage had to be postponed until his future and income were more definitely determined. He had, he wrote, a "conversation with Louisa. Was explicit with her, and obtained her acquiescence." Neither Louisa nor her father was happy, however, with the decision. "The right and the reason of the thing are however indisputably with me, and I shall accordingly persist."

A determining factor would be the election late in the year. If his father became president, it would seem too much like nepotism to accept a new posting. If Jefferson was elected, John Quincy assumed he would be recalled. Regardless, he would resign by the summer of 1797. That would mean Boston and the resumption of the practice of law. He did not know that in April 1796, at the same time that he and the Johnsons were discussing a marriage date, President Washington had nominated him as minister plenipotentiary to Portugal, with instructions to remain at The Hague until his replacement arrived. It was a promotion, doubling his salary to $9,000 a year. "I shall return in all probability to my old profession," he wrote to Charles, "and endeavor to wear off as soon as possible the rust of disuse. But whether I shall again fix myself at Boston or attempt a settlement in a Southern State, remains for my future consideration." Probably Johnson, who had been trying for three years to wind up his London affairs and return to America, had encouraged him to think of Maryland. Louisa might sail with her family while John Quincy was still at The Hague. About to return there at the end of May, he confided to his mother that "upon the whole I have passed a pleasant winter here; the greatest objection that I can make against London is too much society." The Hague would provide an attractive contrast. If Abigail had anticipated, as she had every reason to do, that he would be returning to Holland with his wife, he assured her that he had too much good sense to disregard the prudence that his family preached. "I have, though I own very reluctantly, concluded that I must not yet take upon me the encumbrance of a family." Apparently he assumed that fatherhood would follow soon after marriage. Prudence "tells me that I must return to the Hague alone, and wait for more favorable or more permanent prospects; and although passion has summoned many very plausible arguments . . . it is all to no purpose. Prudence is inflexible, and I go from hence alone."

After an "evening of delight and of regret" and "with sensations unusually painful," he parted from Louisa, though he was happy to be reunited with Thomas, who had had a sickly winter, and whose mother urged him to be proof against the shafts of Cupid's arrow "until you return to your Native Land, and then choose a wife whose habits, tastes and sentiments are calculated for the meridian of your own Country." In essence, Abigail wrote, do not make the same mistake that John Quincy had. Louisa, worried that she would have a stern test to meet when she came directly under the eye of her mother-in-law, existed in a limbo of indefiniteness. In the meantime, all she could do was wait.

During the next twelve months, she and John Quincy exchanged close to seventy-five letters, Louisa developing from an inexperienced, awkward corre-spondent who needed help to a capable and expressive writer. Her dominant theme was her eagerness to be with him. His was patience embedded in eagerness. His expressions of love were inseparable from his guidance, a degree of mentoring that he disguised as the lessons of experience rather than pedagogy. She objected to his occasional ex cathedra comments and advice. Only once did they have a truly angry exchange. She objected to his harsh criticism of her proposal that she visit him in Holland. It seemed to her not only a reasonable but also a loving proposal. And though it had only been implied that perhaps they might marry at The Hague, he took it either as pressure to do that or as an act likely to expose him to accusations of impropriety. She was furious with the implied charge. "You mention the pain it gives you to write me in this style. If thus painful to *you* judge what it must be to *me*, whose mind must be doubly wounded at the idea of having given rise to it. . . . Ah my beloved friend, this boasted philosophy that I have heard so much of is indeed a *dreadful* thing. I have too much reason to dislike it, as I see too plainly that it dictates every action, and guides your pen *I* hope in contradiction to your feelings. When you were here you have often said that you could see no faults in your Louisa, but alas how are you changed, you now charge her with impropriety of conduct, and there is even an indication of want of *delicacy*." He backed down, and they reaffirmed their absolute commitment.

At The Hague, he found himself busy catching up with what he had missed, renewing contacts, and sending off dispatches to the secretary of state and to his father about the changes on the European chessboard, including the game-changing victory in Italy by a new French general known as Napoléon Bonaparte. As the demands on Adams' time gradually decreased to match the usual slow rhythm of life at The Hague, he was able to return to what he had missed so much in London: reading, writing, and translating; long walks in the fields and woods; an early to bed and early to rise schedule; the resumption of his friendship with Bielfeld, the companionship of his brother, and pleasant dinners hosted by Count Löwenhielm. John Quincy's writing was confined mostly to his letters and diary. "The time for original composition has not yet come; I know not whether it ever will. . . . I have no subject, and am far from being yet satisfied with my style."

After learning in October 1796 of his appointment to Portugal, he spent much of the year waiting for instructions. At first he expected that he would be able to depart in the late fall or winter; this plan most likely would include a marriage in London before Christmas and departure together. When no replacement or new instructions arrived, the couple's hopes were dashed. Her disappointment was exacerbated by his constant exhortations to stoic acceptance. "In the intercourse of friends, of lovers, but more especially in the tender and inseparable connection which I hope is destined for us," he wrote to her, "nothing bears so hard upon the ties of mutual kindness and affection as suspicion and distrust. Between you and me may it never rise."

But it did, off and on, with recoveries that each at different moments initiated. A quick learner, Louisa responded that "reason tells me I shall at least acquire resignation if not happiness. . . . Never my beloved friend let my weakness lessen your affection for me but write to me constantly and if possible teach my rebellious heart gently to acquiesce without murmuring." Their bond was strengthened by the process. "You are the delight and pride of my life," he assured her.

But he was stuck in place until spring 1797. The money he saved from his $4,500 salary he transferred to his brother Charles, who touted excellent investment opportunities in New York. "I will not say that I envy your happiness," John Quincy wrote to him, "because I shall always rejoice at your prosperity. But what a sorry figure politics and celibacy . . . make in comparison." Their sister Nabby was discovering that she had a husband addicted to financial extravagance, overextended credit, and land speculation. Bankruptcy seemed imminent. Stuck at The Hague, John Quincy consoled himself that at least he enjoyed the advantage of a disciplined and economical life. He reread *Paradise Lost*, "the admiration of which increases in my mind upon every perusal," and indulged his passion for buying books. He went to the theater, often with Bielfeld, whom he enlisted to give him Italian lessons.

Soon he was reading Torquato Tasso's *Jerusalem*. He worked at his Dutch and refreshed his Latin by translating Tacitus and Cicero. In his end-of-the-year diary summary, now an annual practice, he concluded that the last seven months had "been a time of as steady and constant application as ever occurred in the course of my life." Early in 1797, James Monroe came from Paris for a week of diplomacy and conversation. He continued, in Adams' view, to be too explicitly pro-French at a time when the French government was treating American neutrality with disdain. Still, with discretion on both their parts, they got on politely. "He said nothing to me upon the subject of his recall," Adams wrote to Rufus King, the new American minister in London, "but it was easy to perceive that it was the idea constantly predominant in his mind." Monroe and Adams could not have guessed that they had a future together.

As the winter passed, he chafed at the delay, eager to be married and take up his new assignment, in whatever order seemed best. Tell Louisa, Abigail exhorted him, "that your Mother has for 30 years been tried worn and inured to separations which have torn every fiber of her heart under circumstances the most perilous to those she held most dear." By November 1796, suspecting that he would be at The Hague at least until March and that the Johnson family would leave for America in early spring, he thought the marriage might have to be delayed until he himself returned to the United States. That might not be for years, a prospect that undoubtedly sent a chill through the Johnson household. With Johnson's business affairs in transition, they were all eager to have Louisa settled into her new life. What could they do to expedite the marriage and their own return to Maryland, where Johnson needed to go as soon as possible in order to protect his business interests? Perhaps, they suggested, John Quincy could sail to Portugal via Britain. That would allow the marriage to take place in London, with the Johnsons to depart for America immediately afterward.

Across the Atlantic, December 1796 was a month of suspense for the Adams family. The contested presidential election had been scorching the newspapers, pamphlets, tavern placards, and rumor mills for almost half a year. Jefferson had emerged from his strategic retirement to reclaim the government from Washington and Adams, whom he and his surrogates denounced as pro-British monarchists who favored replacing an elected president with a hereditary king. Actually, both parties embraced government by elites, with the voters to defer to their superiors in wisdom and wealth: large Southern land- and slaveholders, especially in Virginia, the largest state; educated experts in government and commerce, especially in Massachusetts, the second largest. As usual, the public rhetoric and the votes were driven by the extremes. John Adams could face in public, he wrote to Abigail, the possibility that Jefferson would be elected president "with firmness & a good grace that I don't fear. But here alone abed, by my fireside nobody to speak to, poring upon my disgrace and future prospects—this is ugly." What the Federalists most feared was that a Jefferson presidency would mean war with Great Britain and the dismemberment of the Union because of internal bickering or military defeat. They had reason to believe that a loose confederation was indeed what the Southern advocates of extreme states' rights wanted. For John Quincy, who shared his father's view, a dissolution of the Union was the ultimate disaster. If he was not elected president, John Adams vented to Abigail, he would run for the House of Representatives, where he would "drive out of it some Daemons that haunt it. . . . But this is my Vanity. I feel sometimes as if I could Speechify among them, but alas, alas, I am too old. It would soon destroy my Health."

On Inauguration Day in March 1797, John Quincy still did not know who had been elected president. In an election in which there were four candidates, Adams had received 71 electoral votes and Jefferson 68; Adams carried 9 states and Jefferson 7; and Adams collected 35,726 and Jefferson 31,115 votes, though if the slave states had not had the scales tipped in their favor by the three-fifths provision in the Constitution, Jefferson's electoral vote total would have been noticeably smaller. Before the Twelfth Amendment, each candidate ran unyoked to any other and those with the two highest tallies served in the two highest offices, since the creators of the Constitution assumed that none of the candidates would be of different parties or that there would be parties at all. Abigail hoped that Jefferson would now put country before ideology and party. "I trust his conduct will be wise and prudent . . . a means of softening the animosity of Party, and of cementing and strengthening the bond of union." It seemed reasonable for Adams to think that "Jefferson and I should go on affectionately together and all would be well."

At The Hague, John Quincy had reason to think that his father's administration would have multiple challenges. The most dangerous would be an attack on the neutrality policy. In the battle between Britain and France, neither country hesitated to make the United States collateral damage. Would America be forced to fight one or the other or both? That he was the president's son gave John Quincy's position extra exposure. It was rumored that his appointment as minister to Portugal had been heavily influenced by his father. And he still waited for instructions from the secretary of state. "As Events will not accommodate themselves to our desires," he wrote to Louisa early in March, "it is one of the most necessary arts of human life to accommodate our desires to Events. But I am sliding inadvertently again into that terrible thing called Philosophy. Alas! there is no moving a single step without finding the want of it." At last, in late April 1797, letters arrived with the long-awaited news. The new president had appointed William Vans Murray, a congressman from Maryland, as Adams' replacement at The Hague. A law student in London in the 1780s, Murray had been friends with John Quincy; and Murray, who had resided with the Johnsons, had at one time seemed to be in love with Nancy Johnson. "Go to Lisbon," the president wrote to his son, "and send me as good Intelligence from all Parts of Europe as you have done" from The Hague. "I advise you to marry the Lady before you go to Portugal," his mother counseled.

From the Netherlands, John Quincy frantically explored how he and Louisa would get to Portugal. It seemed impossible that an American diplomat could sail from London on an English ship without embarrassing his country's neutrality. Perhaps they could sail directly from Holland to Lisbon. In London, Louisa could hardly believe that their separation was about to end. Could it really be true? "My last disappointment has taught me to fear, and I find it almost impossible to check my apprehensions." A marriage in London and the end of what seemed an indefinite delay depended, it now seemed, on whether her father could arrange for a private schooner to take them to Portugal. As soon as Joshua Johnson learned the good news, he insisted on outfitting a small ship that could safely take them to Portugal and postponed his own and his family's departure.

In late May, John Quincy sent some of his luggage and crates of books directly to Lisbon. When Murray arrived in early June, John Quincy escorted him on his round of formal introductions. Adams also greeted John Marshall, the Virginia Federalist on his way to Paris. The president had appointed him a member of a three-man delegation to resolve the tensions between France and the United States. Word soon came from Johnson that the transportation problem had been solved— they could sail from London. At the end of June, John Quincy and Thomas happily boarded a packet to cross the Channel, though unfavorable winds kept them at anchor for eight days. What John Quincy did not know was that he would not be going to Portugal, ever. There had been a change of assignment. He had been appointed minister plenipotentiary to Berlin.

Chapter 6
Begin Anew the World
1797–1801

Six months later, on a rainy November night in Berlin, John Quincy marked "this month as one of the most unfortunate that have occurred to me in the course of my life." For a man of careful judgment to feel that it had been one of the worst months of his life suggests an abyss of misery. It had indeed been a terrible month. For much of it Louisa Catherine had been fighting for her and her unborn infant's life. That night she miscarried. The ordeal had gone on for eleven days. She dimly heard her brother-in-law Thomas asking the doctor if she were dead. "The mind at least submits," Adams wrote in his diary, "however the heart will rebel."

The calamity had been preceded by a half year of unhappiness that began soon before his marriage on a bright summer day of that difficult year. He had learned a month before his wedding in July 1797 that he had been appointed minister plenipotentiary to Prussia. This unsettled and even angered him enough to initiate a disputatious exchange with his father. Instead of going to Portugal, John Quincy would be sent to renew the treaty of commerce with Prussia and Sweden and provide intelligence. But he had already spent money on his Lisbon arrangements, which would leave him that much less for Berlin. Worse, the Senate had consented by a vote of seventeen to twelve, hardly a ringing endorsement, and John Quincy considered asking for his recall. The anti-Adams newspapers were filled with scorching accusations of nepotism. More sensitive than ever to accusations of favoritism, he vowed to stay in Berlin as short a time as would be consistent with his duty. Berlin seemed to him almost the end of the world, a difficult place in which to be a stranger, and he had no knowledge of the language. "I know not a human being there," he wrote to the president. It did not seem an auspicious place to bring a young wife who had been brought up in comfort. You are going to the Athens of Germany, his father responded, with a wife to accompany you, and your objection to receiving an appointment from your father, given your obvious merit, "seems to me too refined."

As the wedding day approached, John Quincy had no change of commitment, let alone heart. His appointment notice required that he stay in London until his official statement of commission came, which meant that he and Louisa would need to find accommodations. He had already sent twenty-eight crates of books to Lisbon. The house he had rented there would have to be disposed of. Since Johnson's Maryland partners had dissolved the firm, Johnson had decided to do what many had advised over numbers of years: go to Annapolis to secure by his presence his financial interests. He suspected that an examination of the books would reveal that his partners had embezzled money, some of which belonged to him. He had already sent his nephew, Walter Hellen, who had been Nancy's suitor, to collect money owed him. But nothing had been forthcoming, though Hellen prospered. Now Johnson himself needed to press for what was due him, especially since, trading on his own, he had stretched his credit to the extent of being short of cash to honor business and personal debts. He was, though, expecting any day a remittance from his ex-partners from which he would pay his most pressing obligations. The week before the wedding he gave John Quincy "directions concerning his affairs with his former partners, in case [he] should ever have occasion for them," and his son-in-law believed he had reason to expect that Johnson eventually would pay the £5,000 dowry he had promised each of his daughters.

On their wedding day, John Quincy and Louisa took a brief outing. The wedding night goes unremarked in his diary. Louisa stayed at her parents' home, John Quincy both there and in his rooms at the Adelphi. "I have now the happiness of presenting to you another daughter," he wrote to his parents. "The day before yesterday united us for life." He hoped that within a year he would be a father, a happiness he intensely desired. No later than the middle of the next month, Louisa was pregnant. John Quincy did some business: the usual letters to America and a visit with the new ambassador to London, Rufus King, with whom he had "a long conversation . . . upon our public affairs. . . . I perceive something of the talents which have raised him so rapidly to his present situation. His career has been brilliant and will I think continue to be so." He admired King's ability to deal with the British government "without either yielding to or quarreling with them." It was a meeting of minds and a friendship that was to endure. Much of August Adams spent in social activities, reading, attending church and the theater, going to dinners and a ball at the Johnsons'. In early September, still waiting for his instructions, he and Louisa moved into an apartment in the Adelphi Buildings with a number of servants, including John Quincy's Tilly Whitcomb, who had come with him from America, and one of Louisa's maids. Around the same time, the Johnson family sailed for America. The night before they had visited the Adamses. To Louisa, her father looked brokenhearted. She did not know that the next day had been set for departure. "When I arose and found them gone I was the most forlorn miserable wretch that the Sun ever smiled upon." She feared she might never see them again.

Eager to leave for Berlin, the young couple amused themselves as best they could. "She is indeed a most lovely woman," Thomas wrote to his father, "and in my opinion worthy in every respect of the man for whom she has with so much apparent cheerfulness renounced father and mother . . . to unite her destinies with his." From the storm-battered Orkney Islands, where the Johnsons' ship had taken refuge, Louisa's mother wrote to John Quincy "that no event of my life has given me more heartfelt satisfaction than that which enables me to subscribe myself your affectionate Mother." Joshua sent a more foreboding message to his son-in-law. "I need not attempt describing to you my sufferings on this occasion of leaving England. You are a second witness to them. My determination was forced from disappointments and I am persuaded that it is the most proper to enable me to do speedy justice to everyone. However I doubt not but many will censure me. . . . I deposited a Paper with you. . . . Should any accident happen to me, I recommend that to your serious attention." Within days, creditors began knocking at John Quincy's door, demanding payment of Johnson's outstanding bills. The young couple were at first startled, then shocked. Suffering morning sickness, Louisa was bewildered. Harsh voices accosted her. Sleepless, exhausted, she could not believe what was happening. John Quincy directed the creditors to Johnson's bookkeeper, who by now should have received the long-overdue remittance. The bookkeeper claimed that it had not arrived. Suspicion arose that he had been bribed, and Louisa quickly constructed a narrative of betrayal, shifting the blame to someone other than her father. When news of his departure spread, more creditors came to the door. Others wrote threatening letters demanding justice. Louisa dreaded every knock. "Find the affairs of Mr. J. more and more adverse," John Quincy noted in his diary. "This trial is a strong one." Had Johnson not left funds, the creditors asked, in the hands of his daughter and son-in-law to pay these claims? Was Mr. Adams not morally responsible for his father-in-law's debts? Would he not protect the family name and honor? They were not his debts, he responded. "My connection Sir," he wrote sharply, "was with his daughter, and not with his estate."

Regardless, he had to live daily with a wife pained by the thought that other people might believe that her father had known about his insolvency before the marriage and had hastened it in order to dispose of his daughter. Many men Adams knew or knew of had experienced sudden and unexpected business failure for reasons beyond their control, a transformation from riches to rags that every Boston merchant feared. Prudence was only a partial shield, and there was reason to think that Johnson had not been prudent. But there was no doubt in John Quincy's mind that he had been honorable. Still, no matter how much he assured her, Louisa could not, for the rest of her life, get it out of her mind that her husband in his heart of hearts did not believe that. She immediately felt she "had forfeited all that could give me consequence in my husband's esteem or in my own mind." Every attempt at vindication without evidence in her father's favor seemed self-serving. How could her husband not suspect that he had married her under misrepresentations and false pretenses? She had absolute confidence in her father's integrity. But others did not. It was a lifelong nightmare for Louisa.

———

Arriving in Berlin in early November 1797, they were relieved to be done with a difficult journey, the last part over sandy roads between Hamburg and Berlin, a voyage especially hard for a pregnant woman. It rained constantly. At the entrance to Berlin the dapper officer who questioned them had never heard of the United States. Worried about Louisa's fragile condition, John Quincy saw the situation descend quickly from instability to severe crisis. She went to bed, attended by her frightened maid. The hotel "was dirty, noisy, and uncomfortable—The beds miserable; the table execrable," like living in a tavern. With her health varying precipitously from day to day, the pregnancy was in danger. So was her life, and Thomas was also seriously sick. "Three fourths of my time is employed in nursing," John Quincy moaned. "I have never before in the course of my life been afflicted with so many real and severe calamities flocking upon one another at once." Charles Brown, a Scottish physician highly favored by the court and the upper classes, came at John Quincy's urgent request. "This I believe saved my worthless life," Louisa later wrote, "as I lay gasping on my rude bed of agony after eleven days of dreadful anguish." She survived the miscarriage, barely.

After presenting his credentials to the foreign minister, Adams learned that Frederick William II, the son of Frederick the Great, was on his deathbed. Since his successor was not likely to accept credentials that had not been made out to the new king, Adams would have no standing until he could get new credentials from Philadelphia, unless the Prussian government dispensed with the sacred formalities. He argued for an exception, and was delighted when the new king, Frederick William III, "determined to give me an audience as if my credential letter had been delivered to the late king." It was a major concession—and a kindness. It was also by a kindness that they were able to move, in a city short of housing, from their unfurnished apartment at the Brandenburg Gate, where they found the changing of the guards and the military parades unbearably noisy, to a larger, more comfortable apartment close to where other diplomats and Dr. Brown lived. It was convenient to have the doctor close by since Louisa, troubled with headaches and fainting fits, quickly became pregnant again. The courtly doctor and his family entertained regularly; their home was a semiofficial salon for English-speaking residents and visitors. The British ambassador, Lord Elgin, and even Prince Frederick Augustus, George III's son, residing temporarily in Berlin, came regularly. The Adamses were soon close friends with Dr. Brown, his wife, and their three daughters. The Welsh-born Mrs. Brown, "the beau ideal of a good, honest, simple minded, kind hearted old Lady, of the old School," became a much-loved maternal presence. Isabella, the same age as Louisa, became her closest friend. And, to Louisa's delight, two women who came to see her at the doctor's suggestion were soon also good friends, one of whom, Countess Pauline Néale, a Prussian aristocrat and maid of honor at the court, became a confidante and mentor. The language that Louisa shared with the Prussian upper class allowed her to have a fluent voice. An obsessive Francophile, Frederick the Great had made French the official language of the court. And it was in French that Adams conversed with his diplomatic colleagues, some of whom became friends, especially Count Karl von Brühl, who had been military tutor to the new king and whose family joined the Browns and the Néales in the Adamses' social circle.

Although Louisa at first felt too fragile to be introduced to the court, the American minister plenipotentiary had no choice but to do what his status required, including regularly attending what the Adamses referred to as dissipations: lavish balls, dinners, and galas attended by hundreds of people, dressed in the richest fashions, and six-hour tea parties at which the main activity was playing cards for money. "These parties all have a great resemblance to one another," John Quincy complained, "and you meet at them nearly the same company." Soon tongues wagged about Louisa. Why was the minister's wife never to be seen? Was it only her health? Was it an attempt to conceal how ugly she was? Or some other peculiarity or deformity? In the spring, Louisa finally made her royal introductions. Shy, simple in dress, attractive in form and features, pale in complexion, and without any makeup, she was a great success. To the extent that her health permitted, she now accompanied her husband. Twice, when she succumbed to the temptation to use rouge, he bullied her to remove it. She remained pale-faced, a widely noted distinction that made her even more welcome in court society.

But the business for which he had been sent to Prussia was on hold. European politics made it, for the time being, inopportune for Prussia to renew its commercial treaty with the United States. The monarchies in opposition to republican France were finding it difficult to unite into an effective coalition, partly because the enemy of their enemy in some cases was also their enemy. Prussia and Austria especially detested one another. This unstable coalition of Great Britain, Russia, Austria, and Prussia competed with republican France for dominance of the smaller monarchies, the Netherlands, Denmark, Spain, and Sweden. And the shifting scene was made even more complicated when the French Directory, spurred on by victories, especially Bonaparte's in Italy and Egypt, transformed republican France into an oligarchy. "Buonaparte affects great splendor and magnificence everywhere out of Paris," John Quincy wrote to his mother at the end of 1797, "and as great meekness and simplicity there. . . . He has already become almost too great for his masters of the French Directory." Two months later it was clear that "the military power in that Country has become the only real and effective Power." France was no longer a republic. The Prussian government in principle favored a commercial treaty with the United States. But the elephant in the room was France, limited in its worldwide ambitions only by the power of the British Navy. At the moment, Prussia and France were at peace. With a depleted treasury and a deteriorated military, Prussia needed to buy time. And France at the moment was angry at the United States. The Jay Treaty seemed to the French to betray American neutrality and Franco-American friendship. In retaliation, the French were expropriating American cargo ships in French ports. By late 1797, newspaper accounts reached Berlin announcing the probability of war. It was clear that Adams would have to wait until the Prussian government thought the time more propitious for the renewal of the Prussian-American treaty. In the meantime, he had to be resigned to an indefinite residence in Berlin.

At least there would be time for reading and writing, to take excursions, and to learn German. Still, at the start of the new year, with the purpose of his mission balked, he felt "very much dissatisfied. . . . I can do little good, and obtain scarcely any information." Dispatches to the government, no matter how detailed, filled only a few of his work hours, as most of the information was taken from European newspapers. The immediacy of writing made him feel more alive, though a writing plan that he dared not even identify in his diary did not work out. It "presents difficulties which I fear will prove insuperable." For the time being, his diary continued to be mostly a record of daily events, with some personal references, mainly about his professional frustrations and Louisa's health. By early spring, he was almost in despair about his uselessness as a minister. "Find myself in a state of suspense and embarrassment to which I am not unaccustomed. I know not whether I am to remain here longer or return home."

He set himself the challenge of learning German. "Three quarters of my life have been spent in learning languages," he noted, "and I know but two, tolerably well," French and Latin. It was a hard slog, partly because he had no one with whom to practice conversation. Everyone in the diplomatic world spoke French exclusively. He made slow progress, starting with newspapers, then novels, and eventually works of history and philosophy, for which he felt his German inadequate. By June 1798, having spent most of his leisure time at it, he could read with facility. But "I find I can yet understand very little German, when spoken." He appreciated the irony of the Prussian nationalist, Frederick the Great, discouraging the use of the national language. By August, after a hiatus, he hired a tutor. "I find some encouragement in my progress hitherto, and hope to persevere until I can read the language with perfect ease." That never happened, especially when ideas were involved, though specifics of plot, character, and setting made novels and dramas more accessible. By late summer, he had progressed to reading Johann Christoph Friedrich von Schiller and August von Kotzebue, many of whose plays he saw performed, and Gotthold Lessing.

By the end of the year, he was finding German "tolerably easy." He began to read German poetry, especially Christoph Martin Wieland's *Oberon*, "a poetical romance" published in 1780. Having cut his teeth on

Latin, he naturally practiced translation as a learning device. By early November 1799, he had finished reading *Oberon*, "which is one of the best things I have read in German." Averaging a chapter a month, he completed in May 1800 a credible first draft of a verse translation. The poem had multiple attractions for Adams. Set in the medieval world of Charlemagne, it provided escape from his present tedium and anxieties. As a romantic and exotic fantasy and love story in which all the elements take their value from Christian virtues, it allowed him the pleasure of writing rhymed verse, for which he had a talent. It also provided a text for translation that was difficult enough to be challenging but not so difficult as to be discouraging. Although it seemed far below Shakespeare or *Don Quixote*, which he had recently reread, he agreed with what his friend Bielfeld had told him in Holland: it was the best modern German long poem. When his German tutor sent part of the first draft to Wieland himself, who commented favorably, and numbers of readers raised the idea of publishing it, Adams was pleased and flattered. He did not resort to German, in which it would have been a hard slog, but to a translation to read Immanuel Kant's *Project for a Perpetual Peace*. "I wish not to judge of him uncharitably, but from what I have seen of his works and what I have heard from his followers, there appears to me a very short and easy solution to all his mysterious darkness. I suppose his moral system has atheism and his political system revolution for their only bases." Kant seemed to John Quincy a revolutionary fifth columnist in the precincts of royalist Prussia, encouraging in a coded language the destruction of those who employed him. He was in his own way in the service of France. Kant's obscurity seemed to Adams a strategy to evade responsibility for subversive ideas.

His mind was also on matters of state, particularly the threat of war with France. By 1797, the French sale of every American ship it could capture created a crisis for the Adams administration. From Holland, and now Prussia, John Quincy saw Franco-American relations go from poor to bad. In May 1798, the French foreign minister, once again Talleyrand, let the three-man American delegation sent to Paris by President Adams know that he required a large personal payoff and his government a substantial loan in order even to open discussions. When the XYZ Affair, after the code names of its perpetrators, became known in the United States, anti-French sentiment exploded into bellicose anger. "At length Talleyrand the corrupt has come forth foaming and snarling with the symptoms of mania to be expected," John Quincy wrote. "He denies none of the facts stated by the Commissioners, but vents all his rage upon them personally and upon the president." The American public demanded an apology or war. Federalists to the right of John Adams demanded immediate war. Pro-French Republicans counseled moderation. With a minimal army and navy, the United States was totally unprepared. Would a parsimonious and divided Congress vote funds to expand the military? "In Congress," John Quincy wrote to William Vans Murray, "one half of the House of Representatives have to the last moment contested every measure, even of the defensive kind. . . . Even now the most indefatigable pains are taken to throw the blame of a rupture upon our government, or rather upon the President personally." Worse, a war with France would throw the country into the arms of the British. But if President Adams attempted to negotiate a settlement, let alone succeeded, the strong anti-French right wing of his own party would disown him.

By the summer of 1798, the pot was boiling. "A long and terrible War is opening upon us," John Quincy wrote to his mother, "and the more so, as they have the most ineffable contempt for us as Enemies." French hypocrisy appalled him. France maintained that "free ships make free goods" when that policy was in its interest. When not, it acted to the contrary. John Quincy favored arming American merchant ships, and, like his father, believed that Congress should allocate funds to build a strong navy. The only way to sustain neutrality and national integrity was to have sufficient military strength to give both France and Britain a reason for not trampling on American sovereignty. But Jefferson and the Republicans feared more than anything else a standing army. That meant taxes, and a standing army might be used by the executive to dominate Congress and the people. France "can hurt us little by sea," John Quincy wrote to Murray, "and I am sick, heartily sick, of the servile acquiescence with which we have so long received from her buffetings and indignities, and returned her thanks." If America could not raise some martial spirit and defend itself, then it was doomed to be the whipping boy of France and Great Britain, a shuttlecock in the badminton game between the great powers. France had been in an undeclared war against the United States for a year and a half. Much depended on the next Congress. "But if the negro keepers will have French democracy—I say let them have it." It seemed to Adams paradoxical and self-destructive that the Southern slaveholders, who dominated the Republican Party, did not see that the French democracy they extolled, which no longer existed in any event, was incompatible with their own slave-holding reality. It provided the rationale and impetus for slave rebellions.

Two personal anxieties continued to cast dark shadows. John Quincy had transferred to his brother Charles his $4,000 of savings to invest in securities or real estate, with the condition that preservation of capital had the highest priority. Charles was to have the usual rate of commission. By February 1798, John Quincy had cause to be concerned. He had written a number of times for an accounting. "In all my letters I have urged you to write me, constantly and frequently, and particularly to send me a state of the accounts between us at the close of every year. I must in the most pointed manner again entreat you to show this attention to my business." He was shocked to learn that Thomas Welsh, to whom he had entrusted the management of his small investments in Boston, had become bankrupt. John Quincy grieved for the misfortune of his friend. But the amount owed to him was small, and Abigail transferred her son's business to her own trusted agent, Cotton Tufts. It was more worrisome that he had not heard from Charles. By the summer of 1798, he had good reason to believe his worst fears. His mother had learned that Charles had been pressured by Nabby's husband to lend him money. "Happy Washington!" John Adams lamented to his wife. "Happy to be Childless! My Children give me more Pain than all my Enemies." With no security except a note signed by Smith's brother, Charles had lent John Quincy's $4,000 to Smith to save him from imprisonment for debt. "It is hard upon you to have your prudence and economy thus trifled away," Abigail wrote to Berlin.

To his parents it seemed a failure of character, an inability to say no, an absence of backbone. Charles may have thought he would be helping his sister. To John Quincy it was a breach of trust, a greater blow than the loss of the money. "My confidence in him costs me dear," he wrote to his mother. "But what is much worse, I am forced to conclude that it was misplaced." Abigail responded, "He is not at peace with himself." He had behaved dishonorably. "Your Brother Charles is, what shall I say that will not pain us both? . . . He has formed some good resolutions. Could he keep them how would it rejoice us all, but the heart, the principles must cooperate. How sharper than a serpent tooth it is to have a graceless child, may you my dear Son never experience. . . . Blessed be God, I have those in whom I can rejoice." What depths had Charles fallen to by the summer of 1798 John Quincy could only guess. When he later got the full story from Thomas, he was "much more deeply afflicted with the account of my agent at New York than with the loss of all my property."

He had also lost another brother. Thomas Boylston left Berlin for the United States in September 1798, his ultimate destination Philadelphia, where he intended to resume his career as a lawyer, having come to the conclusion that he lacked his elder brother's talent for diplomacy. He did not have his "facility in writing nor his readiness, nor yet his diligence," he admitted to his mother. "There is no equality between us." Concerned for his health and aware that remaining in Europe only postponed the pursuit of a legal career, his parents were eager for his return. They hoped he would settle in or near Quincy. Having made diligent efforts to locate a replacement for Thomas as John Quincy's secretary, Abigail settled on Thomas Welsh's son. His father's bankruptcy had left him financially adrift. On the morning of Thomas' departure, John Quincy reminisced that "it is something more than four years since we sailed from Boston together. . . . He has ever been a faithful friend, and kind companion, as well as an industrious and valuable assistant to me . . . and now it is with an heavy heart that I part with him." Louisa wrote to her much-loved brother-in-law that she had never seen John Quincy so affected by anything as his departure. "He seldom mentions you without tears in his eyes." And it was a loss to Louisa also. Thomas had, from early on, been her friend and supporter. In London, she had opened her heart to him "and assured him of [her and her father's] perfect innocence." He was "a solace in my moments of mental anguish, and . . . contributed to my comfort, and my pleasure, both in sickness and in health. . . . I never saw so fine a temper, or so truly and invariably lovely a disposition. . . . I have always believed that he both respected and loved me, and did me justice in *times* when I needed a powerful friend."

Louisa's health, though, cast the darkest shadow on John Quincy's life. They both wanted a child. She might not have been able to say for which of them she felt the need more strongly. He feared losing her as well as being childless. There was, though, no reluctance to keep trying. Since Louisa was often sick with many of the symptoms she had when pregnant, it was and is sometimes difficult to determine whether an illness or a fainting spell was pregnancy-related. She may have been pregnant again as early as January 1798 when, early in the month, John Quincy noted, "my wife yet very ill." Soon she was "again better, to flatter us again, and make the curse more bitter." She recovered slowly. But from what? And did "the curse" refer to her generally poor health or to being childless? In the late winter or early spring, she was definitely pregnant. By March, she was mostly confined to bed. "My prophetic heart! I have no doubt of the cause," he wrote in his diary. "The cup of bitterness must be filled to the brim and drank to the dregs." In a calm moment he had written to William Cranch that "the blessings of wedded love have been mine, almost a year, but I have not the happiness of being a parent. The partner of my life, like yours, adds every day to the ties of affection which united us, and but for a misfortune of which you will have heard before this time, I might have now . . . participated in all the sources of domestic felicity."

Two days after his thirty-first birthday in July 1798, Adams still had "some faint and feeble hopes of a better event than I had anticipated hitherto." They were soon dashed. "The tortures of Tantalus have been inflicted upon me without ceasing," he wailed. Louisa was in extreme pain. Would she live? Would the baby live? Writing in his diary became "excessively painful," with "nothing but distress to record." After a dreadful night, Dr. Brown came in the early hours of the morning. John Quincy feared that "the case appears . . . similar to that of last November. Patience and Resignation is all that we can have." He stayed up nights with Louisa, whose main suffering came from her awareness that the doctors did not expect her to keep the child. "The anticipation of evils that we cannot prevent is itself a great misfortune," John Quincy wrote. "I have these 8 months been convinced, with scarcely the shadow of a doubt what this Event would be. Yet now that it happens I feel it with no less poignancy than if it had been unexpected. The prospects of futurity that it still holds out to me are horrible. I realize them as if present. I have no more doubt of them than of death. They poison every moment of existence, and when they come will be not less bitter for having been foretasted." The end came toward the close of the month. On July 26 he noted that it was the first anniversary of their marriage. Although there had been two miscarriages in that one year, "from the loveliness of temper and excellence of character of my wife, I account it the happiest day of my life."

To his surprise, Louisa quickly rebounded, gaining weight through the summer and fall. She was able to take excursions and attend court functions. She had friends, admirers, and sympathy, especially from the women, including the queen. A newly married English couple, Elizabeth and George Errington, became her close friends, and Louisa and Elizabeth were inseparable. John Quincy continued to urge the Prussian ministers to begin negotiations for the renewal of the commercial treaty, though he remained convinced that, notwithstanding what they said, "they mean to *delay*." He felt a gnawing sense of failure for which he was not at fault and about which he could do nothing. At best, he was contributing information and analysis of European events to which he gave little value. But, his mother boasted, "you judge and think so accurately respecting the affairs of your country . . . that every syllable you write ought to be made public." In late winter, Louisa became pregnant again. In April 1799, she again miscarried. When Elizabeth Errington gave birth to a daughter, Louisa lamented that "she was blessed with a fine child while I only lived to witness the pangs of disappointment." It "so bitterly distressed my poor husband, and destroyed all the comfort of my life." Worried that she might die in childbirth, she had gotten Elizabeth to pledge to be a mother to her child if it survived her.

By July they had both sufficiently recovered to travel to a fashionable health resort in Saxony. Brown and his medical colleagues, though they remained discreet, believed that Louisa's underlying illness, separate from her pregnancies, was tuberculosis. They did not think she would live much longer. But comfort, rest, and restorative waters might help. At Toplitz, where they stayed for seven weeks, they enjoyed the sociability, international community, tea parties, music, and baths, and the countryside with its magnificent views. John Quincy had reason to be in much better spirits. A week before leaving Berlin, he and the Prussian ministers had at last signed a new treaty of "amity and commerce" between Prussia and the United States. Franco-Prussian relations were now such that it was in Prussia's interest to sign. And, at the suggestion of France and to the horror of right-wing Federalists, President Adams had appointed a new peace commission to Paris. The French were once again ready to negotiate.

From Toplitz, where they had walked to the top of the Schlossberg, "a great proof how much her health has improved since we left Berlin," they went to Dresden. The countryside and the Königstein fortress along the Elbe were riveting. In Dresden, Adams noted masons hard at work cutting stone for buildings. The city still showed the scars that Frederick the Great had inflicted in the Seven Years' War. Along the Elbe, "the views . . . were sometimes wild and sublime and sometimes elegant and cheerful." Friends from Berlin were in Dresden. The Adamses were welcomed by the upper levels of government and the aristocracy, the friendly British minister providing hospitality and introductions. Almost every day Adams went to the state art gallery, fascinated by its collection of German, Italian Renaissance, and eighteenth-century art. At the Catholic church, he noticed Raphael Mengs' striking altarpiece. The national library's immense collection of books and manuscripts impressed him. In the evenings, they went to the opera or theater, where they saw August Wilhelm von Schlegel's German version of *Hamlet*; or they read Shakespeare and Chaucer aloud. By the middle of October, after a detour to Potsdam, they were back in Berlin. Louisa's health after the three-month holiday seemed "as much improved as could reasonably be expected." She was as well as he had ever known her to be.

A new pregnancy had begun about the time of their return from Dresden to Berlin. Until the middle of November 1799, Louisa seemed well, though she deeply regretted the departure of her friend Pauline Néale for England. Suddenly Louisa became very unwell, then seemed better. In early December, at a large party attended by the king and queen, she had a fainting fit. A lady whom she had been approaching fell, broke her leg, and fainted. After helping the doctor cut away the lady's stocking, Louisa began to tremble uncontrollably. Then she herself fainted. "I brought her home," John Quincy wrote in his diary, "and from eight o'clock in the Evening until midnight she had a continual

succession of fainting fits, and cramps almost amount-
ing to convulsions." She was better for a few days,
then became violently ill. At the end of December,
Dr. Brown was in constant attendance. Louisa kept
to her bed, "suffering no pain; and still entertaining
some hopes." At the beginning of the new year, when
the doctor bled her, she "bore the operation well, but
fainted in the Evening." The next week she miscarried. "I can only pray to God," John Quincy wrote in
his diary, "that there may never again be the possibility of another like event. A better hope it were folly to
indulge." Hope was "but an aggravation of misery." He
became unable to sleep. "These violent, long continued
and frequently repeated shocks from the state of my
wife's health must at last have an effect upon mine."
Dr. Brown prescribed opium for his "nervous agitation." As Louisa recalled, John Quincy was "sick from
anxiety, and want of rest."

Their spirits warmed with the approach of spring.
Louisa had recovered enough to seem actually well.
They were soon cheered by letters from America, the
first they had received since the previous December
when John Quincy had been deeply affected by the
news of George Washington's death. Louisa, constantly
anxious about her family, was pleased when late in
the spring she learned that Congress had appointed

Joshua Johnson superintendent of stamps at a salary of $2,000 a year, the only good news from her family since they had parted. Her father's health had deteriorated. He was having no success obtaining from his former partners what he believed he was owed. John Adams had exerted his influence on the appointment, and Johnson's Southern connections had helped. Vice President Jefferson, whose tie-breaking vote carried the appointment, announced that "although the President's nomination was alone sufficient . . . he had a further inducement from being personally acquainted with the candidate and knowing him to be well qualified to discharge the duties of it." Even before the good news, Louisa had perked up. After the first two miscarriages, she had felt so fat and ugly that she would not sit for a portrait. "You know I am very proud," she wrote to her sister Nancy. "Therefore it would not suit me to have a picture worse than common." Her good looks and better health were now obvious.

In the spring weather, John Quincy finished his translation of *Oberon*. To keep himself busy, he translated one of Juvenal's satires. His only remaining business in Berlin was to exchange the final copies of the ratified treaty. After that, he could not see any good reason for remaining. It would be a waste, he told his father, of the American taxpayers' money.

On a hot summer day, high in the Sudeten Mountains between Poland and the Czech Republic, John Quincy and Louisa gathered wild raspberries in the ruins of a castle built in 1292. It was July 1800. They had left Berlin, hot and dusty in the summer, for a three-month holiday excursion. The hope was to keep Louisa healthy in the country air. They traveled east and south to what was then called Silesia, an ethnically diverse region that had been an eighteenth-century battleground between Austria and Prussia. Isolated, poor, and provincial, it had bad to nonexistent roads and few visitors. Frederick the Great had made it a Prussian province. From high in the mountains, they could see on a clear day as far as forty miles. On one side was the source of the Elbe, on the other the Oder; in the distance, they could see Breslau. As they walked in the ruins, it occurred to them both that the castle had been built two hundred years before the discovery of America. All was now desolation. "Five centuries had . . . converted the abodes of social life here into a wild and desolate ruin, while at the same time they had changed in our country a howling desert into flourishing cities." Their minds were on what he expected would be their imminent return and reunion with their families. As they traveled, they were lovers again. They might have hoped the wild raspberries were a good omen.

One day John Quincy rose at 2 A.M. to see the sunrise from the summit of the Risenkoppe, a steep ascent up stone steps designed to make the chapel at the top accessible. Louisa, with a headache, stayed behind. On the whole, he wrote to Thomas, "it would astonish you, as it does me, to see how she supports the fatigues of this journey." Climbing, he saw the dawn about to break. Doubling his pace, intent on not missing the first burst of red light, he was in time by ten minutes. Having heard so much about the splendor of dawn from the top, his expectation exceeded reality. There were low clouds and mist. But still, the moment and the view were splendid. "The spectator has but to turn on his heel, and all Silesia, all Saxony, all Bohemia, pass in an instant before his view. It is therefore truly sublime."

But, like all vast landscapes, it seemed to him chaotic, demonstrating the difference between reality and the ordered arrangement of a landscape painting. Nature and art were different things. For John Quincy, the view from the summit inevitably brought to mind "the supreme creator of the universe," the ultimate artist "who gave existence to that immensity of objects. . . . The transition from this idea, to that of my relation, as an immortal soul, with the author of nature, was naturally immediate. From this to the recollection of my country, my parents, and friends, there was but a single and a sudden step." It was the chain of connections that held his life together. When he returned to the hut in which they had spent the night, he wrote some lines in the guest book:

> From lands, beyond the vast Atlantic tide,
> Celestial freedom's most beloved abode,
> Panting, I climbed the mountain's craggy side,
> And viewed the wondrous works of Nature's God.
> Where yonder summit, peering to the skies,
> Beholds the earth beneath it with disdain,
> O'er all the regions round I cast my eyes,
> And anxious, sought my native home—in vain.
> As, to that native home, which still enfolds
> These youthful friendships, to my soul so dear,
> Still, you, my parents, in its bosom holds,
> My fancy flew, I felt the starting tear.
> Then, in the rustling of the morning wind
> Methought I heard a Spirit whisper fair,
> "Pilgrim, forbear, Still upwards raise thy mind,
> Look to the skies—thy native home is there."

As he traveled, Adams wrote a detailed account of his Silesian journey in a series of forty-three letters to his brother Thomas, without it being settled that they would have readers other than family and friends. "Therefore make up your account to receive patiently all my tediousness, or as I said before, bestow it all upon my mother." From the outset, though, he had in mind a larger audience, starting with his father. On his own initiative, Thomas gave the letters to Joseph Dennie, the pro-Federalist publisher of the *Port Folio*, a new literary magazine in Philadelphia dedicated to the advancement of Federalist values. He assumed correctly that his brother would not object. The first thirty-one letters, a narrative of what John Quincy saw, he wrote while traveling. The remaining twelve he wrote between December 1800 and March 1801 in Berlin; they composed an account, mostly from books, of the history, culture, institutions, and literature of Silesia. As he told Thomas and then a larger audience, he hoped that it would stimulate commerce with the United States. He granted that there were serious obstacles, especially distance. But since, for example, Americans bought English glass at a high price, why not buy Bohemian glass of an equal quality for much less? "Even making every allowance for the necessary difference in the price of transportation . . . an advantageous trade in this article," as well as others, "might be carried on between our country and Bohemia, and I hope it will one day." One purpose of his tour "was to obtain information" that would help "diminish the commercial dependence of our country upon G. Britain." That "ought . . . to be one of the favorite objects of every American patriot." Trade and commerce were essential to America's future greatness— and to its independence.

He also had in Silesia an ugly example of a long, pernicious religious conflict. Its absence in America was essential, he believed, to the country's prosperity. The antagonism in Silesia between the large Protestant and Catholic populations was vitriolic. "There is perhaps no part of Europe where the root of bitterness between the two parties is yet so deep and cleaves with such stubbornness to the ground as here." Everywhere he traveled, its presence was inescapable, the concept of religious tolerance or of a civil religion in a common nation totally absent. Prussia was Lutheran, Austria Catholic. Throughout Europe, rulers determined the official religion of each country. In most cases, it was winner take all. If religious differences were ever to create a politics of religious identification in America, Adams feared that the United States would go down the same road. It would become destructively divided or even theocratic. America, he believed, was a Christian nation. That was simply a statistical and historical fact.

But one Christian sect was not inherently superior to any other. The lessons of the Christian dispensation were available to all Christians. What was important was ethics, not theology. And the way forward for the new republic was tolerance of diverse religious views and a commitment to the highest ethical values, though he had difficulty believing that ethical practice could be separated from belief in a providential God.

Two years earlier, in Berlin, he had puzzled over Gotthold Lessing's controversial play *Nathan der Weise.* "I see not what Moses Mendelssohn," a German-Jewish textile manufacturer and philosopher, "thought so wonderfully fine in it, unless it be that the greatest character is a Jew." Washington had affirmed, in his visit to the Sephardic Jews of Newport, that America accommodated religious diversity. Still, Jews were few in number and rarely in sight. In Europe, especially in Holland, Germany, and Silesia, they were unmistakably visible. Adams had been noticing them in his travels. Some Jews, he observed, were as uncharitable as many Christians. In Amsterdam, he had seen a Jew "apparently at the point of death. Three or four persons were round him, Christians and Jews." They "seemed to throw upon each other the burden of giving him any assistance whatever. They said he had the falling sickness; but upon a piece of bread being held to his mouth, the convulsive manner in which he snapped at it . . . discovered that his only falling sickness was hunger." Many of the Dutch Jews he knew were respectable businessmen. One Sabbath he went to their synagogue. "Heard their Devotions. There were no women present." It came easily to him to use "Jew" as a descriptive adjective, though it was more a mark of difference than a stigma. In late August 1800, visiting a ghetto in Silesia, he described to his American readers "this ridiculous and barbarous regulation" requiring Jews to be segregated. Its history is "represented in a picture, which yet disgraces the catholic church in the town. . . . It relates that about the year 1450, certain Jews obtained possession of a consecrated host, which they treated with contempt and indignity." Two of the Jews were "stabbing the wafer with daggers, and the wafer . . . streaming with blood. For this offence ten Jews and seven of their wives were burnt at the stake, and the town was formally privileged never again to be contaminated with the presence of a Jew." The anti-Semitic blood libel seemed to him preposterous and vicious, and the heavy tax that Frederick the Great had imposed on all Jews illiberal and counterproductive.

In Frankfurt an der Oder, he was appalled by the dirtiness of the Jewish quarter, the squalor of its impoverished life on the fringes of the European economy.

"The word *filth* conveys an idea of spotless purity in comparison." As he traveled, the speech, gestures, dress, and public manner of many of the working-class Jews he encountered repelled him. But in Berlin, he and Louisa moved freely and without the slightest hesitation, let alone prejudice, among the wealthy Jewish merchant, manufacturing, and banking families that had been granted the privilege of residence in the city. Ephraim Cohen, from a wealthy Dutch-Jewish banking family, their landlord when they moved from the Brandenburg Gate to their handsome house in the best section of Berlin, was a neighbor and friend. And while such Jews could not participate in court life, their stone mansions were social and cultural centers, frequented even by the aristocracy. "Mr. and Mrs. Cohen," Louisa recalled, were "very rich Jews; were also very polite to us; and lived in a style of great elegance, entertaining the first and highest persons of rank in the City." Their mansion had a private theater, and they hosted large parties that the Adamses regularly attended. Although the Cohens converted to Christianity, their Jewishness was unmistakable.

There is no indication that Adams connected a public controversy about Jewish emancipation to his private friendships. He was aware that Jews like the Cohens who wished to have the rights of citizenship created a difficult problem for the Prussian state. No non-Christian could become a citizen. In June 1799, he was given a copy of a "new pamphlet, a response to the Jews who have lately proposed to renounce their religion and adopt a sort of Socinian Christianity," which denied the divinity of Jesus, with the intent of qualifying for citizenship. The proposal "has produced some sensation here." That fall, he had much to say about the subject. He believed that there should be no religious test. But the movement among elite German Jews to become citizens was inseparable from the Enlightenment values that had produced the French Revolution and could be a challenge even to the tolerant and rational Christianity that Adams valued. Emancipation was difficult to separate from the rationalism that resulted in atheism. In Berlin, and in reading *Nathan der Weise*, he had become aware of the Jewish Enlightenment, of the efforts of Mendelssohn and others to modernize orthodox Judaism. For some, like Ephraim Cohen, converting to Protestant Christianity was the practical road to citizenship. But were their conversions genuine? the clerical and governmental authorities asked. And did not such conversions compromise all religion? Were they not contaminated by indifference at best, atheism at worst? The movement seemed to Adams inseparable from "atheism and revolution." Liberal Protestantism was being created by Friedrich Schleiermacher and Johann Fichte. Fichte, who "wanders about Germany, scribbling, and holding himself forth as the victim of persecution," Adams noted, formulated a version of Christianity that liberated it from miracle, from the divinity of Jesus, and even from God. The latter eliminations were two steps too far for Adams.

Their Silesian wanderings came to an end in September 1800. In Breslau, they visited churches, the museum, and the university. It seemed to have more business activity than Berlin, its main trade in cloth and linen. Adams made a special effort to satisfy his curiosity about the Moravians to whose Pennsylvania settlements his brother Thomas had been. By early September, they were on their way to Dresden, with the intention to stay a week and then go to Leipzig for a month. The large autumn fair seemed an attraction worth seeing. Now certain that Louisa was pregnant again, Adams thought she might benefit from resting in one place before returning to the social demands of Berlin. In Dresden, he resumed his slightly uneasy relationship with the British minister. Some of their friends from Berlin court society were there. And they were pleased to see Ephraim Cohen and his wife. John Quincy spent most of their six days in Dresden at the

art gallery and library, opening rare volumes and seeing more of one of the great collections of prints and
Renaissance paintings. He spent even the morning of
the day on which they left "at the gallery of pictures,
of which I have now taken leave, probably forever."
His mind was on the future, on his return to America
with his wife and, he hoped, a child.

But he had reason to be pessimistic. Why should
this pregnancy be different from the previous four?
Louisa had written to her father from Breslau that she
had been "very unwell and glad to rest a little." By late
September, they were in Leipzig, where Whitcomb had
arranged lodgings for them. John Quincy was ill with
a severe cough, Louisa unmistakably pregnant and
sick in the usual way. "We have a dismal month before
us," he wrote in his diary. "But her case no physician
can remedy. . . . She is already very unwell, and will
continue so until the severe and inevitable trial has had
its usual end. . . . The misfortune lingers this time, to
make itself more keenly felt." She seemed likely to lose
the child; the danger was "more threatening than it has
been yet." When they left for Berlin toward the end of
the month, she was ill and tired.

With trepidation and surprise, she held the child
through the fall and winter months. "I was sick almost
unto the death," she later wrote, "and sadly wearisome

to everyone; but they bore with me . . . with the patience of Angels." It helped raise her spirits that the beautiful Queen Louise was also pregnant. Dr. Brown and his family resumed frequent evening visits, the ladies chatting, the doctor and John Quincy playing chess. Pauline Néale returned from England. And Louisa had a new intimate friend, Lady Carysfort, the wife of the British ambassador, whom Louisa "clung to . . . as if she had been my own mother." John Quincy kept himself busy reading and writing the usual letters and dispatches and finishing his letters from Silesia, his mind always on the hope that Louisa would not miscarry. Every few days there seemed a likely crisis; each time, to his relief and surprise, the crisis passed. On the last day of 1800, he went to a ball hosted by the Swedish minister. In late January, the Adamses hosted a party and dance at their apartment. Louisa was fatigued but nothing worse.

By late winter, she was still holding the child, at a much later stage of pregnancy than she had ever been before. Most evenings he read aloud to her. During February, she was neither well nor unwell, mostly up and around. Still, she later recalled, "my health was so weak that few expected that I should survive the trial which I had to sustain." When she was confined, King Frederick William ordered that the street should be closed so that there would be no noise to disturb her. Frightened that she would die in childbirth, Louisa elicited from Lady Carysfort the promise that she would look after the infant until John Quincy would be able to take it to America. Every day the queen sent someone from the palace to inquire. On April 12, 1801, John Quincy wrote in his diary, "I have this day to offer my humble and devout thanks to almighty God for the birth of a son." Years later, Louisa wrote to her youngest and only surviving child, "I was a *Mother*. God had heard my prayer!"

Across the Atlantic another struggle in the Adams family had ended in death. John Quincy's brother Charles had died in New York at the end of November 1800. He had been drinking constantly and compulsively, gripped by the alcoholism that the Adamses feared as the most dangerous threat to family and society. He had raved about paying back his brother's money. As he became ill, he probably took opiates; he was incapable of carrying on his legal practice, his family destitute. To his parents and siblings the loss was painful. But in the moral calculus of their lives, it was the inevitable, even righteous penalty for dissolution.

Like his parents, who would look after both grandchildren and support the widow, John Quincy was deeply mournful. Put out of your mind the loss of my money, he urged his father, who he felt had worries enough. "I deplore the unhappy habits of life, which led to the damage I sustained, but at the same time I must acknowledge it might have been much worse." Might he have lost even more, or would Charles have done other things that would have caused them additional pain? It had been hard for the family avatars of moral strength to be of help to Charles. His parents had neither the mind-set nor the tools. For them, and for Charles, the pain of disgrace was powerful and paralyzing. Abigail had "painted . . . the misery he was bringing upon himself, his amiable wife and lovely innocent children. . . . His constitution is nearly destroyed and still he persists in practices which must soon terminate in death."

Damage control could not save him. His wife and children, Abigail lamented, "are the innocent victims of a miserable man, whom I can no longer consider as my Son. Yet am I wounded to the Soul by the consideration of what is to become of him. What will be his fate embitters every moment of my life." John Quincy shared in the pity, grief, regret, and judgment. "Such was the infatuation which had taken possession of him, that he was lost [to life], and rendered every one miserable who possessed a regard and affection for him. . . . He suffered, much, endured much. His mind was constantly running upon . . . making reparation; early principles though stifled now discovered themselves; and mercy I hope was extended to him; but it rends my heart to think upon the Subject. In silence I must submit."

When his infant son was being christened in May 1801, John Quincy may not have had his brother's death on his mind. Still, he recognized that bringing a child into the world was like buying a lottery ticket. What would be the fate of his son with the propitious name George Washington Adams? "I implore the favor of almighty God, that he may live, and never prove unworthy of it." Louisa was too ill to attend the ceremony. During the delivery, she had been mangled so badly that she could not get out bed for almost six weeks. Her health kept him "in a crucifying state of suspense," worried that she would die. Abigail was incensed about the name. Why had he not named the boy after his grandfather? After all, as helpful as the first president had been to John Quincy, the deceased Washington could not know about his namesake. But John Adams knew that his name and family tradition had been rejected. "I am sure your brother," Abigail wrote to Thomas, "had not any intention of wounding the feelings of his father, but I see he has done it.

Had he called him Joshua, he would not have taken it amiss."

It also would not have called attention to a difference between Washington and Adams of serious consequence to the Adams family. Washington had been a two-term president. Probably he would have been elected to a third term if he had chosen to run. In February 1801, the day after he learned of his brother's death, John Quincy saw in newspapers forwarded from Hamburg and London that Thomas Jefferson had been elected president of the United States. His father had become the first one-term president.

He was prepared for the news of his father's defeat. From Berlin, months behind the latest news from America but up to date when the action was hot in Paris, John Quincy appreciated the bold step that his father had taken in 1799. America and France were in a quasi-war. After the XYZ affair, American anti-French anger was intense. Adams sent a three-man delegation to Paris to see if war could be averted. Aware that his cabinet was more loyal to Hamilton than to him and that it was opposed to sending delegates, Adams had bypassed it. The Federalist power structure was furious. It preferred to punish Adams more than it desired to keep Jefferson out of the presidency. In Quincy and Philadelphia, the Adamses waited anxiously for news from France, but it came too late to influence the election. The main charge against Adams was that he was a monarchist, eager to have a surrogate of George III on the throne of the United States. "The conduct of men," John Adams noted, "is much more governed by their *passions* than by their *interests*; the whole history of mankind is one continued demonstration of this axiom." The division between the pro-French and pro-British ideologues determined a scorched-earth policy at both extremes. The middle, where John Adams attempted to position himself, became untenable. With a new Directory and Talleyrand once more foreign minister, the French government acceded to almost all the American demands. But in midsummer and early fall 1800, when elections for state legislators were held, no one yet knew that. In some key states, legislators who elected the presidential electors narrowly favored Jefferson. "You speak of it as a problematical point," John Quincy wrote to Thomas in early December 1800, "whether the federalists will split. . . . By all the accounts from America it appears unquestionable that they will. I consider already the result as perfectly ascertained." The Hamilton wing of the party and the Essex Junto, the right-wing branch of the Federalists led by Timothy Pickering, had not only deserted Adams; they had opposed him. He had won the peace but lost the election.

Still, the election was so close that if Jefferson had not benefited from the additional electors provided by the three-fifths provision of the Constitution regarding the slave population, Adams would have been re-elected. There would have been no Jefferson presidency, until at least 1805 or ever. And for many Federalists like John Quincy, the kind of democracy that Jefferson and his colleagues preached, that made them partisans of France and the French Revolution no matter what, had at its dark heart a glaring inconsistency—more than that, a hypocrisy and a dangerous one. "Those absurd principles of unlimited democracy," John Quincy wrote to Thomas, "which the people of our Southern states, by the most extraordinary of all infatuations have so much countenanced and encouraged, are producing their natural fruits, and if the planters have not discovered the inconsistency of holding in one hand the rights of man, and in the other a scourge for the backs of slaves, their negroes have proved themselves better logicians than their masters."

Late in the same month in which his first son was born, John Quincy received the letter he had been expecting. President Adams had recalled his son. "Possibly," he wrote to his father, "some of my late letters to my mother may lead you to the apprehension that this recall has proved personally unwelcome."

But, he assured his father, he was eager to come home. On May 1, 1801, he requested permission to leave Prussia. Both he and Louisa had been appreciated by the court, the treaty of commerce and amity had been satisfactory to both governments, and the Adamses had been valued by their friends in Berlin's diplomatic and social world. Louisa's long recovery slowed their departure. It seemed at best that she might be up to travel by midsummer, though the sooner they departed the less likely they were to have a rough voyage. Beyond that, John Quincy needed to reassure his parents and himself that he would overcome being once again without a guaranteed income. He had no desire to return to the practice of law. But that was the profession available to him. And if it meant having to play catch-up in the competition for clients, he was ready to do whatever was necessary to provide an income for himself and his family. There were worse things than the Boston bar.

Meanwhile, he kept himself busy reading, especially sermons on the education of children by the English clergyman John Tillotson. In John Quincy's view, his role as a father was inseparable from his role as a teacher. Like his own father, he believed that education begins at home. Keeping his fingers crossed, he made their travel arrangements. As the weather turned warm, he refreshed himself with swimming in the

Spree. When Dr. Brown used a new method to inoculate the baby against smallpox, John Quincy was concerned that it would not be effective. By early June, he determined that Louisa, though still struggling, was well enough for them to leave at the middle of the month. She had her last dinners with the Carysforts, the Brühls, and the Browns. John Quincy packed, arranged to have their furniture sold, read Schiller, and managed to write a letter for the *Port Folio* on a recent book by a German author on the American and French revolutions. Louisa had to be carried into the carriage. Four days later they were in Hamburg, where they found that the ship on which they had engaged passage was too small for comfort. He booked passage on a larger ship, for Philadelphia rather than New York. With a healthy baby in her arms, Louisa boarded the propitiously named *America* for what was likely to be a two-month voyage. At its end was a happiness she yearned for and had many times thought she might never realize: her reunion with her family. As the ship made its slow way down the Elbe, John Quincy celebrated the start of his thirty-fifth year. It was July 11, 1801.

Soon they were into heavy weather on the North Sea and the Atlantic. Winds forced the *America* into a thousand miles of tacks, hardly advancing their westward progress. As he always had aboard ship,

John Quincy felt bored, restless, capable of only sporadic reading. He now chose to tell Louisa about Mary Frazier. Louisa later recalled that his description of Mary excited a desire to meet her. She "was not jealous," though she "could not bear the idea of the comparison that must take place, between a single woman possessing all her loveliness, and a poor broken consumptive creature, almost at the last gasp from fatigue, suffering and anxiety. . . . I had every confidence in my husbands affection, yet it was an affair of vanity on my part; and my only consolation was, that at any rate I had a *Son*." Why John Quincy thought this the right time to tell her about Mary, or why he believed there was any right time, is unclear. Probably it was an expression of his definition of honorable conduct. He had been reading Cicero's *De Officiis*, his cherished handbook of the rules of moral conduct. Perhaps his mind drifted in the torpor of the voyage into the confessional mode. Since he was now bringing his wife for the first time to the places of his past life, it may have felt right to make clear to her what had preceded their life together.

When the wind eventually became favorable, they sailed westward with a sense of expectation. He looked into the diminishing distance toward reunion with his brother, sister, and parents, and to introducing Louisa to her new home, the land of her father and husband, a place in which, with her English accent and French fluency, she would be both a stranger and a native. For John Quincy, it was a resumption of an old and the beginning of a new life. At the close of 1801, expressing gratitude for the blessings bestowed on him, he affirmed his strong sense of renewed futurity. From the depths of his memory, he drew on a phrase from Paine's *Common Sense*. He might have struck out the phrase if he had realized the source. "To begin the world anew," he wrote, "with the common chances of good or ill success. I have only to implore the favor of Heaven for a continuance of health and for the will and power to practice all the virtues which are calculated to promote the happiness of my fellow creatures, and my own." Abigail, writing to Thomas about John Quincy, had used a variant of the same phrase. "I feel very anxious for him in all respects; I pray God send him a safe and fortunate passage to his native land, with his poor, weak and feeble wife and boy. I know and feel how many cares . . . he will have to encounter . . . to have to begin anew the world, and that in a profession which he never loved, in a place which promises him no great harvest."

Abigail was to be wrong about the harvest to come. And in his own use of "to begin the world anew,"

John Quincy may have unconsciously felt that the words applied both to himself now and, as they had in their original application, to his country. Late in August, he noticed a change in the color of the water. The sea became smoother. Land birds were sighted. On September 3, 1801, "with a faint and, irregular breeze," they sailed slowly up Delaware Bay. The next day, at the wharf in Philadelphia, they stepped onto American soil.

Chapter 7
The White Worm
1801–1804

At sea John Quincy had celebrated his thirty-fourth birthday. The country of which he was a proud citizen was twenty-five years old. He had been away almost eight years, and Louisa had never been in America before. The intensely hot early September weather was a shock to both of them. Britain and Northern Europe had been pleasantly temperate climates, and the New Englander had always hated heat. Louisa had never experienced it. Although eager to see his family, he felt apprehensive about his return to Boston and the bar. His brother Thomas, who met them at the dock, was shocked and concerned when he saw Louisa, who had been through four miscarriages and a difficult childbirth. She still could not walk normally. John Quincy seemed to Thomas to have hardly changed, though he had "a sort of fatherly look." But to those who had not seen him for close to a decade, he looked older, with a receding hairline and a gravitas that combined self-control, reserve, and strength. He was tired, and the first blast of hot weather let him know that he was in a different climate.

The heat was also political and cultural. Jefferson believed that his elevation to power represented a revolution, though the revolution was in the main an evolution, changes that would over time probably have occurred anyway. But feelings were raw and anger high, politics a zero-sum game, and the partisan press vitriolic. To its advantage, the new administration came to office with a gift: John Adams' treaty with France, which split the Federalist Party. It also came into office with a gift from its foreign enemies: a new outbreak of war between Great Britain and France allowed American commerce to flourish. The Democratic-Republicans, whom Federalists began to call simply Democrats, had been given a good hand to play. Federalists like John Quincy had reason to assume that they would not have a seat at the table.

After a few days in Philadelphia, he headed north to New York to see his sister, then to the long-anticipated reunion with his parents, parting from Louisa for the first time since their marriage. With the baby, she hastened to Washington, where the Johnsons had settled, eager to embrace her parents and sisters, especially her father, whose financial misery continued and whose health had declined. He was an emotional and physical shadow of the man she had worshipped. When, the next month, John Quincy got to look over Joshua Johnson's papers, his pessimism about Johnson's recovering any money from his business debacle increased. "He has been unfortunate in his trusts and considered as a prey by every man with whom he has dealt. I am strongly apprehensive." The family was living at the edge. So too were the William Stephens Smiths. John Quincy found his sister in good health. His visit lifted her spirits, though her stoic composure could not belie the insecurity that her husband's financial recklessness created. The Revolutionary War veteran, who believed he deserved a prestigious government job, never tired of pushing himself forward as if he were the aggrieved party being treated badly. In New York, John Quincy dined with the vice president, Aaron Burr. "So you see what good company I am getting into," he wrote to Thomas, "and may expect a proportionable improvement in my republican principles." He did not see his fellow Federalist Alexander Hamilton. In the long run, he was not to find either Burr or Hamilton admirable. He thought the former an unscrupulous political adventurer, the latter a manipulative man whose character betrayed his genius. In late September, he was in Providence, then Boston, then at last at his father's house. Within days he felt the embrace of friends and family. He soon had a grand reunion dinner with twelve of his friends, including Mary Frazier's husband.

In Quincy, the Adams household resonated with the resentment the entire family felt at the result of the election, and Abigail worried that Louisa did not know that discretion was the first rule of a political wife. She will be "amongst a people where it will be impossible for her to be too guarded; every syllable she utters will be scanned . . . with . . . carping malice; such is the spirit of party. You too my son," she warned him, "must look for your share of calumny, and arm yourself." In politics, Abigail had learned, that which might seem paranoia was often reality. The former president and his wife felt betrayed by party spirit, envy, and slander, and badly repaid for decades of personal sacrifice. But, John Adams had decided, it was desirable that a former president not engage in partisan politics. Federalist pamphleteers, though, were loudly hostile to the Jefferson administration, especially its early version of the spoils system. Some advocated forceful

John Quincy Adams was born in 1767 in the seventeenth-century Braintree farm-house on the left, next to the house in which his father had been born in 1735. John Adams later owned both of these buildings, ten miles south of Boston, on the busy Plymouth Road. After the death of his father, John Quincy kept both houses in the Adams family. They now are part of the Adams National Historical Park.

The center of the town of Quincy, drawn in 1822 by Eliza Susan Quincy, John Quincy's cousin. The cemetery in the foreground is where John Quincy in July 1787 attempted to read "the inscriptions which love and friendship have written on the simple monuments" that memorialized previous generations of the Adams family.

These pastel drawings of 1764 by Benjamin Blythe depict John Adams, a relatively unknown young lawyer, and his wife Abigail the year of their marriage and three years before the birth of their first son, John Quincy. The relative simplicity of their clothing for this dress-up occasion reflects their lives and personalities. Though idealized, the drawings express their characteristic physical features: one thin faced, with sharp features; the other round of face and build.

JOHN QUINCY ADAMS

Sidney L. Smith did a delicate engraving (1877) of Isaak Schmidt's 1783 pastel portrait of sixteen-year-old John Quincy, who had just returned to The Hague from St. Petersburg. His father remarked that "he seems to be grown as a man; and the world says, they should take him for my younger brother if they did not *know* him to be my son."

The inside front cover of twelve-year-old John Quincy's first journal volume, begun on November 12, 1779, when he sailed with his father from Boston to Spain and France. Already an obsessive writer, on the cover he wrote that this is "a JOURNAL BY ME: JQA." The inside of the cover contains notes written at a later date.

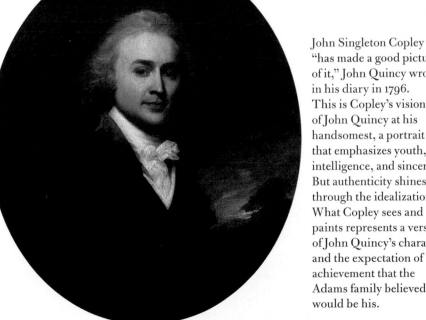

John Singleton Copley "has made a good picture of it," John Quincy wrote in his diary in 1796. This is Copley's vision of John Quincy at his handsomest, a portrait that emphasizes youth, intelligence, and sincerity. But authenticity shines through the idealization. What Copley sees and paints represents a version of John Quincy's character and the expectation of achievement that the Adams family believed would be his.

Thomas Boylston Adams, painted at The Hague in 1795, by a Dutch artist known only as Parker. John Quincy's intimate companion, Thomas accompanied his brother to Holland as his secretary and then to Berlin, where he endeared himself to the newly married Louisa. Without his brother's talents and character, he later struggled with his failed career and with alcoholism.

John Quincy's twenty-year-old sister Abigail, painted in London in 1785 by Mather Brown, who had been a student of John Singleton Copley. Dressed fashionably as the daughter of the newly appointed American minister to the Court of St. James's, Abby at about this time broke her engagement to Royall Tyler and the next year married William Stephens Smith, a marriage that the Adams family came to regret.

Edward Savage's flattering portrait of Louisa Catherine at age eighteen in 1793, four years before her marriage to John Quincy.

Adams believed the Treaty of Ghent, ending the War of 1812, his greatest achievement. It settled the American-Canadian border except for Maine and the Oregon territory and left American territory intact, despite the poor performance of the American military. Amédée Forestier's 1914 painting, *The Signing of the Treaty of Ghent, Christmas Eve, 1814*, is subtitled *A Hundred Years' Peace*, though Adams assumed in 1814 that further British-American conflict was likely. Adams stands at the center, holding a copy of the treaty. Albert Gallatin is immediately to his right. The white-haired man is James Bayard; the man seated to his right is Henry Clay.

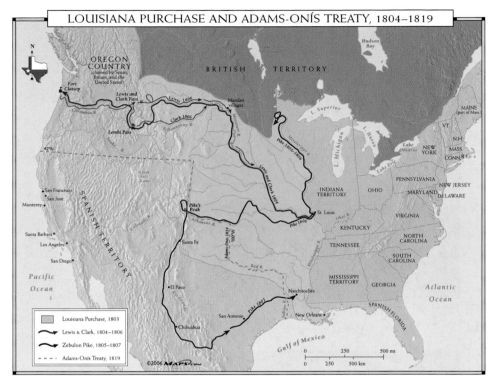

LOUISIANA PURCHASE AND ADAMS-ONÍS TREATY, 1804–1819

Legend:
- Louisiana Purchase, 1803
- Lewis & Clark, 1804–1806
- Zebulon Pike, 1805–1807
- Adams-Onís Treaty, 1819

Adams considered the Adams-Onís treaty with Spain the second-greatest achievement of his diplomatic career. It established the previously contested western and northern boundaries of the Louisiana Purchase and, in the northwest, extended the United States to the Pacific Ocean. On the day of its signing, he was overwhelmed by a feeling of "involuntary exultation."

Charles Robert Leslie, in London in 1816, painted the successful middle-aged diplomat. Adams at fifty had lost the curls that Copley had depicted in 1797, and he had gained weight. Leslie's is a portrait of a man who has arrived, and, as rumor had it, was soon likely to arrive even more fully.

Leslie's 1816 companion portrait of Louisa captures the high fashion and aristocratic pose expected of the wife of the American minister to the Court of St. James's. The facial features are recognizably those of the eighteen-year-old girl whom Savage painted in 1793.

resistance. What nonsense, John Quincy remarked. "Threats of insurrection for so small an affair as subaltern displacements . . . are absurd. . . . There is every reason to suppose that these official revolutions will *for a time* strengthen instead of weakening the President's party among the people." Aware that anything he might say in public would make him enemies on both sides, he was determined "to have no concern whatsoever in politics. There is not a party in this country with which an honest man can act without blushing, and I feel myself rather more strongly attached to my principles than to the ambition of any place or power in the gift of this Country."

He needed to get his feet firmly on practical ground, which meant setting up a law office. He felt pinched, still unable to retrieve what Charles had lent to Colonel Smith, though there was hope that the colonel's brother would honor his cosignature. John Quincy's own savings were banked in London and Holland, with the same firms that managed America's European loans. At his father's urging, he took on the management of his father's money. Cotton Tufts still managed John Quincy's small investments in stocks and his modest property in Boston. Eager to have Louisa and the baby join him, he found a house on Hanover Street, which he purchased for $6,000, providing bills of exchange against his account with the London firm of Bird, Savage, and Bird. It was the house in which John Quincy had lived with the Welsh family years before. Despite the loss of the money handled by Charles, he had saved about $10,000. Although he was by nature and principle set on never depriving Louisa of anything she wanted and he could afford, what he could not provide was better health. In Washington, she collapsed into illness: fever, cramps, disabling headaches, and extreme pain in her hands. It was thought to be gout or inflammatory rheumatism. Despite her thinness and frailty, she was cheered by how healthy little George was. "Every creature that has seen him has been struck with the striking resemblance he bears to his grandfather Adams." How he longed to kiss George's lips, John Quincy wrote to her, and "as for those of his mother I say nothing. Let her consult my heart in her own and all that pen can write or language express will shrink to nothing." They were eager for reunion, and Louisa needed to be introduced into his New England world. In mid-October 1801, he was on his way to bring his wife and baby home.

In Washington, he was happy to find her in better health; she actually looked "better than she has for years," he wrote to Thomas. John Quincy made the rounds, including a courtesy call on the president and visits to Secretary of State Madison and Secretary of the Treasury Albert Gallatin. Aghast to discover that he had been debited for $45,000 more than he had received, he needed to straighten out bookkeeping errors and settle final payment for his expenses in Europe. Not surprisingly, it took some work to correct the accounts, and the audit subalterns were quirky, irritable, and slow to act. He happily dined with his cousin William Cranch, who had been one of President Adams' last-minute appointments as a circuit court judge in the District of Columbia, a position to which he would give long-lasting distinction. With the Johnsons and Madisons, Adams dined with President Jefferson, who had fond memories of the young man from their Paris days. He and Louisa went to Mount Vernon, dining and staying overnight with Abigail's good friend Martha Washington. "The views along the Potomac," he remarked, "are very beautiful, as are the prospects from Mount-Vernon." Accompanied by Louisa's brother, he visited the site of the Capitol building, which was under construction.

Anxious to be home, he had business to attend to in Boston. The night before departure they dined at the Madisons', where the secretary of state was interested in the views of a man who had been more years abroad in the diplomatic service than any other American.

Accompanied by the Johnsons and three of their daughters, including Caroline, who had decided to spend the winter with the Adamses, they left by stage, the Johnsons traveling with them as far as Frederick, Maryland, where a number of Johnson relatives lived. Joshua Johnson had a cold and fever. On route, he had a kidney stone attack, with pain of more "excessive violence than I ever witnessed," John Quincy noted. It came in waves, again and again. He sat up with Johnson most of an entire night. Hesitant to leave, they delayed a few days. When Joshua seemed a little better, they left on a 4 A.M. stage to Philadelphia, where Louisa became sick, "more depressed in her spirits than really ill." She had good reason to worry that she might never see her father again. As their coach creaked forward on muddy and rutted roads, it began to rain heavily. In New York, he had an offer from Justus Smith, William Stephens' brother, to repay the loan with parcels of land in upstate New York, about which John Quincy asked for further information. Meanwhile, not even the interest was forthcoming. On Long Island Sound, as they headed for Providence, it turned cold and stormy. It was raining when they landed and all the next day in Boston. At Quincy, he "had the pleasure of introducing my wife and child to my parents." His days as a European traveler were finally over.

By the end of December 1801, he had resumed his life as a Boston lawyer in a rented office on State Street, a short walk from the Hanover Street house. Unenthusiastic about practicing law, he felt he had no alternative. His mother agreed that her eldest son had little choice but to try to earn a living in a crowded profession in a city that paid lawyers poorly. "I know very well that it has been in compliance with the wishes of your father that all my sons studied law, but it was contrary to my judgment." It did not help John Quincy that he had a sickly wife. "Her frame is so slender and her constitution so delicate," Abigail observed, "that I have many fears that she will be of short duration." Although John Quincy also had such fears, he had begun to accommodate himself to the patterns of Louisa's health, as if it were natural for her to have numbers of sick days each month, whether from a cold or flu or headache or neuralgia or skin eruptions or menstruation or pregnancy, and between these inevitable periods of illness to have days of lively conversation and activity. "I thank God I can yet struggle with the ills of life allotted to me," he noted as he began his new routine. Struggling and dissatisfied, he played with the idea, if the Smiths should pay their debt to him in western New York land, of striking out to the frontier, as an escape from the law and a new start. But soon he had most of the patterns of his Boston life in place.

Weekdays he was at his office or the courts. He attended a few stockholders' and directors' meetings. He negotiated for the purchase of some investment property and for a better house for himself. In the evenings he read, often aloud to Louisa, usually Shakespeare, poetry, history, and philosophy, including John Locke's essay on education. Occasionally they attended dinners and parties, and he spent many weekends with his parents at Quincy, with Louisa whenever her health permitted. They attended a few of the season's balls, where he enjoyed the dancing and remarked on how many new faces there were since he had last lived in Boston. But he was restless. "I feel strong temptation," he wrote in his diary in late January 1802, "and have great provocation to plunge into political controversy." But he would do his best to stick to his resolution neither to write nor to speak on political subjects. "A politician in this Country must be the man of a party. I would fain be the man of my whole Country."

On numbers of nights, awakened by alarm bells, he helped fight fires. In a city whose wooden buildings made conflagrations common, every able-bodied man was expected to be a part of the volunteer fire society. He also helped with household management, too often, he felt, "burdened with the minutest and vilest details of our domestic economy." One evening a week he attended meetings of a society, which he helped initiate, for the study of "experimental and natural philosophy." Another evening he met with a later version of the social club he had belonged to in his earlier Boston years. On Sundays, he attended church twice, the sermons being the main attraction. William Emerson, one of the ten members of the science club, whose son Ralph Waldo was born the next year and whose sermons at the First Church John Quincy admired, became a friend.

Since he did not expect to have much legal business, he resolved at the start of 1802 to devote himself to learning. Still, when a committee from the Charitable Fire Society requested that he give their annual speech, he consented. It would be his first such public appearance. "It is late in the progress of my life to complete my education, and perhaps after my age, a man can learn but little. Yet must I not despair." He even set himself to learn double-entry bookkeeping, having realized how flawed his account keeping had been, particularly when he needed to demonstrate to the State and Treasury departments the credits due to him. He found himself, though, with less time on his hands than he had anticipated, partly because it took him longer and required more application than he had anticipated to pick up the threads of his legal knowledge. He did find time for reading, partly because he had become obsessive about a disciplined schedule, but mainly because, despite self-complaints to the contrary, he learned quickly. And he began to have clients.

He was from the start welcomed into the official and social company of the governing elite of Boston. That he might be offered some sort of appointment was in the air, though he put no credence in the rumors. When the resignation of a state supreme court judge was announced, his name was widely mentioned, and he was happy to be appointed a bankruptcy commissioner: he was paid a small fee for each case, a welcome addition to his income. His name alone put him on lists of possible candidates that the Massachusetts legislature would be considering for election to the U.S. Senate. Abigail was firmly against it. "I hope if any attempt should be made to send him to the National Legislature he will decidedly reject it." She did not see what good could come from his being in a Federalist minority in the Republican lion's den.

But political office pursued him. The Federalist leaders thought it in their interest to nominate him for election from his Boston district to the Massachusetts state senate. The voters chose an entire ticket, not individual candidates. "I have little desire to be a [state]

Senator," he wrote in his diary. Still, he did not with-draw. "You will see my name upon the Suffolk list of Senators," he explained to Thomas, "and perhaps be a little surprized that I suffered it to be run. But as it was extremely doubtful and generally doubted, whether it could be carried, I did not choose either to shrink, or even to have the appearance of shrinking from the trial. So I did not decline. And indeed a man may as well be busy about nothing for the public as for him-self." In April, the Federal list received 2,375 votes, the Republicans 1,498. Ever the cheerleader and the realist, Thomas congratulated him: "The fed's will be glad to use your name and talents, so long as they can give them a lift; I am willing however to give them the credit of intending you a compliment. . . . The triumph is both honorable and consolatory."

Later that month, the dreaded news of Joshua Johnson's death arrived. When they had said good-bye in Maryland, he had seemed unlikely to rebound from pain and depression, but Louisa had continued to hope for his recovery. A world without her father seemed worse than bleak. The love of her life after her husband and child would be gone forever. After four years of sep-aration, father and daughter had at last been reunited. But that one reunion was all they had had. Death, Louisa knew, would close forever the possibility of his redemption; it would close the book on his business failure with that failure still imputed to him. It would deprive Mrs. Johnson and her unmarried daughters of his income, leaving them dependent on the support of family and friends, and there would be no possibility that he would be able to honor his pledge of a dowry. That promise was still in the back of Louisa's mind. It was a debt of honor. His death seared into her psyche her conviction that her husband and others believed John Quincy had been tricked into marrying her, that her father had known of his business failure before the marriage, and that she had been palmed off as damaged goods on an unsuspecting man. Ill and deeply depressed, she was heartbroken.

In late May 1802, Adams took the oath to uphold the constitution of the Commonwealth of Massachusetts. He and others would have been aware that it had been written by his father. Two days later, at the Old State House, he joined the officers and members of the Charitable Fire Society, which raised money for equipment and encouraged new firefighting equipment and techniques. They marched in procession to King's Chapel, where, before an audience that included his parents, he gave his first public address since his July oration in 1793. He had written and memorized the hour-long speech, the usual length since the series had been started ten years before and much shorter than the typical political or commemorative speech. This was a fund-raiser, to loosen purses rather than try the patience of the audience, a call to help Boston become a safer city. The town was discussing legislation that would require that any new buildings higher than ten feet be made entirely of brick. Adams gracefully reviewed the history of such efforts, and the hope that, like Augustus, who found Rome made of brick and left it made of marble, the current generation of Bostonians would be able to boast that they had found their city made of wood and left it made of bricks. It would be an act of Christian charity, which would prevent much misery and save many lives. After a eulogy to the Fire Society's first president, he adroitly segued from the tendency of people to form societies for good works to the larger theme of union and patriotism. They were, in his view, inseparably connected. For "the largest portion of the country is united under a social compact, which makes its inhabitants equal fellow citizens of one great and growing empire. To preserve, to strengthen, to perpetuate this union is the first political duty as it ought to be the highest glory of every American." It was to be the dominant theme of Adams' life.

He soon had another, more impressive occasion on which to develop the theme. Early in August, he received from Plymouth an invitation to deliver the annual speech commemorating the anniversary of the landing of the Pilgrims. He had five months to prepare. It was an honor to be embraced, a recognition that the thirty-four-year-old Boston lawyer and state senator had presence in the Federalist world. His speech to the Fire Society had been well received. His pamphleteering had revealed an educated and politically expressive voice. His translations, letters from Silesia, and original verse in the *Port Folio* called attention to his literary skills. That he had represented his country in Europe distinguished him from almost every other American, and that he was distinctively his father's son, even to their physical similarity, had a particular resonance. From the moment he received the invitation, he set to work. During the "excessively warm and sultry" summer days, he went frequently to Quincy, where he bathed in Black's Creek, walked the hills with his father, and read outdoors, happy to escape oppressive and noisy Boston. He had his Wednesday Club meetings and committee duties as a newly appointed Harvard overseer. His old friend James Bridge visited from Maine. At the invitation of the Boston selectmen, he attended a round of school visits, impressed by how many children of both sexes were being given a basic education. Education was, in his mind, a distinctive New England quality, inherent in Puritan values.

His focus was interrupted briefly when Josiah Quincy, a distant relative on his mother's side and five years younger, declined to try another run for Congress and urged him to be the Federalist candidate. His mother thought this was also a bad idea. "I see no chance for quiet, no hopes for social harmony, the bitterness of party thirsts far more than the cooling stream. The spirit of party is blind and deaf, but not dumb." The Jefferson administration, John Quincy wrote to Rufus King, "rests upon the support of a much stronger majority of the people throughout the Union than the former administration ever possessed since the first establishment of the Constitution. . . . There never was a system of measures more completely and irrevocably abandoned and rejected by the popular voice. It never can and never will be revived." But Adams consented anyway, on the premise that the office had sought him. In November, he lost by only 59 votes out of almost 3,800 cast. "I must consider the issue as relieving me from an heavy burden, and a thankless task," he wrote in his diary.

A few days before Christmas 1802, he was on the stage from Boston to Plymouth, with a small party including William Emerson, over bad roads and in a heavy rain. He had in hand a carefully constructed speech, designed to last a few minutes more than an hour, which he knew would be printed and widely distributed. The history of the Plymouth settlement provided, he believed, an example of American values at their best. "Liberty, Religion and Philosophy," he wrote to his brother Thomas, "are and must ever remain the blessings and ornaments of life, however they may sometimes get ill-sorted." It was his mission to sort them out as best he could in the context of the themes of American history, and the connection between the values of those who had made the first settlement and the challenges that the country now faced. After visiting "the rock of the first landing," he marched in procession to the meetinghouse. Like all such public rituals, the procession exemplified the social compact in praise of which he began his speech. "Man . . . was not made for himself alone—No! He was made for his country." Whatever its flaws, the original Pilgrim settlement was "perhaps the only instance, in human history, of that positive, original social compact, which speculative philosophers have imagined as the only legitimate source of government." But the utopian impulse, whether religious or political, cannot be sustained. "A wiser and more useful philosophy . . .

directs us to consider man, according to the nature in which he was formed; subject to infirmities, which no wisdom can remedy; to weaknesses which no institution can strengthen; to vices which no legislation can correct."

It was the Federalist view of human nature: human beings were not corrupt, but they were deeply flawed, as were the Pilgrims who came with the expectation of a communal utopia. They misjudged, Adams argued, the importance of private property and individual exertion to human nature. "We have seen the same mistake, committed in our own age, and upon a larger theatre. Happily for our ancestors their situation allowed them to repair it, before its effects had proved destructive." National harmony required a realistic view of governance, respect for political disagreement, absolute commitment to justice, and appreciation of the mixed qualities of human nature. This was inherent in the founding social compact, from the Pilgrims to the Declaration and Constitution. Liberty is power, the power to institute and sustain an independent nation devoted to justice. Religion strengthened the moral character; Christianity, as a religion of hope and redemption, instilled hope for the future; and philosophy offered moral guidance, intellectual development, and wisdom. They were and should be the foundation for the American future. "The history and the character of the Pilgrims remind us of these values." And though the Pilgrims were too narrow in their conception of religion, that could be understood in the context of the times, of the persecution they had suffered. It should be forgiven.

But there was another charge against the Pilgrims that Adams believed needed to be put into perspective. That they were not a people intent on enriching themselves distinguished them, and consequently distinguished their relationship with the native tribes from all other American settlements. From the start, they established communities, not outposts for conquest. To their great credit, Adams argued, they "obtained their right of possession to the territory on which they settled by titles as fair and unequivocal as any human property can be held." That process occurred, in his judgment, because it was the only settlement in America based on religious rather than material values. "On the great day of retribution, what . . . millions of the American race will appear at the bar of judgment to arraign their European invading conquerors! Let us humbly hope that the fathers of the Plymouth Colony will then appear in the whiteness of innocence. Let us indulge the belief that they will not only be free from all accusation of injustice to these unfortunate sons of nature, but that the testimonials of their acts of kindness and benevolence towards them will plead the cause of their virtues as they are now authenticated by the records of history." If the word "indulge" expresses tentativeness, or even doubt, it was an expression of hope as well as conviction. And an oration in honor of the Pilgrim Fathers was not an occasion on which to darken conviction with shadows.

Still, whether or not the European settlers had taken land that already belonged to its native inhabitants needed to be addressed. Possession by fraud or forced eviction was reprehensible in Adams' view, a form of criminality widespread and soon to become universal. What justified taking possession of huge tracts of land that had sustained the life of a hunting and nomadic culture, even if the transactions could be defended as fair? It was that

the Indian right of possession . . . stands . . . upon a questionable foundation. Their cultivated fields; their constructed habitations; a space of ample sufficiency for their subsistence, and whatever they had annexed to themselves by personal labor, was undoubtedly by the laws of nature theirs. But what is the right of a huntsman to the forest of a thousand miles over which he has accidentally ranged in quest of prey? Shall the liberal bounties of Providence to the race of man be monopolized by one of ten thousand for whom they were created? Shall the exuberant bosom of the common mother, amply adequate to the nourishment of millions, be claimed exclusively by a few hundreds of her offspring? Shall the lordly savage . . . forbid the wilderness to blossom like the rose? Shall he forbid the oaks of the forest to fall before the axe of industry, and rise again, transformed into the habitations of ease and elegance? Shall he doom an immense region of the globe to perpetual desolation . . . ? Shall the fields and the valleys . . . be condemned to everlasting barrenness? Shall the mighty rivers poured out by the hands of nature, as channels of communication between numerous nations, roll their waters in sullen silence and eternal solitude to the deep? Have hundreds of commodious harbors, a thousand leagues of coast, and a boundless ocean been spread in the front of this land, and shall every purpose of utility to which they could apply be prohibited by the tenant of the woods?

As for Lincoln sixty years later, it was an article of faith for Adams that God had created the earth for cultivation. Those who did not plant and build prevented progress. This was a use-it-or-lose-it argument. It was a definition of the right to property that no white American applied to his own undeveloped tracts of land. It was, though, Adams believed, applicable in this case. And it had another dimension, far removed from, though on the same continuum as, the Indian dispossession. It was an argument that Adams was to maintain throughout his life about one of the main issues of the day: the extent to which the federal government should plan and pay for the development of a national infrastructure to unify the nation and advance its prosperity. Active engagement with the earth and its resources for national growth and prosperity was a civic and sacred duty. For it to be successful, it had to be done on a national scale. What he had to say in December 1802 was in essence and by extension what he would have to say as a senator, secretary of state, president, and congressman. Liberty was power. But power needed to be inseparable from justice. And liberty was compatible with science, intellect, education, roads, bridges, public works, national growth, and national prosperity. All these required a unified nation led by its federal government.

Let "the occasion and the day," he concluded, not "be dishonored with a contracted and exclusive spirit. Our affections as citizens embrace the whole extent of the union. . . . The destinies of this empire . . . disdain the powers of human calculation." Let us "perceive in all their purity, refine if possible from all their alloy, those virtues which we this day commemorate as the ornament of our forefathers—Adhere to them with inflexible resolution. . . . Instill them with unwearied perseverance into the minds of your children; bind your souls and theirs to the national union as the chords of life are centered in the heart . . . that the dearest hopes of the human race may not be extinguished in disappointment, and that the last may prove the noblest empire of time." Although the empire was not to prove as noble as his rhetoric, Adams' belief in this vision of America's future never faltered.

After a sweltering summer, John Quincy, in late September 1803, placed in a large trunk all his letters, letter books, journals, and diaries. Their safety was becoming one of his preoccupations. Having now closed his "residence in Boston," probably "for several years . . . perhaps forever," he deposited the trunk temporarily in his father's office at Peacefield in Quincy. The year since he had delivered his oration at Plymouth had been momentous: in February 1803 he had been elected by the Massachusetts legislature to a six-year term as U.S. senator; on July 4 he became the father of a second son, this one named after his own father; his mother had been seriously ill twice, her life at risk; and, almost simultaneously, the Adams family had received a damaging financial setback for which he felt responsible.

At the statehouse, to which he walked each day, the attempt to have new banks and insurance companies chartered by the state was dominating the agenda early in 1803. There was money to be made, even from disasters. Well-financed interests took the lead. Debt structures sometimes involved the state's participation as a lender or stockholder. As a leading member of the committee to write banking and insurance bills in the state legislature, Adams had his hands full for the first five months of the year with meetings, negotiations, and drafts of legislation. A rumor circulated that he opposed a major bill favored by the financial industry. "Mr. Emerson told me I was reported to be in opposition to the new Insurance Company; and that I had no conception how unpopular my opposition is." He objected to shares' being placed in reserve for members of the legislature. He wanted all stock issued publicly. When the charter for a new bank came before Adams' committee, it was "supported by the principal money men in the town." But it was "opposed by J. Q. Adams, whose popularity is lessened by it," Fisher Ames, an influential Federalist, remarked. "They say also he is too unmanageable."

In February, Adams and Ames had a conversation as they left a dinner party. Adams had heard that both his and Ames' names "were upon the nomination list, now in the Massachusetts House of Representatives, for the choice of a Senator in Congress." In caucus, the Federalists agreed that Timothy Pickering would be their first choice to fill the six-year term. But if he failed to achieve a majority on the first two ballots, they would switch to Adams. Then Pickering would be nominated for a vacant four-year term. Some Federalists worried that Adams and Pickering could not work together; they were political enemies. Adams gave assurances that they could. When Pickering fell short, Adams received an eighty-six-vote majority in the house on the fourth ballot. The influence of Harrison Gray Otis helped turn the vote. Adams was startled that Otis had supported him. "After we adjourned Otis took me into one of the lobbies, to talk with me upon the subject of the application for a new bank in the town of Boston." Adams assured Otis that he did not oppose the bank in principle. In the Massachusetts State Senate, on February 8, he got nineteen of the twenty-six votes. Two days later, he received his credentials as a U.S. senator.

To take advantage of a better exchange rate, Adams had transferred the only substantial cash savings his parents had from a Dutch bank to Bird, Savage, and Bird, which the U.S. government used for its British transactions. He was free to write checks against the account, especially for large transactions, the account to be made whole again as circumstances warranted. Recently, he had asked Thomas in Philadelphia, where there was a better exchange rate, to sell a note for £1,000 drawn against the London account. He had used notes against the account in two real estate transactions. Twice the previous year he had had minor concerns about the bank's solvency, and at the beginning of April he got the shocking news that it had suspended payment of all obligations. Checks he had written would not be honored. The sum at issue was at least $16,000, and there would be hefty penalty charges from creditors.

Huddling with his parents, whose only savings had suddenly disappeared, he took responsibility for the loss. "I feel myself in a great degree answerable for this calamity. . . . The error of judgment was mine, and therefore I shall not refuse to share in the suffering." As soon as he could, he sold some of his and his father's stocks, his Hanover Street house, and his recently purchased Franklin Place house. He could not find a buyer for his father's Court Street building. Especially pressed since the former president had just purchased the Mount Wollaston property, formerly in Abigail's family, he had his Quincy properties appraised, prepared to sell if necessary. "A Catastrophe so unexpected to us, and at a time when we had become responsible for so large a sum," Abigail wrote to Thomas, "has indeed distressed us. . . . I do not dread want, but I dread debt." Boston friends and Rufus King, about to return from his post in London, came to the rescue with loans at the market interest rate. Over the next months, John Quincy covered the notes written against the London account. After the sale of stocks and property, he repaid the loans. In New York, he joined other unsecured creditors in a claim against the assets of the bankrupt firm, the start of a lengthy recovery attempt that took twenty-six years. It was a harrowing experience.

So were the illnesses that threatened Abigail's life. Both occurred soon after Louisa gave birth to John Adams II on July 4, 1803. The birth date seemed a happy omen. Under the care of Dr. Welsh, Louisa came through. Still, her health was a concern, her history of miscarriages inescapably present. John Quincy looked, his mother wrote, "as anxious as though he had the trouble himself to pass through. She is very well for so feeble a body as she is. When you take a wife, it must be for better or for worse, but a healthy and good constitution is an object with those who consider, maturely." He was at Quincy when Louisa gave birth. "For this new blessing, I desire to offer my humblest gratitude to the throne of Heaven." Louisa slowly recovered, with setbacks. In the middle of the month, the baby was baptized by the Reverend William Emerson. That evening, John Quincy read two sermons by Tillotson with titles that turned out to be eerily appropriate for his and his son's life, "Good Men" and "Strangers and Sojourners on Earth."

Good fortune had its counterweight. In early June, Abigail fell backward coming down a flight of stairs at home. She was "much bruised, and inwardly hurt." At the beginning of August, a more drawn-out threat occurred, an extremely painful "large and highly inflamed tumor . . . with a persistent high fever." John Quincy spent much of the first two weeks of August with her. Dr. Welsh thought it likely she would die. Preparing for the end, she made provisions in her will about the Quincy property and a farm she owned in Medford. The Quincy property was to go to John Quincy, the Medford estate to Thomas. She provided gifts of stock to Nabby's and Charles' children. John Quincy spent solemn days with his father, walking and riding, the former president fearing that he was about to become a widower. "The heaviness of heart and of spirits, which I am unable to resist," John Quincy wrote in his diary, "disqualifies me from every proper occupation." He could do nothing, paralyzed by anxiety. Dr. Welsh became more pessimistic. "She has little or no expectation of her own recovery . . . her spirits have sunk with her strength." The family despaired. John Quincy watched, waited, and hoped that there would be a resolution before his scheduled departure for Washington. In mid-September, he went out to Quincy, where he found his mother, "God be praised, much better."

Entering the small city of Washington, dust rising from the dry road, John Quincy and Louisa met another carriage in the early evening light of October 20, 1803. A familiar face greeted them with significant news. Samuel Otis, the secretary of the Senate and Harrison Gray Otis' father, was returning from the President's House to the Capitol. He had brought to Jefferson the formal notification that the Senate had ratified the cession of Louisiana from France to the United States. The vote was twenty-four to seven. John Quincy was one day late. The journey from Boston had taken almost three weeks, lengthened by bad roads and weather, by traveling with a wife and two young children, by illness, and by diversions to avoid

New York and Philadelphia, where yellow fever raged. The Republicans supported the purchase. Many were aware that the Constitution provided no provision for buying territory, but a constitutional amendment would take months to enact. The sale might have collapsed, so these strict constructionists rationalized. That the purchase price would have to be borrowed was a scruple also readily overcome by an administration whose avowed principle was minimal government and the eradication of all debt.

How would Adams have voted? Most Federalists opposed the purchase. New Englanders feared the dilution of their influence as population and territory spread southwestward. In the next months, as Louisiana dominated the legislative halls and national discussion, it became clear that Adams would have voted in favor. That vote would not have pleased his fellow Federalists, and he would soon have the opportunity for others that would confirm Fisher Ames' view that he was "unmanageable." The vote that he had missed ratified the purchase but dealt with none of the controversial subsidiary issues. What had defined the colonies and the country from the start was the availability of land. Here was more of it, a vast amount. Americans became better off by either obtaining real estate or marrying up or both. That was what George Washington had done. So had the Jefferson family and the Madisons. So too had most of the notable families of Virginia and many of the signers of the founding documents. In his small way, John Adams had expressed his land hunger by his Quincy purchases. Some wealth, of course, came from commerce. The closer to New England, the more that was so. Still, real estate and marriage as sources of wealth were a national preoccupation. By 1803, most of the untapped land was to the west and southwest: vast tracts to be bought, sold, settled, hunted, mined, cleared, tilled, planted, and developed into towns, estates, farms, and ranches, to be settled by a growing population, to attract foreign investment, to send its animal skins, silver, gold, and agricultural products eastward. At the end of the trail, there were sparkling Pacific waters, which Jefferson soon sent Lewis and Clark to gaze upon. From that moment, like Keats' Cortez when, "with eagle eyes / He stared at the Pacific," in its imagination the United States was already a transcontinental nation.

Adams granted that New England interests needed to be protected. But since the purchase contributed to national security by further distancing Spanish and French neighbors, and gave the United States control of the Mississippi River, with a thriving port for international trade, he had no doubt that it would benefit the country. Considerable tension, though, seethed about the thorny subsidiary issues. Did the purchase require an after-the-fact constitutional amendment? How was this vast territory to be governed? Were the French, Spanish, Indian, American, and black populations of New Orleans and its adjacent areas to have by virtue of the purchase all the rights of American citizens? It seemed to Adams incompatible with the Declaration and Constitution to take possession of not only territory but people without their consent. It was obligatory to ask, by referendum, for their approval. Without it, you were simply buying people. And would it continue to be permissible for slaves to be imported from Cuba and Africa into Louisiana, or should Louisiana be an advance instance of the exclusion of the slave trade that the Constitution stipulated could not be enacted before 1808?

Startled that not even the strictest constructionists were inclined to offer an amendment to provide authority for acquiring the new territory, Adams went to see Secretary of State Madison. "I asked him whether the Executive had made any arrangements with any member of either house to bring forward the proposal for an Amendment to the Constitution to carry through the Louisiana Treaty. . . . But if not, I should think it my duty to move for such an Amendment."

Madison responded that "he did not know that it was universally agreed that it required an Amendment of the Constitution." But he had "no difficulty in acknowledging, that the Constitution had not provided for such a case as this. That it must be estimated by the magnitude of the object, and that those who had agreed to it, must rely upon the candour of their country for justification." Adams agreed "but urged the necessity of removing as speedily as possible all questions on this subject, to which he readily assented." Madison was not being entirely truthful. The Jefferson administration had decided that the Constitution should be disregarded, an answer with which Adams, who was soon reading *The Federalist*, did not agree. The next day, when he added the words "consistently with the Constitution" to the Louisiana enabling bill, the addition was ruled out of order. When he and others proposed that approval be made conditional upon the passage of an amendment, the Republican Congress and administration united against the proposal.

At the end of November 1803, Adams moved an amendment allowing "Congress to admit into the Union territory that had not been part of the US and to give all the inhabitants full rights of citizenship." It was sent to a special committee to which he was appointed, though he opposed its creation. It had, he knew, been formed to support the administration's wish that enabling legislation be passed as soon as possible. Having spoken in favor of the purchase itself, he soon got a lesson about the danger of not being in lockstep with the Federalists. "The Hon. John Quincy Adams will certainly be denounced and excommunicated by his party," the *Worcester Aegis* remarked. His father's enemies held him in distaste, and the Federalists would denounce him as an apostate whenever he voted with the government. Since his election, he had anticipated that his road in the Senate would be a rough one. His mother had urged him not to go into the lion's den. It was more than impolitic—it was politically suicidal, and he knew from the start that it would be only a matter of time before the Massachusetts legislature would replace him, whether Federalists or Republicans held the majority. In the meantime, he had "already seen enough to ascertain that no amendments of my proposing will obtain in the Senate as now filled. . . . Firmness, perseverance, patience, coolness and forbearance" were called for. But "the prospect is not promising." All his resolutions, he noted, were feathers against the whirlwind. Conscience and politics did not mix well. He was "between two rows of batteries directly opposite to and continually playing upon each other, and neither of which consider me as one of their soldiers." When, early in the new year, he introduced a resolution opposing taxing the inhabitants of Louisiana without their consent, he touched a raw nerve. Still, the bill that taxed without representation passed. In fact, "my warmth of opposition against these measures," he noted in his diary, "has reconciled some persons to it who hate me rather more than they love any principle."

The consideration of Louisiana dragged on until every member of both houses seemed to Adams sick of it. "If any gentleman," he argued to the Senate in early January, "can controvert the principle that by the laws of nature, of nations, and of God, no people has the right to make laws for another people," including requiring the payment of taxes, "without their consent unless it be by right of conquest, I shall be glad to hear him. . . . I cannot prevail upon myself to vote for a law with a clear and undoubting conviction . . . that Congress have not the shadow of a right to pass it." By early March 1804, he realized that it was the natural course of a legislature to take final action on a bill when a majority was sick of the subject rather than when the legislators had its provisions right. It gradually dawned on him that most of the opposition to his position resulted from the widespread belief that he was motivated by a desire to abort the purchase rather than make it constitutionally valid. Many Republicans assumed he was maneuvering to protect New England interests. "But their suspicion of me is totally groundless," he wrote to Thomas. "I sincerely believe that every State would ultimately have agreed to the Amendment." His expectation that he would be "stigmatized on all quarters" was correct. But "it is the price of Independence as things stand; and I must pay it."

At the end of January, he had voted against making it illegal to import slaves into Louisiana, claiming that Congress had no right to make any laws there at all without the representation of its inhabitants. He did not favor the importation of slaves, and the bill had no chance of passing anyway. But without a constitutional amendment, he believed there should be no enabling laws passed, even ones that he favored. The politics of slavery, he knew, were dynamic and complicated, even more so in the South than in the North. Some Southerners favored the indefinite extension of importation. Others looked to the increase in the value of their slaves when imports ended after 1808. Observing the machinations of senators to kill the bill that prohibited importing slaves into Louisiana by adding to it an abolition clause, Adams was repelled. "The workings of this question upon the minds and hearts of these men," he noted, "opened them to observation as much as if they had had the window in the breast. . . . This is now in general the great art of legislation at this place. To *do* a thing, by assuming the appearance of *preventing* it. To *prevent* a thing by assuming that of *doing* it." Like most New Englanders, he was realistic enough to know that, despite widespread unease about slavery, abolition was not even remotely feasible. The Constitution sanctioned slavery. Since the South believed its economic and social interests to be inseparable from perpetuating slavery, an abolition proposal would be perceived as a call for the breakup of the country.

When he attended a dinner celebrating the ratification of the Louisiana Purchase, he had no regret about either his support or his reservations. One of the toasts was to a "Union of Parties," echoing Jefferson's inaugural address. That, John Quincy wrote to Thomas, "is like drinking the Millennium. I suppose they will come together." Since he had nothing more to lose or gain from either party, he had decided that he would accept invitations from Republicans as well as Federalists when the event interested him, whether it was a political or social occasion. Political differences rarely affected social activities. Washington society mostly managed to be hospitable, and John Quincy made a few friends, especially three Federalist senators, William Plumer of New Hampshire, James A. Bayard of Delaware, and Uriah Tracy of Connecticut. Jefferson continued

Washington's practice of inviting members of all factions to his table, though he practiced an informality that would have been antithetical to the first president. By himself, and sometimes with Louisa, John Quincy attended dinners and other entertainments. Some were at Republican homes, the president's and James and Dolley Madison's especially. At Jefferson's table, he was startled by the president's tall tales, some of which he knew to be outright fictions. They contributed to his sense of Jefferson as living in a world of fabrications, many of which the president believed to be absolute truths. Adams never quite brought himself to say that Jefferson was a liar, but he came close. And though he would support any policy he believed in the best interests of the country, his childhood admiration for Jefferson as a man of honor had long disappeared, destroyed by Jefferson's self-serving actions as John Adams' vice president and his sponsorship of anti-Adams slander in the 1800 campaign.

The political pot kept boiling. In the late winter and early spring of 1804, two bitterly divisive issues erupted, both with long-term consequences. Adams and other Federalists were appalled by what suddenly developed as a full-scale attack on the independence of the judiciary. Congress was consumed from March

1803 to April 1805 by aggressive, painfully controversial attempts to remove a New Hampshire federal
district court judge and a Supreme Court justice from
office. To Jefferson, the federal judiciary was an inferior branch of government, which should be subordinate to Congress and the state legislatures. They, not
the Supreme Court, should determine whether an act
of Congress was constitutional. Behind Jefferson's and
his colleagues' desire to limit the reach of the federal
courts hovered their concern that those courts might
one day make slavery illegal. Jefferson's antidote to
Federalist domination of the courts was to inter-
pret the clause in the Constitution that judges "shall
hold their offices during good behavior," and the
clause that "all civil officers of the United States, shall
be removed from office on . . . conviction of treason,
bribery, or other high crimes and misdemeanors," to
mean that judges could be impeached for a wide range
of behaviors, including the expression of political
views. His purpose was to replace Federalist with Republican judges.

The first case, against New Hampshire's John
Pickering, came to trial before the Senate in March
1804. A mentally unstable alcoholic, Pickering was incapable of performing his duties. Since it was clear that he
needed to be removed from office, Pickering's family

sought an honorable resolution. Adams did not believe the charges rose to the level of impeachable offenses. "Motions were made to assign him counsel who upon the plea of not guilty should give in evidence, insanity, by way of mitigation; as if a madman could either plead guilty, or not guilty." The motions were denied. Eager for a useful precedent, the Republicans argued that an impeachment trial did not give the defendant a right to counsel or to call witnesses. When the Senate convicted him, Adams thought the proceeding cruel, dishonorable, and a travesty of justice.

The Supreme Court had already become a palpable presence for Adams, who had been admitted to practice before it in February. It was not improper for federal legislators to argue cases on behalf of clients. In late winter, he was cocounsel on behalf of the wealthy Boston businessman Peter Chardon Brooks. "Mr. Adams is so much engaged he scarcely allows himself time to eat, drink or sleep," Louisa wrote to her mother-in-law. In spare moments, he went to watch and learn. One of the well-known justices was Samuel Chase, an outspoken opponent of Jefferson and the Republicans. Here was an opportunity to purify the court and create a vacancy. The Republican House voted for impeachment, charging "high crimes and misdemeanors"—including Chase's assertive, blustering expression of political views from the bench. The trial was high drama. It came to the floor of the Senate in February 1805. No defense was allowed, only a plea. "Impeachment is nothing more than an enquiry by the two houses of Congress," the leading Republican senators claimed, "whether the office of any public man, might not be better filled by another . . . and on the same principle, any officer may easily be removed at any time." The impeachment of a judge or justice "need not imply any criminality or corruption in him. Congress had no power over the person, but only over the Office. And a removal by impeachment," Adams observed, "was nothing more than a declaration that 'you hold dangerous opinions, and if you are suffered to carry them into effect you will work the destruction of the Nation. We want your Offices; for the purpose of giving them to men who will fill them better.'"

But Chase and his team of lawyers forced a defense, maneuvering, with the consent of an increasingly uneasy Senate, to a full rebuttal in response to the House prosecution. In late March, John Randolph, the House manager of the prosecution, whose arrogant bungling further poisoned matters, "began a speech of about two hours and a half with as little relation to the subject-matter as possible. Without order, connection or argument: consisting altogether of the most hackneyed common-places of popular declamation, mingled up with panegyrics and invectives upon persons, with a few well expressed ideas, a few striking figures, much distortion of face, and contortion of body, tears, groans and sobs, with occasional pauses for recollection, and continual complaints of having lost his notes." It seemed a mockery of justice, sometimes even comic in its pettiness and absurdity, yet also immensely consequential. "Sir Gravity himself," John Quincy wrote to his father, "could not keep his countenance . . . at the puerile perseverance with which *nothings* were accumulated, with the hope of making *something* by their multitude. . . . Judge Chase . . . contented himself with observing to the Court, that he expected to be judged upon the *legal* evidence in the case." The future of the Supreme Court was being determined. Of the thirty-four senators, twenty-five were "political opponents of Mr. Chase." In the end, to Adams' surprise and satisfaction, Republican defectors denied the House managers the necessary two-thirds vote on every one of the counts. Chase was acquitted.

At the same time, debate on an international issue introduced the possibility of war. It was an issue on which Adams felt passionately. It soon pushed him into his own "furnace of affliction" and the country, nine years later, into an unnecessary war. When, in 1803, war resumed between the two major European powers, Great Britain began to assert more than ever the right it claimed to remove from American ships sailors believed to have deserted from the Royal Navy. British law did not allow voluntary renunciation of citizenship. And Britain desperately needed sailors for its huge navy. Any Englishman, at home or at sea, was subject to impressment, an early form of drafting men into the military. Some British deserters found employment on American merchant ships, which provided better conditions than the harsh regime on British warships. In labor-short America, shipowners needed crews. Britain, fighting against Napoléon, dependent on its fleet for survival, needed every sailor it could get. And some of its sailors were on American ships.

There were two related concerns: how to define neutral trade and whether American ships could trade in British colonial ports, particularly in the West Indies. For Adams and his countrymen, "neutral ships made free trade" unless the cargo was explicitly war related. For Britain, the lumber, hemp, and wheat that American vessels carried to European ports empowered its enemies as much as guns and cannons. Britain believed that necessity gave it the right to impound ships carrying war materials, as it defined them. But, Jefferson and Madison responded, Britain needed

American products. If it were prevented from getting them by an embargo, it would see the folly of its ways. That was America's best defense. It did not need a navy or an army. American reaction to impressment, the British attitude about neutral trade, and whether American ships could enter British West Indian ports were inseparable from widespread Anglophobia. Many Republicans hated Great Britain, intellectually and emotionally. Its American allies, the Federalists, were always scheming, Jefferson believed, to return the United States to its prerevolutionary subjugation. And no matter what crimes the French revolutionists and then Napoléon committed, France and America were natural allies. It seemed to Adams, who tried to be realistic and objective, that the United States needed to be wary of both countries. Britain could not be trusted. It would always act in its own interest. But France was a disaster, Napoléon a tyrant. And France also would always act in its own interest. How to steer a safe course between the whirlpool and the flood? Neutrality continued to be the best policy. War would be disastrous for an unprepared America unwilling to spend the money to prepare itself.

In January 1804, passions were inflamed on the impressment issue. Backed by the Jefferson administration, the Republicans introduced a bill claiming that British authorities had no right to board American ships in British ports, even on the Thames River in London. The ostensible purpose was to protect American seamen from impressment. In effect, the purpose of the bill was to prevent British authorities from removing British subjects, including deserters, from American ships in British territorial waters. It was less about protecting American seamen than about encouraging British deserters. Adams felt the bill was motivated by an Anglophobia so intense that it disregarded the law of jurisdiction and the law of nations, which sanctioned the right of a government to inspect foreign ships in its own ports. "If there is any one thing that we can do better calculated to plunge us into a naval war than any other, I think it is this very project." When Adams, after waiting in vain for someone else to take the floor, acknowledging that he did not have the good sense "to be silent upon anything," spoke out against the hypocrisy, illegality, and dangerousness of the bill, he was pilloried. "I wish you had seen," he wrote to his brother Thomas, "the hornet's nest that burst down upon my head on the first day's debate. Not a soul supported me in my principles and half the federalists declared against me. However they at last thought it was worth thinking a little more about, and at the second day's debate the federalists rallied a little, and the others began to stagger." To his relief, Congress reconsidered and decided that the bill was too inflammatory. In mid-March, Adams wrote to his father that "we have but one week more for this drudgery. I shall hail the moment of release."

When, in early April 1804, John Quincy left Washington, he traveled northward alone. To stay through the summer was a horrid thought. He wanted the restorative air of Quincy and the company of his parents. But he missed his family. "I feel already to use a vulgar phrase, like a fish out of water, without you and my children," he wrote to Louisa, "but I will not complain. I hope to hear from you very soon, and to be assured that you are all well." He did have a complaint, though. They had parted on a misunderstanding. Louisa felt distressed by her husband's comment that their finances required that she and the children spend the full year in either Quincy or Washington. She assumed she could choose to stay in Washington or go with him, but, if she went, not to return to Washington until the following fall. She chose to stay. He saw it as an unconstrained choice on her part. "I preferred," she responded, "passing the summer months with my family to living alone at Quincy through five dreary winters. I do not think my beloved friend you do me justice when you say I prefer a separation from you rather than separation from them. . . . I am ready when you please to relinquish their society and reside at Quincy to insure your affection and esteem."

That she had thought him unkind pained him. He thought himself practical and prudent. "The first wish of my heart is to make you happy as far as it is in my power, and it is a subject of deep affliction to me, that my means of accomplishing this wish are not more adequate to its ardor and sincerity." He wanted his wife and his sons with him. But it was also his duty and his wish to be with his parents. She thought his claim that they could not afford to be together for the full year in both places unconvincing. "Adieu, my dearest friend," he wrote. "May you never feel a pang imparted from your husband's hand; and may his feelings of the warmest and tenderest affection ever meet with equal and correspondent sentiments in return."

As the spring blossomed, he more and more missed his wife and sons. "I think of scarce anything else. . . . And I count every day. And every hour until we meet again." Although the edge of his desire remained, his practical and stoic temperament asserted itself. It was too late to make her trip worthwhile. But he committed himself to their being together in Washington and Quincy the next year. He still left open whether she would then have to remain in Quincy. He did indeed have other things to occupy his mind and time, and their letters during the next five months reveal how effectively they each adjusted to the obligations of the moment, mostly of family on her part and, on his, of literature, intellect, politics, and gardening. But they were also love letters between a husband and wife who were happily intimate. "Good Night my best beloved. I send you the most tender kisses of love," he wrote at the end of May. Committed to the happiness of her "most beloved friend," she hoped he would "learn to read my heart more correctly than you do at present. I meant not to make conditions in my letter but to act solely for the future as you thought fit. However painful a separation must ever be to me your interest alone must be my consideration and everything else must give way. . . . I am ready to do anything you please for the future." Together or apart, they worried about one another's health. In August, Louisa became frantic at the news that he was ill. A racking cough that he had had for months had returned. The thought of not being with him when he was sick "is torture and I shall know no peace until I hear. Oh this separation. Life is not worth having on such terms."

An important person from Adams' past suddenly became a vivid presence. In July 1804, Mary Frazier died at the age of thirty. "Sargent has had the misfortune of losing his wife," he wrote to Louisa. "She has left a child about 6 months old." Sargent was never to remarry. John Quincy said nothing more, but Louisa, who had met Mary twice in social company, expressed herself spontaneously and generously: "inexpressibly shocked at the melancholy news. . . . Poor Mrs. Sargent. I most sincerely sympathize with you my beloved friend in grief for her early death. Amiable and lovely as she was everyone who has seen her must deplore her loss, but you my best friend who have known her so long and once loved her so well must indeed mourn her untimely fate and bury her faults (if faults she had) in eternal oblivion." Louisa instinctively found the words to let her husband know how much she valued him. When she had first seen Mary, she could not help concluding that Mary "still retain'd her affection for you which she could not conceal and I pitied her from my Soul, convinced as I am she never ceased to lament the folly she was urged to commit and to deplore the blessing she had lost." Yes, he responded, he was deeply affected by the news of her death. He had known she was ill but had assumed that her illness was not life threatening. The next he had heard was that "she was dead. But without intending to affect either indifference or sensibility, I must assure you that I lamented her loss as I should have done that of any other young woman the wife of my friend. The sentiments I had felt with regard to her for at least ten years were those of a common acquaintance, coupled perhaps with a peculiar coldness of reserve. . . . I never felt the wish to see her, nor was I conscious of a wish to avoid her." His inconsistent account of his feelings seems an attempt to convince both of them of what was obviously not true.

When Louisa attended the July 4 reception at the President's House, Jefferson asked after Adams. He "was very anxious to know when you returned." The president seemed to Louisa "so altered I scarcely knew him. He is grown very thin and looks very old." In April, Jefferson's daughter Polly had died. The grieving father responded to a heartfelt letter of consolation from Abigail, who in 1797 had briefly provided a home for Polly when she was on her way to France. The complicated relationship between the Adams family and Jefferson had taken a more outspoken turn recently, and a polite communication had resumed. But Abigail could not resist referring to the Adams-Jefferson alienation. In the few long letters she and Jefferson exchanged, she chastised him for pettiness in having forced the removal of John Quincy as a commissioner of bankruptcy. Jefferson could not resist defending himself. When the change was made, he explained, it had been part of a clean sweep, dictated by impersonal considerations. He had not even known that John Quincy held that position. The exchange turned politely cold. John Adams was not aware of the correspondence, and, when he learned of it, he kept his distance. But, despite his lingering bitterness, he had some words of praise for Jefferson, though he thought many of Jefferson's views crazy— for example, that it would be better if France would return to the condition of 1791. And Adams expressed unequivocally to his son that he had not lost faith in the American people, despite recent events: that over the long haul, good sense would triumph, though, like John Quincy, he had no doubt that a sea change had occurred in American politics and culture. "I take it for granted that public Virtue is no longer to rule: but Ambition is to govern the Country. . . . All we can hope for hereafter is that we may be governed by honorable, not criminal Ambition."

A shocking instance of "criminal ambition" had occurred that summer. When John Quincy learned, in July 1804, that Vice President Aaron Burr had killed

Alexander Hamilton in a duel, he shared the widespread horror and disgust. He had met Burr a number of times in New York and Washington, though he had still been in Europe in March 1801 when Burr temporarily benefited from the flaw in the Constitution that did not distinguish between candidates for the presidency and those for the vice presidency. Electors voted for each candidate as if he were running for president, even when everyone knew which office each was ostensibly slotted for. In a statistical and political accident, Jefferson and Burr received the same number of electoral votes, requiring that the House vote by state to determine who was to be president and who vice president. To everyone's shock, Burr refused to withdraw. Jefferson thought Burr's conduct detestable. Adams, who had many opportunities to observe Burr in the Senate, noticed the coldness with which Burr was treated by the Republican establishment. Although Adams had no personal animus against Burr, he thought him "a man of very insinuating manners"—politically clever, intellectually sharp, with a gift for impromptu and courteous eloquence.

But he also seemed an epitome of ruthless ambition, with a self-regarding willfulness that in the duel with Hamilton rose to the level of criminality. A grand jury was soon to indict Burr for murder, though, to

John Quincy's astonishment, he continued to function as the president of the Senate. "I cannot conceive any possible circumstances, which can justify the conduct of Mr. Burr, either preceding the fatal day, or immediately subsequent to it." Burr had precipitated the duel. There were no words, let alone conduct, on Hamilton's part that justified Burr's challenge. "The conduct of Mr. Burr through the whole affair appears to me strongly to corroborate that opinion of his character which his enemies have long ascribed to him." John Quincy was not an enemy, however, but a moral critic, and of Hamilton also. He believed that Hamilton, he wrote to Louisa, "was a man of considerable, but over-rated abilities, openly and scandalously vicious in his private character. . . . His tragical end, I lament as much as any man. The distress of his family, I feel for in common with the warmest of his friends; but in very deed I do not think he was either a demi-god or a Saint."

In the hot summer weather, John Quincy attempted a holiday from all political matters, including provocations. He refreshed himself regularly at Black's Creek, and spent much of his time reading the *Iliad* in various translations, a portion of the *Odyssey* in Greek, and Aristotle's *On Rhetoric.* Some volumes of French memoirs and letters entertained him. He began to read in their entirety the laws passed from the first Congress to the present. In the evenings, he read *The Faerie Queene*, often aloud to his parents, delighted that Abigail was in better health than she had been for some time. He responded to Louisa, who was reading Madame de Staël on marriage and divorce, that he would read it too, if he could find a copy. "As to Madame de Stael's opinions upon the subject of divorce, and the marriage vow, they are such as might be expected from her history and her character. After having sacrificed all decency as well as all virtue in her own conduct, it is natural enough to find her torturing her ingenuity to give infamy itself a wash of plausibility." John Quincy had the frequent company of Thomas, who had finally capitulated to reality, relocating from Philadelphia to Quincy and occupying a farm in Braintree that had been a gift from his father. The brothers hunted together, stalking the usual birds for the dinner table, and John Quincy considered alterations to the unoccupied house at the foot of Penn's Hill in which he had been born. It had also been a gift from his father. Abigail and John were eager to have both sons close by.

As he walked his farmland, John Quincy discovered a dormant passion for gardening. He was pleased to find that all the apple and peach trees he had planted were thriving. "I pay so much attention to the poor plants from hour to hour, that the only danger is of my killing them all with kindness." His interest in trees had surfaced years before, and the previous summer he had planted some peach stones that produced saplings in May. He now grafted apple and peach trees. He had begun, he half-joked, his apprenticeship as a farmer. To improve the land by planting trees soon became a passion. They would be, he believed, a gift to posterity. And reforestation would contribute to personal and national wealth. He also got an object lesson in how unfriendly nature could be to human desires. He soon sang to Louisa a gardener's lament: "A most pernicious insect . . . devours the grapes, the peaches, apricots, nectarines, and the leaves of every plant that grows. But another much more diabolical is a white worm which attacks the peach trees at the roots, and has destroyed more than half the trees in the garden, within these six weeks. After blossoming as full as I have ever seen . . . they began of a sudden to droop. Leaf after leaf, and peach after peach, drops in melancholy succession, till the whole garden is strewed with dead vegetation, as at the close of autumn, and the trees stand bare and blasted as if to reproach the skies with their fate." His own fate was much on his mind. When he attended the funeral of Harvard president Joseph Willard, he was startled to be asked by Josiah Quincy whether he would allow himself to be a candidate for the office. He thought Quincy was joking. The next week his cousin repeated the request. Although he knew how unlikely it was that he would serve more than one term in the Senate, he responded, "It will not answer."

Having been instructed by the Massachusetts legislature to introduce in the next session of Congress a constitutional amendment to strike out the three-fifths rule, he spent the rest of the summer thinking about what to say and writing. In late July, he began work on an essay, "Serious Reflections, Addressed to the Citizens of Massachusetts." It was to be published in four installments between late October and mid-November 1804 in *The Repertory*, a Boston weekly newspaper, under the pseudonym Publius Valerius. Its central theme was the threat to the union posed by the three-fifths rule and the moral viciousness of slavery. Brilliantly written and unsparingly frank, it dissected the hypocrisy of the previous Congress, the unequal distribution of power among the states, and the likely catastrophe to which the three-fifths compromise would lead.

Working assiduously through August and September, he finished well before he left in mid-October for Washington, where he at last embraced his wife and children again and paid his usual and well-received visits to Jefferson and Madison. Both talked freely

with him about foreign policy, especially impressment and whether to allow trade with the new slave
republic in Haiti. Jefferson was determined to suppress
any American assistance to blacks. Such suppression,
Adams knew, would offend New England's principles
and its pocketbooks. His own attack on the three-fifths
rule took Southern sensitivity about blacks far beyond
the Haiti problem. "I miss you, beyond expression,"
his father wrote to him. And "I pity you, in your
Situation. But . . . I have great Confidence in your
Success in the Service of your Country, however dark
your prospects may be at present. . . . I should think
it an honor to go with you to the Guillotine." John
Quincy's future seemed unpromising, his country's
dangerous. With a loving evocation of the dangers they
had lived through together, John reminded his son of
how, in ferocious Atlantic storms, they had "clasped
each other together . . . and braced our feet against the
Bed boards and Bedsteads to prevent us from having
our Brains dashed out against the Planks and Timbers
of the Ship."

Chapter 8
Fiery Ordeal
1805–1808

Adams arrived in Washington in October 1805 with a bad chest cold, worrying his mother and wife. He also arrived with a sense that the Senate session would have political firestorms hot enough to temper the cold weather. As an independent associated with the minority party, he participated in the debates on numbers of issues. But he was aware that the paucity of his appointments to committees, in which the Senate made its crucial decisions, reflected the indifference or mistrust in which he was held. Hypersensitive to what he considered his lack of debating skills, and particularly aware that he could not, on his feet, create a cohesive argument that contained a useful balance between the details and his main thesis, he felt awkward to the point of embarrassment, as if he could be effective only if he spoke from a written text. Otherwise the supporting details slipped his mind at exactly the point at which he needed them. Uncomfortable pauses resulted as he worked to remember what he well knew. Although he feared this limitation was a part of his personality that he could never remove, he vowed to work at improvement.

Fortunately, the issue that most preoccupied him through the winter–spring congressional session allowed him to make his case in writing. Whether or not Jefferson had read Publius Valerius' "Serious Reflections," he never could have had any doubt about Adams' views on the three-fifths provision in the Constitution, slavery in general, and Southern domination of the federal government. Although Jefferson was softer in tone than many of his Southern colleagues, he did not have to read Publius to know that Adams was the literary voice behind those who felt themselves oppressed by a provision of the Constitution that placed them under the Southern lash. The larger the slave population, the greater the political power of the South. On the one hand, Adams could participate in social pleasantries as the president's guest, as he did early in January 1805 when, at a dinner, Jefferson indulged "his itch for telling prodigies," this time the tall tale that in Paris he had seen the thermometer never rise above twenty degrees below zero Fahrenheit for sixty consecutive days. On the other, he and the president knew that the argument Adams had made in Publius, and in the constitutional amendment he and Pickering had placed before the Senate in December, was a direct attack on Jefferson's policies. In the Publius articles, Adams attacked the Massachusetts Republicans who voted for the interests of the national Republican Party ahead of those of their region: New England would pay most of the cost of the Louisiana Purchase but suffer from the shift in power that would inevitably result since the new states would be slave states.

The three-fifths rule, he reminded his congressional colleagues, had been a compromise requiring that federal direct taxes be based on representation. The South would get extra representation; the Union would get tax revenue according to the value of that extra representation, the slave property. The more slaves one owned, the more direct taxes one would pay. But the South had successfully opposed the levying of any direct federal taxes at all. So, in the end, Adams noted, "the whole revenues of the US, as far as they are a burden upon the people, are now collected from the commercial states alone," despite the intent of the constitutional provision about slavery and direct taxation. In effect, the North had swallowed both a monetary and a moral poison pill. "The President of the United States belongs to that part of the State of Virginia which, by the effect of the iniquitous mode of representation now established, sends at least two representatives to Congress, where upon principles of equal rights, they ought to send but one. His personal and local interests are, of course, in opposition to the proposed amendment. . . . If we judge of the party which now governs this Union by their acts, it will appear that their whole political system centers in personal attachment to him and his views."

Since "change is the only unchangeable characteristic of our Governments," John Quincy wrote to Thomas, it might happen that one day a president and legislators with views other than those of Jefferson and the Southern tyranny would be in power. Still, since the Republicans were better organized and the Federalists divided, the Republicans were likely to remain in power for some time to come. But one could hope. Adams agreed with Aristotle that "Man is a social animal. His passions, and imaginations," Adams remarked, "are contagious. Popularity and unpopularity are as catching as the small pox or measles. They are taken by a whole congregation from a cough in a church. They are tides that ebb and flow. They are gales of wind that blow down the stoutest oaks."

The president and his cabinet pleaded principle when they opposed amendments for political or ideological reasons. Such strict constructionists, Adams concluded, were strict only when it was in their interest to be so, and the Jefferson-Madison government did not practice what it had preached before it came to office. No one could miss the reference to Jefferson, who believed that each generation needed to rewrite or even create a new Constitution, and to the self-serving amendments that congressional Republicans regularly proposed. Publius had no hesitation in speaking in print about Jefferson as a slave owner. Our "fear of giving offense by the exercise of an indisputable right . . . is an appeal to weakness, a plea to cowardice, an argument fit only for slaves to utter and to hear. It discovers a mind prepared for every degree of submission. It is language of a negro driver on a plantation to the wretches who tremble under his lash; but it can find no accessible corner in the heart of a New England farmer." Soon and quickly, Adams' congressional colleagues, some as if brushing away a pesky fly, others as if squashing a disease-bearing mosquito, disposed of his amendment.

He was relieved to leave Washington in April 1805. Having been ill through part of the winter, at home in Quincy he continued to feel drained, feverish, without appetite, and unable to focus on the reading to which he had been looking forward. His own nature, often in conflict with messy reality, could not always accept what was beyond his control. In the Senate, he kept trying in circumstances that left a bitter taste in his mouth, though he exerted every effort to remain even-tempered, forgiving, and realistic; and to be critical of policies, not their proponents. His mother kept reminding him to stay well dressed, mind his appearance, and be kinder to himself. Was he not too stiff-tempered? she asked. As a diplomat, he had been obliged to be guarded and discreet. He had become habitually reserved, Abigail told him, and he studied too hard. The advice his mother gave him he gave to Thomas: "There is more merit in bearing up against depression, and keeping a placid and cheerful heart in ill-fortune; than in commanding all the votes in the world. In recommending these reflections to you, I am sensible they are equally necessary to myself." But he knew that the physician could not cure himself. Self-awareness was not self-cure.

To his surprise, the administration signaled that it viewed him as not only useful but valuable. His independence represented possible support. By 1805–1806, Jefferson and Madison were facing difficulties beyond their control; their own party was dividing along lines of ideology and self-interest. At dinner, Jefferson hinted to Adams that he had exactly the qualifications necessary for the position of governor of Louisiana, for which the president could find no other qualified candidate. Adams avoided any expression of interest. But could a Republican administration appoint a Federalist without offending its own base? Jefferson and Madison believed that this depended on particular circumstances. It was obvious to everyone that Adams' Senate days were limited. And Madison, who respected John Quincy's qualifications for a diplomatic post, mentioned that possibility to a close friend of the Adams family, Benjamin Rush, the Philadelphia doctor and avid Republican, John Adams' colleague at the Continental Congress. Rush told John Quincy that it was Jefferson's "wish to employ me on some mission abroad if I was desirous of it." Adams sent back word that he would seriously consider such an appointment if it were offered. When the rumor that he would be appointed minister to Great Britain made the Washington rounds, he thought it improbable. Perhaps webs were being spun. If so, he would not be pulled in.

In Quincy in April 1805, he moved with his family into the small house in which he had been born, practicing the economy to which he had committed himself after the failure of Bird, Savage, and Bird. He was "entering upon a new mode of life . . . dictated by necessity." Its discomforts, he hoped, would accustom his family to live in a way compatible with his reduced income. When his Senate term was up, he would return to Boston. He would have his plants, his trees, his orchard, his parents, and Thomas, who had just married, close by. "In these measures, my intention is to do all in my power, for the preservation of my family. The issue must be beyond my control. May it be auspicious!" But he still had three years of his term remaining.

And he had another prospect. In late June, Josiah Quincy called on him. He "informed me that the Corporation of Harvard University had the day before yesterday elected me Professor of Oratory," the Boylston professorship established by a grant from Nicholas Boylston, a first cousin of his father's grandmother. The subject had been mentioned to him several times before, "but I yet know not whether it will be proper or expedient for me to accept this situation." The Boston and Harvard world was still a kinship society, and the Boylston money spoke loudly in such deliberations. The bequest stipulated that Adams be offered the position. His age, reputation as an orator, and government service were in his favor. The Harvard establishment thought well of him, and he was known as an assiduous scholar and writer. It was not, though, a full-time position, and the salary would be only a supplement to his other income. Still, he could not occupy his Senate seat in Washington and at the same time teach during the full academic year. Also, he objected to the religious oath that Harvard required. A nonsectarian avowal of Christian belief, he insisted, should be sufficient. Negotiations continued into the fall. In early November 1805, an agreement was reached. He would deliver an inaugural lecture the next June, the start of a seven-year commitment. While a senator, he would teach from the spring to the late fall. From November to April, he would be in Washington.

In New York, his sister Nabby's life had become even more distressing. Who could have predicted in the 1790s how vainglorious and financially irresponsible William Stephens Smith would turn out to be? His brothers, Justus and James, irregularly repaid William's debt in response to pressure, even the implicit threat of legal action, from John Quincy. No more than barely holding his own financially, he resented having to spend time writing letters to get small sums in dribs and drabs. Smith did one foolish thing after another. He deluded himself into believing that the highest levels of government had given him the go-ahead to help General Francisco de Miranda, a Venezuelan patriot fighting for independence from Spain, secretly purchase, arm, and man a ship in New York Harbor. Miranda would then take to sea against Spain, an enemy of the United States and of South American independence. Smith hoped to sail with Miranda. But federal law forbade arming a foreign vessel in an American port. The United States, which coveted Florida and even Cuba, was engaged in what turned out to be a decade-long negotiation with Spain that numbers of times threatened to collapse into a war the government did not want. And it certainly did not want individual citizens, whatever their motives, determining foreign policy.

Miranda had urged Jefferson and Madison to help his cause. They had given him a polite but cautious reception. Smith was influenced by Miranda's misleading report of these Washington interviews, and his enthusiasm overcame his prudence. "If you were induced to believe it was the President's intention to countenance any purpose of hostility to Spain," John Quincy later wrote to Smith, "I very much regret both the fact and its consequences. That he had no such intention is obvious from the course of policy which has prevailed, and has always appeared evident to me." What Smith had done was illegal. Even worse for Nabby, when Miranda sailed from New York in February 1806, her eldest son, William Steuben

Smith, was aboard, with his father's permission, as an American volunteer to fight against Spain. His nephew, John Quincy lamented, had "a very sorry prospect before him. . . . Bred to nothing—possessed of nothing—Having Nothing to expect. This quixotic adventure at the outset" will "make him fit for nothing upon this Earth." In May, on his way to Boston, John Quincy visited Smith in debtors' prison. "The Colonel," he noted, "keeps up his Spirits, with apparent cheerfulness; and still seems to flatter himself with prospects, which to my eyes have not so much rational foundation as a fairy tale." In the spring, the Jefferson administration removed Smith as surveyor of the port of New York. John Quincy urged his sister to spend at least the summer in Quincy. But her Adams sense of duty required that she stick by her husband, who was awaiting trial. In Quincy that summer Nabby's fate weighed heavily on the hearts of all the members of the Adams family.

Awakened by loud noise and bustling activity at the home of Walter and Nancy Johnson Hellen in Georgetown, where he and Louisa were living, John Quincy asked if anyone had been ill. Betsey, the maidservant, he was told. He was surprised to learn that up until that night there had been two pregnant women in the house. Louisa had become pregnant, probably in October 1805, before leaving Quincy for Washington. Betsey had just given birth, apparently concealing her condition until the last moment. Such pregnancies had the potential to be, at a minimum, domestic disturbances, at worst, Calvinistic night-mares. This melodrama turned into a comedy with a happy ending, well-managed by Louisa's elder sister and her husband. John Quincy was amused rather than moralistic, and they were all used to marriages after pregnancies or childbirth, especially in the servant class. "As the adventure was followed in the course of a few days by a wedding and a christening," he wrote to Thomas, "all improper breaches were repaired, and the Lady being again an honest woman must be respected in her reputation." Plain English rhyme, he told Thomas, rather than serious Latin or Greek, was appropriate for the event. He enclosed a comic poem, "The Misfortune":

> Poor Betsey was a maiden pure
> Declined in years, but so demure
>   That Man was her aversion;
> And night by night her door she barred
> With treble bolts, her fame to guard
>   From Slander's foul aspersion.

When lo! all in the dead of Night
Came Mary, breathless with affright,
                    Wringing her hands and crying
"Oh! Mistress! Mistress! Rouse! Awake!
To Betsey come, for Heaven's sweet sake!
                    Poor Betsey!—She's a dying!"
The Lady, tender and humane,
Starts from her bed, and flies amain,
                    The wonder to unravel.—
Flies to where Betsey lays and moans
And straight perceives what caused her groans
                    —Poor Betsey!—was—in travel!!

Since his election to the Senate, John Quincy had found little time for writing poetry. In Washington, and especially during his months in Quincy, he assigned himself heavy reading, particularly in American history and official documents. Before 1803, he had had a burst of more literary reading, including Michel de Montaigne's essays, Edward Gibbon's *Memoirs*, and Blaise Pascal's *Provincial Letters*, and he wrote a biographical essay about Pascal for the *Port Folio*. He would have liked, in his thoughts about an alternative life, to be solely literary, devoting himself to filling in the deep chasm of American ignorance of foreign literature. Even if it were more than a fantasy, it was obviously an impossibility. To improve his Greek, he read chapters in the Greek New Testament. As both a moral and a literary commitment, he devoted an hour on first rising to what became a lifelong effort to read continuously through the Bible, starting again once he had finished. He also kept up his French and German. Occasionally, for amusement, he read a novel. He began reading through the works of the Roman playwright Plautus and became increasingly engaged with the Roman satirist Juvenal. Much as he loved Cicero, it was Juvenal who now preoccupied him. Juvenal's satires became the focus of Adams' fascination with Roman history, his desire to improve his Latin, and his need to express himself in poetry. The Roman poet had written sixteen verse satires, averaging two hundred lines each in unrhymed dactylic hexameter, the same meter as Greek and Latin epic poetry. Between 1800 and 1805, Adams translated three or four of them in full and portions of others, and he read professional translations, the best-known published in 1800 by the English satirist William Gifford. Although Adams flirted with undertaking a complete translation, he soon decided that existing translations were superior to his efforts.

In 1800 in Berlin, he had translated Satire XIII for his own amusement and sent it off to Thomas, who saw to its publication in the *Port Folio.* When he tried

VII, he found it hard to understand, let alone translate. Much of Juvenal's urban grittiness and sexual frankness were beyond John Quincy, or at least too distasteful and colloquially Latinate for easy translation. He gave up, then tried again in 1803. The challenge gave him great intellectual and aesthetic pleasure. It engaged and expanded his capacity for language and art. Like Gifford, he translated Juvenal's six-beat hexameter lines into rhymed couplets, the traditional five-beat line of English poetry. Dryden and Pope, the two eminent poet-translators of the seventeenth and eighteenth centuries, looked over his shoulder, and his translations have merit in the eighteenth-century tradition of transforming the past into a version of the present. They deal with substantive essence, not literal reality. And, like other contemporary translators, Adams viewed Juvenal mainly as a moralist, though of a different kind from the one he actually was. Adams thins out much of Juvenal's wit, cynicism, anger, coarseness, prejudice, and sexual frankness. His closed couplets provide some of the slash but none of the bitter humor of the Roman poet's attacks on corruption. And his Juvenal is a moral gradualist with a belief in human redemption, a pagan with a Christian sense of providence and ultimate justice, embodying Adams' own moral and aesthetic values. The daily life of a raunchy and nasty Rome, with Juvenal himself one of the participants, is elided into an overview much like Pope's *Essay on Man.* Juvenal's harsh and caustic irony disappears. So too does his pagan mind.

When he accepted the Boylston Professorship, Adams knew he would have even less time for poetry. But rhetoric and oratory had always interested him. Now their theory and practice would have to become a preoccupation. Since there had been no such course before at Harvard, he would have to invent it from scratch. That meant intensive reading in the literature from classical times to the present. While he had always been a regular attendee at sermons, he now would have to study sermons as object lessons, with a critical and analytic attention that he had only irregularly applied previously. He began to make detailed analyses in his diary of the sermons he heard in Quincy, Boston, and Washington. His lectures would themselves be public performances to be judged by the very standards the lectures proposed.

In the summer of 1805, he devoted most of his reading hours to classical writers, from Demosthenes and Aristotle to Cicero and Quintilian; then to modern writers from the Renaissance to the late-eighteenth-century theoreticians and the best-known modern practitioners in politics and the church. He read two

well-known Anglican clergymen, the late-seven-
teenth-century archbishop John Tillotson and the
eighteenth-century churchman and novelist Laurence
Sterne, whom Adams admired for their writing skills
and emphasis on Christian ethics rather than theology.
Speeches and sermons existed as a genre of literature
that at its best, he believed, deserved a place in the literary canon. During the next three years, he created
thirty-six lectures, as if he had undertaken a sacred
trust and his self-worth depended upon its fulfillment.
He was determined to give Harvard his best effort.
Rarely more than a lecture or two ahead of his teaching
schedule, he often worked at the next lecture right to
the day of class. From the start, he had two connected
questions in mind: in what way was oratory useful for
good rather than evil, and why was it especially crucial to a republican government run on democratic
principles?

With his teaching scheduled to start in July 1806,
exactly on the day that began his fortieth year, he felt
compelled to go to Quincy, where he would have the best
conditions for writing. Louisa, about six months pregnant, would remain in Washington, where she would
have the assistance of her family, give birth in her sister's
house, and await his return in the late fall. Until April
it had been a comparatively easy pregnancy, with only

brief moments of illness. But in late April, she became "extremely ill" with "a "violent head-ache," her "health . . . so precarious that I cannot leave her without feeling great concern. But I am compelled to go, and it is impossible for her to accompany me." John Quincy believed he had no choice. From the start of the separation, Louisa missed him terribly. "The loss of Mr. Adams's society is to me irreparable," she wrote to Abigail. "His tenderness and affection . . . throughout this winter have been inexpressible and I have enjoyed almost perfect happiness. I already look forward to his return with the most anxious impatience and should it please heaven to spare me and my infant [I] anticipate with delight the moment when I shall present it to receive a father's blessing." For much of May, Louisa seemed better, except for the swelling of her feet. She could not walk.

Early in June, Louisa again became very ill. She was still able, though, to keep up her end of their correspondence, and she soon seemed better. From May to July Adams worked at writing lectures, lodging in Cambridge with Benjamin Waterhouse. By late July, he had written the seven lectures he needed to carry him through commencement in August and to give himself a head start for the new term in October. On June 12, Adams was inaugurated as the first Boylston Professor. A few days later in Boston, he watched a total eclipse of the sun, fascinated as always by astronomical phenomena. On June 9, Louisa had written him a letter, which probably arrived about June 20: "I have continued very ill ever since I wrote you and am still unable to leave my chamber. More than half distracted" with pain, she had "small abscesses" at the back of her throat. They "break and form continually." Her legs and thumbs were so swollen that "it is with difficulty I can hold my pen. . . . I shall not be a very regular correspondent until I have got through my difficulties." She got through them on June 22, unhappily. On the last day of the month, William Shaw, searching for John Quincy, found him at the Suffolk Insurance office. The letter Shaw gave him was "a message of Misfortune," he wrote in his diary, unself-consciously echoing the title of his comic poem about "poor Betsey." After twenty hours of difficult labor, their son had been born dead. For the moment he did not know what he was feeling. He walked in the rain to Cambridge. Quietly, alone in his room, he cried.

As soon as he could control himself enough to hold a pen, he wrote to Louisa. Seeking consolation in resignation, he hoped to emulate her example. Her letter, he wrote, had affected him "deeply by its tenderness, its resignation, and its fortitude." It was "the most excellent though the most painful letter I ever received from you." Why, he asked himself in his diary, had he allowed himself in the past to express "too bitterly the pangs of such disappointments." After all, "Heaven has compensated me for all those sufferings by my two boys, who promise all that a parent's heart can wish from children of their age. I had given up my Heart to Hope, and Joy in the Hope of a third. It is gone. Let the rigor of the stroke help to purify my soul by affliction, and may my never ceasing gratitude flow for the blessings which remain to me by the bounty of Providence," especially that his wife was still alive. If his cup did not overflow, it was at least half full. His affirmation of the bounty of providence was more than formulaic. It was emotionally sincere and consolingly ritualistic. By late July, Louisa felt well enough to travel. "My anxiety to see you and my children is almost insupportable. I cannot feel one moment of happiness until I again behold and clasp you to the heart." In mid-August, she arrived with her sister Caroline. They moved—with the children, who were recalled from their great-aunt and grandmother—into the Old House at Quincy. The long separation was over. Three months later, however, John Quincy was on his way back to Washington, alone.

In much of the three years from his inaugural address in the summer of 1806 to his final lecture in July 1809, Adams gave half his time to the challenge of his teaching duties and half to his political career. Actually, he had no sense of the latter being a career. There was no chance that the Massachusetts legislature would re-elect him. In Washington, much as he was an insider by heritage and connections, he was a Federalist outsider who dissatisfied his own party. He felt himself a man in transit, probably back to the legal profession, perhaps to continue indefinitely as the Boylston Professor. He learned to be a teacher as he taught. Gradually he got over feeling "very awkward in the business." His own skill as a writer in his inaugural address and the lectures is unmistakable. His spoken voice, like those of his contemporaries, can only be guessed at. With the formal humility of an oratorical tactician, he confessed his lack of preparation for the task, and identified himself as a novice teacher who had been only a casual student of the subject.

But he would learn along with his students. Oratory, he would demonstrate, was not a frivolous ornament. And it was useful much beyond success in the professions in which public speech is essential. It was connected to the soul of the nation. It had a special role to play in the United States. Technical skills were important—a central part of the course activity, both in student declamations and in his lectures. But to be highly skilled was not enough. The utility of oratory, he argued, was inseparable from the moral values of the orator. And the mission of the orator was to use the art of persuasion for moral ends, whether in the pulpit, courthouse, or legislature. To be of value to society, the orator had to have total respect for the integrity of language and a message grounded in the highest values of the Judeo-Christian tradition. Service to others was at oratory's core, and the grossest disservice in oratory resulted from the misuse and corruption of language for self-serving ends.

What had happened to oratory since its glory days in Rome and Athens, Adams asked, "when eloquence produced a powerful effect, not only upon the minds of the hearers, but upon the issue of the deliberation?" Oratory could flourish, he argued, only in a free society that welcomed debate. Its strength depended on the service it provided to the pursuit of justice. It had not thrived and could not thrive in the European world of arbitrary power. The United States, in contrast, provided fertile ground for oratory. Effective public discourse, he told his students, was essential for the success of democracy. "Consecrate, above all, the faculties of your life to the cause of truth, of freedom, and of humanity. . . . Under governments purely republican, where every citizen has a deep interest in the affairs of the nation, and . . . has the means and opportunity of delivering his opinions, and of communicating his sentiments by speech; where government itself has no arms but those of persuasion; where prejudice has not acquired an uncontrolled ascendency, and faction is yet confined within the barriers of peace; the voice of eloquence will not be heard in vain."

In Washington, as he settled in for the winter of 1806–1807, it was clear to him that the values inherent in the kind of oratory he commended to his students were absent, with minor exceptions, in the halls of Congress. These values did not exist in the political culture. Oratory could indeed be heard in vain. From the start of Washington's second term, American public discourse had become debased by linguistic distortion and dishonesty. The glory years of the revolution and the first years of the republic had been tarnished by brutal infighting, the lust for power and patronage, the disdain for compromise, the passionate defense of slavery, the expansion of the nation in a way that made New England feel victimized, and a coarsening of public life in which the uneducated, even the illiterate, had power much beyond their ability to use it wisely. The elite leadership sustained a winner-take-all power game, the object of which was to wipe out the other party. Politicians were motivated by ambition for their own rather than for the country's success.

In April 1805, leaving Washington, burdened with
this disillusionment, Adams had found himself traveling in the same coach with Aaron Burr. Adams had
been present at "the last act of his political life," Burr's
farewell speech to the Senate the previous month, in
which he had urged the Congress to honor and exemplify "dignity, order, and morality." Adams had been
impressed by his eloquence, modesty, and sound sense.
But was he sincere? Was it not possible that there were
two Burrs: the opportunist who had murdered Hamilton
and the modest speaker of the farewell address offering
good counsel to the nation? Burr seemed a man who
could play whatever role was advantageous to him at
the moment, revealing a pervasive insincerity, a lack
of honesty and moral consistency that typified the use
of oratory for evil ends. Adams had kept his distance
from Burr during the congressional session, outraged
that a man charged with murder had the gall to act as
president of the Senate. Here was a character, Adams
had concluded, who embodied talents and vices that
exemplified the art of persuasion in the service of evil.
"But now, when he was a private citizen again and in
circumstances of obvious difficulty, struggling with
misfortune and maintaining a deportment superior to
it, I had not strength of mind enough to retain in their
full inflexibility the resentments." As they traveled

and talked, Burr was "the most amiable, attentive, and complacent of travelling companions." To his surprise, Adams "felt a degree of compassion for the man, which was almost ready to turn to respect." He was easy to disapprove of but harder to hate, a testament to the complexity of human nature.

In late 1806 Burr was back in the news. When Adams returned to Washington, the rumor that the former vice president had instigated an armed rebellion—though to what extent and with what intent was unclear—alarmed the city. What was known was that, after purchasing land from the Spanish government, Burr had raised a small number of men in Ohio and Kentucky, purchased flatboats for transport down the Mississippi, and suborned some officials and private individuals to join him for the purpose of occupying Spanish territory in the Southwest. His intent might have been to form a government under his own authority. His initial destination seemed to be New Orleans, where the U.S. military commander and governor, James Wilkinson, himself secretly in the pay of the Spanish, had been playing a double game, encouraging and betraying Burr. Wilkinson had alerted the Jefferson administration. But whether Burr had gone west to settle, as he later claimed, or to conquer needed to be clarified. As rumor and fact slowly made their way to Washington, the Jefferson administration became frightened and furious. The president personally detested Burr, whose attempt to deprive Jefferson of the presidency could not be forgiven. Eager to have Burr convicted, even before all the facts were in and even when the facts seemed damaging but inconclusive, the president publicly accused him of treason. Burr's activities, whether treasonous or not, seemed likely to have international consequences. The last thing the administration wanted was war with Spain. The country was not prepared for war with anyone, even the weakest of European powers.

But Congress, the president, and public opinion outside New England wanted Florida. The administration had requested in early 1806 that the House, in secret session, allocate $2 million for its purchase. Congress complied. The hope was to repeat the success of the Louisiana Purchase. If Florida could not be bought, many Americans, especially in the South and West, urged military conquest. By early 1807 it appeared unlikely that Spain would ever sell Florida. But Spain, Adams recognized, was not a threat to the United States. On the contrary, Spain had to worry that the United States would take the Floridas by force. Nonetheless, in the view of the American government and of public opinion, Spain was entirely at fault. Behind all the public bluster, however, Americans understood that the real threat came from Great Britain and France, not that it was believed that either would attack the United States. Still, America's reaction to the confiscation of its merchant ships and the impressment of alleged British subjects might escalate into a war that neither country desired. At the start of 1806, as the Jefferson administration began its attempt to pressure Britain and France by trade restrictions, war was exactly what Adams feared. "The hurricane of animosity and of hostility against Spain," he predicted, "will gradually subside and die away, until scarce a distant murmur of it will be heard; while the moderate breeze which first blew against Britain will freshen to a gale, until the tempest will try the strength of our timbers to a degree which I cannot look forward to without concern." He had no doubt that resistance was necessary. To sidestep it would "be equivalent to a total abandonment of all rights of neutral commerce. I feel no disposition for my own part to abandon those rights; and I believe by a firm and determined opposition to her pretensions . . . we can bring her to relinquish them." He was to learn that he was wrong, though from the start of the embargo policies he had little faith they would work. And he doubted that the American government had the skill, flexibility, and capacity to manage trade restrictions successfully.

No American, not even Adams, believed that Great Britain and France had benign intentions. Either of the two major powers would have been delighted to coerce the United States into dependency or, in the case of France, total subjection. Britain, though, was gradually limiting its expectation of North American territorial dominance to Canada only. Consequently, it had a clear path to reasonably good commercial relations with the United States once it had favorably concluded its European war. In the meantime, Britannia ruled the waves. Its concern was the United States' helping Napoléon to rule the European mainland and conquer Britain. From 1806 to 1809, Jefferson, Madison, and the Republican Congress instituted a series of nonimportation acts and embargoes, the point of which was to show Britain that it had such great need of American raw materials, and Americans so little need of British manufactured goods, that Britain would have to renounce impressment and recognize the right of neutral trade. From the start, Adams, like his father, believed that the medicine, if continuously applied, would be worse than the disease. For New England "to have committed the command of [its] Commerce to Virginians was to commit the Lamb to the Guardianship of the Wolf." The business of New England was business in ships, stores, offices, and banks. The business of the

South was agriculture. When England and France confiscated American ships, they hurt New England most. When the United States closed American ports, it hurt New England even more. That would plunge the national economy into depression and motivate otherwise law-abiding citizens to find ways to break the law in order to survive. Jefferson and Madison, blindly, with total conviction, insisted that their policies would work. Between 1806 and 1809, a series of nonimportation acts and total embargoes were passed by a Republican-dominated Congress. The government hoped that the trade restrictions would last only a short time; Jefferson predicted rapid British capitulation. When he was proved wrong, he urged continuous doses of the strongest medicine possible.

Like most Federalists and some Republicans, Adams thought it likely that only time would resolve the conflict. After all, Britain and France could not keep fighting indefinitely. But for how long and at what cost to America could the United States delay taking action? And since America had refused to arm itself, what actions could it take? If necessary, it could impose a total embargo, Jefferson and Madison argued, which should bring, they claimed, one or both of the perpetrators to the table. To Adams, his own issue was simple: should he vote for acts that had some chance, no matter how small, of success, or vote against them when the only alternative was the view that patience was America's best weapon? And was he not committed to vote for measures on their merits rather than their party origins? But if he voted in favor of the embargo, he would be in even deeper trouble than he already was with his own Federalist constituency.

In April 1806, he became the only Federalist to vote in favor of the first nonimportation act. He then voted for the series of restrictive measures that followed. At the same time, since he believed they were unlikely to succeed, he urged that the country arm itself. Neither Britain nor France would take a toothless America seriously. In December 1807, he was the only Federalist to vote for the most extreme form of trade restriction, a total embargo on imports from and exports to Britain and France and a total embargo on trade with any other country in order to prevent the movement of goods through a third party. To many New England Federalists, Adams was a traitor to his class, his state, and his region.

It was a painful session for Adams. His political situation was untenable. Worse, he felt inadequate to the task, with "a distressing consciousness of my own weakness of capacity, together with a profound and anxious wish for more powerful means. I lament the want of genius, because I want [to be] a mighty agent for the service of my country." In the secrecy of his diary he pledged himself to compensate for his lack of influence and oratorical talent with honest forthrightness, though he well knew that it was a weak currency in political transactions. When New England investors requested that he introduce petitions on their behalf to compensate them for the cancellation by the Georgia legislature of their land claims in the Yazoo River area of the Mississippi territory, he felt his personal voice would be ineffective and the petitions unsuccessful. He had nothing to add, he felt, to the debate on eliminating the importation of slaves. A senator from Kentucky ten years his junior, newly appointed to fill the remainder of a vacated term, impressed Adams when he spoke against the importation of slaves. "Mr. Clay the new Member from Kentucky made an ardent Speech. . . . He is quite a young man, an orator, and a Republican of the first fire." When Adams introduced a proposal for the federal government to subsidize the creation of canals and roads and to dredge harbors, the Senate voted it down with almost no discussion. Although there were constituencies in both parties for federal funding for infrastructure, the Republicans mostly held back. If the Constitution could be interpreted to allow even partial federal funding, would this not strengthen the central government at the expense of the states? And the money would have to come from either taxation or borrowing, both of which the Republicans opposed. Still, it disappointed Adams, who strongly believed that roads, canals, and harbors were essential to the future prosperity of the country and the promotion of national unity. Nothing in the Constitution, as he read it, prohibited such activity. It was a vision for America's future that Adams was never to relinquish.

Although he had the company of the Johnson and Hellen families in the winter of 1807–1808, he missed Louisa and his sons. They were constantly on his mind. The death of the Hellens' two-year-old-son intensified Adams' tendency to worry about his own family. Was George forgetting all his French? How was John doing at school? Would the end of the congressional session drag on beyond March? Louisa urged him to be especially attentive to her grieving sister. She thought he too often gave the appearance of being insensitive to other people's feelings, especially women's. "For your sake and for that of your Sisters," he responded, "I have often wished that I had been that man of elegant and accomplished manners, who can recommend himself to the regard of others, by *little attentions*. I have always known however that I was not, and have been sensible that I could never be made that man." But, though he thought it unlikely, he hoped that at least those who knew him well would not read his heart in his face. He had no shortage of invitations to the usual Washington events. One of the "little attentions" he paid to his wife's family was to escort them to dinners and parties that he might otherwise have happily skipped. To make amends to his sister-in-law, he presented her with a consolatory poem in the sentimental tone and language that assured her that her innocent infant had ascended to heaven. Its theological pieties were sincere and uninspired, a textbook expression of Christian consolation. Whatever good it did the grieving mother, as poetry it had to make do with the merit of sincerity and good intentions.

In mid-January 1807 he wrote a poem of a very different sort, "Lines, To Miss In Full Un-dress at a Ball." It had verve and humor and, he explained to Louisa, tongue in cheek, a moral purpose. It was written "in consequence of the writer's meeting in company a young lady rather more than usually undressed," displaying a low neckline in the French fashion. He assured Louisa that he had written the poem for her eyes alone, though, with logical inconsistency, also to excite "Ladies to *Reflection*" about "the effects which their exhibitions have a tendency to produce upon *Sensation*." Louisa herself had no need of the lesson.

It was obvious that, despite the moral gloss, he had enjoyed the sensation and that the poem was meant to share with her, despite her absence and in anticipation of their reunion, an expression of long-distance love-making. It contained, she responded with delight, "the sauciest lines I ever perused. I have a great inclination to have them published in the next Anthology," she teased, "for the sake of my sex in general to whom I hope they would prove a serious advantage." The advantage clearly was entirely to the two of them.

> *When first, in Eden's flowery dell*
> *Our Grandam Eve was framed,*
> *She was, as holy legends tell,*
> *Though naked, not ashamed.*
> *But when the Serpent's subtle head*
> *Had brought her to disgrace;*
> *When Innocence and Bliss were fled—*
> *The fig-leaf took their place.*
> *Still the more guilty she became,*
> *She shewed herself the less;*
> *And thenceforth nakedness was shame,*
> *And Innocence was—Dress.*
> *But soon, her daughters shall again*
> *Like Eve, before the fall,*
> *Concealment of themselves disdain*

And stand displayed to all. . . .
Already, Sally now reveals
To view, Neck, Arms and Breast;
While a bare spider's web conceals,
And scarce conceals the rest.
Dear Sally! let thy heart be kind—
Discover all thy charms—
Fling the last fig-leaf to the wind,
And snatch me to thy arms!

"God bless you my best beloved," she responded, "for my heart throbs to behold you. . . . Mon Ami, I count the hours till you return."

As her thirty-second birthday approached, he was again in a poetic mood. Echoing in the title one of his favorite Shakespeare plays, he composed fourteen eight-line stanzas to celebrate "Louisa's Birth-day 12. Feby: 1807. A Winter's Tale to Louisa." It was an everyday tale, an account of his daily life, poetry as an alternative to diary, a love letter in verse, and an affirmation of his emotional attachment.

1. Friend of my bosom! wouldst thou know,
How, far from thee, the days I spend?
And how the passing moments flow?
To this short, simple tale attend—

When first emerging from the East
The Sun-beam flashes on my curtain,
I start, from slumbers ties released
And make the weather's temper certain.

2. Next, on the closet's shelf I seek
My pocket Homer, and compel
The man of many wiles, in Greek,
Again his fabled woes to tell. . . .

4. Then forth I sally for the day;
And musing Politicks or Rhyme,
Take to the Capitol my way,
To join in colloquy sublime. . . .

9. As Eve approaches, I ascend,
And hours of Solitude ensue—
To public papers I attend,
Or write, my bosom's friend, to you.
Gaze at the fire, with vacant stare,
Suspended pen, and brow contracted;
Or, starting, sudden from my chair,
The chamber pace, like one distracted.

10. I see the partner of my Soul,
I hear my darling children play;

Before me, fairy visions roll
And steal me from myself away.
Not long the dear delusions last—
Not long these lovely forms surround me—
Recovered eyes too soon I cast,
And all is Solitude around me. . . .

14. *Thus, in succession pass my days,*
    While Time with flagging pinion flies;
    And still the promised hour delays,
    When thou shalt once more charm my eyes.
    Louisa! thus remote from thee
    Still something to each Joy is wanting;
    While thy affection can, to me
    Make the most dreary Scene enchanting.

As Adams traveled from Washington to Boston in early March 1807, he was relieved to have the congressional session behind him. He had in mind the lectures he still needed to prepare, the situation of his sister in New York, and the pleasure of imminent reunion with Louisa, his children, his parents, his brother Thomas Boylston, and his Boston friends. John and Abigail were reasonably well. Thomas, who had married in 1805, now had a one-year-old daughter. The previous autumn, a few weeks before leaving for Washington,

John Quincy had taken five-year-old George on a tour of his Quincy farm. His object was "to familiarize him at this season of his life with the scenes upon which my own earliest recollection dwells." First impressions are the most lasting, he believed, and he wanted his eldest son to have the same emotional allegiance as he had to this landscape and community. In New York, he had persuaded Nabby, with her two younger children, John and Caroline, to travel with him to Quincy to stay at least the summer, as William Stephens Smith struggled with his financial difficulties, including the possibility of resettling in western New York state. The notion of Nabby situated in a frontier existence horrified the Adams family.

What John Quincy apparently did not know was that Louisa was four months pregnant. It seems likely that she kept it from him to prevent his worrying about her health, concerned that this might prove another disappointment. In mid-August, she gave birth to a healthy son. Although her labor was severe, it was short. "He is born to be lucky as you say," John Quincy wrote to his mother-in-law. At first, the baby had shown no signs of life. Louisa thought he had been born dead. "In about five minutes however, while preparation was making to set its lungs in motion, it commenced respiration of itself, and very soon appeared to be in full life." He was baptized in mid-September by William Emerson, whose four-year-old son Ralph Waldo was probably at his father's church that day. The infant was named "Charles Francis . . . in remembrance of my deceased brother, and . . . as a token of honour to my old friend and patron Judge Dana." Louisa, not well enough to attend, waited outside in her carriage. Charles Francis was indeed to be their lucky child.

When he managed to sell the house he owned in Half-Court Square, Adams moved into one of the two newly built adjoining buildings at the corner of Nassau Street and Frog Lane. He also completed the purchase of a house on Hancock Street, another rental property. He had some law business, including serving on a committee with Harrison Gray Otis, with whom he was on reasonable terms. "Though systematically a man of great politeness," he "is quick, and hasty in temper, and at such times very irritating in his manner. Without his accomplishment, I am afflicted with his infirmity, and we catch like fire and tinder." Adams dined with friends, including Daniel Sargent, and spent time with John Hall; James Bridge, who visited from Maine; and especially John Gardner, who remained his loyal friend at a time when most of Federalist Boston was frowning on him. Boston had many familiar faces, including Adams' law mentor, Theophilus Parsons, whom he visited on the day Parsons was appointed to replace Dana. But John Quincy had days on which he was so exhausted that he despaired of getting serious work done. Grateful for commencement and the August holiday, he stumbled on. Despite periods in which he felt too tired or distracted to write, and days on which he felt he had given dull performances, by the time of his departure for Washington he had written and delivered through lecture twenty-six, a total of fourteen from April to October 1807. He had completed more than half the course, as he conceived it.

He and the country, though, had no way to determine when they might be done with the subject from which there seemed no viable escape: British and French mistreatment of American ships and sailors. An explosive incident further exacerbated the situation. News reached Boston at the end of June that a 40-gun wooden frigate, the USS *Chesapeake*, had been fired on and boarded off Norfolk, Virginia, by the 50-gun British warship *Leopard*. When the *Leopard*, at sea, requested to board the *Chesapeake*, just starting a long voyage, the American commander refused. Since the British Navy had previously restricted impressment to American merchant vessels, the captain had no reason to expect he would be attacked, though the judgment that led to his conviction at a court-martial applied the traditional standard that a naval vessel should always be ready to defend itself. The *Chesapeake*'s gear and provisions were on deck; its guns were not in firing position. Almost defenseless, it was heavily damaged and boarded; four sailors were removed and eighteen wounded, including the captain. Three of those removed turned out to be British deserters. American public opinion exploded in outrage. At first, the Federalist elite in Boston thought silence the best response. Great Britain was a danger but not an enemy (ideologically); France was an enemy (ideologically) but not a danger. While the Republicans organized a meeting of protest and solidarity, the Federalists hesitated. Adams, with no Federalist alternative, attended the pro-administration meeting at the statehouse. Although he declined to act as moderator, he accepted appointment to a committee to draft a protest resolution, which was unanimously approved. The next day, his fortieth birthday, he was told by a friend that "I should *have my head taken off*, for apostasy, by the federalists. I have indeed expected to displease them, but could not help it. My sense of duty shall never yield to the pleasure of a party." Adams had been an unreliable Federalist. Now he was an apostate. The rumor spread that he had changed parties. The next week, belatedly, the Federalist leadership sponsored a town meeting to draft its own protest resolution, which Adams helped to compose. Republicans and

Federalists were now on the same page, and, for the time being, Adams was aligned with both.

In October 1807, after he had walked out to Cambridge to deliver his twenty-sixth lecture, John Quincy and Louisa, with two-month-old Charles Francis, left for Washington. George was sent to boarding school with Abigail's sister Elizabeth and her second husband, Stephen Peabody. John was left again with Mary Cranch. In Washington, the Senate session started slowly. Early in November, John Quincy dined at the President's House with a large group of congressmen. A few weeks later, at a smaller dinner, the president noted to the British minister, who was being replaced after a negotiation fiasco, that he hoped "in the meantime your nation will make peace, and leave us nothing to dispute about." That confirmed Adams' view that procrastination "includes the whole compass of Mr. Jefferson's policy." The conversation at both dinners was lively, including much interest in Robert Fulton's newly invented steamboat and torpedoes. The entrepreneurial Fulton was seeking government support for his projects and was "very anxious to make an experiment of his torpedoes before both Houses of Congress. . . . On the whole it was one of the agreeable dinners I have had at Mr. Jefferson's." The next week he dined at the President's House with most of the cabinet officers, including Vice President George Clinton and the secretary of the treasury, Albert Gallatin. Adams was to develop over the years a cordial and admiring relationship with Gallatin. Still, despite the inevitable social and professional engagements, he had associates but few friends. Uriah Tracy had died. William Plumer had not run for re-election. Adams took bemused delight in the inventive ideas of William Thornton, the head of the Patent Office, who had designed the Capitol building and was eagerly promoting a scheme for "universal education." They met frequently, and years later were to be neighbors and friends.

Much as he did not want war, Adams found the evasion of war to the point of national humiliation abhorrent, including the administration's response to the *Chesapeake* attack. It seemed clear that the president and Congress had no intention, despite bluster to the contrary, to prepare for war, though it had previously seemed to Adams "that this country cannot escape a war." He now saw that he had been wrong. There is "an obvious strong disposition to yield all that Great-Britain may require to preserve peace, under a thin external show of dignity and bravery. . . . The national representative pulse never has beaten so slow as it does at this time. The Cotton of the South is a great peacemaker. . . . Every man seems to tremble least he should do something rash." Congress, whose temperament and principles would not permit it to legislate any taxes besides a modest tariff, would not fund a navy, even with European guns at its head. At the end of 1807, Adams added his voice and vote to the repeal of the nonimportation act, which was a selective embargo. "The experiment has been made. Whatever good effect it could produce it seems to me is past. It is too much, or too little, for the present state of things; and its effects will distress our own people, without producing the effect for which it was intended." Britain clarified its policy: it would not impress British deserters on American warships. But merchant ships were fair game. Locked in mortal struggle with Napoléon, it would rather suffer the tensions created by impressment and the confiscation of neutral shipping than sacrifice valuable manpower or allow life-sustaining materials to reach French ports. It was a version of total war. And Great Britain had no hesitation in repeating, with additional emphasis, that a British national was a British national forever.

At the end of 1807, at the administration's request, Congress passed a total embargo bill, which Adams reluctantly supported. It could be quickly revoked when conditions changed. He acceded to the argument that an embargo would save American ships by ordering them home, would help in negotiations with the new British minister, and might pressure Britain and France to change their policies. Both were confiscating American merchant ships in large numbers. Britain claimed it had the right to stop any trade by a neutral power between French colonies and France that had not been allowed by France at the time of the Seven Years' War. France retaliated by claiming that the ships of any country that accepted the British claim or traded with Britain were subject to capture and confiscation. In effect, for both countries, there were no neutral ships and there was no freedom of the seas. At the same time, fortunes were being made. Each country allowed trade with the other when it had an interest to serve. Profiteering was widespread—and some Americans profited handsomely.

But the American economy also suffered grievously. Every aspect of commerce was affected. "I do not believe indeed that the Embargo can long be continued," Adams wrote to his father. The administration denounced those who opposed the embargo as traitors, those who defied it as criminals. Still, public resistance was so great, and opportunities for smuggling were so many, that the administration could not enforce its edicts. When out-of-work American sailors rioted in the streets of Boston, John Adams pointed out to his

404 · JOHN QUINCY ADAMS

son what his son well knew: "You may as well drive hoops of wood or iron on a barrel of gunpowder to prevent its explosion when a red hot heater is in the center of it as pretend to enforce an embargo on this country for six months. It would be utterly impracticable if you had a regular army of ten thousand men employed with all their bayonets and balls to keep the peace." The law and its enforcement were felt to be as repressive as Jefferson had believed the Alien and Sedition Acts to be. Just as in 1799 he had claimed that each state had the right to nullify federal law, now the Essex Junto claimed that individual states had the right to withdraw from the Union.

Like many others, including increasingly nervous Republicans, Adams recommended that Congress discuss lifting the embargo and implementing alternatives. In January 1808 he introduced a resolution "for the appointment of a Committee to consider and report when the Embargo may be taken off, and whether merchant-vessels shall then be permitted to be armed, and in defence of their lawful commerce to resist foreign aggression." A nervous Congress, caught between loyalty to Jefferson and electoral politics, retreated to a mixture of delaying tactics and poison-pill conditions. Should it be total embargo or selective nonimportation? Jefferson and Madison still had absolute faith in total embargo, if only the screws could be tightened and the leaks plugged. Most of the country agreed with Adams. It was also aware that a new president would be elected in late 1808. An outgoing president, Jefferson believed, should defer all major decisions to his successor. In essence, the embargo policy would stay in place at least until the Republican congressional caucus met. Madison was the likely successor, but there was opposition. Extreme states'-rights Virginians, led by John Randolph, supported James Monroe. Each Republican member of the House and Senate would have one vote, followed by an electoral college vote. Since most of the legislatures were in Republican hands, the Republican nominee with the highest number of electoral votes would succeed Jefferson.

If the choice were to be between Madison and Monroe, John Adams wrote to his son, he would prefer Madison, and "if our present rulers would not garble the Constitution and would let us have a reasonable revenue, fortifications, ships and troops I would not quarrel with them at all. . . . Mr. Madison's bias towards the French has always been too great but I never suspected him of corruption." Father and son granted that many Federalists, including their friend John Jay, had been biased toward the British. The former president no longer thought of himself as a Federalist: "I have long since renounced, abdicated and disclaimed the Name and Character and Attributes of that Sect as it now appears." Like his father, John Quincy now thought of himself as a man without a party. He regretted that neither he nor the Federalists, unless he rejected party solidarity, could have any influence on national policy. That seemed self-defeating. He took seriously the argument by leading Republican senators that the country would be best served if at least some Federalists supported sensible Republican policies. In late January 1808 he received an invitation from Stephen R. Bradley, a Republican senator from Vermont and Madison supporter, to attend the Republican congressional caucus.

"Some days after the meeting, when it was generally known that I had been the only federalist invited," John Quincy explained to his brother, "I asked . . . why he had invited me. He said it was because I had received marks of confidence from the Republicans among my own Constituents, and referred immediately to [the] meeting at the State-House in Boston last Summer, and to my name appearing on the Committee appointed upon that occasion." The invitation recognized his status as a semi-independent. It may also have reflected the respect in which he was held by some Republicans and their hope that someday he might more fully identify with them. The caucus nominated Madison for president and Clinton for vice president. Adams voted for Clinton for both offices, though he never explained, even later, why Clinton rather than Madison. Voting in the Republican caucus was a defining act that he never seems to have regretted, though his presence, let alone his participation, is inconsistent with his claim that he attended only as a "mere spectator." How was it possible for Bradley and others to take at face value his claim that he had no horse in the race? Or for most Federalists not to think his actions incompatible with his position? Was this not absolute evidence that he was no Federalist at all?

But his own party, he believed, had deserted him by advocating policies placing the interest of party above the welfare of the country. When an influential Republican told him he would be offered a high position if he publicly stated his approval of Republican principles, he nevertheless declined: "Deeming it inconsistent with my duties ever to shrink from the service of my country, I have always adhered to the principle that I should not solicit any of its favors." But rumors circulated: he had been offered the nomination for vice president by the Monroe faction; he would be offered a judicial or diplomatic position by Madison. The claim that he had made a self-serving deal was false, he told his brother. On the contrary, those Federalists who made the accusations were themselves "making advances" while "I was receiving them. While they were soliciting pledges of personal support and local influence, I was rejecting them—while they were offering themselves to market, I was explicitly declaring, that . . . I had no personal views whatsoever. Nothing to ask—nothing to wish." When Josiah Quincy told him that his "principles were too pure for those with whom [he] was acting, and they would not thank me for them, I told him I did not want their thanks." He would gladly sacrifice his political career to prevent the worst evil he could imagine—civil war and the dissolution of the Union, which would result in the total subservience of New England to Great Britain.

Now under Federalist control again, the Massachusetts legislature passed resolutions attacking the administration, the embargo, and its supporters. Adams had sold his soul to Jefferson, Madison, and Napoléon, a trio of devils plotting to bring French troops and Napoleonic rule to the streets of Boston. Led by Pickering, the Essex Junto fired away. Britain could do no wrong. The United States ought to submit with gratitude to its protection and do everything possible to help it achieve victory, including declaring war against the atheistic republic turned tyrannical dictatorship. "On your return home," Thomas wrote to his brother, "you must expect to find yourself in a strange land. The sour looks and spiteful leers will not be few that you will have to encounter, but they will be seen on the faces of those who have long harbored a secret heart-malignity against yourself as well as against your father." To the Adams family, this was an extension of the same right-wing Federalism that had attacked John Adams in 1799. It preferred an independent New England as a separate republic. And it was so thoroughly anti-Republican that it rejected the republicanism of the revolutionary generation, not simply specific policies of the Jeffersonians but the basic principles of the Declaration.

When Pickering, in early 1808, urged the legislature to condemn the embargo and pass laws to nullify it in Massachusetts, Adams published a response, soon in print as a pamphlet. He anticipated that it "will bring upon me the fury of all my former enemies and a host of new ones. . . . State legislatures should not," he argued, "become forums for debating and judging the decisions made by the US Congress." Foreign policy was the province of the federal government, not the states. And Pickering had misrepresented the history of the embargo act, as if British policies had had nothing to do with it. "The embargo was the only shelter from the tempest," Adams wrote, "the last refuge of our violated peace. . . . The embargo . . . is . . . always under our own control . . . open to a repeal . . . justified by the circumstances of the time. . . . I am ready to admit that those who thought otherwise may have had a wiser foresight of events . . . than the majority of the national legislature, and the President." Still, the ballot box, not the state legislature, Adams argued, was the proper place for opposition to the embargo policy. "Confidence is the only cement of an elective government. Election is the very test of confidence, and its periodical return is the constitutional check upon its abuse." He granted that New England commercial interests were badly battered by the embargo. But the interests of national security came first. And to the extent that New England had legitimate grievances, it needed to respect the constitutional division of responsibilities. It needed to go to the voters rather than to battle stations.

In Boston, when he called on the state chief justice Parsons, Adams got from his Newburyport legal mentor a full exposition of the views of the Essex Junto. The conversation was polite but definitive. "I found him as I expected totally devoted to the British policy, and avowing the opinion that the British have a right to take their seamen from our ships. . . . And a right by way of retaliation to cut off our trade with her Enemies altogether." France was the real enemy.

"The only protection of our Liberties, he thinks, is the British Navy." At a dinner to celebrate the ordination of a new minister, he was attacked by a clergyman in "a rude and indecent manner. . . . I told him that in consideration of his age, I should only remark that he had one lesson yet to learn, of which I recommended to him the study as especially necessary. And that was Christian Charity." When the legislature voted that its two senators be instructed to urge Congress to repeal the embargo, Adams believed he had no choice but to resign, not only because he believed the legislature had overstepped its legitimate powers in the national scheme but also because he anticipated that capitulation to such instructions would deprive him of the exercise of his own judgment. Were elected officials to be mere mouthpieces of those who elected them? Were they to be only conduits of the opinion of their constituents and of public opinion? Both roles were important, he acknowledged, but leaders in a republican and democratic government needed to be able to exercise their own judgment; to take positions determined by their information, insight, and values; and to be held accountable to the ballot box, not to the state legislature or the public opinion of the moment.

In June 1808, his long-held belief that he would not be re-elected was confirmed. The principal object of the Federalist Party, Governor James Sullivan wrote to Thomas Jefferson, "appears to be the political and even the personal destruction of John Quincy Adams. They have yesterday come to the choice of a senator in Congress to succeed him next year. James Lloyd had 248 votes, Adams 213. It is of great consequence to the interest of Mr. Adams, and to that of your administration, to rescue him from their triumphs. I know not how this can be done otherwise than by finding him a foreign appointment of respectability." Ordinarily this election would not have been held until December. Aware that it had taken place early in order to mark how strongly the Federalist majority in the legislature disapproved of him, Adams submitted his resignation. It was ten months before the end of his term. Why wait? The election and the anti-embargo resolutions indicated that he could from that moment on no longer be of service. Perhaps, he hoped, though not optimistically, the new senator could help devise new policies to "enforce the means of relieving our fellow citizens from their present sufferings, without sacrificing the peace of the nation, the personal liberties of our seamen, or the neutral rights of our commerce. . . . I now restore to you the trust committed to my charge, and resign my seat as a Senator."

Chapter 9
Paradise of Fools
1809–1814

From shipboard, he could hear the bells of Boston ring out one o'clock in the afternoon as the merchant ship *Horace* sailed from the wharf at Charlestown for cold northern seas, its destination St. Petersburg. It was Saturday, August 5, 1809. As the *Horace* disappeared into the fog of the Grand Bank, John Quincy had in mind more what he was leaving behind than what he was going to, and he could not have guessed that he would be absent from the people he loved so much for over eight years. He would also be absent from people who now at best disapproved of him, at worst hated him. "I have lost many friends and have made many enemies." He did not take the clash between politics and principles lightly. The tension caused him considerable anguish, though he had no doubt that his duty required that principle always weigh heaviest in the balance. His frame of mind and vocabulary reflected the interaction between an idealism rooted in family psychology, civic idealism, and Christian values and the experience of being pilloried. He found some relief in thinking himself persecuted by enemies and honored by martyrdom.

Between the resignation of his Senate seat in June 1808 and his departure, he had the painful experience of witnessing how many of those he had thought his friends could neither accept his politics nor honor him for his principles. Even among those who did not turn their faces away there were some who expressed their disapproval with awkward silence. He recognized that "it is not magnanimous and certainly not wise to quarrel with human nature for being weak. That a man should be deserted by his friends in the time of trial is so uniform an experience in the history of mankind that I never had the folly to suppose that my case would prove an exception to it." But he was still bitter. There were notable exceptions: Judge John Davis, a pro-Adams Federalist who was a constant supporter and friend; Reverend William Emerson; his intimate friends John Hall and John Gardner. At Harvard, his colleagues and students made him feel his mission there was valued.

Everyone else seemed to have deserted him, and, as he elevated himself to the ranks of the persecuted, he had the satisfaction of taking his place alongside his father in the family psychology. On his forty-second birthday, he noted that "the year of my life now expiring has been marked by a continuance of that persecution which the combined personal enemies of my father and myself had unrelentingly pursued the year before." Some connections, though, were too strong to be broken, even if they were strained. And there remained a few Adams loyalists among the Boston Federalists who gave him little bits and pieces of comfort. Some gave him legal business, which along with his real estate investments and dividends allowed him to think he might make enough not to have to dip into his small capital. "Your exile will be to your office and chair and you may soon be called again to higher cares," his father counseled. Ever the lover of the law, the former president thought the bar a desirable place.

But few among the political cadre in Boston or Washington believed that John Quincy would remain there for long. From the moment of his resignation from the Senate, he discovered that what had alienated many Federalists now made him attractive to the Massachusetts Republicans. A week before his resignation, a Massachusetts Republican and friend of

Jefferson's requested clarification of the widespread assertion in Washington that President Madison had offered him a foreign embassy if he would join the Republicans. Adams denied it. There had been no such conversation and no promises made. Undoubtedly, his friends believed him and his enemies did not. But the Massachusetts Republicans saw an opportunity. His candidacy as a Republican would be popular in their constituency. It would be a triumphant display of a new recruit and a likely electoral victory. In the election for his replacement, the large number of votes he had received had come from Republicans. And it was likely that the anti–Essex Junto moderates among the Federalists would vote for him, especially in a secret ballot.

For Adams, his resignation had brought genuine relief. He would no longer be in the public spotlight, targeted by both parties. His brother thought he had never looked so relaxed. But he noted that "politics threaten me again." The first threat, in September 1808, came from a representative of the local Republican committee. Would he not run on their ticket "to represent this District in Congress?" Silence would be taken as assent. He was tempted. It would be a platform on which he could affirm the views that had lost him Federalist sponsorship. But he declined. He had confidence in the current congressman, his friend Josiah Quincy. If he opposed Quincy for no reason other than ambition, it would be a personal betrayal. When the offer was repeated by three other members of the committee, he stuck to his refusal. In November, the Republicans tried again, this time with the offer of support if he would become their candidate for governor. William Eustis, a prominent Boston Republican to whom Adams had lost an election to the House of Representatives in 1802, led the effort to enlist him. Late in the month, in a long conversation with John Gardner, Adams "came to a determination to remain politically inactive at present, from a conviction that by now taking a part, I could do little or no useful service to the public." With a Federalist legislature, a Republican governor of Massachusetts would have limited powers and unlimited ceremonial duties. Having been burned so badly by voting independently in the Senate, he found that alternative political infernos had little attraction. And what if he should lose? Gardner agreed. So did his father. "Let me hope," John Quincy wrote in his diary, "I may not . . . suffer my feelings to be sported with, or my imagination to be deluded, by the electioneering intrigues of any party. In the dreams of other fancy's, may the reality of my own situation still be present to myself, and teach me the steady possession of myself." How to prevent becoming the projection of other people's desires? That was the greatest challenge, especially to a man in public life: to keep control of his own emotions, thoughts, and decisions; to create harmony between his inner life and his worldly activity; not to allow ambition to overcome integrity; to ground himself in the reality of self-knowledge. The effort was an unending process.

The electoral vote late in 1808 had made James Madison president. Both Adamses were comfortable with his ascension. Federalism had swallowed the poison pill of the Essex Junto, and John Quincy had become convinced that Pickering and his associates had made plans to take New England out of the union if circumstances gave them the opportunity. "For among themselves, I know that they . . . chuckle and exult as much at the operation of the embargo, as in public they whine and rage about it." In January, traveling by sled from Boston to New Haven, Adams had mostly in mind his legal business. But the Boston rumor mill was rife with speculation that he was going to Washington for appointment to a high position. At an overnight stop, he met a young Massachusetts congressman, Joseph Story, a Republican whose legal skills were becoming widely acknowledged. Neither Adams nor Story could have guessed that within two years Story would be appointed a Supreme Court justice. They talked until past midnight, "almost entirely upon the situation of public affairs. I have found a great agitation of mind everywhere as I came along. Much alarm and dissatisfaction, with an expectation of strong internal commotions." It was frightening. "Politics—the rage of politics, boiling everywhere with a fierceness I never knew before," he wrote to Louisa from Baltimore. "In Massachusetts, and Connecticut all was questioning— Had the General Court declared the division of the states?—Had they recalled all their members from Congress? . . . All this, and I know not how much more such stuff had been reported, and seriously believed."

Still, this Washington visit was about legal rather than political matters. He was on his way to try three cases before the Supreme Court, two with significant constitutional considerations. *Fletcher v. Peck* was a case about the Yazoo land claims, which raised the issue of the powers of the Supreme Court. The other case raised the question of whether a corporation had the right to sue or be sued in federal court. Adams had thought long and hard about *Fletcher v. Peck*, though when he arrived in Washington in early February 1809, he felt keenly that he was insufficiently prepared: "I find the pressure upon my mind very great. My anxiety relative to the discharge of my duty, excessive."

As a senator, he had done his best to have as little as possible to do with the attempt by New England speculators to convince Congress that the Georgia legislature had illegally confiscated their property on the grounds that all but one member of the legislature had been bribed. The corrupt legislators had sold the 35 million acres of state-owned real estate for less than a penny and a half an acre. Organized as the Mississippi Land Company, the buyers claimed they had a legal right to ownership. One of them, John Peck, who had sold some of his land to Robert Fletcher, acting on behalf of the interested parties, was asking the court to validate that he had legally owned the land he had sold. The state of Georgia wanted its cancellation of the initial sale confirmed as legal. Could the state arbitrarily cancel a contract? Or was a land grant a contract at all? Georgia argued that it was not and that, anyway, the state was the sovereign judge of such matters, not the Supreme Court.

Adams spent much of the next six weeks in court and making visits. He called on the president, the president-elect, numbers of secretaries and Supreme Court justices, and the British and French ministers. He dined with notables, including his former colleagues. He met a new senator from Georgia, William Harris Crawford. On the day of Madison's inauguration, Congress lifted the embargo, though it made clear the administration's intent to put new but unspecified sanctions in place thereafter. But who would implement Madison's policies was the subject of the day. The rumor that Adams would be appointed secretary of war or secretary of state bounced between Washington and Boston. For a short time all Boston believed he was to be appointed minister to Great Britain. None of the rumors had "the slightest foundation," he wrote to Louisa. When he visited Jefferson, he "found the President reading newspapers, which appeared to contain something offensive to him. He told me that one of his greatest enjoyments was in the prospect of being released from the necessity of reading them at all."

Two days before Madison's inauguration, Adams argued for a full day that Peck had a valid legal claim to the ownership of the land at issue: the sanctity of the contracts clause in the Constitution made the revocation of the original land grants illegal. He thought his presentation was poor. But "the Court did . . . hear me through." The arguments continued for two more days, and Adams thought it likely the judges would decide against him. Actually, they decided not to decide at all. At an evening dance party, Associate Justice Henry Brockholst Livingston intimated to him that the court would postpone its decision on *Fletcher v. Peck.*

Chief Justice John Marshall read the court's opinion. Since there was a technical defect in the presentation of the case, it would not at this time rule on the substantive issues. But could the defect in the pleadings be waived, Adams asked, if both parties agreed? The court consented. He immediately wrote and submitted a waiver request signed by both attorneys, though he knew that the court would not return to the case until the next term. But now at least the technical matters were in order.

On March 4 he witnessed Madison's inauguration. That afternoon he attended the reception and in the evening the inaugural ball. "The crowd was excessive, the heat oppressive, and the entertainment bad." The next morning, he received a note requesting that he call on President Madison. Emperor Alexander I of Russia had strongly urged an exchange of ministers. Russia and the United States, the president and emperor agreed, had important mutual trading interests. Both countries were concerned about the seizure of neutral ships. A closer relationship would be advantageous. To Adams' surprise, the president asked if he would agree to be nominated to be the first U.S. minister plenipotentiary to Russia. What, he asked Madison, would be the subject of negotiation? His mission, Madison responded, required that he be prepared to negotiate a commercial treaty if favorable circumstances should develop. How long he would be expected to serve there would depend on events but perhaps three or four years. He should depart as soon as his "convenience would admit. . . . I told him that upon the little consideration I was able to give the subject upon this sudden notice, I could see no sufficient reason for refusing the nomination." The next day there "was a whisper circulating that the Senate had not concurred in the nomination . . . and resolved that it is not expedient to send a Minister there at this time." He would not be going to Russia after all. "In respect to ourselves and to our children it would have been attended with more troubles than advantage," he wrote to Louisa, with dramatic understatement, "and although I feel myself obliged to the President for his nomination, I shall be better pleased to stay at home than I should have been to go to Russia."

In private, Chief Justice Marshall intimated that Adams need not do any more preparation for the corporations case, *Bank of the United States v. Deveaux*. It also would be postponed for continuation in the next session. About to leave Washington, he got the unexpected news from his cousin Billy Cranch, now chief justice of the District of Columbia Circuit Court, that the Supreme Court had seen the light and come to a decision on the corporations case. It had decided against

Adams' client. A corporation, "as such, cannot be a citizen," Marshall wrote. It cannot as a corporation sue in federal court, though individual citizens as members of a corporation may sue the citizens of another state in federal court. In this indirect way, a corporation's interests may be represented in federal litigation. A year and a half later, six thousand miles away, Adams learned the court's final decision in *Fletcher v. Peck.* The contract clause in the Constitution applied. A state could not arbitrarily void a contract. In the eyes of the law, Peck owned the land and had a right to sell it. "I was . . . delighted to learn from you," John Quincy wrote to Thomas, "that Mr. Peck attributes the success of the cause to my exertions at the preceding session of the court." He was to appear before the Supreme Court only one more time, thirty-two years later.

When he returned to Boston in March 1809, Adams found himself restless and unfocused. His tendency to divide himself among numbers of tasks made him feel he was not giving any one of them its due. His left eye continued to bother him, with heavy tearing and inflammation. Thomas Welsh took him to a specialist who diagnosed an obstruction, which could be removed by surgery. He put that off until the end of his teaching term. Unexpectedly, his move from the

Washington public world to a more limited personal community deflated his spirits, and he found he could not stay away entirely from political matters. The substitution of a selective rather than a total embargo seemed acceptable but unlikely to work. The president and Congress still refused to arm the country in any meaningful way. Great Britain and France continued to seize and confiscate American ships at will. What most irritated Adams, though, was the publication, in January 1809, of the *Works of Fisher Ames*. Adams was infuriated because it seemed a travesty on the recently deceased Ames' life. It was Essex Junto propaganda and a desecration of privacy to have published personal letters that the deceased did not have the opportunity to disallow. Indeed, Ames, a member of the Essex Junto in his last years, had freely expressed his antidemocratic and pro-British views in private letters. But during the revolutionary period he had served his country with great distinction, a patriot to be admired. The introductory biographical essay and the selection from his writings emphasized only the last ten years of his life. That seemed to Adams a distortion, even a dismembering, of the helpless body.

He argued his position in a series of six articles published in the *Boston Patriot* from late April to mid-June 1809, though he confessed to himself that "this is itself a sort of dissipation, and I know not whether it will do any good." It was indeed a poor performance: repetitive, rabid, and too long to be readable, an ineffective vehicle for his anger at his treatment by the Massachusetts Federalists, his hatred of their extreme antidemocratic views, their subservience to Britain, and their disloyalty to the union. While supposedly salvaging Ames' reputation, Adams attacked him mercilessly. Even a friendly reader might have thought it tedious and unfair. At the same time, John Adams had begun publication of letters from his presidency in the same paper. When a well-known Federalist clergyman preached the Election Day sermon in May, John Quincy was "informed that he was very abusive upon my father and upon me. No doubt he was chosen for that purpose. . . . Mr. Emerson told me after coming out, that there were two or three Sentences in it" for which the preacher "ought to be hanged." Friends of Ames probably thought that Adams ought to be hanged, or at least deported.

When Adams received a letter from the secretary of war, William Eustis, expressing the hope that the former president would publish no more of his letters at this delicate time, the brief postscript sent John Quincy's world spinning again. "I think," William Eustis wrote, "the nomination to St. Petersburg will be made this session." By July 4, the Senate had confirmed Adams' nomination by a vote of nineteen to seven. His father, offended by Eustis' letter, had advised John Quincy that if its hint turned out to be correct, he should refuse the appointment. Now the reality had to be confronted. As he listened to the July 4 oration at the statehouse, he had the arguments on either side of the balance sheet in mind. All who wished him well preferred that his appointment be "on American ground." In mild weather, letters would take two months to reach St. Petersburg, in the winter as many as six. Such a mission, a political friend wrote, "is to a man of active talents somewhat like an honorable exile." Some well-wishers thought that it would be in his interest to be sent out of the country. An appointment at home would be too controversial. And after all, as President Madison thought, who was better qualified to undertake this important European mission? No one else both had been to St. Petersburg and spoke French, the language of the court. "My personal motives for staying at home are of the strongest kind; the age of my parents, and the infancy of my children both urge to the same result. My connection with the College, is another strong tie . . . and by refusing the office I should promote my personal popularity more than by accepting it. In my own opinion also,

I could do more public service here in a private station than abroad."

But he had no doubt what his decision would be. It was a simple formula: if called, he would serve. No matter what the objections, no matter what the personal sacrifices, he was compelled to accept the responsibilities that he had been taught from earliest childhood were his sacred duty. With "a firm conviction," he wrote to the secretary of state, Robert Smith, "that the first object of the President's administration is the welfare of the whole Union," he had "an ardent desire to contribute whatever aid" he could. Wherever he was asked to serve, he would unhesitatingly go, a response that neither Madison nor his successor would undervalue or forget.

One voice of opposition was especially outspoken. "This embassy to Russia sits heavy at my Heart," his mother wrote to her sister Elizabeth, as John Quincy hurried to arrange passage, transfer his large library to the Boston Athenæum, and attend to the details of departure. Abigail feared the dangers of the voyage, especially so late in the sailing season. "Both his father and I have looked to him as the prop and support of our advanced and declining years," she continued. "His judgment, his prudence, his integrity, his filial tenderness and affection, his social converse and information have rendered his society peculiarly dear to us." Perhaps, she feared, he might never return. Or they would be dead before he did. His mother was to scheme and work for the next four years to arrange his recall.

But three other important voices did not have the same perspective or warrant to express themselves. John Quincy thought it would not be in his two older sons' best interests to go. There would be no school for them in St. Petersburg, it would be difficult to find tutors, there would be no playmates, and the weather would be unfriendly. Instead, George and John would board with their aunt Elizabeth. They had no say in the matter. But they would remember the long separation and be affected by it for the rest of their lives. For their mother, it would be a devastating separation. One-year-old Charles Francis, still at her breast, would go with them. But that was not enough to alleviate Louisa's horror that she had to go abroad without George and John. She had nightmare visions of never seeing them again. Like Abigail, she had no quarrel with subordination to her husband. But she had not anticipated this sacrifice. The blow sent her reeling. "O it was too hard! not a soul entered into my feelings and all laughed to scorn my suffering. . . . Every preparation was made without the slightest consultation with me and even the disposal of my Children . . .

was fixed without my knowledge until it was too late to Change. . . . Oh this agony of agonies! can ambition repay such sacrifices? never!!—And from that hour to the end of time life to me will be a succession of miseries only to cease with existence—Adieu to America."

John Quincy also had to say a premature good-bye to his Harvard students. In early July 1809 he gave his letter of resignation to the president of the college. At the end of the month, he concluded his final lecture to a large audience of students and guests with a personal farewell. He had faith in the intelligence and ideals of his students, he told them. They had been loyal to him, and forgiving. "Youth is generous," he wrote to Thomas, "and although the majority of the students were made to believe that I was a sort of devil incarnate in politics . . . yet they never could be persuaded that I was the ignorant impostor in literature" some had claimed. His farewell words, personal and eloquent, beating "with every pulsation" of his heart, spoke for all humanists past and present. "I would entreat you to cherish and to cultivate in every stage of your lives that taste for literature and science, which is first sought here. . . . I would urge it upon you, as the most effectual means of extending your respectability and usefulness in the world. I would press it with still more earnestness upon you, as the inexhaustible source of

enjoyment and of consolation. In a life of action, however prosperous may be its career, there will be seasons
of adversity, and days of trial. The trials of prosperity
themselves, though arrayed in garments of joy, are not
less perilous or severe than those of distress. . . . At no
hour of your life will the love of letters ever oppress
you as a burden or fail you as a resource." Although he
had hoped to be their teacher for years to come,

from these dreams . . . I have been awakened, by a
destination, of uncertain continuance, to a distant
country. It is not without reluctance that I have
yielded to this call and to the public service of the
country, as that from which an unsolicited call will
admit of no refusal. . . . Finally, gentlemen, though
my inclination still lingers at the word, I must, however reluctantly, bid you one and all, adieu. I have
heard of two lovers, who, upon being separated
from each other for a length of time, and by a distance like that to which I am bound . . . mutually
agreed, at a given hour of every day, to turn their
eyes towards one of the great luminaries of heaven;
and each of them, in looking to the sky, felt a sensation of pleasure at the thought, that the eyes of the
other at the same moment were directed towards
the same object. . . . Whenever the hour of studious

retirement shall point our views to those luminaries of the moral heavens . . . when the moralists, the poets, and the orators of every age shall be the immediate objects of our regard; let us in the visions of memory behold one another engaged in the same "celestial colloquy sublime." Let us think of one another as fellow-students in the same pursuit. Let us remember the pleasant hours in which we have trod together this path of wisdom and of honor; and if at that moment the sentiment of privation should darken the retrospect, may it be your consolation, as it will be mine, that the only painful impression which resulted from our intercourse arises from its cessation; as the only regret, with which the remembrance of you can ever be associated, is that which I now experience in bidding you FAREWELL!

As the *Horace* slipped down the harbor on August 5, 1809, the cannon from the Charlestown Navy Yard boomed a farewell salute. So did the cannon at Fort Independence on Castle Island, whose garrison was on parade in his honor, and the guns of the revenue cutter *Massachusetts* and the frigate *Maryland.* Louisa's sister Kitty and John Quincy's nephew, his sister's eldest son, William Steuben Smith, employed as his private secretary, accompanied them. He had taken on two others as unofficial secretaries. One was Francis Calley Gray, the son of the owner of the *Horace*; another was Alexander Everett, from a family with whom Adams was to have close ties. There were also John Quincy's valet, a black man from Trinidad named Nelson; and Louisa's maid and the baby's nurse, Martha Godfrey. Thomas Welsh and Thomas Boylston Adams were on board as far as the lighthouse. In his hand, John Quincy held a farewell note from his mother, "which would have melted the heart of a Stone." It was also addressed to Louisa. "My dear children, I would not come to town today because I knew I should only add to yours and my own agony, my heart is with you, my prayers and blessing attend you." To his mother, "the separation appeared like the last farewell." They all knew there was a chance that her son and his wife would never return or, if by good fortune they did, one or both of his parents would not be there to welcome them back. By nightfall, the *Horace* was out of sight of land. "At this Commencement of an enterprise, perhaps the most important of any that I have ever in the course of my life been engaged in," he wrote in his diary, "it becomes me to close the day by imploring the blessing of Providence."

At Kronshtadt, at the eastern end of the Gulf of Finland, the Adams entourage set foot on Russian soil. With a fair wind behind them, they sailed down to St. Petersburg. "We landed on the quay of the river Neva just opposite the magnificent equestrian Statue of Peter the Great." It was eighty days since they had embarked from Charlestown. Mounted on a twenty-five-foot-high stone pedestal, itself another twenty feet high, the bronze Peter the Great looked westward; his gaze represented his obsession with modernizing Russia. It also expressed the vision of the current emperor. Adams, who had begun to read Russian history, knew that his mission was the result of a long train of historical forces that made cooperation between the United States and Russia likely. When Emperor Alexander looked westward, Napoléon was directly in his line of sight, as much threat as friend. Beyond, he saw Great Britain, the naval power that controlled the oceans. Neither country respected Russian or American neutrality, though Russia was, for the time being, allied with France against Britain. Farther westward, he saw America, whose minister to Russia stepped onto the dock at St. Petersburg with an anxious sense of what a difficult mission he had undertaken. It seemed simple: negotiate an agreement with an emperor who desired active trade between their two countries. In reality, the war between the two main powers and their allies, with the balance of power always in flux, made a commercial agreement subordinate to changing fortunes and national interests. Adams would have to be patient.

The voyage itself had had its difficulties. Fog and clear weather, tedious calm and brisk breezes, alternated. Most of the party was seasick until the middle of August. Adams was not. An experienced sailor, he calculated that this was the fourth time he had crossed the Atlantic in an age in which it was unusual even for diplomats to do it more than once. As always, he found shipboard life tedious, its routines, crowded quarters, and constant distraction creating a feeling of disconnection that he struggled to overcome. It was also irksome and worrying that Kitty's flirtatious presence sparked a rivalry between Everett and Gray. Her sister and brother-in-law had tried to discourage her from coming. The voyage was a harbinger, they feared, of the likely trouble of having an attractive single woman with them. Sleepless nights made writing and even reading difficult for John Quincy. He managed some of Plutarch's *Parallel Lives* and Dryden's poetry, a French edition of the Koran, and sermons by the early-eighteenth-century French bishop Jean-Baptiste Massillon, who joined Tillotson and Sterne among Adams' most esteemed Christian orators. He especially admired Massillon's emphasis on ethical conduct and "the baseness of the mere terror of Hell." Writing, though, was more difficult than reading. Still, he kept up his diary entries and began a series of letters to the sons he had left behind. These letters, modeled more on Cicero than on Massillon, included lectures on education and morality; general principles about purpose, work, duty, civic responsibility, happiness, and the good life; and hortative advice about civic republicanism and Protestant responsibility. They were an attempt to perform a duty that he regretted he could not fulfill in person.

In mid-September 1809, the voyagers entered the North Sea off the coast of Essex. When stopped, they were congratulated by British officers on not having been blown out of the water. British warships blockaded every northern European port, intent on preventing any trade with Napoléon, though Britain and France granted special licenses whenever it suited their needs. Profiteering, bribery, forged papers, and the misuse of neutral flags flourished. The Danish Navy, which the *Horace* next encountered, had an additional incentive to confiscate all ships and cargoes. Under Napoléon's edicts, Denmark was starving. In southern Norway, Adams found that about thirty American ships were being held for adjudication in Copenhagen. As the *Horace* sailed down the Øresund Strait connecting the North Sea with the Baltic, and came to Elsinore, he would have preferred to focus on Shakespeare and *Hamlet*. When the castle cannons fired on a nearby British ship, the *Horace* had to maneuver between the suspicious demands of British warships and Danish forces. In Copenhagen, "the sight of so many of my countrymen in circumstances so distressing is very painful, and each of them has a story to tell, of the peculiar aggravations of ill-treatment which he has received." Although he had no basis in his instructions, he made representations to the Danish government to release the American ships. He was not successful. That he was on a private vessel and had no official passport complicated his ability to claim diplomatic immunity. Having feared that if he waited longer he would not get to St. Petersburg before the ice closed entry, he had sailed before the news arrived that Madison was sending his passport and the USS *Essex* to take him to Russia. Later, he recommended to the secretary of state that a minister be sent abroad only on a government ship. Otherwise, speed, security, and status would be compromised.

As the *Horace* crossed the Baltic and entered the Gulf of Finland, the weather became dreadful. The experienced captain, who had made eleven voyages to St. Petersburg, wanted to turn back. He dreaded the gulf more than any other body of water. They should winter over in Kiel, he recommended. At first, reluctantly, Adams consented. "I had objects on board, more precious to me than my own life, and there was some reason for shrinking from a risk of the ship and cargo, which was not mine, and which was the special trust of the Captain." They turned back. When the weather suddenly became favorable, he insisted that they turn again. They beat eastward and northward, arriving in St. Petersburg just before the ice closed the city.

Overnight in Kronshtadt, Louisa, constantly ill and frightened during the voyage, got her first taste of how difficult winter in Russia would be. As she tried to sleep in a cold stone room, she discovered that it was "so full of rats that they would drag the bread from the table by my bed side . . . and fight all night long and my nerves became perfectly shattered with the constant fright least they should attack the Child." The water made them all sick. Diarrhea. Smells. In St. Petersburg, they hoped their stay at a hotel would be brief; the walls were so thin that whenever Kitty and Louisa played and sang they were applauded by the unseen hands of strangers. There was no house available. "Everything here as to price exceeds anything you can form an idea of," Louisa wrote to her mother-in-law. How would his wife and sister-in-law, John Quincy worried, adjust to a totally foreign culture? How could an American minister survive honorably on his modest salary, with nothing extra for expenses, given the high cost of living, if he desired anything even close to American comfort? And how could an American minister make a dignified impression, given the style of life the Russian government expected of a foreign minister, as he socialized with well-supported ambassadors and a Russian nobility that delighted in displays of imperial wealth?

It was also cold beyond anything they had ever experienced. Indoors they were protected by the Russian custom of double windows, hot stoves, and three-foot-thick outer walls, which kept out the cold but also any fresh air. Despite the temperatures, John Quincy insisted on daily walks for his health and to see the city. The fur on the edge of his cape, which he wrapped around his ears and face, became "powdered with a white frost, from my own breath, which as it issues, settles and freezes on the hair. The same effect gives a singular appearance to the men of the country who wear whiskers on the upper lip," as if they had been powdered. "Horses which have been driven fast enough to produce a free perspiration are entirely covered with it, and appear white." Louisa felt besieged with difficulties, including her health and Charles Francis'. As she chastised herself with unremitting self-condemnation for having abandoned her two eldest children, as if her integrity as a mother had been irredeemably forfeited, from the start what she wanted most was to go home. To her shock, she discovered that she was the only diplomatic wife in St. Petersburg, with one exception. She would have few opportunities to make female friends. In the spring, they were at least able to move into a house vacated by a departing minister, though at a barely affordable cost.

They both were challenged to adjust to the daily regimen of court life, which meant rising late; having tea in the late afternoon; and attending dinners, parties, masquerades, and balls, often with three hundred guests, until long after midnight, and often without release until four in the morning. "I can take no pleasure in parties, the sole amusements of which are dancing and gambling," he noted, "and which begin at 11 at night. . . . At these parties, I am reduced to the necessity of dancing, to avoid gambling." The elite, from clerks to aristocrats, lived beyond their means, with "extravagance and dissipation . . . a public duty." Expenses were so severe that John Quincy soon realized he could pay them only by supplementing his salary from his savings and Boston income, especially since he had to sponsor all the expenses of the American mission. To their shock, they had no choice but to employ about fourteen servants. The French ambassador had nearly eighty. It was expected as a matter of status and as a contribution to the Russian economy. And the house servants believed they had a right to supplement their salaries with a percentage of the household resources. They felt they were taking what custom entitled them to. But to the Adamses, who could not accept that they had this obligation, they stole. A depressed Louisa proposed that she go home in the spring. "You can form no idea of the morals the manners and the people of this country," she wrote to Abigail, "and I am so conscious that we cannot in any degree assimilate with them that I foresee nothing but perpetual mortifications of every kind to be heaped on us during our stay." But, John Quincy pointed out, the cost of supporting two households would probably be more than the cost of St. Petersburg alone.

By late January 1810, they had at last gotten "through the continual series of invitations, which have so long kept us in a State of dissipation, and absorbed my time." He had been introduced formally at court, where etiquette required that he wear a wig. Court activities, from levees to religious rituals to lavish dinners, required his attendance. Louisa's introduction was delayed. "This mode of life is dreadful to me and the trial is beyond my strength," she wrote.

When presented to the emperor, she was questioned as to why she had been so reclusive that the American minister had had to attend court events by himself. By late spring, Louisa was three or four months pregnant. In mid-July 1810, she lost the child. "I am just recovering from another severe indisposition," she wrote to Abigail, "which has deprived me of the pleasure of presenting you with another little relation it is only four days since and I am so weak I can scarcely guide my pen." In November, she became pregnant again. Still, she was a hit at court, as was John Quincy. The emperor and his entourage took a liking to them both, showing patriarchal concern for, and deference to, Louisa's pregnancies.

Soon after Adams arrived, the American consul, Levett Harris, had introduced him to the emperor. Prim and self-satisfied, or so Louisa thought, Harris ran his own commercial empire. Like many unpaid American consuls, he paid himself handsomely with bribes and business deals. To be permitted to do business in Russia, American ship captains had to pay Harris a percentage of their cargoes. Although Adams knew from previous service in Europe that many consuls exploited their official positions, he would only gradually learn the extent of Harris' profiteering. Louisa found Harris' condescension offensive and laughable.

When, in 1811, American ship captains and merchants brought complaints about Harris to Adams' official attention, his investigation revealed what verged on illegal conduct. Relations between Harris and Adams cooled. Eventually Harris, who provided unconvincing explanations, had the opportunity to defend himself in the appropriate Washington forum. Still, they carried on cordially with their overlapping obligations in St. Petersburg.

Adams was also concerned about Francis Gray, not about business but about relationships. In February, he noted that Kitty "this day gave me information with respect to herself and Mr. Gray, which places me in a situation of some delicacy." When they moved from the hotel into a house, there was no room for Gray and Everett, which must have been a relief to John Quincy. Whatever had happened between Gray and Kitty, it had no long-term consequences. Neither did the emperor's interest in her. A handsome, imposing, and courtly man, with a receding dark blond hairline and engaging blue eyes, thirty-two-year-old Emperor Alexander had inherited the throne from his murdered father. Educated in the values of the European enlightenment, he ruled a feudal, faction-ridden country dominated by aristocratic families and the Orthodox church. Sensitive and evasive, he delighted in being liked. A seductive personality, he needed to be wily and suspicious, to charm as well as command. His purse and position dominated St. Petersburg and Russian society. Curious about America, he took formal and informal opportunities to query Adams about American things. Like Adams, he walked for exercise, usually on the promenades by the Neva, where they frequently met. He was solicitous about the health of Adams and his family. He was attentive and gracious to Louisa, dancing with her at imperial parties, and he took a special liking to Kitty—part curiosity, part gallantry, and part flirtation.

When the emperor wanted to see more of Kitty, he overruled her exclusion from court functions, creating much gossip. Most likely it was thought that the emperor wanted her for a mistress, or that she already was. Since Kitty, Louisa wrote in her diary, "was a great Belle among our young Gentlemen this circumstance though customary with the Emperor towards many Ladies whom he met gave umbrage to Beaux and occasioned so much teazing and questions that we left off our promenades for some time." Kitty was flattered and excited, Louisa worried. John Quincy was sure there was nothing to fear. He worried more about Kitty and Gray; then, later, about Kitty and William Steuben Smith. When Charles Francis' nurse, Martha Godfrey, became "attached to a young Russian . . . she expected to marry," the Adamses discovered "unpleasant facts about him." They were able to prevent what would have been an unhappy marriage for Martha, whose talents and personality they thought well of.

Most of Adams' official business was with the fifty-five-year-old Russian chancellor and foreign minister, Count Nikolai Rumiantsev. When Adams had a formal request to make, the foreign minister conveyed it to the emperor. When he urged the Russian government to pressure Denmark to release impounded American ships, the request was eventually honored, signaling the emperor's eagerness to show goodwill to the United States. Rumiantsev and Adams developed a personal rapport. Like Adams, Rumiantsev had a passion for books; he collected old Russian manuscripts. Like the American minister, he was passionately patriotic and intellectually engaging. His grandfather had been a well-rewarded associate of Peter the Great. His father, Catherine the Great's field marshal, had served his country, as had John Quincy's, at the highest level. Having lived in France, Rumiantsev knew the French language well. Although he took Napoléon seriously, he viewed the French emperor as a transient phenomenon, the product of unusual circumstances and personal genius. Great Britain seemed to him, as it did to Adams, the dominant world power, a much-admired model of social, monetary, political, and commercial achievement, the ruler of the seas from whose dominance both Russia and the United States needed to be liberated. He made it clear to Adams that there was no possibility of a commercial treaty at this time. France forbade it. Powerful Russian Francophiles argued against it. Powerful Russian Anglophiles promoted a Russian-British alliance. For the time being, the emperor mediated between these pressures, trying to keep the Russian state secure. American matters were secondary.

To Adams' American eyes, what seemed most distressingly noticeable, as newspapers and couriers brought European news to St. Petersburg, was the contrast between what ten years before had been Europe's emerging republicanism and its recent descent into assorted tyrannies. In 1776 America had proclaimed itself the leader of a worldwide revolution, its founding documents the lights by which other nations would be guided. But in Europe the lights had been mostly extinguished: Napoléon was an absolute tyrant attempting to stock European thrones with relatives and offspring, from the Netherlands to Spain, Scandinavia to Austria and Italy, Germany to the Russian border. In St. Petersburg, the French ambassador, the Marquis de Caulaincourt, who lived "in a style of magnificence scarcely surpassed by the Emperor himself," dominated the diplomatic corps with his charm, wealth, and hospitality. The Adamses enjoyed his masquerades, ice parties, parties for children, and literary entertainments, especially the performance of scenes and soliloquies from classics of the French theater. But such entertainments reminded Adams that Europe remained what it had always been: a feudal world ruled by an elite, a dispiriting contrast for a Massachusetts republican from a better place four thousand miles away.

In the two friends he made among the foreign diplomats, Willem Six van Oterleek of the Netherlands and General Benito Pardo de Figueroa of Spain, Adams had compelling examples of the influence of Napoléon on the European psyche. Napoléon represented liberation from the old tyrannies, the attraction of individual genius, the rise of talent to power, the breaking of William Blake's oppressive "mind-forged manacles," and the rejection of rule by divine right. But the liberties he granted could be taken away as a matter of policy or whim. He created the drama of divided allegiance in local patriots, who were dazzled by his person and promise. The liberator could also be the enslaver. The Chevalier Six van Oterleek had been plain Citizen Six when Adams first met him in Holland in 1795. When Six was transformed from radical republican to the minister plenipotentiary of the king of Holland, Napoléon's brother Louis, "his subjugation of soul [was] complete," Adams noted. "But he is a man of great political information; of long experience; of better principles than most statesmen, of this or any other day; of good intentions; of good disposition, anxiously desirous of the esteem of others, and especially of those whose judgment he fears. . . . When his chain galls him he looks at it, and takes comfort in the thought that it is gold." When Napoléon in 1810 incorporated the Netherlands into France, he eliminated his brother Louis' throne. That eliminated the office of Dutch minister. When Six left for Paris, John Quincy missed him. He was shocked when he learned that his friend had drowned in an Amsterdam canal, apparently a suicide.

General Pardo, who had sided with the French invaders against the Bourbon dynasty, identified with the Patriot Party that fought against Napoléon, as if one could be both a loyal Spaniard and a supporter of Napoléon's invasion. A delightful companion and classical scholar, he spent hours in conversation with Adams. Intemperate, irresolute, his brain and his heart rarely in harmony, Pardo seemed to Adams torn between two impossibilities. He identified with the patriot victories against Napoléon's army but represented

Joseph Bonaparte at the court of St. Petersburg. By the autumn of 1812, Pardo's tensely balanced equilibrium became unbalanced by Napoléon's losses in Spain, the end of Joseph Bonaparte's brief rule, and the events that brought Russia, France, and Great Britain into a different alignment. Pardo was without salary or position. With a son in Paris, he left St. Petersburg, took ill, and died "of a broken heart [in] a small, mean hovel of an inn, upon his journey from this City." Such was life in the diplomatic corps, "which of all the movable sand banks in the world of mutability," John Quincy wrote to his mother, "is perhaps the most given to change And here ends my canto of mutability . . . how they pass like Chinese shadows before us."

As always, he had an insatiable interest in the customs and institutions of the country in which he served. His diary is a record of what he saw, heard, and felt: mainly court and diplomatic life; the daily interactions with servants and merchants; and street life, including St. Petersburg's fairs and festivals. He often took Charles Francis with him for walks, and Louisa and Kitty, when weather permitted, on carriage rides. During 1810 and 1811, he went on every major religious holiday to the prominent Russian Orthodox and Roman Catholic churches to learn about Russian religious rituals. Russian Orthodoxy was "indeed encumbered with innumerable superstitions and armies of saints and miraculous relics, and images and trivial formalities." Although its core seemed to him as truly Christian as his own New England Congregationalism, its frequent festivals and elaborate ceremonies seemed "suitable to a people of slaves, and no other." Everyone, except the emperor, waited for orders from above; the country's public spectacles embodied the mind-set of a slavish society. That kind of religion, sincere as it was, was a substitute for freedom. And at the bottom of the hierarchy were the serfs, heavily bearded, all dressed in caftans, the Russian equivalent, he noted, of American slaves.

His own life and Louisa's, they were aware, were passing like "Chinese shadows," transient silhouettes in the puppet shows they saw at St. Petersburg street fairs, thousands of miles from where they wanted to be. "How ardently," Louisa wrote, "I wish the time was come for us to return [to] the quiet and easy life I led in America." Adams felt frustrated at how little real work there was for him to do, "the stagnant political atmosphere and the Scythian winters of St. Petersburg," and the onerous diplomatic obligations, which sometimes had him at four court functions in a day. He read as much as his schedule would permit, though recurrent eye inflammations interfered. He could not escape a racking cough and head colds, which he blamed on the climate. When he worked as hard as circumstances allowed to teach Charles French, he began to appreciate how difficult it was to teach children anything. "The best of all possible educations I know is but a lottery, and without a corresponding disposition in the child, all that you can do for him is but labor lost. . . . The most essential part of education after all is to teach a child to think. Perhaps, too, it is the most difficult." Learning Russian, from which he could expect only limited returns, began to seem unnecessary. His conversations with the emperor and with every Russian functionary were in French. He was, though, pleasantly surprised to discover that the emperor spoke English. When they met one day along the Neva, Alexander, having heard Adams had been sick, expressed in English his concern about his health.

By late 1810 Adams, like the Russian government, had almost no doubt that Napoléon would attack in either 1811 or 1812; which, Louisa feared, would imprison them in St. Petersburg. "This is an exile which I fear will not shortly be terminated," she wrote to her mother-in-law. Since her son could not afford the high cost of living in St. Petersburg, Abigail told President Madison, he should be recalled. "I hope you may have already received through the Secretary of State his own request to this effect." But John Quincy had no intention of making such a request. It was against his principles. As much as he and his family wanted him home, he believed, as did the president, that to return prematurely would abort an important mission. It would also perhaps offend the emperor. Apparently neither her son nor Madison was offended by Abigail's intercession. After all, it came from the heart. But John Quincy would return home only if recalled at the government's initiative. "As no communication of your wishes, however, has yet been received from yourself," the president wrote to him, "I cannot but hope that the peculiar urgency manifested in the letter of Mrs. Adams was rather hers than yours; or that you have found the means of reconciling yourself to a continuance in your station." It was too important a mission to leave without a replacement in place. "We end this year in bad health and in worse spirits than ever," Louisa wrote. "God help us these are honors dearly bought." The next month he had in hand a letter from Washington giving him permission to come home at his discretion. Abigail's lobbying had, after all, been successful. When he told Count Rumiantsev that he had been given consent to leave, the foreign minister urged him to stay. A few days later the emperor, who had also heard the news, told him that "he was sorry for it. I told him that at least I hoped it would not yet be for some time; probably for some months."

John Quincy did not yet know that Louisa had become pregnant. That knowledge soon ended discussion of an early return. "Nothing short of the extremist urgency could induce me to embark with a wife and child later than the first days of September." But Quincy and Washington did not learn of the new reality until spring 1811. In late February, Madison, having been turned down twice, nominated Adams to be a Supreme Court associate justice. The political stars were aligned for the appointment of a New Englander. When the Senate confirmed the nomination unanimously, the Adams family in Quincy was overjoyed. Always an enthusiastic partisan of the legal profession, John Adams would have the happiness of seeing his son occupy a seat on the highest court in the land. "There is great anxiety about your return," he wrote to John Quincy. "Some are impatient to have you here, and hope to have you for their gladiator . . . others dread your appearance lest you should be a candidate . . . a third sort wish you here that you might be an upright enlightened Judge; and a fourth that you might be made a Judge in Spite, to put you away . . . out of sight of the searches for candidates. It is a terrible thing to be a man of so much importance." He argued strongly for acceptance.

But John Quincy had significant doubts. He was as qualified as most of the likely candidates, though he urged that his better-qualified Boston friend Judge Davis deserved the appointment. Was he himself, he asked, suited to spend his life listening and making judgments rather than persuading and acting? "I always shall be too much of a political partisan for a judge," he wrote to Thomas, "and although I know as well as any man in America how and when to lay the partisan aside, I do not wish to be called often and so completely to do it as my own sense of duty would [require] were I seated upon the bench," a view that Chief Justice Marshall also privately expressed. If he had no better way to earn a living, John Quincy concluded, he would return to the practice of law rather than sit on the bench or submit to "the servile drudgery of caucuses, the savage buffeting of elections, the filth and venom of newspaper and pulpit calumny, and the dastardly desertion of such friends as Anthology critics and Boston legislators." The bench did not suit his temperament. Politics was a mean profession. Anyway, he could not leave St. Petersburg. The young Joseph Story was appointed instead.

The rumor spread that Adams was being transferred to Paris. Carrying on with his diplomatic conversations, he tried to convince the Russian government to cease cooperating with the French embargo. Count Rumiantsev granted that it probably worked to the benefit of Great Britain. Its main advantage to France was the money it brought in through confiscations of American ships and fees from licenses. But the conversations produced no change in policy. As Adams reported to Quincy and Washington, the Russians were buying time. Napoléon would be gone, eventually. Britain was the long-term rival. Under pressure from France, Russia signed treaties with Sweden, Denmark, and France; France and Russia signed treaties with Austria. None of these, the emperor believed, would hold. But they would buy time. The key was Prussia and Poland, both under French rule. Prussia hoped to regain its national autonomy. If Napoléon revived Poland as a militarized vassal state, Russia's national security would be at risk. By late 1810 it become clear that Napoléon intended to use Poland as a launching pad for an invasion of Russia. Inevitably, then, Russia and Britain would resume their interrupted good relations. They would be allies against Napoléon. And it would not be in Alexander's interest, as he explained to Adams, for the United States and Britain to be at war.

A weak America, John Quincy wrote to his mother, "debilitated by internal dissensions" is "the common foot-ball of Europe." Would the football attempt to kick back? The notion that the United States would declare war on Britain seemed possible but far-fetched. It would be untimely, since an intensification of the European conflict would probably be to America's advantage. The European "war is . . . likely to be long," he wrote to Thomas. "Should we join in the conflict, we could scarcely hope for a better fate than to be sacrificed as one of the victims at its close."

As St. Petersburg held its breath during 1811, Adams hoped that the turmoil of invasion would be delayed until the next year. By late spring it seemed likely that Louisa would not miscarry, though miscarriage remained a fear until the last moment. The ghost of their stillborn child would give them no peace until they held a living child in their hands. And news from America brought distressing shocks. Although the information took many months to reach St. Petersburg, in April 1811 Francis Dana, ill and irritable for twenty-five years, died, to be followed in May by William Emerson. John Quincy had first set foot on Russian soil accompanying Dana to the court of Catherine the Great. He had been a mentor and friend. As a Harvard student, John Quincy had sat by Dana's bedside after his first paralytic attack.

And John Quincy had admired Emerson's intellect and religious liberalism. Most important, his friend and pastor had not deserted him when he had run afoul of Federalist approval. "I felt a strong attachment to him and lament his loss." When, also in April 1811, his sister Nabby was diagnosed with breast cancer, medical advice urged surgery. Abigail feared she would lose her only daughter, John Quincy his only sister. In late May, when John Quincy and Kitty were in his study reading newly delivered letters, some almost half a year old, six-months-pregnant Louisa walked in. She "immediately saw by their distressed countenances that bad news had come to us." Her thirty-seven-year-old sister, Nancy Hellen, had died in childbirth. So had the child. "The shock was sudden, unexpected and violent." Louisa was "alarmingly ill." Was she going into early labor? Would she miscarry? After a few days of headaches and "nervous agitation," with the help of laudanum she was able to sleep. The crisis passed. Depressed and sick at heart, they focused on the hope for a safe childbirth.

In June, Louisa was ill again with a high fever. Erysipelas, a skin disease in one of her ears, recurred, as did her usual sick headaches. Medical care, as it had in Berlin, required bleeding. Leeches were applied to the wrists to relieve the fever. In July, on its fourteenth anniversary, John Quincy gave thanks for his marriage. He had qualifications: "the ill health which has afflicted [Louisa] much of the time"; their different views about running a household and educating children; their shared inclination to be "quick and irascible." His temper was "sometimes harsh," he recognized. They were both thin-skinned. Quick to take offense, she was sensitive to what she perceived as his coldness. Such natural frailties led to arguments and alienation. "Our union has not been without its trials; nor invariably without dissensions between us." For Louisa, the circumstance under which they had married was never out of mind. She had no doubt, though, that she loved and honored him, and he had no reservations. "She has always been a faithful and affectionate wife, and a careful, tender, indulgent and watchful mother to our children. . . . My lot in marriage has been highly favored." He could not imagine that he could ever have had such happiness outside marriage.

Late in the month, lingering on the riverbank near the house they had rented on Apothecary's Island, he listened to the music played by a military band, about twenty-five clarinets, horns, and drums. His mind was on Louisa. She had been sick much of the day, with her confinement approaching. He was finding it difficult to sleep, bothered by the constant daylight. He worried about diplomatic matters, especially the likelihood of war between Russia and France, the United States and Great Britain. He felt at the mercy of dispatches, letters, newspapers, and visitors who took up much of his time, resolved nothing, and intensified his feeling of uselessness. He worried about his sons in Massachusetts. And he worried most about Louisa. "May the Mercy of God, grant her a safe and joyful deliverance!" She went into labor at seven in the morning on August 12, 1811, attended by the nurse and midwife. John Quincy stayed with her most of the day. In the early evening, when he had left her for a few minutes, Kitty came to tell him he had a daughter. The doctor arrived two hours later. The mother was fine, "the child . . . as healthy and lively as we could wish. . . . I think this will convince you," he wrote to Abigail, that "the climate of St: Petersburg is not too cold to produce an American." They hoped to return to the United States the next year or the year after with a little girl who would bring delight to the hearts of the Johnson and Adams families.

Some of those hearts had stopped beating in the fall of 1811. The most shattering news arrived in St. Petersburg in February 1812. Louisa's fifty-four-year-old mother had died at the end of September, the victim of "a malignant bilious fever" that swept through Washington. "My Poor Mother! After ten years of poverty dependence and severe suffering which at this great distance it was so utterly out of my power to mitigate or assuage—How different will home appear should we live to return." The same letter brought the shocking news that Louisa's brother-in-law, youthful Andrew Buchanan, who had married her sister Caroline, had died a few days after Catherine Johnson. News from Quincy made it an even fuller season of misfortune for both families. Abigail had "consigned to the tomb in one day, my dear and venerable brother [Richard] Cranch, and my beloved sister [Mary], after a sickness of four months. . . . O he has only stepped behind the scene, I shall know where to find him." She had intimations that she and her daughter Nabby might one day soon also have to search for one another there. At the Old House in Quincy, Nabby's cancerous tumor and her entire breast were cut away. The letters to St. Petersburg in February 1812 brought a tale of "the frail tenure of human existence." They were all transient characters, Adams acknowledged, in the same eternal story.

John Quincy and Louisa had reason to be anxious about their infant daughter, named after her mother. "You have a sweet little Sister," Louisa wrote her sons. She looks like "Grandmama Adams. She is very handsome . . . she plays all day long." But to herself she noted, "Every one who sees her stops her in the Street. . . . All say 'that She is born for Heaven.'" The Russian superstition read an early ascension in the face of the child. Although her teething had started normally, an infection developed. In July 1812, the eleven-month-old infant developed dysentery. On some days she was wretchedly ill, on others better. The dysentery became severe, with high fever and dehydration. Her mother began to wean her in the hope that the change in diet would be beneficial. In severe pain from teething, the baby cried incessantly. In August, her distressed parents prepared for the worst. So did the doctors. "Warm baths, and injections of laudanum and digitalis" were "tried . . . with no favorable effect." The doctor "ordered the . . . hair to be cut off from her head; and . . . as a last resource a blister was applied on the back of the head." Early in the morning of September 15, little Louisa died. As always, Louisa blamed herself. If she had not stopped nursing, she was convinced, the child would have survived. "I pray that the calamity which has befallen me, may produce no unsuitable impression upon my heart or mind." But she could not get her baby's "last agonies" out of her mind. "My babes image flits forever before my eyes and seems to reproach me with her death." If only she had died instead of the child.

She stayed depressed for months, dreaming about death and hoping for her own. Her life seemed "cold blank and dreadful!" Ill herself, she thought it likely that people would think she had lost her mind. "I am a useless being in this World and this last dreadful stroke has too fully convinced me what a burthen I am become—surely it is no crime to pray for death." Her severe grief lasted for six months. For the rest of her life, the loss was an ache in her heart. John Quincy tried to contain his pain. "Her last moments were distressing to me and to her mother, beyond expression. . . . The Lord gave, and the Lord hath taken away." During the deathwatch, he was too agitated "for the ordinary occupations." But he did not desert his diary, and it did not desert him. Each day he wrote his account of events and feelings, bringing to bear the power of language to express and contain what otherwise would be inexpressible and uncontainable. The child has no more suffering to endure, he wrote. But "how keen and severe" is the pain of the parents. "She was precisely at the age . . . when every gesture was a charm, every look delight; every imperfect but improving accent, at once rapture and promise. To all this we have been called to bid adieu . . . stung by the memory of what we already enjoyed." Gradually he mustered the stoic resignation and consolation an active life required. "If there be a moral Government of the Universe," he wrote to his mother, "my child is in the enjoyment of blessedness, or exists without suffering, and reserved for unalloyed bliss hereafter." It was an "if" that he occasionally allowed himself, particularly in his struggle to deal with personal tragedy. But he usually kept it at bay. Although his Christianity served him well under most circumstances, it was force of personality, constructive willfulness, and a sense of what he owed himself and the world that carried him through.

And in his world there were now two wars with immense consequences. In June 1812, at President Madison's urging, spurred on by war hawks such as Henry Clay and a new congressman from South Carolina, John Calhoun, Congress declared war on Great Britain. About twenty thousand American and British citizens were to die in a war made ludicrous by the fact that the United States was unprepared to fight and the declaration of war came a few weeks after the British had repealed the policy that had been its cause. Slow trans-Atlantic communication added irony to folly. The United States had much cause to complain. But impatience, political pressure, bad judgment, and greed for real estate, particularly for Canada, resulted in a monumental mistake. In July, Napoléon invaded Russia, another mistake, this one with devastating consequences for France. Gradually, Alexander's strategy had become clear. Napoléon's army would be allowed to advance westward to Moscow, living off and ravaging the land, without a major effort to halt its progress. On the day little Louisa died, Moscow burned. Without supplies or shelter, backtracking across a desolate countryside, pursued by Russian forces, Napoléon would barely survive the inevitable retreat. By the late fall of 1812, "it has become a sort of by-word among the common people here," Adams wrote to his mother, "that the two Russian generals who have conquered Napoleon and all his Marshalls are General Famine and General Frost."

For the Adamses, "literally and really sick of this climate," there also seemed no escape. Napoléon was discovering, Adams commented, that it is "contrary to the course of nature for men of the south to invade the regions of the north." The American invaders of Canada should have understood that too. Undertrained, undersupplied, and badly led, the American armies suffered military defeats on the Canadian-American border through 1812 and much of 1813. In frigid St. Petersburg Adams could only watch from a great distance and hope for better news. Although the war seemed to him an untimely mistake, he recognized that it was a culminating, even if irrational, response to ten years of British insults. He shared the national anger. "If all the people of America . . . were of my sentiment, the last drop of American blood and the last dollar of American property should be staked, rather than flinch an hair's breadth from our whole ground in this quarrel. It is pure unmingled tyranny that constitutes the whole British claim."

In London, the declaration of war was interpreted, even by long-standing friends of America, as pro-French. In Quincy, John Adams consoled himself, as did his son in St. Petersburg, that now Congress would have to spend money on national security. America's best hope was its navy, its best manpower sea power, its large pool of experienced sailors. On land, it had a minimal army, untrained militiamen, incompetent officers, and a remedial command structure. Like Russia, though, it had a huge landmass that would cost the British immense resources to conquer, a reminder of the major British liability in the Revolutionary War. It would cost more than even the richest country in Europe could afford. That country had its priorities: to defeat France, reestablish a balance of power in Europe, and maintain British commercial supremacy. Now the retention of Canada was added to the list.

Early in 1813, toward the end of an abysmally cold winter, the worst he had ever experienced, John

Quincy felt hopeful about an imminent return home, though he had no specific reason to be sanguine. He had not yet received his recall, and how to get home in the midst of a war was uncertain. But there was reason to anticipate that his St. Petersburg years would soon be over. World events were increasing that likelihood. News, though, came slowly. Information about American and European events arrived most quickly from London via courier to the British embassy and the Russian government. For the moment, Adams' views about British turpitude were frozen into angry pessimism: "It is now for the God of Battles to decide. . . . They have in short everything against us but a righteous case." Realistic Americans knew that Britain would never renounce impressment. The practice would end only when the British had no more need for it, which would be when its war with Napoléon ended in victory. For clear-eyed Americans, the best hope was that the American military, the size of the country, and war weariness would bring Britain to the negotiating table. For the time being, "the only point indispensible on our part," an unrealistic Madison had told Albert Gallatin, was "to obtain a stipulation which should protect our Seamen from impressment." But how to get negotiations started? The Canadian campaign was a disaster, the East Coast was blockaded, and the embargo policies were inciting New England to threaten not only noncooperation but also disunion.

In March 1813, Emperor Alexander, unhappy at the continuation of a war between his two main trading partners, offered to act as mediator. Madison immediately dispatched two delegates to join Adams in St. Petersburg: Albert Gallatin and James Bayard, the Federalist senator from Maryland. Gallatin was to head the peace commission and trade negotiations with Great Britain, Adams the trade discussions with Russia. In April, the secretary of state confided to Abigail that in case of peace with Britain, John Quincy would be offered the mission to London, "which would at least bring him closer." Perhaps Alexander's mediation would be successful, John Quincy wrote to his mother. Otherwise there would be no peace. Britain, which had no desire to add to the Russian emperor's glories and viewed its differences with the United States as a family quarrel, summarily rejected Alexander's offer. Since it was scoring points on the battlefield, Britain felt no urgency, though it had, behind the scenes and in its own councils, already decided to agree to negotiate directly. Madison was surprised and disappointed. His critics snickered. In St. Petersburg, Gallatin and Bayard impatiently awaited either a renewed effort by Alexander or conditions that would allow them to leave. They had no further instructions from Washington. Would the attempt to mediate be revived? If not, what next? The Americans were eager to move the process forward.

It was, in Adams' view, a solemn mission of the highest importance, an opportunity to salvage an honorable peace and fulfill his lifelong desire to be a peacemaker and nation builder. "This day thirty years ago," he wrote to his father, "you signed a definitive treaty of peace between the US of A and GB, and here am I authorized together with two others of our fellow citizens to perform the same service." There was, though, "little prospect of a like successful issue." He felt as impatient, frustrated, and restless as Gallatin and Bayard. And the intensity of his frustration was exacerbated by already having spent five years in St. Petersburg with nothing to show for it. Prayer and poetry helped.

> But from my Heart, spontaneous may arise
> A prayer sincere and fervent, to the skies,
> That all Earth's choicest favors may attend,
> And all the joys, upon my bosom's friend!
>
> That thou wouldst bless, with ever bounteous hand
> My Parents, Children, Friends, and native Land.
> Nor be my Vows, to these alone confined;
> Forgive my foes, and bless all human kind.

John Quincy found it difficult to bless the marriage in mid-February 1813 of Kitty to his secretary and nephew William Steuben Smith. It was hastily arranged, perhaps with concern that Kitty, a few years older than the twenty-four-year-old Smith, might be pregnant. A relationship had developed that prompted John Quincy to have a "very serious conversation with Mr. Smith, who finally avowed a disposition to do right." The couple had no resources or financial prospects, and it seemed to the Adamses that Smith had some of the propensities of his father, who had burdened John Quincy's sister with his impecuniousness. The marriage "looks dark and . . . imprudent," Abigail wrote to her sister. At the same time, news reached St. Petersburg that the mother of the groom had died six months before, emaciated and in great pain, the victim of an aggressive cancer that spread rapidly despite a mastectomy. John Quincy, who had been for months preparing himself for news of his sister's death, was nevertheless stunned into inactivity and grief. "May no lesson of the great teacher Death, be lost upon any one of us. May we all learn to be also ready!"

It was a season of waiting. On New Year's Day 1814 it was so cold in his study that he could not hold his pen without pain. "A nervous agitation, an unsteady hand . . . a flutter of the Spirits, and a perpetual sensation of hurry" expressed his impatience and low spirits. Louisa felt terror "at the idea of what another winter may produce." If only, she thought, her husband could "pass a few months in any mild climate I am confident he might be rapidly restored" from his frequent coughs and head colds. But he was not then and hardly ever seriously ill, though still prone to eye infections and now a hand tremor. He busied himself with the little work he had, especially conversations with Gallatin, for whom his respect grew, and with Bayard, with whom he had had a contentious relationship when they were both senators. They restlessly waited for Alexander, who renewed his mediation offer, and the English to make their intentions clear. "Mr A is even more buried in study than when he left America," Louisa wrote to her mother-in-law, "and has acquired so great a disrelish for society that even his small family appears at times to become irksome to him. . . . The melancholy prospect of public affairs all over the world preys upon his spirits." He immersed himself in astronomy, reading about and observing the planets and stars. And he pursued his interest in historical chronology, weights, and measurements. He and Louisa read poetry separately and together, particularly George Crabbe, Sir Walter Scott, and a new star that had burst onto the literary horizon, Lord Byron, whose *Childe Harold*

Adams found compelling but morally suspect. The poet's ego seemed outsize, the poetry to have more rhetoric and passion than sense. "The plan of the Poem is original, the versification spirited, the character of *Childe Harold* purposely bad."

For wisdom, he resumed reading Sterne's and Massillon's sermons. Reading Thomas Malthus, he thought the argument that population increases geometrically and food supply arithmetically irrefutable. "The chapter on emigration and that on the Poor Laws are highly important, and expose the erroneous policy of all the European governments on these subjects." The Bible was sustaining, the book of Job capturing his full attention. It was, he decided, "a philosophical Romance," not a history of facts, "full of profound and admirable instruction." The narrative exemplified the truths that defined Adams' belief system—there is a just ruler of the universe, "the afflictions of the righteous are trials of their integrity," human beings are frail and infirm, and it is our duty to be resigned to God's will and have "humility under affliction." He regretted that he could not read Hebrew. In mulling over the claims for Christian belief, he read William Paley's *Natural Theology*, containing the influential metaphor of God as the perfect clockmaker from whose creation his existence could be deduced. But it

bothered Adams that Massillon, "among the most distinguished and most virtuous defenders of the Christian
Cause . . . incessantly preaches unrelenting Persecution
to Heretics" and that "Paley curtails the jurisdiction of
Christianity within narrower bounds than of a Justice
of Peace." Since, for Adams, love and charity were the
highest Christian virtues, there could be no such thing
as a heretic. Tolerance was an indispensable virtue,
and though human reason was obliged to test religious
claims, there were claims whose merits were beyond
the power of reason to determine.

By New Year's Day 1814 he knew he was about
to resume his wandering life again. The diplomatic
trio had struggled with their awkward situation for
six months. Could they leave St. Petersburg without
new instructions from Washington? Should they go
to London, where maybe negotiations would begin?
Possibly there were new instructions awaiting them
there. But it seemed to Adams that he had to stay at his
post in St. Petersburg until new instructions arrived.
That, he surmised, would freeze him out of the negotiating process, leaving it all to Gallatin and Bayard.
But since word had reached them that the Senate had
declined to confirm Gallatin, would that not leave as the
only official negotiator Bayard, a man in whom neither
Adams nor Gallatin had confidence? In late 1813 word

came unofficially from London, then from the British ambassador in St. Petersburg, that Great Britain desired direct negotiations. The site chosen was Göteborg, Sweden. At the end of January 1814, Gallatin left for Switzerland, with the possibility that his resignation as treasury secretary would at a later date provide Senate approval for rejoining the peace mission. But that now made Adams head of the commission.

Having survived a cold, restless, frustrating winter, he waited in St. Petersburg for instructions. At the end of March it seemed likely that the allies had entered Paris. "The Empire of Napoleon is in the Paradise of Fools," he wrote to his mother. But it was, he knew, one of many such paradises. He hoped that a treaty could be negotiated that would put an end to American involvement in the British Paradise of Fools. And he hoped that his own country could, in its coming majesty, avoid that delusional state. At the end of April, he said good-bye to Louisa and Charles Francis, who were to remain in St. Petersburg. He implored God's blessing for them all and set out by coach across the frozen landscape.

Chapter 10
My Wandering Life
1814–1817

Two months later Adams was surprised to find himself traveling from Amsterdam to Ghent through familiar scenery that touched him to the heart. When he left St. Petersburg in late April 1814, he knew that he might never see it again, though he had left behind his wife and son as hostages to a climate they had come to hate. For all its imperial splendor and the emperor's solicitations, it had been a place with few friends and much personal heartache. But he was still minister plenipotentiary to the Russian court. He might have to return, and he doubted that his assignment as the first of the commissioners named to negotiate a peace treaty with Britain would be successful. With the end of their fifth year abroad approaching, what he most wanted was to return to America, to his parents, and especially to the sons he had left behind. With a severe cold caught on the North Sea, he stayed in Amsterdam only long enough to realize how exhausted he was. In the meditative mood triggered by revisiting familiar places and scenery after a long absence, he noted that the hotel he lodged at was "the same house where I have always lodged on my visits to this City, from 1780 until now." How many times had he been in Holland since he had first arrived there as a boy of thirteen? How many of the significant events of his education and young manhood were associated with its places and scenery? As he traveled by coach from Amsterdam to Ghent, eager to be the first commissioner to arrive, his quick passage through Haarlem and The Hague slowed the personal timetable of his mind with recollections and strong feelings.

Startled at the overflow of emotion he felt as The Hague came into view, he anchored his feelings in the safety of fact, a brief chronology of his times in the Netherlands, as if getting the facts right was inseparable from identifying what he felt. He had lived at The Hague during "several of the most interesting periods" of his life: in July 1780 when he had come with his father and brother from Paris to Holland; in 1783, when he had traveled alone from St. Petersburg and resided at The Hague for four months; and in 1784, when he was studying for his

Harvard entrance examination. That "was the precise time of my change from boy to man, and has left indelible impressions upon my memory." Ten years later, appointed minister to Holland, he had returned. Much was behind him: a Harvard education, his legal training, and a love affair that had ended under family pressure. He then made his first "entrance on the political theatre as a public man." And it "was here that the social passion first disclosed itself with all its impetuosity in my breast," the feeling of sympathy and identification with others, the desire to make the world a better place. "It is not in my command of language to express what I felt," as he passed the house where he and Thomas had lived, "and thence through the town, along the road . . . to Delft. It was a confusion of recollections so various, so melancholy, so delicious, so painful, a mixture so heterogeneous and yet altogether so sweet, that if I had been alone I am sure I should have melted into tears."

At Ghent, he also had strong feelings: resolution, stubbornness, and pessimism. Prussian troops were withdrawing from the city, and British troops were arriving, the start of a military occupation as the allied armies began sorting out the political and territorial changes resulting from Napoléon's defeat, including the division of spoils. For the time being, the two de facto monarchs of Europe were Emperor Alexander and the

Duke of Wellington. The direction of diplomatic energy was toward Vienna, where a pan-European conference was soon to determine the arrangements for a permanent peace. In Adams' view, it was not likely to be permanent at all. What was certain was that Great Britain now had even stronger muscles for striking the only country with which it was still at war, a war the British resented as unnecessary, initiated at its time of trial by a puny nation that sought to benefit from Britain's preoccupation with Napoléon and that had always, anyway, favored France. Adams had good reason to be pessimistic. Why should Britain offer terms the United States could accept? Although America had not been defeated, it had revealed itself to be badly led, irrationally volatile, internally divided, inept if not cowardly, with an untrained militia and a tiny navy, and with little prospect of righting the foundering ship of state. If Britain cared to pour major resources into the war, no doubt it could inflict additional heavy punishment on a semi-defenseless nation. To Adams and the other negotiators, it seemed sensible to be pessimistic. He settled into his hotel, awaiting the arrival of his colleagues and the British commission.

In late June 1814, a short time after Adams' arrival in Ghent, Emperor Alexander, en route home from London, passed on horseback, in constant rain, through Ghent. His passage was so quick that there was no chance to have a personal moment with the American minister. Adams merely watched from a distance. The emperor was in plain dress, identifiable only because Adams knew exactly what he looked like. He had most recently seen Alexander in the city of Reval, now known as Tallinn, where the American minister had arrived in early May on his way to The Hague. Four days out of St. Petersburg, as he entered the Russian port on the Baltic Sea, he heard bells ringing and saw rejoicing in the streets. Paris had fallen. That night at the theater, he was startled to see Emperor Alexander materialize on the stage, at the climax of a prologue to *Europe Delivered: An Allegory*, in a symbolic representation. Europe was "a fair Lady" in chains, subjected to "an evil genius." The Old Year appeared, accompanied by "War, Poverty, Famine, and Pestilence. . . . Then came another Fair Lady, representing Russia, who broke the chains of Europe; and at whose command the evil Genius vanished." Suddenly the scene changed. A reborn world rose from the Ocean, representing the peaceful and prosperous New Year. The play concluded, Adams wrote in his diary, "with a Chorus from Mozart's Opera of Titus . . . in honour of the

Emperor Alexander." The emperor's bust, "crowned with laurel, appeared at the back curtain of the Scene." Russia and Alexander, enabled by "General Frost and General Snow," had conquered Napoléon. If only Napoléon, Adams noted, "with his extraordinary genius and transcendent military talents, possessed an ordinary portion of judgment or common sense, France might have been for ages the preponderating power in Europe." Now "the sufferings of Europe are compensated and avenged in the humiliation of France." For a brief moment all Europe idolized Alexander, "the darling of the human race."

Crossing Sweden from east to west, on the same road between Stockholm and Göteborg he had traveled more than twenty years before, Adams learned, to his regret, that the site of the peace negotiation had been changed. He would have liked to argue against holding the talks in a city occupied by the enemy. His delay at Reval, as "the prisoner of the ice and the winds," had made that impossible. When the British offered Ghent or London, Bayard and the newly confirmed Gallatin, who had resigned as secretary of the treasury, thought the Belgian city preferable to the enemy capital. In St. Petersburg, Adams had been impressed by Gallatin's "quickness of understanding, his sagacity and penetration, and the soundness of his judgment."

Six years older than Adams, his accent revealing his Swiss birth, Gallatin had risen to high office by virtue of talent alone. Handsome, slight of build, soft-spoken, a master of statistics and monetary policy, with a sense of humor that lightened argumentative discussion and an ability to persuade inoffensively, he maintained in his American life a touch of European courtliness. To his critics, he was too much the foreigner. To his admirers, his abilities and personality were perfectly matched. Gallatin remembered seeing Adams studying law in Newburyport in 1788. An immigrant traveling in New England, he had thought Federalist Boston even more dull and puritanical than his native Geneva.

Two additional commissioners had been appointed, Henry Clay and Jonathan Russell. A loyal Jeffersonian from Massachusetts and a war hawk, Russell had also been appointed minister to Sweden. Four years younger than Adams, he lacked the dedication and talent of his colleagues. Bayard, almost exactly Adams' age, had taken the harsh ultra-Federalist line during Adams' Senate term. An impressive orator, he was intemperate with words and wine, with a talent, Adams noted, for slander and mischief making that he seemed to practice impartially. In St. Petersburg, he had bad-mouthed Adams to Gallatin and Gallatin to Adams. At first strongly opposed to the war, he became one of the few

Federalists to support the government once it had been declared. During his fifteen-year career in the House and Senate, he had had the good sense, which Adams respected, to remind his colleagues that, since they refused to provide funds for the military, a declaration of war was undesirable.

A decade younger than Adams, Clay had served in the Senate and House, twice as speaker. A mercurial personality and gifted orator, he was an idealistic patriot with an immense ego. Like Bayard, he had little intellectual curiosity and the politician's gift of not seeing the slightest gap between his own ambition and his country's well-being. In 1806, when Adams had heard Clay speak against the importation of slaves, he had been impressed by Clay's rhetorical skill and commanding platform presence. "His school has been the world," Adams was to write in 1821, "and in that he is a proficient. His morals, public and private are loose, but he has all the virtues indispensable to a popular man." Lanky and tall, he had sparkling eyes and light blond hair. He dressed well, reveled in his own charm, and had unlimited confidence in his powers of persuasion. His "temper is impetuous and his ambition impatient," Adams noted. As speaker, Clay had taken a leading role in encouraging Madison to approve and Congress to declare war. An inveterate gambler who loved cards, whiskey, and late night entertainments, he was a Westerner both in essence and in pose, as if there were no confusion between who he pretended to be and who he was. Clay seemed to Adams a brilliant but limited politician, his assets charm, oratorical skill, and political shrewdness.

When by early July 1814 the British commissioners had not yet arrived, it became clear that the delay was purposeful. Military realities favored the British. The Atlantic Coast was blockaded, American commerce and revenue were severely curtailed, and the Eastern seaboard cities were easy targets for the British fleet. New England was keeping its cooperation to a minimum. That Canada was ripe fruit about to fall had proved to be a delusion. For over two years, most of the news from the United States had been "mortifying." If another British incursion could take New Orleans, the United States would be at a further disadvantage. The American commissioners had good reason to be pessimistic. Why should the British rush into negotiations? And, once they began, why not keep progress dependent upon news from the battlefront? The delay seemed a bad omen. Six weeks after the Americans, the British finally arrived. What Adams feared, as the negotiations began, seemed a likely reality: the process would drag on for months. Or it might end abruptly with no agreement. And what was there to negotiate about? Impressment. But it was, for the British, a non-negotiable item. For the Americans, it was the main item, raised to the top of the list when Congress, the president, and the American press found themselves deprived of their first rationale by the British withdrawal of the Orders in Council that had sanctioned seizing American merchant vessels. To withdraw the demand that the British renounce impressment would be a bitter pill to swallow. The half of the country, Adams realized, "which approved the war would never approve or be satisfied with a peace which should give up the point of impressment." But there could be no peace treaty if the United States did not.

The British negotiators immediately put the Americans on the defensive. "Divided among ourselves," Adams wrote, "with half the nation sold by their prejudice and their ignorance to our enemy, with a feeble and penurious government, with five frigates for a navy and scarcely five efficient regiments for an army, how can it be expected that we should resist the mass of force which that gigantic power has collected to crush us at a blow?" Having assumed that Ghent's proximity to London would be an advantage to the British, Adams soon realized that it limited their flexibility. In contrast, the Americans had to make decisions on the spot, with the hope that these would be approved later. If they held firm on the important issues, they would be supported. And while the Americans needed to negotiate among themselves, with an eye toward what Washington might eventually think, the British commissioners—James Gambier, Henry Goulburn, and Dr. William Adams— had to clear every move with their superiors in London: Lord Castlereagh, the foreign minister; Lord Bathurst, the colonial secretary and secretary of war; and the Earl of Liverpool, the prime minister. A naval admiral, Gambier's claim to infamy was that he had commanded the British bombardment of Copenhagen in 1807. William Adams was a maritime lawyer. Only Goulburn, previously undersecretary of state for war and colonies, had any significant presence in London. The smartest and best of the British negotiators were preoccupied with Vienna, not Ghent. Still, though London wanted delay, it also wanted a peace treaty. From the start, the three Tory ministers had in mind to squeeze as much out of the Americans as possible but also to achieve a treaty that took into account the stressed British treasury and its European obligations. Castlereagh, a brilliant strategist and the council's intellectual leader, envisioned a mutually profitable long-term commitment to peaceful free trade with the United States and the emerging South American republics.

To their shock, the Americans were presented in August 1814 with a list of outrageous demands. There could be no peace treaty, the British maintained, unless the United States agreed to a new boundary with Canada, including a buffer zone consisting of almost the entire Northwest Territory to be reserved for the Indian allies of the British; the Indian tribes were to be recognized as independent nations and the British allowed to trade with those remaining on American soil; all American naval forces were to be withdrawn from the Great Lakes; areas of northern Maine would be turned over to the British as a transit corridor; the British right to navigate the Mississippi River, guaranteed in 1783 by the Treaty of Paris, would be affirmed; and the right of Americans to fish in Canadian waters and dry fish on the shoreline, also part of that treaty, would be canceled. In effect, the British negotiators presented as the condition of a new treaty a revocation of a substantial part of the 1783 peace treaty. It had been nullified, they claimed, by the American declaration of war. London was pushing and testing.

When news reached Ghent that in August a British landing party had burned much of Washington, the American commissioners were appalled. But when the British were forced to abandon an attack on Baltimore, it was clear that significant offensive gains were unlikely.

It had been all along for the British mainly a defensive war, London's highest priority being the retention of Canada. The attack on New Orleans, planned for the end of the year, was to be only another incursion, an attempt to elicit better peace terms. The same was true for a planned thrust from Canada into northern New York state at Lake Champlain, especially since the Great Lakes still remained mostly under American control. Week by week, as the Americans angrily rejected the entire list, they expected the talks to collapse. Any agreement that consented to the British rejection of the 1783 treaty and the loss of American territory would be denounced and renounced. Its perpetrators would live in infamy. And the American commissioners were still impaled on the cross of impressment. Point by point, the Americans beat back the British demands. Clay performed well, his gambling and bluffing skills useful; Bayard kept his cool and made sensible suggestions; Adams provided expertise, knowledge, stubbornness, and writing skills; Russell was a minor contributor; and Gallatin diffused tensions with his good-humored negotiating temperament. Although they made tactical concessions to keep the negotiations alive, the Americans held firm on all the major points. Though there were disagreements, on the important matters the commissioners remained united.

Meanwhile the Americans waited for further instructions, especially about impressment. Both sides looked for news from the battlefield. Clay, Bayard, and Russell, to Adams' tacit disapproval and bafflement, often played cards and drank heavily into the wee hours of the morning. But differences of temperament did not undermine their working sessions. When Castlereagh passed through Ghent on his way to Vienna, he drily tamped down the obduracy of his negotiators. The negotiators had a more stringent view than their London superiors. They had the incorrect impression that their instructions were absolute: either the Americans would agree to these harsh demands or the war would continue. By early November 1814, both sides had received important news from America. At first the British newspapers reported that the American military had been defeated at Plattsburg, New York, but the report proved incorrect. The British had been defeated, with the U.S. Navy the decisive factor. The British press and popular opinion reacted furiously. The war must continue; the United States must be destroyed. But Liverpool's cabinet, which had already transformed demands into negotiating points, had no intention of being bullied into further misery. "The continuance of the American war," the prime minister wrote to Castlereagh, "will entail upon us a prodigious expense, much more than we had any idea of." The costly expeditionary force on its way to New Orleans probably would not produce a victory soon enough, he assumed, to affect the terms of the negotiation, unless it was to be dragged on into the new year. The naval force would influence the war or peace only if the negotiation at Ghent failed.

By mid-November, the British had decided to move quickly. "I think we have determined, if all other points can be satisfactorily settled, not to continue the war for the purpose of obtaining or securing any acquisition of territory," Liverpool wrote to Castlereagh. Because of the "unsatisfactory state of the negotiations at Vienna . . . the alarming situation" in France, "the state of our finances, and . . . the difficulties we shall have in continuing the property tax . . . it has appeared to us desirable to bring the American war if possible to a conclusion." Castlereagh agreed. New instructions went to the commissioners. At the same time, the ministers played with an alternative policy: would the Duke of Wellington lead an army against the Americans in order to force the United States to cede territory along its northern border? Although he at first thought favorably of the idea, Wellington soon concluded that his participation, let alone the continuation of the war, made little sense. Even the attempt to retain the American territory now occupied by British forces had no basis in reality. Without naval superiority on the Great Lakes, further military action was unsustainable. "I confess that I think you have no right," he wrote to Liverpool, "from the state of the war to demand any concession from America." He advised that they settle for a return to the situation prior to the declaration of war.

That is exactly what the American commissioners had just proposed. Clay thought the proposal likely to be insultingly rebuffed. "I was earnestly desirous," Adams wrote to the former Georgia senator William Harris Crawford, now the American minister in Paris, "that this offer should be made, not from a hope that it would be accepted . . . but from the hope that it would take from them the advantage of claiming that our proposed articles" showed the Americans were not inclined to make peace. Gallatin and Bayard agreed with Adams. Clay and Russell disagreed, partly because one of the inducements allowed the British to continue to trade with the Indians. Fearing the reaction of his Western constituency, Clay argued that it was against instructions. It would leave impressment unaltered. But it was not expressly against instructions, Adams and Gallatin responded. Less than a week later, when new instructions arrived from Washington, Adams, Gallatin, and Bayard were vindicated. Madison authorized concluding "the peace on the basis of the Status ante Bellum, precisely the offer which we have made in our last Note, and of which I found it so difficult to obtain the insertion." At the same time, the British negotiators received new instructions, withdrawing the demand for a buffer between Canada and the United States. A disappointed Goulburn remarked that he had had no idea until he came to Ghent "of the fixed determination which prevails in the breast of every American to extirpate the Indians and appropriate their territory; but I am now sure that there is nothing which the people of America would so reluctantly abandon as what they are pleased to call their natural right to do so."

By the end of November they were back at the negotiating table, the tone good humored. Although there was more sparring to be done, a peace treaty was imminent. It took a month to dispose of a few nasty details. The British still wanted navigational rights on the Mississippi and the end of American fishing and drying rights in Canada, which they defined as a "privilege." And they wanted Moose Island, off the Maine coast, which they occupied. Clay envisioned a damaging threat to his Western identity and political career. Adams, loyal to New England, would rather have died than give up American access to the sacred cod. There was an angry altercation between them about which, if either, should be given up. "Mr. Gallatin brought us all to union again by a joke," Adams appreciatively noted in his diary. Gallatin "said he perceived that Mr. Adams cared nothing at all about the Navigation of the Mississippi, and thought of nothing but the fisheries. Mr. Clay cared nothing at-all about the fisheries, and thought of nothing but the Mississippi. The East was perfectly willing to sacrifice the West; and the West was equally ready to sacrifice the East." But, he argued, they would lose both if they agreed to lose either. Clay rightly sensed that the British would not break off negotiations. Adams worried that they had been led down the garden path, that the British strategy had been to insist on their Mississippi and fishing positions in order to force the Americans to take the blame for failing to end the war. Clay turned out to be right. Both sides badly wanted a peace treaty.

The process of working out the details had occasional touches of rancor. But they followed the cardinal strategy of diplomacy: everything that could be settled, they settled; everything that could not, they postponed. The negotiators agreed to make no reference to the two issues over which Adams and Clay had fought, an agreement that protected them both. Although commissions would be established to determine boundaries and the ownership of disputed territory, the overall Canadian border would be guaranteed, the Great Lakes kept free of navies, and the Indians on American territory remain under American jurisdiction. Mediation would settle outstanding differences. None of the British principles or practices to which the United States had objected was renounced, and America got no territorial advantage from the war it had started. At best, the United States had earned the right to boast that it had survived a war against the greatest power in the world. It had entered the lion's den and come out intact, though not unscorched. How many had died? Been crippled? How many families had lost loved ones and breadwinners? How much national treasure had been wasted? Still, for Adams and many Americans, even the New England antiwar faction, there was every incentive to look forward, not back, to emphasize national pride. When the peace treaty was signed on December 24, 1814, Adams said to Gambier that he "hoped it would be the last Treaty of Peace between Great-Britain and the United States." Copies were immediately dispatched on their month-long trip to Washington. In Ghent and London, there was suppressed awareness of the now superfluous British expedition to New Orleans. But not even the fastest ship could get there in time to stop it.

The American commissioners tidied up. Clay wanted, for political advantage, to sail home with the mission's official papers, a flourish of assertion and mastery. As head of the commission, citing precedent, Adams refused to give them up, then offered to do so if Clay would sign a statement detailing the process that had resulted, over Adams' objections, in his obtaining them. Clay blustered, swore, and then refused to sign. Adams had checkmated Clay's strategy. Clay retreated into pleasantry and compromise, asserting that he had always been and continued to be on the high ground. Both he and Clay were prone to be irritable and overbearing, John Quincy admitted to Louisa, but whether Clay had made it, as much as John Quincy had, "the study of his life to acquire a victory over it, and whether he feels with as much regret after it has passed every occasion when it proves too strong for him; he knows better than I do." How would an impartial judge distinguish between them? "I think [he] would say that one has the strongest, and the other the most cultivated understanding; that one has the most ardency, and the other the most experience of mankind." Much of the tidying up was pleasant, even triumphant; if no victory was gained at least a war was not lost.

Remaining in Ghent, they basked in the plaudits of a population that got great pleasure out of what seemed a British retreat. With the city occupied by British troops, the citizens delighted in lauding everything American. "The hatred [here] of the English is so universal, and so bitter." You could not "easily imagine," John Quincy wrote to Louisa, "how *Hail Columbia* has become the most popular song in Ghent. . . . Now it is everywhere played as a counterpart to *God Save the King.*" "Thus ends the most memorable year of my life," he noted in his diary. Each of the commissioners looked toward his next assignment. From Ghent, Bayard and Clay planned to return home via Paris. Gallatin intended to remain in Europe, with the possibility that he might replace Crawford as minister in Paris. Adams had two main commitments. One was not to return to St. Petersburg, to "the endless night of a Russian Winter." The other was to be reunited with Louisa and his children.

Their common destination was Paris. In Brussels, en route to Paris in late January 1815, he went to the theater, was an honored guest at the literary club, and attended Mass on Sunday, interested, as always, in church practices. "Whenever the kneeling time came an [altar boy] gave notice of it by the ringing of a bell. This is their form of worship. A mass every hour, from seven in the morning until noon; and the

worshippers come in and go out when they please."
What intrigued him even more was a visit to the best
private collection of paintings he had ever seen, the
great Dutch and Flemish painters, with four portraits
by Holbein, and paintings by Van Dyke, Rubens, and
even Leonardo da Vinci. After the work of the last six
months, this was an aesthetic holiday. He stretched
his legs on the streets and in the museums. A few
days later his recollections were stirred when he entered Paris. "After an interval of thirty years," he
wrote to his mother, "I revisited that great city, where
all the fascinations of a luxurious metropolis had first
charmed the sense of my childhood, and dazzled the
imagination of my youth."

The city still dazzled him. It was where, as a young
boy, he had fallen in love with a young actress. It was
the first great city he had ever seen, the first European
capital he had lived in. "The tendency to dissipation . . .
seems to be irresistible," he noted, a limpness "against
which I am as ill-guarded as I was at the age of twenty."
Apparently he found it pleasurable not to resist. When
he had last been in Paris, "its streets, its public walks,
its squares, its theatres, swarmed with multitudes of
human beings as they do now. And in walking through
the streets now they present so nearly the same aspect
as they did then, this Rue de Richelieu . . . looks so

exactly like the Rue de Richelieu which I first alighted [at] with my father in April, 1778, thirty-seven years ago, that my imagination can scarcely realize the fact that of its inhabitants certainly not one in a hundred, probably not one in a thousand, is the same."

It was a city that was also referencing its past. Its government had been reconfigured by the allies, who preferred a sclerotic monarchy to a military dictatorship. Napoléon was a semi-prisoner on the island of Elba, formidable as a former emperor even in his defeat and abdication. The Bourbons had been restored to the throne. For the first time in over twenty years France was at peace as the allies attempted to steady the various factions, hoping to arrange for decades of European stability. Adams, though, expected that differences among them would set them to war again, and again make America the victim of events beyond its control, voiding the advantages of the Ghent peace treaty. He sensed that there was a powder keg in place. And though France seemed in some ways as it had been thirty years before, some of those ways were distressing. Crowds of beggars—"objects of great wretchedness"— besieged his carriage as he entered the city. Poverty, hard times, unemployed veterans, and class antagonism characterized the tensions among those who had benefited from the revolution and those who had been victimized. In Paris theaters, pro-royalists played monarchical tunes and cried *vive le roi*. Adams noticed how cynically pliable Parisians were, singing whatever was the patriotic song of the moment. Self-interest, he observed, trumped ideology.

At the Tuileries, the king asked if he "was any relation to the celebrated Mr. Adams." John Quincy called on French associates from former days, especially General Lafayette. His French social circle included Count Barbé-Marbois, to whom, as a twelve-year-old boy, he had taught English on shipboard sailing to America. He had a warm reunion with Baron von Bielfeld, his friend at The Hague from 1794 to 1797. He enjoyed the National Museum and especially the theater, which he delighted in, endlessly, whether the plays were good or bad, well or ill performed. "The taste for frequenting the theatre grows upon me, and must be controlled." But he was in an indulgent mood. At the Odeon, he enjoyed Mozart's *Figaro*. He found the music "charming" but not equal to a wildly successful comic opera by Domenico Cimarosa, *The Secret Marriage*, which over the years had become his personal favorite, "the most enchanting music that I ever heard."

At the Théâtre-Français he was particularly engaged by Molière. As always, his judgment was both aesthetic and moral, though these criteria were not always in harmony, especially when Molière's comic genius provided plots and characters in which values sacred to Adams were ridiculed. "It is the contempt shed upon old age, and particularly upon the paternal character; the irreverence of children to their parents; the complacency with which fraud and swindling is represented," which he disliked in *Scapin's Deceits*. They were "moral and literary defects." He attended a performance of *Le Misanthrope*, the house "crowded as if it had been a new Play." He had a seat in the pit, "which at Paris is the criticizing part of the audience, and it is amusing to observe how sensitive they are to every incident that occurs upon the stage. I was surrounded by persons who at almost every first line . . . would repeat in muttering the second. Who censured and approved the manner of the actors at every passage, and as I thought almost always correctly." These months in Paris, he later recalled, were "in many respects the most agreeable interlude . . . of my life."

In contrast, Louisa's journey from St. Petersburg was formidable. John Quincy had written to her from Ghent to join him in Paris as soon as she could. Lonely in St. Petersburg, overwhelmed by the need to make daily decisions, she was also capable and determined, eager to be reunited with her husband and her sister Kitty, who with her husband and infant daughter had moved to Paris. Without a job or an income, Smith consulted his uncle about future employment, hoping to collect money he believed was owed him for official duties. His uncle worried about his character. Louisa hoped once again to have Kitty's company, though, to her disappointment, she was to discover that the Smiths had left for the United States one day before her arrival in Paris. Russian officials and the French ambassador provided Louisa with introductions and passports. She had cash in hand in the form of gold and letters of credit. On a freezing day in mid-February 1815, her Russian-style carriage, on sleigh skids, slid out of St. Petersburg into a white landscape, followed by a smaller vehicle with three servants. The route was via Riga, Königsberg, Berlin, Leipzig, Frankfurt am Main, and Strasbourg, almost two thousand miles of iced-over rivers, precipitous hills, muddy roads, vehicle breakdowns, greedy merchants, hostile local populations, and a landscape still strewn with the skulls of Napoléon's battles.

She soon encountered an unanticipated danger. Napoléon had left Elba in late February. By early March, he was moving toward Lyon, gathering troops and support, his destination the same as Louisa's. The countryside exploded into political and military activity. Getting fresh horses and drivers at post stops became increasingly difficult. Sometimes she had to exercise all her self-control and cleverness to resist panic when exploited by greedy vultures and threatened by the desertion of her French-born servants, who had expected to return to a monarchical France. They all feared they would fall into the hands of pro-Napoléon Jacobins. Ahead were revenge murders, summary executions, and merciless soldiers. As they passed through excited crowds and by marching troops, their carriage provoked rumors, including that she was Napoléon's sister on her way to join him. At Château-Thierry, they found themselves in the midst of the Imperial Guards on their way to meet the emperor. "Presently I heard . . . 'tear them out of the Carriage; they are Russians take them out kill them.'" When she showed her passport to an officer, he shouted that she "was an American Lady, going to meet her husband in Paris—At which the Soldiers shouted 'vive les Americains'—and desired that I should cry vive Napoleon! which I did waiving my handkerchief." Charles, who "seemed to be absolutely petrified," sat by her "side like a marble statue." As they approached Paris, drunks lined the road, shouting "down with Louis XVIII, long live Napoleon." On March 23, 1815, forty days after leaving St. Petersburg, her carriage rolled onto the Paris streets, through the Porte de Saint Martin and into the Rue de Richelieu, to the Hôtel du Nord. Napoléon, already ensconced at the Tuileries, had beaten her to Paris. But that she had gotten there at all was a small miracle.

As he waited for Louisa and Charles, John Quincy had a ringside seat at the return of Napoléon to Paris. Imprecisely and with many self-interested distortions, the news of Napoléon's progress reached the capital. On March 11, Bielfeld told Adams of a reliable French report: Napoléon had taken Lyon, less than three hundred miles away. Stagecoaches were mounted in every section of Paris. A flight to safety had begun. Concerned, John Quincy bought a map indicating all the official post stops. Which was the best road from Berlin to Paris? Was it possible that Louisa's route and that of the royalist troops or Napoléon's supporters might intersect? At the theater, he overheard an agitated man remark that the king had left Paris, which turned out not to be true. Like almost everyone else, Adams had not thought it possible that Napoléon, with a few men, could represent a military threat to the French government. But the rainy streets, as he walked home, were crowded with soldiers, with sentinels at every corner. People were gathered around placards, reading them by lamplight. The city seemed increasingly agitated. A handbill announced that there would soon be very satisfactory news from the king: Lyon would be speedily delivered. Gallatin and Lafayette seemed optimistic.

Baron von Bielfeld "concurred in the opinion prevailing that the Government will be maintained. . . . The number of volunteers who have offered themselves at Paris to march against Buonaparte is greater than the government could accept. . . . It is ascertained that a part of the troops as well as the highest officers are faithful to the king; and Napoleon's soldiers will probably desert him in the end." But the crowded theaters of a week ago were now almost deserted, though the pro-monarchical audiences demanded royalist cheers and music. On March 15 he had a ten-day-old letter in hand from Louisa.

After dinner on March 19, John Quincy walked around the Tuileries and the Place du Carousel through large crowds shouting their support of the king. "If the slightest reliance could be placed upon the most boisterous and unanimous expressions of public feeling," he wrote to his mother, "the only conclusion would be that here are twenty-five millions of human beings contending against one highway robber. In private conversation the universal expectation is that Buonaparte will enter Paris . . . without opposition. . . . It was but last Thursday," he noted, "that the king . . . talked before the two Legislative Chambers of dying in defence of the Country." Soon the same crowds had another view. "The cries of Vive l'Empereur had already been substituted for those of Vive le Roi." Crawford told him that when the royalist officers of the Paris garrison had urged the troops to fight for the king, "the soldiers would say . . . Oh! yes! Vive le Roi, and laugh." On the evening of March 21, the emperor marched into Paris at the head of "the same garrison . . . which had been sent out against [him], and who entered the City with him." The spectacle and the ironies were irresistible. Mingling with the crowds, Adams heard people "laughing and joking . . . swearing vengeance against the Prussians." That night he went to the opera, but the leitmotif of his consciousness was expectation of Louisa's arrival. Where was she? On March 23, when he returned home at 11 P.M., he had, as he had numbers of times already, expected to find her there. But her carriage was not in the yard. A few moments later she arrived. "She and Charles are both well, and I was delighted after an absence of eleven Months to meet them again." His usual understatement concealed his immense relief and great pleasure.

Adams once again had a front-row seat at an enticing and complex political spectacle. He had no doubt that the Paris populace and the army, who shared a "sympathy of sentiment," welcomed Napoléon's return. They preferred the emperor to the king. "If the people of Paris had been seriously averse to his government, the national guards of the city alone would have outnumbered five times all the troops that [he had]." Having imposed heavy taxes and repossessed expropriated state property, as if nothing had changed, "the government of Louis XVIII had rendered itself more odious to the mass of the nation than all the despotism and tyranny of Napoleon had made him in ten years." Adams had no doubt that Great Britain and its allies would attempt to put Louis back on his throne, and much as the Parisians preferred Napoléon, their first thought "will be to save themselves. . . . They will submit quietly to the victorious party, and will do nothing in support of either. If the same Spirit should prevail throughout France, Napoleon will be soon overthrown."

In his downfall, rise, and likely second fall, Napoléon seemed to Adams a more interesting phenomenon than ever before. He had offered the allies peace if they would leave France in peace. But a Europe with Napoléon in power was unacceptable to Britain, Prussia, Austria, and Russia—an ideological insult and a danger to their security. "Napoleon," Adams granted, "is no fit person to built a temple to the name of the Lord." But, as with Milton's Satan, " 'neither do the spirits reprobate *all* virtue base.' " The comparative baseness or questionable virtues of a national leader needed to be evaluated in comparison with the alternative. Who had less virtue, Napoléon or the royalist regime? "Had the name of Napoleon Bonaparte remained among those of the conquerors of the earth, it would not have been the blackest upon the list." After all, he was not a mass murderer, a tyrant who ruled by cruel tortures and wholesale executions. Americans had no reason to think less well of him than they had thought of George III, and more reason to think well of him than of a long list of equally absolutist but more brutal European monarchs.

The next month, in the crowded Tuileries gardens, when the emperor "stood about five minutes at one of the windows and was hailed with loud and general acclamations of Vive l'Empereur," Adams "had a more distinct view of his face than when [he] had last seen him." When he heard from Crawford that Napoléon had requested a performance of one of his favorite plays, Luce de Lancival's *The Tragedy of Hector*, John Quincy took Louisa to the Théâtre-Français, where the house was so crowded that the orchestra was forced to forfeit its seats and play from behind the stage. After the first scene, the emperor arrived to shouts of *Vive l'Empereur*. The theater swelled with republican music; the "Marseillaise" was played over and over again. "Never at any public theatre did I witness such marks of public veneration, and such bursts of enthusiasm for any crowned head as that evening exhibited for Napoleon." Since he could get seats only "on the same side of the Theatre as he was seated," neither John Quincy nor Louisa could actually see the emperor. The play, closely adapted from Homer's *Iliad*, with lines from the original translated into classical French verse, "turns . . . upon the interest of a heroic character," Adams noted, "who deliberately sacrifices his life to the defence of his country." Napoléon's identification with Hector distinguished him from his royalist enemies. "Where is the head or the heart among them capable of rising to the admiration of such a character . . . ? Their Hector belongs not to tragedy but to comedy; not the champion of Troy, but the knave of diamonds." Although he had never been an admirer, Adams concluded, as he calculated degrees of baseness, that Napoléon had at least a few virtues that an American could respect.

By early April 1815 Adams knew that he would not be going home. President Madison had appointed him U.S. "Envoy Extraordinary and Minister Plenipotentiary . . . at the Court of His Royal Highness the Prince Regent of the United Kingdom and of Great Britain and Ireland." In May, after the ministers from countries hostile to Napoléon had left, Adams had a long conversation with his former St. Petersburg colleague, the Marquis de Caulaincourt, now Napoléon's foreign minister. He was surprised to find Caulaincourt optimistic. Wellington, who had been the de facto ruler of France, and the allies, Caulaincourt maintained, were responsible for Napoléon's return. They had put the incompetent Louis XVIII back on the throne; the Bourbons were Britain's stooges and self-aggrandizing reactionaries "so odious that if the Emperor had not come, there would nevertheless have been within two months a Revolution against them." But, Adams knew, no matter what Napoléon's intentions, any Napoléon at all was intolerable to the allies. There would be war. When he returned from a walk with Charles, the long-awaited certainty of change had arrived from Baring Brothers, America's banker in London. They had in hand and held for him his commission to Britain. "I determined to proceed to England with as little delay as possible."

There were some good-byes to be made, particularly to General Lafayette, whom he had seen much of in Paris and at the Grange, Lafayette's country home. At the middle of the month, he delayed departure a few days in order to see Lafayette one last time, an act of respect and affection for a man he assumed he would never see again. He also wanted to sound Lafayette out.

Would he use his influence to prevent America's being victimized again by French and British blockades and seizures? And what kind of government did he anticipate for France, whatever the outcome of this new conflict? He was not surprised to find Lafayette ambivalent about Napoléon and uncertain about what was best. "He wants Napoleon, as a military commander, to defend the country, and yet fears the consequences of his success. He thinks much of the Duke of Orleans. Much of a Republic. . . . At bottom there is an ardent love of his country, and a sincere desire that it should be governed by a free and liberal constitution; but with regard to the means he has nothing settled, and waits for events."

Adams' Paris interlude was over. The family left for Le Havre through a countryside covered with orchards in full bloom. At Dieppe, he arranged for passage to England on a Danish ship and, with a wine merchant, "for a sample of Champagne wine." He was determined "to take a stock of that and other French wines" with him to England. He had developed a taste for Bordeaux especially. Having first set foot in wine country almost forty years before, he had enjoyed superlative wines. In America, wine was a rarity. It had to be imported, which Adams, like Jefferson, was to do. Before the *Olga* sailed, he bought "a small assortment of wines," and added the cases to the family's dozen trunks. "The custom house officers made some difficulty about embarking the wines . . . until I produced my passport from the Duke de Vicence, with which they were entirely satisfied." He himself was happy to be crossing the Channel. It was not the moment to open the champagne, but there were celebratory possibilities. He was to occupy, as his father had done, the highest American position in Britain. Although they were not going home yet, they were moving closer. And an important part of home was moving toward them: they had sent for fourteen-year-old George and twelve-year-old John. Perhaps the boys were already in London.

On May 24, "Dover Castle was in sight." At the York Hotel, the hotelkeeper "recollected my father and mother and Coll. Smith and my Sister, who he said had all repeatedly lodged at his house. I remembered having once been there myself in August 1784." The countryside from Dover to London seemed just as he remembered it. "I had travelled the whole way twice, and the greater part of it, many times." It had been eighteen years since he had last been in England. Outwardly, it looked much the same. Inwardly, and in relation to America, there had been great changes. When he and Louisa drove up to their lodgings at 67 Harley Street, Cavendish Square, London, their two sons were waiting for them.

———————

Suddenly John Quincy was busy, harried by responsibilities. He needed to move quickly to set up an office and residence, and he needed to keep economy in mind. Also, he had no secretary. Within days of his arrival, he had a long discussion at the Foreign Office with Castlereagh about discriminating duties and impressment. When, in early June, he presented his credentials to the prince regent, who, as usual, was not well informed, "the Prince took the letter, and without opening it, delivered it immediately to Lord Castlereagh; and said in answer to me, that the United States might rely with the fullest assurance upon his determination to fulfill on the part of Great-Britain, all the engagements with the United States. He then asked me if I was related to Mr. Adams, who had formerly been the Minister from the United States here. I said I was his Son." A few days later, at a dinner hosted by Lord Carysfort, whom he had known at St. Petersburg, he listened to Sir Humphry Davy talk about his recent visit to Italy and his chemical discoveries. One of the guests predicted that Napoléon "would beat them all." Adams disagreed. When a few days later Wellington defeated Napoléon at Waterloo, the Adams family drove through the streets of London ablaze with spectacular celebratory illuminations.

The search for a house and servants pressed his schedule. They had brought with them John Quincy's valet, Antoine Giusta, a Piedmontese whom he had hired on his way to Ghent, and Louisa's maid. When Adams discovered that the cost of a suitable house in London was beyond his means, he rented "Little Boston House" in the village of Ealing, eight miles west of central London, an hour and a half away by carriage. Convenient and comfortable, with a coach house and garden, it was available for immediate occupancy on an open-ended lease with a one-month-notice requirement. In London, at 25 Charles Street in Mayfair, he set up a small office with a place for him to stay overnight. Americans in London needed an address to come to with requests, particularly for passports. A clerk would act as copyist and staff the office.

When he and Louisa found the two sons they had left behind waiting at Harley Street, they were overwhelmed with emotion. In the excitement, Louisa fainted. The reunion left them all dazzled. The two older boys had last seen Charles Francis when he was two years old. George had become unrecognizable. In July, from their Harley Street lodgings, they took long walks to get fresh air and exercise in Kensington Park, where the boys "played bat and ball." John Quincy took cold baths with his sons. They all went to museums and the theater, where they saw *Romeo and Juliet* and Milton's *Comus*. It was a happy time, a renewal of and new introduction to family life. Marking his forty-ninth birthday, John Quincy had reason, he felt, to be more than satisfied. "It has in relation to public affairs been the most important year of my life, and in my private and domestic relations one of the most happy years." English weather seemed blissful, "a paradise compared to that of St. Petersburg, where life itself scarcely deserves to be called existence. I left that country a skeleton," he wrote to his mother, "and verily believe that before the end of another winter I should have left the skeleton there. Since the days that I quitted the banks of the frozen Neva I have been steadily redeeming flesh."

The Adamses moved to Ealing five days before they marked the anniversary of their sixth year away from the United States. Charles and John were enrolled at nearby Great Ealing School. Since a pew at St. Mary's came with the house, they attended Anglican services. When, after a few weeks, George reported that his brothers hated the boarding school, Adams investigated. He found the boys wanting, not the school. "They have never before been accustomed to the restraints of an English School; and both . . . have been several months released from all study, so that a return to it is irksome. . . . I did not however discover that there was any solid ground of complaint." George joined him in the early morning for Bible reading and Latin and French lessons. When it became clear that home schooling had its limits, he joined his brothers, though as a day student. With John and Charles, John Quincy played backgammon. They all went on fishing parties to the River Brent, a Thames tributary, and the Grand Junction Canal, sometimes with Peter Pio, the butler, and with Louisa, who enjoyed the sport as a happy way to spend time with her sons. It was an activity that came naturally to the family of a Massachusetts father who, like his own father, had the highest appreciation for what fish had done for his state. He wanted his sons to have the same sporting childhood that he had had. But study came first.

After his semi-holiday in Paris, Adams plunged into a full work schedule. He headed the negotiations for a trade agreement. He also needed to pry money from and manage loan repayments to Baring Brothers. And there were constant applications from American sailors, often reduced to begging, who had been impressed into the British Navy, and from American prisoners of war released from the infamous Dartmoor prison. The Ghent treaty had not established a protocol or source of funds. British poverty was palpably visible.

Hungry wanderers begged for food at the door of Little Boston House, "each with a different hideous tale of misery." On one of his walks from Ealing to Charles Street, he "saw a man, decently dressed, lying stretched upon the ground . . . his face downward, and apparently asleep or dead. . . . I then spoke to him and he answered. Said he was not in liquor; but had a bad leg. . . . I asked him if he was in want. He said he had eaten nothing for two days. . . . I gave him a shilling. . . . The number of these wretched objects that I meet in my daily walks is distressing. . . . It is not a month since a man was found dead, lying in a field by the side of the road. . . . The extremes of opulence and of want are more remarkable and more constantly obvious in this country than in any other that I ever saw." He and Louisa helped, with food and medical care, an out-of-work Staffordshire black-smith whose son was dying of smallpox. That there were beggars in America John Quincy knew from experience. But the numbers in Great Britain made American poverty seem minuscule and easily managed. In the United States there was more work to be had, even in hard times. It pained Adams to see starving Americans, many able-bodied, some damaged by war and imprisonment, forced to beg in the streets of a foreign country. He worked tirelessly, and stretched his own and the embassy's resources to assist them.

Adams felt the irony of the prince regent's flippant assurance that his country would fulfill all its treaty obligations. The crucial issues remained unresolved. Through the summer of 1815, with Clay and Gallatin, Adams persistently kept at the attempt to negotiate a trade treaty. The British had appointed two of the same trio that had negotiated at Ghent, a sign that Castlereagh and his colleagues did not attach enough importance to, or have high enough hopes for, the outcome to send in a more highly regarded team, though they did add the vice president of the Board of Trade to provide expertise. Britain, the Americans soon discovered, was not ready to open its West Indies ports to American trade, except with severe restrictions. The British offer limited most favorite nation status to trade with European ports, allowed limited American trade with India, and required that Canadians be allowed to trade with Indians on U.S. territory. Believing they had been misled, the Americans—especially Clay—were disgusted, even angry.

In the end, the agreement provided very limited trade gains. Adams insisted to Clay and Gallatin that he would not sign unless the more restrictive word "Convention" was substituted for "Treaty" throughout, and the proper procedure for alternating signatures between the British and the American copies was followed. Clay and Gallatin, eager to have it done with,

516 · JOHN QUINCY ADAMS

believed the British would not sign with those changes. To their surprise, the British signed. It was agreed that Adams would, upon receiving new instructions, take up the unresolved issues. "With this assurance the British Plenipotentiaries expressed themselves satisfied. On taking leave of them Mr. Goulburn said to me, well, this is the second good job, we have done together. Yes, said I; and I only hope we may do a third; going on from better to better." On July 4, 1815, Clay left for Liverpool to join Bayard, "who was extremely ill . . . and said that he wished to go home to die among his friends and family." Gallatin joined them. They were soon, as Adams put it, "wind-bound" for America.

Soon after moving to Ealing in August 1815, he was unnerved by disabilities that threatened to end two activities that defined his life. His writing hand had developed a tremor, an ordinary affliction, especially for someone who wrote a great deal, now called an "essential tremor." It was aggravated by stress and exhaustion but was not degenerative. For Adams and his contemporaries, there was no telling how serious it might be. In mid-October, he test-fired a pistol bought for his sons to learn the proper use of firearms. The target was a tree in the garden sixty feet away. When he fired, "the pistol flew out of my hand,

and fell at least ten feet distant from me." It had been overloaded with powder. His right hand and its fingers were wounded in four places, particularly the forefinger, which soon became infected. He could not write. "My whole course of life has been changed, and I am yet uncertain how far I shall be able to return, to my occupations." What would happen to his diary? And his reports to Washington? It took almost half a year before he could write normally again. In late October, his left eye became bloodshot. It soon became inflamed, with heavy discharges. He blamed this recurrence on even longer than usual hours of reading. The predisposition for bacterial infection to lodge in his eyes may have been congenital; his father suffered from variants of the same symptoms. This time, when John Quincy awakened, he was unable to open his badly swollen eye. He could neither write nor read nor "bear the light of day." His doctor feared he would lose his vision. Reluctantly, he allowed leeches to be applied. "I was nearly delirious. It seemed to me as if four hooks were tearing that side of my face into four quarters." The condition got better, then worse again. "Thick purulent matter oozed out slowly." He was confined for weeks to a dark room. "It was now that I knew how to estimate the blessing of eye sight, and the wretchedness of being bereft of it." By early

November, there was a turn for the better. Well, we have saved your eye! the doctor announced.

But then the right eye became infected, and the infection returned, though less virulently, in the left. He went to be examined by a London specialist who assured him "that the disorder was altogether superficial, and the vision in no sort of danger." The doctor recommended diet, a constant wash of rose water and alum, and more leeches to disperse the inflammation. Having had enough of the leeches, Adams refused. He also concluded that his illness had been misdiagnosed, though he had at best a vague and inaccurate idea of what might have been a better treatment. Still, the eyewash helped, and the infection ran its course. By mid-November 1815, he could see almost normally again. But daylight still hurt. "After eighteen days of confinement . . . I . . . took a walk for a quarter of an hour in the garden; wearing a pair of green eye glasses, and a green silk shade over them." Boredom and inactivity tormented him almost as much as illness. In a darkened room, unable to read or write, frightened that he might lose his sight, he strained for patience and resignation. Inactivity was a state adverse to his personality. Without writing, thought lost firmness. It required the precise language and shape writing provided. Louisa helped, the start of her role as amanuensis.

In early November she began taking dictation: letters and short minutes for his diary, aids to his memory for a full entry at a later date. At first he found dictation awkward. He soon got used to it, relieved that he could begin working off the large stack of accumulated letters. Reversing their usual pattern, she read to him, novels by Maria Edgeworth and Scott, particularly *Guy Mannering* and *Waverley*. By the end of the month, though he still could not read or write, he was enough himself to refresh his view that, despite their many merits, Scott's novels were too romantic and too inclined to reinforce the widespread human addiction to superstition. "There is more of nature in Miss Edgeworth's characters and incidents. More of real life, and less of romance." By December, he was regaining his writing hand. "My eyes and hand are yet weak, but I hope recovering." But it was slow going. It took him time and effort to do what seemed very little. Still, he was at least up and around and active. By the start of the new year, he was at work again, at Ealing and in London. With the departure of Clay and Gallatin, he had full responsibility for negotiating a trade agreement and implementing the Ghent treaty. The trade negotiation, to his disappointment, was reduced to minor adjustments, and the treaty discussions soon were transferred to Washington. He sent detailed accounts of his discussions with Castlereagh in a series of almost a hundred official letters. But, during 1816, his London responsibilities, other than reporting to the secretary of state and discussing American loans, were mainly pastoral, especially finding ways to help stranded American sailors and prisoners of war, and what he called "Table-Cloth Oratory," the toasts to which he was obliged to respond at charity dinners sponsored by royalty. As the American minister, his participation was semi-required.

Private dinners with British notables were more to his taste. He found himself a happy guest of Whig aristocrats like Lord Holland, "distinguished by his love of literature and the cultivation of science," and the Duke of Sussex. Less antirepublican and conservative than the ruling Tories, they had a positive interest in the United States, whose minister was "an object of more willing and friendly notice to the city and opposition parties than to the court and the cabinet." Adams inevitably had diplomatic and social relations with the Tory regime: Castlereagh, Bathurst, Liverpool, and the ultra-Tory Wellington faction. When Wellington revealed that he had forgotten they had previously been introduced, Adams noted that "this is one of the many incidents from which I can perceive how very small a space my person or my station occupy in the notice of these persons, and at these places." His presence may have been less small than he thought. Castlereagh and his colleagues, in private moments and to some extent in trade considerations, shared Adams' prediction that "Europe, which has already felt us far more than she or we ourselves are aware, is destined yet to feel us perhaps more than she or we expect."

The talented anti-American Tory minister George Canning sought Adams out; he and Liverpool were especially interested in who would be the next American president, though by the end of 1816 that had been determined. Probably they had heard the rumor that Adams might be appointed secretary of state. Having met James Monroe in the Netherlands in 1796, Adams had not been impressed. Monroe seemed neither intellectually sharp nor personally distinguished, a career politician who had risen by service and durability rather than talent. Sensitive to the criticism that Virginia controlled the federal government, President Monroe decided it would be best to appoint a New Englander secretary of state. Adams, an ex-Federalist, began to seem the best choice available, though Monroe did not have in mind that he would be appointing a likely successor. That, he assumed, would be eight years off.

Early in 1817, as Monroe moved toward a decision, Adams determined to return home, whether appointed secretary or not. "That there is nothing to be obtained here," he wrote to his father, "I am fully convinced. That they now strongly grudge what they have conceded, is likewise evident." Anyway, he had had enough of Europe, no matter what the diplomatic post, especially considering the complicated financial calculation. His $9,000 salary was minuscule compared with what European ambassadors were paid. No matter how carefully he operated, he was out of pocket for costs, which might or might not be reimbursed at a later date. But if he were not in public office, what would he do and how would he earn a living? That seemed less on his mind than the uncomplicated insistence that he had an obligation to his wife and children to return home, and that he had a duty of love to once again embrace his Quincy family, who deeply felt his absence and longed for his return. What if one or both of his parents should die while he was away? "May it please God to spare their lives, and to grant me the mercy of beholding them once more in this world!" From London, he could be only as close to them as letters would allow.

Letters did provide pleasure, especially his exchanges with his father about the conflict between the New England post-Calvinist Trinitarians and the Unitarians who denied the divinity of Jesus and believed the Trinity superstitious hocus-pocus.

John Adams had become a freethinking Unitarian. His son tended to keep faith with traditional claims, though not with the tone of their adherents or the doctrine of original sin, which seemed to him damaging nonsense. In London, he enjoyed the company of Unitarian friends, as he had in Boston. Although "the bias of my mind is toward the doctrines of the Trinitarians and Calvinists, I do not approve their intolerance." Father and son continued this discussion for over two years. "I find myself growing more strongly Trinitarian [with] every new pamphlet that I read on either side of this dispute," John Quincy wrote. That did not prevent him from attending Unitarian services. But he doubted that either position could be supported by reason. It was insufficient to the task. "You caution me against commencing to be the champion of orthodoxy," he wrote to his father in 1817. "I think I shall neither commence champion of orthodoxy, nor as your old friend Franklin used to say, of any man's *doxy*. . . . You and I are competent . . . *to hold opinions*, but not to obtain perfect knowledge." He could not give up his faith that Jesus was indeed divine. Without that, he thought the entire Christian structure would collapse. But the subject had more intellectual than emotional attraction. As always, he was more interested in moral and ethical conduct than in theological claims. And both sides often seemed to lack Christian temperament and tone.

As he engaged in intellectual conversation with his father, whether his sons would become learned or not preoccupied him. He alternated between high expectations and realistic acceptance. He wanted them to model themselves after him: his love of learning and his rigorous self-discipline. He did his best to inspire them "with the sublime Platonic idea of aiming at ideal excellence." And he regretted that he had become so pressured by work that he had little time to tutor them. It did not occur to him that they might be better off without his interference, though Abigail, who knew George and John well, advised him to let the boys be boys. Still, he insisted that they all master French and continue the subjects that would qualify them for Harvard. It seemed to be work against the grain of their personalities. None of the boys showed a predilection for hard and disciplined study. George loved desultory reading and began to write poetry. John enjoyed sports more than school. Charles Francis had curiosity and sharpness, but also irresolution and lack of focus. "They are all, thanks to god, boys of good tempers and good dispositions," their father acknowledged, "but the great and constant effort of them all . . . is to escape from study." His exhortations had no effect. "I have given up all opposition as vain. I comfort myself with the reflection that they are like other children, and prepare my mind for seeing them . . . get along in the world like other men." Actually, he did not want them to be like other men. "But I am aware that no labor will ever turn a pebble into a diamond if the pursuit of knowledge . . . is not a passion which will seek its own gratification. I know how useless it is to impose it upon youth as a task." But he could not resist offering advice, and he still hoped that one day his sons would be his intellectual companions.

Over time, the travel between Ealing and London became tiresome. But, as they looked over an eighteen-month period for a London house, rents were too high in desirable locations. They stayed in Ealing. In late 1815 London streets became illuminated by gaslights. "They are remarkably brilliant and shed a light almost too dazzling for my eyes." His discussions with Castlereagh made almost no progress. The Treaty of Ghent, Adams argued, required the British to return or pay compensation for American slaves whom British officers, the United States claimed, had captured during the War of 1812 and sold in the West Indies for personal profit. The British distinguished between slaves who had sought asylum and those who had been forcibly removed. Great Britain was set on the path of suppression of the slave trade and emancipation, with British abolitionists increasingly active in Parliament and the press. There were no influential abolitionists in America. Although this was an awkward argument for the slavery-hating Adams, he pursued it with professional skill. And it was soon to have another complicated dimension: Britain argued for cooperation between the U.S. and British navies to board likely slave ships. American pro-slavery interests strongly objected. When Adams met with William Wilberforce in June 1817, at the British emancipator's request, "the suppression of the slave trade was the subject . . . and we had an hour's conversation relating to it. His object is to obtain the consent of the United States and of all other maritime powers that ships under their flags may be searched and captured by the British cruisers against the slave trade." But Adams could not support granting British warships the authority to board American ships, though American warships would have the reverse right. The impressment experience and its insult to American sovereignty prevented such an agreement.

A new issue arose, though it had been on the horizon for some time. Spain clung to its claim of sovereignty over its South American colonies. By 1816, the degree to which the United States would support

Spain's ownership and when, if ever, it would recognize the independence of Spain's South American colonies had become a serious issue. Venezuela and Argentina had already claimed national independence. Britain feared the precedent of South American independence with regard to its own colonies. It also feared that the "Holy Alliance" of Russia, Prussia, Austria, Spain, and France, which supported Spanish royalty's claim of perpetual ownership, would intervene for its own benefit on the side of Spain. Europe was now "reshackled," in Adams' view, "with the manacles of feudal and papal tyranny. . . . She has . . . returned to her vomit of Jesuits, inquisitions, and legitimacy of Divine Right." What the British, who believed they would dominate South American commerce, most wanted was a favorable trading climate. Spanish rule restricted trade. Castlereagh worried that the United States was set upon seizing additional territory; establishing colonies of its own; and making another attempt perhaps at Canada, certainly at Florida and Cuba, and perhaps at Central and South America. Would the United States like to enter into an arrangement with Britain to act together to protect their similar interests?

Washington was cautious. Many Americans wanted to arm the emerging South American nations. After all, since they were republics, like the United States, why should there be a legal ban against arming them? Castlereagh, though, was right to be concerned. The United States had coveted Canada. Every administration since Jefferson's was committed to obtaining Florida. To support South American independence openly would make Spain less likely to part with East Florida. Given the tensions between the two countries, official American support of South American independence might ignite a war. Some, however, especially in the South, argued that such a war was desirable. Young General Winfield Scott, who visited Adams a number of times in London and who "has much of the ardor of youth combining with a large share of military ambition," believed war between the United States and Spain was inevitable. America should invade and take Florida. Adams and the Monroe administration wanted to tread lightly on this complicated matter. The end was clear: American possession of Florida and the recognition of the independence of Spain's South American colonies. But how to get there was not.

In a retrospective and catch-up mode, Adams devoted the day of the start of his fiftieth year to bringing up to date the arrears in his diary. The acts of self-fashioning that the diary embodied required that it be kept current, a record of the synchronicity between the past and present. Each entry had its own field of energy. When he fell behind, it was as if part of his life was missing. Birthdays and anniversaries often activated efforts to make things right with his diary, an aspect of making things right with his life. Later that month, noting the nineteenth anniversary of his marriage, he made a long entry about the history of his relationship with his diary, one of the two deepest bonds of his life.

Sitting for a portrait by Charles Robert Leslie, he was handsomely dressed for the occasion, though he teased his mother that "he is certainly at fifty what he was at twenty, at thirty, at forty . . . excessively careless of personal appearance, and utterly incapable of ever rising to the dignity of a dandy." In truth, Leslie painted a man who looked strikingly different from the person Copley had portrayed twenty years before. Both are idealized depictions, one Romantic, one Regency. But Adams at fifty had lost the curls that Copley had depicted, and he had gained weight. He was no longer the Romantic opportunist searching for life and love. Leslie's is a portrait of a man who has arrived, and, as rumor had it, was soon likely to arrive even more fully.

But in London in 1816 and early 1817 he lived the continuity between past and present. Much of it was imaginative and literary. In the summer and fall he wrote poetry in imitation of Thomas Gray—odes or fragments of odes and lyrics to Fortitude, Temperance, Prudence, and Justice, intended as moral lessons for George even if they failed as poetry—a letter in poetry to his mother, and some comic verses for Ellen Nicholas, who was a daughter of the Ealing School headmaster and had become a family intimate and a close friend of Louisa's. Adams discarded most of what he had written in this "poetical paroxysm" as poor stuff, imitative, stiff, mere versifying. The lines seemed dead on the page. "Could I have chosen my own Genius and Condition I should have made myself a great Poet. As it is, I have wasted much of my life in writing verses; spell-bound in the circle of mediocrity." But it was an obsessive preoccupation, partly to be judged as a self-fashioning activity, an exercise of the imagination, a way to have an even more intense engagement with literature, to walk and work in the shadow of Shakespeare and Pope. Adams kept a full sense of the gap between his aspirations and his achievements as he watched with interest another poet emerging in the family. His parents probably were pleased and amused to read George's poetic apology for some misbehavior, a poem continuing the family tradition of pen portraits. George was himself "a wayward boy, / His Mother's pride, his father's joy"; John, "open, cheerful, gay, / And happy-thro' the live long day"; and Charles Francis, "a thoughtful mind displayed / A noble heart, a reasoning head. / Deep and reflective he appears / And learned much above his years." The lines contain an eerie anticipation of future realities.

British culture and some of its distinguished people also kept Adams' attention through 1816 and the spring of 1817. Jeremy Bentham, the seventy-year-old writer on social issues and institutions, arguably the founder of modern social science, sought Adams out to facilitate sending his books to leading Americans. Bentham, who queried Adams about his views on social issues, was disappointed at his conservatism. Adams was sorry to discover that Bentham, an "eccentric," though of "great ingenuity and benevolence," was probably an atheist. But they liked one another and took long walks, talking about politics and government. Each took pleasure in the other's learning and intellect. At the end of 1816, Adams read the recently published third canto of Byron's *Childe Harold*, the Byron mania a preoccupation of British society. At a large party everyone noticed the "very notorious" Lady Caroline Lamb, best known for a "scandalous novel" and her affair with Byron. *Childe Harold* did not appeal to Adams. But he was not blind to Byron's power. "The train of thought is original, wild, romantic, and bordering on distraction. The versification singular but habitually running the sense from line to line; to the utter discomfiture of harmony. The sentiment sometimes tender, but always with a mixture of gloom and fierceness." Trained in eighteenth-century poetic propriety, Adams could not feel that Byron's poetry embodied the qualities he most valued. He heard conversation about Wordsworth and Coleridge, but he apparently did not read *Lyrical Ballads*, though his own attraction to the ballad form and his commitment to the quiet virtues of everyday life might have made this collection congenial. He preferred his literature to be inseparable from a moral vision with which he could identify. The Romantic poets were a reach too far. Handel's oratorios, wildly popular in London, had what was for Adams the right combination. They were characterized by a "union of harmony and devotion . . . more perfect and delightful . . . than in any other music that I ever heard."

But he stuck most of all to his highest model of poetic genius, Shakespeare, who dominated his reading and his evenings at the theaters. He read Shakespeare at home, often aloud. He had only a little discomfort with the extravagant liberties taken with performances of *A Midsummer Night's Dream*, distorted by the widespread practice of rewriting Shakespeare's plays to make them more attractive to current tastes. "All which was necessary," he granted, "for there is no interest in the play itself, though it contains some of the finest flashes of the authors genius." He was less comfortable with a performance of *Richard the Third*, which had been, not "improved, and amended," as claimed, but badly rewritten. Adams had a good critical eye for the luminaries of the stage, particularly Edmund Kean, whose performances disappointed him. "There is too much of rant in his violence, and not smoothness enough in his hypocrisy. He has a uniform fashion of traversing the stage from one side to the other when he has said a good thing, and then looks as if he was walking for a wager. At other times he runs off from the stage with the gait of a running footman. In the passages of high passion he loses all distinct articulation, and it is impossible to understand what he says." In compensation, he has "a most keen and piercing eye, a great command and expression of countenance, and some transitions of voice of very striking effect." When he saw John Kemble, from a famous eighteenth- and nineteenth-century acting family, in a performance of *Coriolanus*, Adams noted sadly the change since he had first seen the performer thirty-four years before. He was "still a favorite with the public, and retains the faculties of his mind; but the flesh is weak and exhausted.

His lungs are no longer equal to the labors of a theatrical life. He is but the shadow of what he was." The London theater, though, was alive and vibrant. John Quincy had the time and opportunity to enjoy it.

That soon changed. Shortly after the middle of April 1817, the long-anticipated letter from President Monroe arrived. "Respect for your talents and patriotic services has induced me," it began. Adams had been appointed secretary of state. He had, he believed, the character and moral fiber for the job. But he had less confidence in his ability. Early in the month, he had felt ill and depressed. "Every man knows the plague of his own heart. Mine is the impossibility of remaining where I am; and the treacherous project of the future." Still, he could not say no. This was the service to which he had been dedicated from his birth, and a responsibility he had embraced from his earliest consciousness of self, family, and country. He both welcomed and feared this destiny. "It is probable that this will be my last removal from Europe; at least it is my wish to pass the remainder of my days in my own country."

Louisa and the boys could not be happier. They were all going home, and they could not be prouder of the head of the household and of themselves at the position of prominence to which he had been elevated. Louisa, who had been sick for much of the winter, looked forward to the fullness of spring and a summer crossing. John Quincy gave notice at Ealing, where he had been more comfortable than at any other place before, and took his last walk around the Common. Packing took up much of his and Louisa's time. In May, they left for London lodgings, preoccupied with moving arrangements. He had time in early June for a visit to the newly installed Elgin marbles at the British Museum. On Tuesday, June 10, he wrote in his diary, "London. Farewell!" He had reason to assume it was final. They sailed from the Isle of Wight into the open Atlantic in an aptly named American ship, the *Washington*. He was never to see Europe again.

Chapter 11
The Terrible Sublime
1817–1821

When Adams returned to the United States in August 1817 after an eight-year absence, he had an aura and presence that he had not had before. He had left as a man of distinction. He returned as a man of national prominence. By virtue of the office to which he had been appointed, he was to be the president's chief foreign policy adviser, with responsibility for implementation. But there was more to it than that. Jefferson had served as Washington's secretary of state. Monroe had served as Madison's. Now Adams was to serve as Monroe's. There was reason to speculate that he might be the next president. His influence as secretary of state and the power he would have if he should succeed Monroe were inseparable from his presence.

There were friends who had expectations. There were enemies in waiting. And Adams had no illusions. He would, he soon wrote to his replacement in London, be riding "in the whirlwind."

The voyage home took fifty-one days. Adams alternated between boredom and seasickness. He played chess, mostly with a Bostonian from the prolific Otis clan. He had occasionally played before, always annoyed that he had not learned to play well, competing mainly against his own sense that he had not worked hard enough to do better. Mortified by his loss of composure when he blundered, he lashed himself with harsh self-criticism. "There is a capacity of combinations in chess which is a painful test of intellect. It occupies the understanding and works upon the feelings too powerfully for sport." Making the same mistakes repeatedly hurt his pride. But chess kept him preoccupied during what seemed an endless voyage when the motion of the boat made it difficult to read or write. With a manual in hand, he practiced model moves. Encouraging his sons to play, he whittled chess pieces for Charles. On calm days, he read Bacon's essays. During a stretch of good weather, he brought his diary up to date. Convinced that America's future prosperity and rise to power depended on its growth in population, he was returning with a wife, three children, and three European servants who were likely to become Americans. Louisa was attended by her French maid and by Mary Newell, their English servant; John Quincy by his Swiss valet, Antoine Giusta. Adams' attitude toward employees exemplified the New England view defining people mostly by performance. Class was accidental, a matter of birth and category. Performance was individual, partly under the control of character. He was interested in people of every class. Steerage and cabin passengers mingled in a twice-a-week political discussion group in which he took part. When a poet from steerage sent up complimentary verses in honor of his birthday, Louisa returned the compliment with several bottles of wine. When Adams overheard one of the steerage passengers singing the "Marseillaise," he engaged him in conversation. After twelve years "roving about the world as a mariner" on a French ship and being captured by the British, the man was returning to Wilmington, North Carolina, his birthplace, eager to see his parents once again. John Quincy realized that the man's story had something in common with his own.

This foundational narrative was his unceasing preoccupation. His father's generation had founded a country. Could his maintain it, or would its growth be stunted or even its existence aborted by internal divisions or external enemies? The threat from enemies was in fragile remission. It had been in Great Britain's interest to come to terms. The United States had been fortunate, Adams concluded, to have maintained its territorial integrity and ended the war no worse off than before it began. "Oh! the voracious maw, and the bloated visage of National vanity," he noted in his diary. "If it were true that we had vanquished or humbled Britannia it would be base to exult over her, but when it is so notorious that the issue of our late war with her was at best a drawn game there is nothing but the most egregious national vanity that can turn it to a triumph." The war had indeed redeemed initial stupidity by plucky resistance and geographical advantage. But the unresolved issues might, he feared, produce a new conflict. The United States still had no army or navy of consequence. "Shall we . . . be radically and forever cured of the reliance upon embargoes, non-intercourse, and restrictions, as weapons either defensive or offensive against GB? Shall we perceive that our only effectual defence is a naval force?" He would do everything he could as secretary of state to resolve the boundary issues and advocate a strong national defense. His worst nightmare was another European war. It would again confront the United States with the challenge of how to respond to British impressment of American sailors. But if the impressment issue remained quiescent, other differences with Britain, such as the boundary with Canada on the Northwestern Coast and in the Northeast, seemed unlikely to lead to armed conflict.

The next chapter in the narrative, Adams recognized, was being written in the South and West, with Spain as the primary antagonist. The border of the Louisiana territory was in dispute. How far west did the American purchase go? Did it include Texas, and, if so, to what point? The Spanish argued that the American purchase itself had been illegal. Napoléon had had no right to sell the territory to the United States. Would the Americans also want California and Mexico? For over a decade, the United States had been demanding that Spain sell it East Florida. If not, American war hawks screeched, it should be taken by force. But if the United States seized Florida, would Spain forcefully resist? With its meager resources devoted to combating insurrections in South America, it was already bleeding heavily. Still, Adams worried, an American invasion would put the United States in the moral wrong. It would invite European censure. And what role would Britain, Spain's ally and the world's mightiest military force, play, since it was in Britain's interest to restrict American territorial expansion? The United States had two trump cards: one was geography, the other morality. Spain could not prevent Indian tribes and renegade

Americans from attacking Americans north of the Florida border and then using Florida as a sanctuary. Did the United States not have the right to pursue these criminals into Florida?

On shipboard, he pondered his and his country's future. As much as he worried about foreign affairs, he had an even more pervasive anxiety: the internal health of the nation. Like both his parents, he believed as deeply in the union as he believed in his life. During his father's presidency, both Republicans and Federalists had threatened disunion. After the passage of the Alien and Sedition Acts, Jefferson and Madison had claimed the right to secession. Now the South was demanding that new states allow slavery. Whether they had moral objections or not, Northern states objected to the additional congressional and electoral votes that the South would have. The British could be handled, though there would be difficulties. So too could Spain, which Adams had reason to believe would in the end capitulate. There were trade issues with France, though at their worst they were not likely to result in armed conflict. To Adams, domestic tensions seemed most threatening in the long run. The Declaration of Independence and slavery were, he believed, incompatible. Still, maintaining the union was paramount. "Much as I must disapprove of the general tenor of Southern politics I would rather even yield to their unreasonable pretensions, and suffer much for their wrongs, than break the chain that binds us altogether. . . . If it is once broken, we shall soon divide into a parcel of petty tribes at perpetual war with one another." The War of 1812 had "revealed our darkest side, our tendency to faction, sectionalism, and disunion." The future contained the threat of "the dissolution of the Union and a civil war."

Landfall was Montauk Point. In the distance, his eye tight to a spyglass, Adams had his first sight of home. The next morning at daylight he saw the two old Sandy Hook lighthouses and the impressive new one, their lights burning, an image of human ingenuity and American progress. As he stepped ashore on a steamy afternoon in early August 1817, the entire Northeast broiling in high temperatures of the sort to which he had grown unused, he felt a mixture of relief at their safe arrival and concern for the future, "an anxious forecast of the cares and perils of the new scene upon which I am about to enter. The latter of these feelings was greater and more oppressive than it had ever been on returning to my country heretofore." Eager to be of service, he worried that he would prove inadequate for the service required. He would, he promised himself as he prayed for divine help, do his best to be a good steward of the president's policies and the trust he had undertaken. But the issues transcended personal anxieties and individual talents. "The seeds of the Declaration of Independence are yet maturing. The Harvest will be . . . the terrible sublime."

He had met most of the elite of New York and Boston before. But this time it was he who was the center of attention. The extent to which both cities had grown astounded him. The influx of immigrants to New York crowded the streets and tenements. Boston, Adams noticed, had doubled in size, with the building boom "contributing equally to [its] elegance and comfort," though he disapproved of the leveling of Beacon Hill. In New York, he was briefed by the former head clerk of the State Department and feted at a Tammany Hall dinner sponsored by the political elite, including Governor Clinton, and the business community, especially John Jacob Astor. He visited the Academy of Art, the Historical Society, and the Museum of Natural History, early way stations in America's growth to cultural maturity. He hastened to the Hudson River waterfront to observe the latest milestone in American ingenuity and mechanical ambition, "the magnificent Steam Boat, Chancellor Livingston, which arrived this morning from Albany. It is a travelling City." At the Brooklyn Navy Yard he went aboard the first steam frigate, combining sail and steam power, and the USS *John Adams*.

It took Adams only forty hours via the steamboat *Fulton* and stagecoach to reach Quincy a little after the middle of August. He permitted himself in his diary only the phrase "inexpressible happiness" to describe his feelings on his reunion with his parents. He had for years feared he would be too late to hold them in his arms again, or even just one of them. But they had come through the long separation, and they seemed, given their ages, "in perfect health." Quincy family and friends quickly assembled. The Old House almost immediately filled with Thomas, his wife, and their five children; and John Quincy's uncle, the elderly Peter Adams. "Ancient friends and acquaintance" eagerly came. In Boston, a public dinner in his honor was inevitable. He embraced his old friend John Hall and joined the governor, the state chief justice, and others in a visit to the navy yard, where he boarded the USS *Constitution* and the USS *Guerriere*. Even the remnant of the Essex Junto had faded into silence on the issue of his apostasy, though memories on both sides were long and bitter. For relief from the heat, he bathed at a familiar spot at one of the wharves. After seeking advice about a school for John and Charles, he decided on the Town Grammar School, later to be called Boston Latin; the boys were to lodge with Dr. Welsh and his wife. To his father's disappointment, George proved not proficient enough for immediate entry to Harvard. Arrangements were made for tutoring. John Quincy felt some of the same disappointment he had experienced when, returning from Europe in 1795, filled with anticipatory enthusiasm, he had been told that he too had not yet mastered the entrance requirements.

At Cambridge he had a happy reunion with Harvard friends. He discovered, when he attended a meeting of Phi Beta Kappa, that he had been elected president, though he could not see how he could fulfill the duties of the office. And though his face was not readily recognized outside the elite circles of government, his name was. He now had his first taste of what national office imposed on men of influence. He had been used to occasional pressure from applicants for diplomatic positions. But his new post made him the primary target for job seekers in the State Department. It was as if there existed a vast pool of poorly employed or unemployed candidates for jobs who relied on him to lift them from imminent or actual poverty. There were also young men from good families who needed a start in life. His first rule was that he would not recommend any member of his family, which did not please his mother or Louisa. They thought him too rigid. His second rule was that qualifications were essential. Most applicants thought themselves sufficiently qualified by general education, social position, and need. From the start, he refused to use political patronage to make and retain supporters. High office was a gift to be bestowed, not sought. Need and ambition were not appropriate standards. When, in September, he and Louisa left for Washington, he had an increasingly clear sense that he would face formidable demands on his time, skills, patience, and integrity. In this "new world" he would have "to ride in the whirlwind." And though he had a clear sense of his values and personality, he accepted that he would have days of anguish. The reality of life in the political whirlwind would challenge his deepest beliefs about public conduct.

In the cramped quarters of the State Department, he took the oath of office on September 22, 1817, sworn in by a justice of the peace for the District of Columbia. No one attended other than the acting secretary of state. Introduced to the four subordinate clerks, he agreed to the reappointment, urged by Richard Rush and the president, of Daniel Brent as chief clerk. He would prove loyal, capable, and effective. That was the entire State Department. It was an administrative shambles. Records were haphaz-ardly kept, documents difficult to retrieve. Many had been lost or never filed. Rush, who had kept reason-able order during his brief tenancy, had done his best not to add to the mess. But Jefferson, Madison, and Monroe, the longest-serving previous secretaries, had paid little attention, for personal and ideological rea-sons, to establishing a rational bureaucracy. At the same time, Adams would have to deal with visitors. All government offices could be accessed by citizens at any time, many of them office seekers. One only had to knock at the president's door or walk into the office of the secretary of state.

He would also have to become conversant with the Washington power players, some already familiar, like Henry Clay, the House speaker, and William Harris Crawford, the treasury secretary. Some he knew only by reputation: William Wirt, the attorney general, and John Calhoun, the South Carolinian who, after Clay had turned it down, accepted an appointment as secretary of war. He knew and liked the British minister, Charles Bagot. He had never met Don Luis de Onís, but he had good reason to believe Spain would be troublesome. And he soon met the French minister, Jean Hyde de Neuville, who demanded a long interview that put his mercu-rial character on exhibition. What would it be like to be a member of a cabinet of forceful personalities, with Crawford ambitious to succeed Monroe, and a Congress led by a speaker with the same desire? Adams' appointment, he knew, had alienated Clay and Crawford, each of whom had expected to be offered the position. He also worried that the influence of politics on diplomatic policies would hinder his ability to function effectively. "The path before me," he noted, "is beset with thorns, and it becomes more doubtful than ever whether I shall be able to continue long in it."

He soon saw that the president thought and acted deliberately; that he preferred delay to action, prudence to spontaneity, and cabinet consensus if not unanimity; and that he was sincere in his desire to hear out each cabinet member in full, with a commitment to the predominance of Congress and a narrow construction of the Constitution. More so than Monroe, Adams believed that American prosperity required greater power for the federal government to promote internal improvements and a modern banking system, encourage manufacturing, advance education and science, and strengthen the military. He had an ally in the secretary of war and a partial ally in the secretary of the treasury—less so in the prudent president. Like Crawford and Calhoun, Monroe was a Southerner and a slaveholder. Love of state might be stronger than love of union. And though it was never a matter of discussion between Adams and his cabinet colleagues, they all knew that Adams opposed the three-fifths provision and thought slavery a threat to the union. Neither matter, though, was on the congressional or presidential agenda. On key matters of foreign policy, Adams believed that he and the president were in agreement: it was desirable to remain neutral though sympathetic to the South American revolutionaries; it was preferable to obtain East Florida by negotiation rather than by war; the western boundary of the Louisiana territory should bring the United States as much of Texas and as close to the Pacific as possible; the attempt of France to obtain trading privileges should be denied as a matter of fairness and equity; and the outstanding issues with Great Britain would most likely, if handled firmly but diplomatically, be settled satisfactorily.

What Adams did fear was that the president and the public "have always over-estimated, not the goodness of my intentions, but the extent of my talents." Monroe's performance as secretary of state had impressed him. "I have known few of his opinions with which I did not cordially concur." But "how much more delicate and difficult a task it will be to conciliate the duties of self-respect and the spirit of personal independence with the deference of personal obligation and the fidelity of official subordination under the new station assigned to me." As a minister in Europe, he had implemented policy. Now he was part of a team that made it. There had been dramatic and divisive lapses in cabinet teamwork during the first four presidencies. Only Jefferson "had the good fortune of a Cabinet harmonizing with each other and with him through the whole period of his administration." He would enter into office "with a deep sense of the necessity of union with my colleagues, and with a suitable impression that my place is subordinate." His duty would be "to support, and not to counteract or oppose, the President's administration." But how would his cabinet colleagues, especially Crawford, who had narrowly lost the presidential nomination to Monroe, perform?

No sooner had Adams gotten to his desk than Crawford, a manipulator of patronage, proposed that he take one of his Treasury clerks, a Crawford loyalist, into the State Department. That would make room at Treasury for Adams' nephew, William Steuben Smith. For Smith, who had made himself responsible for his father's debts, "as if his own were not enough to satisfy him," it would be a promotion. Adams said no. "Smith and his wife and all the family were up in arms against me about it, but I thought my resolution right and adhered to it." When Crawford pressured him, he repeated "that it was impossible." In Crawford, self-interest and patriotism were aligned. Adams accepted that self-interest and principles often made a tight fit, especially among politicians. Admiral Stephen Decatur's widely publicized toast in 1816, "our country, right or wrong," struck Adams as not only discordant but immoral. "I cannot ask of heaven success, even for my country, in a cause where she should be in the wrong. My toast would be, may our country be always successful, but whether successful or otherwise always right. I disclaim as unsound all patriotism incompatible with the principles of eternal justice." Adams' response was intellectual and visceral. "This sacrifice of all consideration of right and wrong to state policy always disgusts me." Justice came before patriotism, a principle that he knew would not always make him politically viable, let alone popular. Election to high office was partly earned and partly bought. He found the purchase part, instinctively and intellectually, offensive.

"Business crowds upon me from day to day," he wrote to his mother, "requiring instantaneous attention. . . . I am endeavoring gradually to establish a regular order in the course of business. . . . For myself I can only assure you that I have found the duties of the Department to be more than I can perform. Some of them therefore are not performed," no matter how many early mornings and late nights found him chained to documents and reports. Everything that did not fit under the rubric of the Treasury, War, and Navy departments or the office of the attorney general was shoved into the small rooms of the State Department. Since Congress had a strong prejudice against creating additional cabinet posts, Adams found himself responsible for a host of areas that had nothing to do with foreign affairs, including the census and government pensions. The workload of the secretary of state seemed not only impossible but absurd, especially when the Senate, in March 1817, directed Adams to provide information and make recommendations about weights and measurements. The Constitution required that Congress legislate a uniform system. The decimal system had been selected for coins. But Congress had failed to act on what Jefferson had recommended in 1790: the adoption of the French metric system. Adams also preferred the metric system. It had many advantages. "Yet when I look at the other side of the question, and observe the obstacles and resistances of every kind which stand in the way and make the practicality of so great a change questionable, I shall have some hesitation even in disclosing the opinion that I entertain."

The report would have to answer a series of questions bearing on domestic and international commerce.

Should the United States move toward a system that would create uniformity throughout the country? When a customs official in New York determined that an item weighed one pound, how could one be certain that an official in Baltimore would make the same determination? Should the United States retain its variant of the English system? Should it make its system identical to England's? Or should it embrace the French metric system? Was it desirable and, if so, was it practical to have one international system, or were multiple systems viable as long as equivalents from system to system were sound and consistent?

It was a daunting assignment, sloughed off on Adams as if it were a minor task. By personality and training, he had no choice but to do it thoroughly, to make it a thoughtful, informative, and perceptive contribution to the scholarly discourse on a subject of importance. It proved a nightmare of work. "The subject is great and complicated, and in treating it I find hills over hills and Alps on Alps arise." Gradually, over time, as he collected information, did experiments with weights, made calculations, and considered the issues, he concluded that familiarity outweighed the benefits of change. Better to improve the current system than to challenge an entire nation to change its mental and practical standards for something so basic. It was a conservative conclusion. In the end, in 1821, after four years of intermittent work, the Senate and House acknowledged the book-length report and tabled it without discussion. Congress made none of the improvements he recommended. His colleague John Calhoun tactfully told him, when Adams gave him a draft for comment, that perhaps it was a bit too long for a report to Congress. He meant, between the lines, that most congressmen would not have the time, the interest, or the intellect to read it.

By the end of the year, foreign policy had moved into high gear, with a sideshow that proved troublesome. Jefferson and Madison had felt free to have substantive discussions with foreign ministers at offhand moments, without the presence of the secretary of state. President Monroe announced that he would have no such discussions, formal or informal. All communications would come through the secretary of state. Adams approved. But the foreign ministers thought less well of the change. There was also confusion about how the wives of ministers were to engage socially with the president and his family. Should they leave cards at the President's House? Should Elizabeth Monroe, who was often ill, or her daughter Eliza, the wife of George Hay, make the first visits and leave cards? The same issue arose between cabinet members' wives and the wives of members of Congress. Louisa hated the expectation that she would leave a card at the home of every member of Congress. Adams decided that they would return cards only for every card first left for them. When, in retaliation, many congressmen refused to accept invitations from the Adamses, their guest list reflected the disaffiliation. Was the secretary of state a snooty European-influenced elitist? Was his British-born wife an arrogant snob? Did Adams not believe in making political friends?

Each of the three main foreign ministers had an agenda: Bagot wanted to move ahead with the unfinished business of the Ghent treaty; Hyde de Neuville had French trade interests to advance and a special interest in playing a role in the negotiations between Spain and the United States; Onís wanted American cooperation about Amelia Island, which was connected to Spain's vital interest in maintaining its territory in the New World. Anti-Spanish Americans from Charleston and Savannah had occupied the island, Spanish territory off the coast of Florida, and proclaimed it the "Republic of Florida." After Spanish forces reoccupied the island, American irregulars and French-born pirates wrested back control from the Spanish. To Spain, they were all pirates and privateers, criminals and insurrectionists. To the Monroe administration, they were a problem.

American public opinion was anti-Spanish. It favored South American independence and American possession of Florida. Clay, as Adams noted, had "mounted his South American great Horse," which he hoped to ride to the presidency. Onís demanded that the United States use force to return Amelia Island to Spain. But since Spain had no capacity to govern the island, would that not require the United States to keep a military presence there? And how would that affect negotiations for the purchase of East Florida and the establishment of a western boundary for the Louisiana territory? Eager to put pressure on the administration, Congress requested that the president turn over all relevant documents. Why, it was widely demanded, should Amelia Island not immediately be declared an American possession? That would undermine the negotiation with Spain about more important matters, Adams responded. Monroe, Crawford, Wirt, and Benjamin Crowninshield, the new secretary of the navy, favored immediate withdrawal. Adams and Calhoun argued for temporary possession, as a bargaining chip. Monroe sided with Adams and Calhoun, who had proposed that the United States hold the island in trust for Spain. The justification was self-defense:

the pirates were freebooters who attacked American as well as Spanish shipping.

When Onís protested, Adams argued that the United States "had not taken the possession from Spain, nor committed any hostility against her. We had been obliged in our own defence, to take the place, in defense of our laws, of our commerce, and that of nations at peace with us, Spain included." But, he told Onís, none of this would be relevant if Spain would make a comprehensive proposal "upon which we could in a few days agree." The Spanish minister, three thousand miles from the source of his instructions, had a difficult challenge. Bankrupt and politically chaotic, Spain was ruled by a combination of intrigue and autocracy. Royal rights and the Spanish empire required vigorous defense. His instructions were to make minimal to no concessions, pressure the United States not to assist the South American revolutionary governments, and retain as much of Florida and the Louisiana territory as possible. Adams responded: make me an offer we can accept; otherwise, no more discussions. Both knew that eventually Florida would be put into American hands and that the western boundary of Louisiana would be west of the Mississippi River. But how long would it take to negotiate an agreement? And would an impatient United States take Florida by force?

In the spring of 1818 a new situation threw Monroe and his cabinet into a crisis. Seminole Indians from northern Florida had been raiding American settlements in southern Georgia. Action had to be taken. But what effect would that have on the Adams-Onís negotiations? The Seminole issue itself seemed relatively uncomplicated. American citizens on the Georgia side of the border were entitled to protection from marauding Indians who added insult to injury by incorporating runaway slaves into their ranks. Since Spanish authorities exercised no control over them, should American troops be allowed to pursue these criminals across the border for the purpose of capture and punishment? Was this not the only way to prevent such raids?

In late December, Monroe and the cabinet decided it was legally sound for the United States to take action in self-defense. Major General Edmund Gaines, who had been representing the government in discussions with the Creek Indians, was ordered to assemble forces in Georgia and turn them over to General Andrew Jackson, who was "to repair immediately to the seat of war and take the command." The fifty-year-old victor of the battle of New Orleans, who had recently defeated the Creeks at the battle of Horseshoe Bend, enthusiastically embraced the order. From Nashville, Jackson wrote to President Monroe "that the arms of the United States must be carried to any point within the limits of East Florida, where an enemy is permitted and protected. . . . All opposition must be put down." Jackson defined all Indians as enemies, at best to be tolerated when pacified, at worst driven away or destroyed if unresponsive to the needs of American expansion. Law and order were at issue. So too was valuable real estate.

The best-known American general since Washington, Jackson twenty years before had served a year in the U.S. Senate and one in the House. Did he have political ambitions now? To most of his contemporaries that seemed unlikely. But he was a force to be reckoned with. Willful, self-confident, easily offended, he had a strong sense of personal honor and a violent temper. As a Westerner, he embraced the life-and-death vision of the frontier, the importance of trial by fire. He believed that God was an American and had a special commitment to Andrew Jackson. The highest test of character was personal loyalty. Those who agreed with him and served him were friends forever; those who did not were enemies. Disciplined, relentless, and brave, he was appreciated in Washington as a man suited to the rough work of the battlefield, not diplomacy or administration. When Monroe had enquired of Jefferson "how it would suit to appoint Genl. Jackson to the Russian

Mission . . . his answer was why, Good God! he would breed you a quarrel before he had been there a month!" Jackson had recently quarreled with the secretary of war, who had issued an order directly to an engineer in Jackson's division. Furious, Jackson had ordered his officers to refuse to obey any order not sent through him. At issue was the chain of command. But Monroe sided with Calhoun, who was instituting a much-needed reorganization of the military bureaucracy. "The President's letter," Adams noted, "is very kind and conciliatory, but urges Jackson that he was wrong" to issue an intemperate order in direct defiance of the president and the secretary of war. In the end, Jackson accepted a reasonable compromise. But his penchant to act first and consult later had been put on display.

In Washington, Adams and Onís made little progress. When dispatches from Jackson contained startling news, the relationship was put to an even more severe test. The president believed that his orders through General Gaines and directly to Jackson limited Jackson's incursion into northern Florida to the pursuit and destruction of the Seminoles. They forbade an attack on Pensacola or the occupation of any Florida territory. But the initial dispatches revealed that Jackson had captured the Spanish fort of St. Mark, executed Indian prisoners, and threatened to execute two Englishmen who had been doing business with the Seminoles. Washington soon learned that he had indeed had the two Englishmen hanged. He had also taken Pensacola, where he set up his command, occupying Spanish territory. His actions immediately became national news. The war hawks declared it the perfect opportunity to take the rest of East Florida. An angry Congress protested that only it had the right to declare war, certainly not a general on a limited mission. Jackson responded that he had needed to take Pensacola to protect his troops. How, he implied, could those who issued his orders not know, given his record, that he would take bold measures to do so? Some in Washington concluded that he had envisioned his incursion into Florida all along as an opportunity to force the government to proclaim East Florida U.S. territory. It seemed a usurpation of authority, an act of insubordination, an attempt by a general in the field to make foreign policy.

Whether or not Adams approved of Jackson's actions, he saw good reason to defend them. Natural law and international law, he believed, permitted the incursion. Military necessity justified actions required to protect the safety of the U.S. Army. One could quibble about the appropriateness of any one action, but the overall pattern absorbed all but the most outrageous deviations. And, so far, there was nothing that the law of self-defense did not justify, or at least excuse. The president and every other member of the cabinet disagreed. Onís angrily protested, Hyde de Neuville tried to play constructive mediator, and Bagot expressed his government's likely condemnation of the execution of the two British subjects, though London, eager to avoid an unnecessary quarrel, decided that the men had been operating outside the protection of Britain. Monroe, who had been forced into a difficult situation, could neither fully defend nor fully condemn Jackson without paying a price. His political enemies, especially Clay, were waiting with long knives. The administration, Adams noted, had fewer and fewer friends in Congress, and Monroe seemed vulnerable. In a series of cabinet meetings, Adams proposed a way out. The general had technically acted beyond his orders because of misunderstandings, not in purposeful opposition, and he had, in the most recent dispatches, made clear that he had always intended to return Pensacola to Spanish control, consistent with his original orders. Everything else he had done in self-defense.

Gradually, Monroe and the rest of the cabinet saw the expediency of Adams' argument, except Calhoun, who still demanded a strong public rebuke. Crawford reluctantly acceded, for reasons of politics, not principle. For much of the next year, Monroe's and Jackson's enemies, some of whom were the same, kept up a drumbeat of attack: the Constitution had been violated; Jackson should be censured by Congress. Through all of this Monroe probably had in mind Jefferson's comment that Jackson was a man who bred quarrels. Adams responded informally to a perturbed Onís and a glum Hyde de Neuville that he believed that Monroe "would approve General Jackson's proceedings. That we could not suffer our women and children on the frontiers to be butchered by savages . . . and that when the governor of Pensacola threatened . . . to drive him out of the province by force, he left him no alternative but to take from him the means of executing his threat." Adams formally rejected the Spanish minister's protest. In late November 1818 he composed a state paper in the form of a lengthy letter to George Erving, the American minister in Madrid. It provided the definitive statement of the U.S. position. The letter was for Erving's eyes but for the ears of the Spanish government. With literary skill and sharp logic, Adams argued that the security of the United States required that it possess Florida.

Spain must immediately make her election, either to place a force in Florida adequate at once to the protection of her territory, and to the fulfillment of her engagements, [or] to cede to the United States a province of which she retains nothing but the nominal possession. . . . The duty of the government to protect the persons and property of our fellow-citizens . . . must be discharged. And . . . if the necessities of self-defense should again compel the United States to take possession of the Spanish forts and places in Florida, [we] declare, with the frankness and candor that becomes us, that another unconditional restoration of them must not be expected; that even the President's confidence in the good faith and ultimate justice of the Spanish government will yield to the painful experience of continual disappointment; and, that, after unwearied and almost unnumbered appeals to them for the performance of their stipulated duties in vain, the United States will be reluctantly compelled to rely for the protection of their borders upon themselves alone.

The statement to Onís and the letter to Erving were widely circulated. Jefferson enthusiastically complimented their author. "They are without exception," he wrote to Adams and the president, "the ablest State papers he ever read."

Despite Onís' claim that the Florida incursion had set back treaty negotiations, the Spanish minister and the secretary of state knew that it only gave further emphasis to Spanish vulnerability. Within months, they were bargaining again, though with little progress beyond Onís' statement that the United States could have Florida for practically nothing if it were to agree to the Mississippi as the western border of the Louisiana territory. Attempting to take the high moral ground, he continued to express outrage at Jackson's actions. Frustrated with Spain's unacceptable demands, Adams devoted much of his energy in late 1818 and early 1819 to repeating with blunt clarity the American position. To each of Onís' concessions, offering a border slightly farther west, Adams said no. The United States required that the border be as close as possible to Mexico—the target was the Rio Grande—as far west as possible in the Southwest, along the north-south axis of the Great Plains, and then westward to the Pacific.

Early in 1819, the administration made a final take-it-or-leave-it offer. It would agree to a southwestern border at the Sabine River, the boundary between the state of Louisiana and Texas, rather than at the Rio Grande, if Spain would agree to American terms on every other matter. It was a political gamble. Many Americans had set their hearts on Texas. When Adams, at Monroe's request, had two discussions with Jackson in which he previewed the treaty's provisions,

Jackson remarked that the enemies of the administration would object to the Sabine rather than the Rio Grande. But "the vast majority of the Nation would be satisfied, with the Western boundary . . . if we obtain the Floridas." Jackson gave no hint that he was among those who would be dissatisfied.

If he could have continued negotiating, Adams believed he might have gotten a boundary even closer to the Rio Grande. But "we are now approaching so near to an agreement that the President inclines to give up all that remains in contest." Onís hoped for some minor adjustments in Spain's favor. On the morning of February 20, 1819, he called on Adams at home. He would accept the treaty as is, he conceded, including the provision that all land grants made after January 24 were void, though he still thought the United States should make two minor concessions. Adams was not in a conceding frame of mind. "I observed there was no time left for further discussion; and we had yielded so much that he would have great cause to commend himself to his court for what he had obtained. He said I was harder to deal with than the President." The next morning Adams went to his office for the first time ever on a Sunday. He encountered an old friend who remarked how unusual that was. Suddenly John Quincy was overwhelmed by a feeling of "involuntary exultation." It was a meeting and a feeling he would remember for the rest of his life. The treaty was signed the next day.

On the same day that the Senate ratified the new treaty, a select committee recommended censuring Jackson for insubordination. At a dinner for Jackson that the Adamses hosted, Louisa noted that "our hero looked depressed and dejected and appears to be more severely wounded than one would suppose." He "seems to be very sensitive." The general and the secretary of state had a long discussion about politics, especially about Crawford, whom Jackson believed to be his enemy and whom he detested. Adams passed along Jackson's views to the president. Aware that there would be dissent, Monroe believed, as did Jackson, that the treaty provided ample compensation for the concession of Texas. Florida would be ceded to the United States. Any Florida land grants made by Spain after January 24, 1819, the date on which Onís had first proposed ceding Florida, would be invalid. All monetary claims by American citizens that had accrued since 1795 would be paid by the United States, about $5 million, which would be, in effect, the cash cost of Florida. The western boundary of the United States would follow the Sabine River and then the Red River to Arkansas; it would continue in a jagged line to the forty-first parallel, then directly westward to the Pacific Ocean on a line bordering the southern boundary of the Oregon territory. The United States would be a presence on the Pacific rim. Texas would be an issue for another day.

When, in 1817, Adams settled in Washington, he looked much as his father had at the age of fifty. Early morning summer swims in the Potomac and long winter walks across the frozen city and countryside kept him healthy. He was temperate and disciplined in his life, plain in dress and manners, talented and hardworking. As George Watterston, the librarian of Congress, noted in his *Letters from Washington*, Adams combined "an intuitive and natural" perception with "intellectual power." In appearance, he was an Adams male, not short but below medium height, not fat but stocky. And what his outward appearance did not reveal, his relationship with Louisa did: his capacity to feel love and loyalty. When his workload weighed heavily, his temper was short, and his tongue was sharp, they had difficult days. But not days that denied their commitment to one another. At their best moments, he and Louisa were a happy match.

They both continued to write poems, and in Louisa's case a comedy of manners, contrasting French vanity with British sincerity in matters of love and courtship. She too loved the theater and felt she had a talent for writing drama. In response to Leslie's recent portrait of her husband, she described in poetry what the painter had not depicted fully enough: his wit, love of justice, and intellectual power, and his "harmony of nature where sense and sweetness joined with ease." In return, in a poem called "To My Wife," he acknowledged his good fortune in having married Louisa. He would marry her again if that were possible. Given the tense circumstances that had followed their marriage, this indeed must have been poetry to Louisa's ears.

> Nigh twenty years of lengthened span
> On earth their destined course have run,
> Since, in the face of God and man
> Thy lot and mine were linked in one.
> Louisa! could a mortal hand,
> Break for a moment, thy nest's chain,
> Before the altar I would stand,
> And thou shouldst be my bride again.

After twenty years of marriage the patterns were inseparable from those of daily life: his work schedule; their literary and social evenings; their concerns about their children; Louisa's colds, headaches, and erysipelas attacks; and the need to entertain that Washington life required. If she had major complaints, they were that she was separated from her children and that he spent too many hours at work. His was that there were not enough hours to do what he needed to do. She filled some of her time with a diary of her own, initiated as entries for the eyes of her father-in-law, her "confidential correspondent." She regretted that John Quincy's reticence would not allow him to be as emotionally expressive as she would have liked. His face and voice sometimes conveyed more disapproval than he actually felt, let alone desired to convey. He too often gave the impression of being censorious, "a sort of coldness to his manners which . . . make him seem severe and repellant. . . . He is continually reproached for want of sociability and I for pride hauteur and foreign manners," Louisa wrote. They were both aware that this worked against him. His patience often wore thin. There were provocations, especially from the hordes who sought patronage or recommendations or handouts. When he was harangued by a stranger about "the duty of a man in high office to patronize and recommend poor and ingenious persons like him," he "bore all this with composure. . . . The result is that I am a man of reserved, cold austere and forbidding manners; my political adversaries say a gloomy misanthropist, and my personal enemies, an unsocial savage." Louisa knew better.

Much of Washington society was formal, shifting, and tense. It did not lend itself to intimate friendships. Louisa, fortunately, had two of her sisters and their families as nearby neighbors. In November 1817, her eleven-year-old orphaned niece, her elder sister's daughter, came to live with the Adamses. Louisa at last had a daughter of sorts, the pretty and lively Mary Catherine Hellen. But Louisa worried constantly about her boys, especially their health, and about her husband's overly strict standards. Washington was not a good place for someone in Adams' position to make new friends. He soon, inevitably, had an active social life, with a large number of acquaintances, many in political life or government service. Some were interesting, like the architect, inventor, and longtime head of the Patent Department William Thornton. An impassioned supporter of South American independence, he could not understand why Adams and the president thought him too partisan to be an effective minister to Argentina. Adams did not have the heart to tell him that he lacked two of the essential qualities for a minister: objectivity and discretion. John Quincy and Louisa occasionally hosted large receptions of up to three hundred people at their house on F Street, which he bought and into which they moved in the fall of 1820—his Washington residence for most of the rest of his life. He was, despite the perception that he lacked social graces, essentially by training and culture a social man, as was his father. Solitude, study, and introspection came naturally to him. But he was the child of a culture that defined man as a social creature. His primary obligation was to other human beings.

From the start of his Washington residence, mediating between the two poles of his life had its difficulties. There was little attractive about the still-primitive capital city, and Adams never felt fully at home in the Washington landscape or weather. It had no resemblance to the New England world of long-settled towns, Boston culture, Cambridge education, and the bracing Atlantic rather than the sluggish Potomac. When the president and his cabinet colleagues left in July 1818 for cooler places, Adams forced himself to stay into late August, hoping he could work. But he had never before experienced a Washington summer. He found it impossible to sleep; the heat made him lethargic; the screenless house was invaded by insects. Tossing and turning from what he thought prickly heat, he discovered that "it was caused by a nest of Spiders just from the egg-shell, so small that most of them were perceptible only by their motion. It was like the continual titillation of a feather passing over the skin at a thousand places at once. It was a night of exquisite torture without pain. My linen and body were covered with them." The next summer, "Antoine killed a brownish snake two feet long, in the house at the foot of the staircase."

In Quincy and Boston, there were different discomforts. John and Charles were making progress at school. Amiable and kindhearted, John appeared to be doing well enough. But he had little interest in ideas or books, and a quick temper. Charles Francis was much fonder of reading than study. Reserved and introspective, he was not a committed student. George, having passed his examination, had entered Harvard. His passion was poetry, and he seemed to John Quincy to have inherited his mother's sensitivity but not his father's intellect and discipline. Adams had anxious reservations about all three, whom he judged against exemplary models: his father and himself. "Among the desires of my heart, the most deeply anxious is that for the good-conduct and welfare of my children. In them, my hopes and fears are most deeply involved. None of my children will probably ever answer to my hopes. May none of them ever realize my fears!" He worried enough about his sons to anticipate the consolation if at least they turned out to be of good character. If his sons could not be men of achievement, let them, he prayed, be honorable human beings.

He discovered to his dismay that his brother Thomas had been drinking heavily and perhaps gambling. John Quincy, who had entrusted Thomas with some of his financial matters, was "deeply concerned." It "is of itself a great relief to my mind, from the concern I had felt for your family, and yourself on that account," he responded when Thomas "implicitly and distinctly" denied that he had gambling debts. But he was far from certain he was being told the whole truth. The companion of his youth, his closest friend and associate in the Netherlands and Germany, had not been able to sustain a career. Thomas had tried literature, the law, elective office, government service as a local judge, and a touch of farming. Now there were issues of character. Family help had been forthcoming, especially since he had a wife and five children. Their future weighed heavily on Abigail's heart, and John Quincy had reason to worry. He needed to be "firm, prudent, and affectionate." But it was a dangerous situation.

He soon had serious worries about George too. A student rebellion had resulted in disciplinary action. John Quincy received a badly written, semi-illegible, evasive letter from his son. "I am waiting with extreme anxiety to know," he responded. "I write to you and

THE TERRIBLE SUBLIME · 575

entreat you to relieve my anxiety. But you have a stain upon your face and I wish neither to see you nor hear from you till you have discovered it yourself and done all that remains in your power to wipe it away." At Louisa's urging, he did not send this letter. The letter he did send two weeks later to "my dear son George" was based on information that revealed that his son had not himself been an instigator. Out of loyalty to classmates and friends, he had followed them into acts of insubordination. "I shall use with you no crimination or reproach. I am willing to flatter myself that you have had a lesson . . . which will not be lost upon you." But he was not willing to forgo a sermon on a moral lesson his mother had taught him as a child: "In all cases throughout life, when a difficult choice was to be made . . . to put the question to myself which was right and which was wrong—and if I could answer immediately that question, to inquire no further. To take the right side, and then to be moved from it by nothing upon earth." John Quincy was to discover, over time and to the pain of both, that George would be the master of his own vices, and that no sermon from his father could, in the end, make a significant difference. For the time being, John Quincy added punishment to sermon. "Your mother and I had wished and intended that you should come and pass the winter with us here, but as . . . you may have a opportunity to make yourself in some degree useful to your grandfather, we are willing to forgo the pleasure we had promised ourselves in seeing you here until the next season." Louisa thought the punishment too severe. She wanted the company of her favorite son.

In Boston during the summer of 1818, Adams was compelled, he complained, into "a perpetual course of visiting and banqueting and utter idleness." At Quincy, he talked frequently with his father, and they walked together on paths and roads that he knew from childhood. Nine years older than his wife, John Adams remained in remarkably good health. Seventy-four-year-old Abigail had had numerous brushes with death. Still, when John Quincy left for Washington at the beginning of October, his mother had seemed in good health. In late October, he learned that she was "dangerously ill" with a high fever, nausea, and diarrhea. A letter from Harriet Welsh, the daughter of Dr. Thomas Welsh, prepared him for the worst. At last a letter from Charles confirmed what now seemed inevitable: his mother had died. "Oh, that she could have been spared a little longer. She had known sorrow, but her sorrow was silent. She was acquainted with grief," but she kept it to herself. If "virtue alone is happiness below, never was existence upon earth more blessed than hers." And she had been more than a mother to him. She had been a beneficent presence, "the comfort of my life." Now, "the world feels to me like a solitude. . . . But oh! my father! my aged and ever venerated father! what solace is now left that can attach him to life?"

Louisa turned to pious poetry, "Lines on the Death of Mrs. Adams," invoking Abigail's spirit to

fondly hover o'er thy lonely friend,
In nightly visions resignation find,
Cheer his great mind attune his Soul to peace
Till in this world his hopeless griefs may cease:
And when his spirit quits this mortal clay
To bliss eternal guide him on his way.

Ill with her usual miseries, she took to her bed. John Quincy turned to his diary, his private chamber for recuperation—and in this case for elegiac memories. "She was always cheerful; never frivolous, she had neither gall nor guile. Her attention to the domestic economy of her family was unrivalled. Rising with the dawn, and superintending the household concerns with indefatigable, and all-foreseeing care. She had a warm and lively relish for literature; for social conversation; for whatever was interesting in the occurrences of the time, and even in political affairs. She had been during the war of our Revolution, an ardent patriot, and the earliest lesson of unbounded devotion to the cause of their country that her children received was from her." He knew that much of who he had become and what he was now for himself and his country was her gift to him.

Having feared that the treaty with Spain was too good to be true, Adams waited anxiously for news of Spanish ratification. The February 1819 treaty required that land grants made before the January 1819 date were to be deemed valid. To his shock, Spanish documents now revealed that the king had announced numbers of grants in December 1817. Adams had trusted Onís' word and George Erving's dispatch from Madrid, which had implied that the February date in the translation was the date of the land grants, and this in turn meant that the grants had been made after Onís and Adams had completed the negotiations. Adams had not read the original Spanish documents. If Spain had grounds to claim that the grants were in fact valid, a treaty that the Senate had unanimously approved contained a fatal flaw.

What had been a triumph for the secretary of state now had the appearance of incompetence. He had not been thorough enough in his examination of the documents Erving had sent him. He had trusted Erving's descriptions of their contents. Had Onís, in Washington, been part of the deception? Did he know at the time of the signing that the land grants had been made at an earlier date and concealed that information in order to finalize a treaty to which the United States would never have otherwise agreed? Or had the deception been perpetuated by the Spanish court without his knowledge? In either case, "an infamous fraud has been practiced upon us." The Spanish government needed to accept an emendation to the treaty disclaiming the grants. Otherwise the United States would renounce the treaty, although the president murmured to his inner counsel that even with the outrageous land grants, it was probably worth retaining. After all, there would be ways, once Florida was U.S. territory, to invalidate them.

At a dinner on March 8 hosted by General Jackson a number of guests remarked that Adams "had a care clouded countenance." He could not conceal his "anxiety and mortification." He had committed a primary sin. He had relied on someone else's word instead of closely examining the documents himself. Since the best defense was attack, he immediately called in Onís and Hyde de Neuville. Had Onís not represented to him when the treaty agreement was made that there were no recent land grants that preceded the January 24 date? And had not Hyde de Neuville, who had facilitated the agreement, believed the same on the basis of Onís' representations and the available facts? Neither disagreed, though Onís averred that if he had known of the existence of grants made in 1817, "he should have insisted upon their validity." But he had not known. The American position was clear: Onís had been authorized by his government to sign a treaty. He had not acted outside his instructions. Consequently, the treaty was valid. Madrid needed to sign the treaty expeditiously and provide a statement acknowledging that the land grants were invalid, in keeping with Onís' and Adams' understanding at the time of the signing.

At first Adams suspected that Onís had engaged in a purposeful deception. He soon concluded that he had not. Still, "of all the transactions in which I ever was concerned this negotiation is the most treacherous— not a step of it can be taken without meeting with delusion and fraud." He regretted that Onís, of whom he had come to think better, was soon to be replaced. So too were Hyde de Neuville and Bagot. Each, despite his flaws, had been a capable minister. Bagot especially had been "a minister of peace . . . wise and benevolent . . . temperate and respectful," with a "conciliatory disposition." In fact, Adams concluded, "the mediocrity of his talents has been one of the principal causes of his success." It was a conclusion that staggered his "belief in the universality of the maxim, that men of greatest talents ought to be sought out for Diplomatic Missions. . . . But a man of good breeding, inoffensive manners and courteous deportment, is nearer to the true diplomatic standard than one with the genius of Shakespear, the learning of [Richard] Bentley, the philosophical penetration of Berkley [referring to George Berkeley], or the wit of Swift." Did Bagot have more of the desirable qualities for a diplomat than Adams? Or was it that Bagot did best with ordinary diplomatic activities while he himself had qualities that made him fit for the harder circumstances, including these difficult negotiations with Spain?

When Onís' replacement, General Francisco Vives, arrived in April 1820, Adams believed the treaty was dead. The cabinet discussed contingency plans, including asking Congress to empower the president to take temporary possession of Florida in trust for Spain, as had been done with Amelia Island. Although the treaty is "gone forever," Adams told the president, "the ground upon which we stand is safe. Some convulsion may take place in Spain, upon which we may be obliged to occupy Florida, or some chance may again occur upon which we may receive it by Treaty." But Vives,

Adams was pleased to learn, came with conciliatory instructions. Spain, Vives claimed, had always intended to ratify the treaty. It considered the land grants null and void. In fact, Spain had only one demand. In order to ratify, Spain required a positive stipulation that the United States would not recognize the independence of Argentina. "I told him that our system between Spain and South America was neutrality. That a stipulation not to recognize the South Americans would be a breach of neutrality, and as such we could not accede to it." In fact, Adams and Monroe had every intention of recognizing Spain's South American colonies, but at a time when it served American interests. This was as obvious to Spain as it was to Adams. And, as Vives and Adams discussed the issues in April and May 1820, political turmoil and change in Spain raised more uncertainty: whether the new government would confirm or alter Vives' instructions. A revolt in Spain had forced Ferdinand VII to share power with liberal forces. The new regime needed time to organize itself and its policies. "This will undoubtedly make a new difficulty even if all the others are abandoned."

When it became clear that Vives did not have authority to ratify the treaty, but could only pledge that his government would ratify if the United States responded satisfactorily to the new Spanish demand, Adams was left in a bleak mood. Once again the treaty seemed lost. He was "so immersed in the Spanish business," Louisa wrote in her diary, "that he can scarcely give himself time to eat drink or sleep. . . . He grows very thin and his mind is kept upon the rack night and day." Calhoun told him that Jefferson had recommended to Monroe that the attempt to get Spain to ratify be abandoned. Instead, "we should look to the occupation of Texas." But Monroe and Adams preferred to wait to see what the Cortes, the Spanish legislature, would do. At first it claimed that it had no constitutional authority to "alienate" Spanish territory. But rumors reached Washington in late November 1820 that, seeing the light of realism, the Cortes had indeed ratified the Adams-Onís Treaty. The rumor was true. In mid-February 1821, Vives delivered a copy, signed and ratified by Ferdinand VII. It contained "an express declaration of the annulment of the grants." Adams' long ordeal was over.

When the Senate, in February 1821, ratified the Adams-Onís Treaty for the second time, the vote was not unanimous. Why had the southwestern border been set at the Sabine River? Why not at the Rio Grande? What weakness or incompetence or sectional trade-off had resulted in the United States' not acquiring Texas? After all, what interest had Adams, as

584 · JOHN QUINCY ADAMS

a New Englander, in Texas? And why was Monroe so weak as to allow Texas to slip away? A Virginian and a slave owner, he should have known better. What everyone knew, starting in early 1819, was that the competition between the North and South for political control of Congress, inseparable from the issue of whether slavery would be allowed in the new territories, had suddenly intensified into an explosive conflict. The focus was Missouri. "The excessive curiosity upon the subject of this [Spanish] Negotiation . . . is qualified only by the agitation of a question in the House of Representatives on a Bill for admitting the Missouri territory into the Union as a State. A motion for excluding slavery from it has set the two sides of the house, slave holders and non slave-holders, into a violent game against each other." Between 1819 and 1821, while Congress demonstrated its inexhaustible divisiveness over Missouri, Adams worked to conclude the treaty with Spain. But he could not avoid becoming increasingly aware, with a sinking heart, of the Missouri conflict.

In Massachusetts, Adams saw only free blacks. As the author of the state constitution, John Adams had provided the legal basis for abolition. Slavery, he recognized, was a threat to the Union. "There is a darkness visible upon all our national prospects, which cast a gloom upon my declining days," Abigail wrote to her son Thomas in 1802. "It will soon involve us in a civil war: or a lethargy and stupor render us fit subjects for Southern despotism; the rising generation will have more dangers to encounter than their fathers have surmounted." In Washington, John Quincy, between 1803 and 1808, and again starting in 1817, resided within a living paradox, a city that represented a nation based on the Declaration of Independence and the Constitution but was built by slaves, had a large slave population, and boasted a thriving slave market. Neither John Quincy nor Louisa had a passion for abolitionist views or ideological outrage. But they had no doubt that the climax of this racial nightmare would be violent if the nation did not eventually awaken. In the meantime, though, the Constitution was the law. Adams had written against the three-fifths compromise that kept New England and the nation under the thumb of Southern rule. But how to change that? And, eventually, how to eliminate slavery? He saw no practical path.

As a peace commissioner in Ghent and minister to the Court of St. James's, he had not taken a moral position on the British claim that whatever slaves it had removed from the United States during the War of 1812 had voluntarily sought liberty. As an agent of the government, he had the obligation to argue the official position:

British soldiers had seized American property, then sold it in the West Indies for personal profit. Compensation should be paid. Without conclusive evidence, the British government declined to compromise, let alone accede. After all, it had settled the War of 1812 on terms that many Englishmen thought too generous, and the government had to respond to an increasingly powerful anti-slavery coalition in Parliament and widespread abolitionist sentiment. In the United States, slave owners were the dominant force in determining national policy. In Ghent, London, and Washington, Adams had the duty, lawyer-like, to represent the official position. His own views on the larger subject, the morality of slavery, were another matter. He could deal with the claim that slaves had been abducted for personal profit as a legal issue only. But he could also distinguish between his responsibility to implement the Monroe cabinet's policies and his personal views. And he could keep at moral arm's length two other slavery issues that had become imminent by 1819. One was the international slave trade. Enforcement was often far from strict. Large amounts of money were still being made illegally by British and American citizens, and legally by the French until 1818, in slave trading. Although Spain abolished slavery in 1811, except in Cuba and Puerto Rico, trans-Atlantic slave smuggling flourished, much to the distress of the

British and a mixture of indifference, embarrassment, and disgust in the United States.

For Adams and his colleagues it was an exasperating issue. Anti-British sentiment and resistance to the interdiction of American ships by a foreign power made it impossible to accept the British proposal that the two major maritime powers allow their warships to board vessels of any flag to determine if slaves were being transported. International courts, Britain argued, could be established to guarantee that national sovereignty would be respected. Anyway, slave ships were considered pirate ships—they could be captured at will. Why would the United States not commit itself to joint action? But how, Adams asked, could it be determined that a ship on the high seas, flying whatever national flag, was actually a slave ship without stopping and boarding the vessel? If a British warship stopped an American ship that was not a slaver, there would be hell to pay in Congress and in public opinion. It would be considered an assault on freedom of the seas. The Monroe cabinet pondered and pondered. In the end, it put its emphasis on enforcement of interdiction in American waters only and interdiction on the high seas of only readily identifiable slavers. Bagot and then his successor, Stratford Canning, indifferent to the degree of American sensitivity created by the long conflict with Britain about impressment, argued against the American position, which seemed to them shortsighted and hypocritical. Time after time Adams had to repeat that the United States insisted on compensation for the stolen slaves and that it would not enter into an agreement to stop the illegal slave trade.

Adams also would not accommodate another attempt to deal with slavery. In October 1817, on the steamboat between New Haven and New York, he read a pamphlet about a society recently formed to establish a colony of free blacks in Africa. In Washington, Elias B. Caldwell, the clerk of the Supreme Court and a friend of Francis Scott Key, spoke enthusiastically to Adams about the American Colonization Society, sponsored by a coalition of Northerners and Southerners, including Henry Clay, Richard Rush, and Bushrod Washington, the Supreme Court justice and nephew of George Washington. Adams had no doubt that "the intentions of the Society are benevolent." But he believed the "project to be impracticable and dangerous." Free blacks were its main target. In the South, free blacks were troublemakers, if only by their existence in a place where most blacks were slaves. In the North, free blacks were often the target of prejudice. If Congress would allocate funds to make an exodus and resettlement possible, would not the South and the North be better off? For many, repatriation would eliminate the paradox of free blacks in a slave society and in a society that white people felt was exclusively theirs. A homeland in Africa for free blacks would be the first step in the eventual repatriation of all American blacks. The country would be liberated from the burden of slavery. There were, though, two formidable obstacles: money to pay for the purchase of territory, the cost of transportation and settlement, and the purchase of slaves from their owners; and the absence of political will among the Southern elite, essential for peaceful emancipation.

From the start, Adams had sympathy for the intent. But he thought the project doomed to failure. It had inherent in it such inconsistencies, contradictions, and impracticalities that to hope that it could be successful in any of its stages seemed the delusion of fools and frauds. And even if it were successful in establishing a territory in Africa to which free blacks would migrate, the new "nation" would be so dependent on the United States that it would be in effect a colony of the sort that many of the European powers possessed in Africa, Asia, South America, and the West Indies. It would be the first step in transforming the American republic into an American empire. It would be a betrayal of the ideology of the revolution and the principles of the Declaration. It would be an unintended but perverse consequence of good intentions. The society's members, he told Monroe and the cabinet, "are men of all sorts and descriptions . . . some exceedingly humane, weak-minded men, who have really no other than the professed objects in view; and who honestly believe them both useful and attainable. Some speculators in official profits and honors which a colonial establishment would of course produce. Some speculators in political popularity, who think to please the abolitionists by their zeal for emancipation, and the slave holders by the flattering hope of ridding them of the free colored people, at the public expense. Lastly some cunning slaveholders who see that the plan may be carried far enough to produce the effect of raising the market price of their slaves." Monroe disagreed. The Colonization Society might be a good start toward gradual emancipation. He then enlarged, Adams noted, "upon the great earnestness there was in Virginia for the gradual abolition of slavery, and upon the excellent and happy condition of the slaves in that state, upon the kindness with which they were treated, and the mutual attachment subsisting between them and their masters." Monroe saw beneficence; Adams saw the whip.

When he told two representatives of the society who asked for his support that he believed "that the mass of

[free] colored people who may be removed to Africa, by the Colonization Society, will suffer more and enjoy less than . . . if they should remain in their actual condition in the United States," it was not the message they wanted to hear. "Their removal will do more harm than good to this Country, by depriving it of the mass of their industry." Free blacks were American citizens and an asset to the United States. Adams had the "highest respect for the motives of the Society," and he hoped that his refusal to support it would not become a political weapon to be used against him. "I apprehend the Society," he wrote privately, "which like all fanatical Associations is intolerant, will push, and intrigue, and worry, till I shall be obliged to take a stand, and appear publicly among their opponents. Their project of expurgating the United States from the free people of Color, at the public expense, by colonizing them in Africa, is so far as it is sincere and honest" like the "project of going to the North Pole" by "travelling within the nut shell of the earth." When one of his visitors attempted to convince him, stubborn hope met stubborn realism. The Negro was here to stay, Adams had no doubt, even if simply as a practical matter, and why any free black would voluntarily export himself to a primitive African environment seemed beyond common sense and experience. So why the willingness

among otherwise smart, experienced, and even learned
people to indulge in the colonization fantasy? Behind
it all, Adams believed, were fear, desperation, and the
dark chasm of irreconcilable differences between the
North and South.

As the conflict over the admission of Missouri played
out, it seemed to Adams that, even if some compromise
was patched together, the North and the South would
be at war eventually. He had no direct role to play in the
riveting congressional conflict about admitting Missouri
as a slave or free state. But the issue haunted his sense
of self and nation. "The Missouri Question" has taken
deep "hold of my feelings and imagination," he wrote
in early 1820. Clay had once said, as he and Adams
walked from the Capitol to visit with the chief justice,
that though "it was a shocking thing to think of . . . he
had not a doubt that within five years from this time the
Union would be divided into three distinct confedera-
cies." Adams attempted to sort out his own views in his
diary. Convinced that slavery was inimical to republican values and natural law, which gave every person the
right to life and liberty, he deplored the inconsistency
between the fact of slavery and the country's founding
principles. As a New Englander, for whom the distribution of political power seemed weighted unfairly toward
the South, he believed that the North and its values

would be put at further disadvantage if pro-slavery political power was increased by the admission of new slave states without offsetting free states.

His strong preference, though, was for no new slave states at all. What would be the point, if there was to be gradual emancipation, of extending slavery further and strengthening those who favored it? Without gradual emancipation, there would be an inevitable disaster. "I have within these few days begun to commit [my thoughts] to paper, loosely as they arise in my mind. There are views of the subject which have not yet been taken by any of the speakers or writers by whom it has been discussed. Views which the time has not yet arrived for presenting to the public; but which in all probability it will be necessary to present hereafter. I take it for granted that the present question is a mere preamble; a title page to a great tragic volume." But "the time may, and I think will come, when it will be my duty . . . to give my opinion, and it is even now proper for me to begin the preparation of myself for that emergency."

Adams watched and listened, aware, like many Americans, that a future-defining debate was in progress. "By what fatality does it happen that all the most eloquent orators," he asked, "are on its slavish side," except for Rufus King, who spoke "with great power, and the great slaveholders in the house gnawed their lips, and clenched their fists as they heard him." Those on the side of freedom had "cool judgment and plain sense . . . but the ardent spirits, and passions, are on the side of oppression. Oh: if but one man could arise with a genius capable of comprehending, a heart capable of supporting and an utterance capable of communicating those eternal truths that belong to this question, to lay bare in all its nakedness that outrage upon the goodness of God, human slavery, now is the time, and this is the occasion upon which such a man would perform the duties of an angel upon earth." There was no such leader. Even if Adams had been in the legislative arena rather than the executive, he had no sense that he himself might be that man.

Monroe's prediction that "this question will be winked away by a compromise" proved accurate. In March 1820, a compromise was patched together. Missouri and Maine would be admitted to the union, one slave, one free. And slavery would be prohibited in any other new states formed from the Louisiana territory, north of the 36° 30' latitude line, with the exception of Missouri. When the governor of Indiana wrote to Adams about an attempt to reinstitute slavery in his state, Adams responded, emphasizing that "I consider slavery as the misfortune but not the fault of the states

THE TERRIBLE SUBLIME · 595

where it exists, and exemption from it as the happiness but not the merit of those where it does not exist. The abolition of slavery where it is established must be left entirely to the people of the state itself. The healthy have no right to reproach or to prescribe for the diseased; but that slavery should dare to claim legislative sanction in an American state where it has once been prohibited passes my comprehension." It would be like the legislature of Massachusetts bringing in a bill "for a nursery of rattle-snakes for the purpose of propagating the breed, or to import the yellow fever for the benefit of the infection."

But how much time was there? Gradual emancipation? It had begun to become clear that the dominant opinion in the South was increasingly against any emancipation at all. And the Missouri Compromise had by late 1820 itself become a source of further conflict. The compromise required that the Constitution, just as it protected the right of slave owners to bring slaves into Missouri, also protect the right of free blacks to travel into and reside in Missouri. In November, as Adams was riding with his colleague John Calhoun to a cabinet meeting, Calhoun expressed "great concern at the re-appearance of the question upon the admission of Missouri as a state into the Union. After all the difficulty with which it was compromised at the last session of Congress, the Convention which made their Constitution has raised a new obstacle, by an article, declaring it to be the duty of the legislature to pass laws prohibiting free negroes and persons of color from coming into the state; which is directly repugnant to the article in the Constitution of the United States which provides that the citizens of each state shall be entitled to all privileges and immunities of citizens in the several states." Calhoun did not know how this difficulty could be surmounted. It seemed to the North a deal breaker.

In February 1821, after furious legislative maneuvering, wheedling, threatening, and bribing, facilitated by House Speaker Clay's sleight of hand, a second compromise barely passed. Its grease was language so vague that the South could swallow it. It allowed Missouri to implement the provision in a way that was consistent with its own interpretation of the Constitution. For the next forty years, no free black could enter Missouri except at risk to his freedom. Adams expected that this second compromise was "not destined to survive" Monroe's "political life and mine." He was wrong. But he predicted accurately that the situation's ultimate resolution, whenever it would come, would not be without bloodshed and civil war. It was only a question of when.

Chapter 12
The Macbeth Policy
1821–1825

As Adams mounted the speaker's platform in the House of Representatives to deliver the annual July 4 speech in 1821, he had in his hand the parchment original of the Declaration of Independence, signed by his father and at least ten others he had known personally, all dead but three. Stored in a drawer at the State Department, it showed the wear of casual hands and handling. Any reputable person who asked to examine it would not have been denied. Concerned about its deterioration, he had commissioned a facsimile copy to be engraved on copperplate. Two years later, Congress ordered that two hundred copies taken from the copperplate be printed on vellum; this method was the origin of all modern photographic images.

Adams had pondered whether to decline the invitation, concerned that what he said might be viewed as a statement of official policy. He decided that it would be understood that he spoke in his private capacity. But he knew, as did his large audience, that his public position would influence the reception of his words, however distinctively individual his claims. In fact, the oration was to be perhaps his single most memorable and prescient speech. He had had little time to prepare it, and he wrote out of his memory, his heart, and his convictions to an audience that gave him its riveted attention, alert to the possibility that he might be the next president.

Was the United States to be a nation that embodied power or liberty? he asked. It could not be both, he maintained, in the ordinary sense of how those words were understood. The nation that most embodied the inevitable imbalance in the effort to combine power and liberty was Great Britain. From the start, its liberty had been corrupt because it had been bestowed on the people as a limited gift from its rulers. It had been further weakened by the establishment over the centuries of the British empire. Empire and liberty were incompatible. At home, limited freedom; abroad, servile and shackled dependency. And that was what the American colonists had rebelled against, having given Britain every opportunity to eliminate or at least decrease its colonial misrule. Let me remind you, he told his audience, as he had reminded his audience in 1793 in Boston and in 1802 at Plymouth, of the narrative of our founding. And let me respond to those in Britain who provokingly and disparagingly ask the question, What has the United States as a nation ever contributed to civilization, to the arts and the sciences?

The country has contributed, he argued, something extraordinary. "It demolished at a stroke the lawfulness of all governments founded upon conquest. It swept away all the rubbish of accumulated centuries of servitude. It announced in practical form to the world the transcendent truth of the unalienable sovereignty of the people. It proved that the social compact was no figment of the imagination but a real, solid, and sacred bond of the social union." A dedication to justice and human rights was "the only legitimate foundation of civil government," Adams continued, and the gift of America to the world is that it is the first country ever created on the basis of this claim. The universally valid social compact that founded the United States proclaimed that all human beings have by the law of nature and of nature's God the innate right to govern themselves. That is what America has "done for the benefit of mankind." She has spoken, "though often to heedless and often to disdainful ears, the language of equal liberty, of equal justice, and of equal rights. She has, in the lapse of nearly half a century, without a single exception, respected the independence of other nations while asserting and maintaining her own. She has abstained from interference in the concerns of others, even when the conflict has been for principles to which she clings, as to the last vital drop that visits the heart." Her commitment is to liberty, to the power of an idea, not to the power that comes from guns and cannons.

The greatest threat to America, he argued, came not from abroad but from within. The United States is in danger of forsaking the policies that have embodied its dedication to its own liberty and to the principle of the liberty of others. Its foundational history proclaimed it the champion of anticolonialism and national independence. If the nation is to be true to its principles, whatever new territory the United States gains must come from moral leadership, from its commitment to justice, from fair and honest dealings, and from the consent of the governed, not from the power of intimidation or conquest. "Her glory is not dominion, but liberty. Her march is the march of mind. She has a spear and a shield; but the motto upon her shield is, Freedom, Independence, Peace." She may, at her peril, become one of the great rulers of the world. But "she would be no longer the ruler of her own spirit." And the neutrality policy, initiated by President Washington, is the right policy to maintain America's moral self-identity and leadership.

> She well knows that by once enlisting under other banners than her own, were they even the banners of foreign independence, she would involve herself beyond the power of extrication in all the wars of interest and intrigue, of individual avarice, envy, and ambition, which assume the colors and usurp the standard of freedom. The fundamental maxims of her policy would insensibly change from liberty to force. . . . Wherever the standard of freedom and independence has been or shall be unfurled, there will her heart, her benedictions and her prayers be. But she goes not abroad, in search of monsters to destroy. She is the well-wisher to the freedom and independence of all. She is the champion and vindicator only of her own. She will recommend the general cause by the countenance of her voice, and the benignant sympathy of her example.

Her blood and treasure, Adams continued, should be expended only when her own national security was directly attacked. All else was a matter of diplomacy, not arms. Under his and Monroe's leadership, the United States had been wise, he implied to an audience that could fill in the examples, not to interfere in the wars for South American independence—and not to seize Florida by force or invade Cuba and Texas, despite the hawkish drumbeats.

Six months later, writing to Edward Everett, the coeditor of the *North American Review*, Adams responded to criticism of his speech. Some New England Federalists believed his characterization of Britain had been too harsh. They saw in the speech the same Adams they had refused to re-elect to the Senate. To many in the South and West for whom liberty meant the right to conquer new territory, Adams' message was unpalatable, a lecture by a moralist who did not share their desire for additional real estate. He also responded to those who criticized his passionate tone and figurative style, as if he had indulged in emotional self-expression at the cost of good taste. "I have considered that the first object of a writer for the public was to obtain as many readers as he could, and that something *remarkable* in the style of composition was among the most attractive lures to readers." It was "not an oration constructed according to rhetorical rule; it is a continued tissue of interwoven narrative and argument . . . in answer to the question, what has America done for the benefit of mankind? and with a peroration of ten lines at the close."

The address, he told Everett, had done six things. It had again vindicated his father's generation, whose rebellion was justified. It had demonstrated that the South American rebellions also were justified, though whether the new nations would be able to establish republican governments remained unclear and perhaps unlikely. The address looked "forward prospectively to the downfall of the British Empire in India." That was inevitable. And it "anticipates a great question in the national policy of this Union. . . . Whether we too shall annex to our federative government a great system of colonial establishments." For the address demonstrates, Adams continued, "that such establishments are incompatible with the essential character of our political institutions." Even worse, "that great colonial establishments are but mighty engines of *wrong*." Like the slave trade, they must be abolished.

It was a powerful address, though a minor distraction from the political preoccupation of the day: who would be the next president? As early as the summer of 1821, Washington was obsessed with the question. Ideology was one of the concerns. So was real estate, particularly the pricing of government-owned land, whether it would be set high enough to pay for new infrastructure or low enough to make its purchase cheap for settlers and speculators. So too, though with less intensity than from 1819 to early 1821, were slavery and the claim of the South and its Northern allies that the federal government should be subordinate to the states; that the Constitution prohibited Washington from making laws regulating the internal activities of the states; and that an infringement of states' rights was just cause for secession. The issue was not abolition of slavery but its restriction or expansion into the new territories, and much of the intense focus was on what role the federal government could play in sponsoring internal improvements, such as roads and canals. Did the Constitution allow that? And where would the money come from? Internal taxes? The tariff only? The sale of Western land?

Adams had come to respect Monroe. He admired the president's fair-mindedness; his amiable, mostly unruffled doggedness; and his obvious incorruptibility. Virginians, Adams noted, make states' rights and jealousy of the executive the basis of their "political fabric, because they are the prevailing popular doctrines in Virginia." They "occasionally render service to the nation by preventing harm; but they are quite as apt to prevent good; and they never do any." In Adams' view, Congress specialized in doing little.

Its dominant mission was to keep spending to a minimum, even when that did not address domestic problems and threats to national security. Monroe, too, attempted to do good by doing little, in basic agreement with the Virginia philosophy that the executive should be subordinate to the legislature, except in foreign affairs. "The President is often afraid of the skittishness of mere popular prejudices," Adams noted, "and I am always disposed to brave them. I have much more confidence in the calm and deliberate judgment of the people than he has." Unlike Jefferson, the behind-the-scenes master of a compliant Congress, Monroe confronted in his second term a fractious Congress eager to put the president in his place. By personality, he disliked confrontation. Often the secretary of state found he had to tell the president things he preferred not to hear. "He rather turns his eyes from misconduct, and betrays a sensation of pain when it is presented directly to him . . . and in the way of censure or punishment, if an order that he gives should not be executed, I doubt whether he would ever notice it, unless by having it again called to his attention." Monroe seemed "more governed by momentary feelings, and less by steady and inflexible principle. . . . But his failing leans to Virtue's side. He is universally indulgent, and scrupulously regardful of individual feelings."

Like Madison, Monroe had learned during the War of 1812 that a national bank was essential to national security and prosperity. Without a cohesive banking system, the government had no efficient means of borrowing and depositing money. Even foreign loans had to be handled through clumsy mechanisms. Hostile to banks and eager to minimize federal power, the Madison administration had let the charter of the First National Bank expire. In 1816, a chastened Madison signed a bill to create the Second Bank of the United States, though he had previously believed a national bank unconstitutional. It would control lending and borrowing, keep the currency sound, serve as a depository for government funds, and fund government borrowing. To widespread surprise, Madison, just before he left office, vetoed an infrastructure bill, which seemed to many no less constitutional than a national bank. President Monroe made it clear that, without an amendment to the Constitution, he would do the same. When, in 1819, a blistering depression overwhelmed the country, Monroe and Adams agreed that reckless state banks needed to be purged. "The Virginian opposition to implied powers therefore is a convenient weapon," Adams remarked, "to be taken up or laid aside as it suits the purposes of state turbulence and ambition." Adams had total certainty on both issues. The country needed to be brought together by a national transportation network. Its future prosperity depended on that. And the proper role of the federal government was to be forward looking, to become the instrument and agent of infrastructure, education, science, and the arts.

What Monroe lacked, Adams thought, was the will to take initiatives when actions could and should be taken. He disagreed with Monroe's strict interpretation of the "necessary and proper" clause of the Constitution. Like Adams, Monroe thought roads, canals, and the dredging of rivers and harbors desirable. His Republican predecessors also had. But only if the Constitution were amended to say that expressly. Jefferson and Madison had come to power, Adams noted, by attacking Washington and Adams "under the banners of state rights and state sovereignty. They argued and scolded against all implied powers, and pretended that the government of the Union had no powers but such as were expressly delegated by the Constitution. They succeeded. Mr. Jefferson was elected President of the United States, and the first thing he did was to purchase Louisiana. An assumption of implied power greater in itself and more comprehensive in its consequences than all the assumptions of implied powers, in the twelve years of the Washington and Adams administrations put together." William Wirt, the attorney general, rejected Adams' narrative. Adams provided a verbatim account of their disagreement in his diary:

WIRT: Not so!

ADAMS: Why not?

WIRT: I'm too much a Virginian to grant that.

ADAMS: Constitutional scruples are accommodating things. Whenever there was an exercise of a power that did not happen to suit Jefferson and Madison they would allow of nothing but powers expressly written; but when it did, they had no aversion to implied powers. Where was there in the Constitution a power to purchase Louisiana?

WIRT: There is a power to make Treaties.

ADAMS: Ay! a Treaty to abolish the Constitution of the United States?

WIRT: Oh! No! No!

ADAMS: But the Louisiana purchase was in substance a dissolution, and re-composition of the whole Union. It made a Union totally different from that for which the Constitution had been formed. It gives despotic powers over the territories purchased. It naturalizes foreign nations in a mass. It makes French and Spanish laws a part of the laws of the

Union. It introduces whole systems of
legislation, abhorrent to the spirit and
character of our institutions; and all this done
by an administration which came in, blowing
a trumpet against implied powers. After this,
to nibble at a bank, a road, a canal . . . was
but glorious inconsistency.

Important as were such issues, people and personalities mattered equally. All eyes were on the succession, and the secretary of state was an inevitable target.
Adams knew that his personality worked against him.
Widely respected, he was less widely liked. Clay could
charm the birds out of the trees. Crawford had a bluff,
bearish, embracing friendliness. Calhoun was handsome, articulate, and visibly energetic. Jackson always
had hovering over his head the golden halo bestowed
on America's most prominent living war hero. These
were Adams' competitors, all Southerners and slave
owners. Clay and Crawford each believed he should
have been offered the State Department. Crawford had
even immediately begun a behind-the-scenes campaign. With the advantages of the speakership, Clay
had been as active as Crawford. Calhoun had his eyes
on the highest office, though he was aware that 1824
might not be his year. Jackson, who had not previously
shown any political ambition, seemed to many unfit

to be president; his actions in Florida had not been forgotten. Jackson and Adams, though, were on good terms between 1819 and 1823. The general appreciated Adams' support in the Seminole War, and Adams continued to defend Jackson, who "had rendered such services to this nation that it was impossible for me to contemplate his character or conduct without veneration." Soon there was talk that Calhoun might settle for the vice presidency, and that Jackson would be a helpful running mate for any of the other candidates.

Would it not be likely, Adams' opponents murmured, that he would be hostile to Southern interests? "I have lived but little in my own state," Adams wrote to a friendly newspaper editor. He had worked with statesmen of both parties from all areas of the country. That he was hostile to the South was nonsense. "I have been entirely and spontaneously indebted to the Virginian Presidents, from all four of whom I possess testimonials of personal esteem and confidence." Prejudice against Virginia "would add the crime of ingratitude to the wrong of illiberality. . . . My relations with all her Presidents have been those of independence, candor, and confidence. I have experienced nothing else from them." Then there was the question of how the next president would be elected. That might help determine the outcome. The Constitution mandated some of the features of the process, but the details were left to the states. Some had the state legislature choose electors. In others, voters in congressional districts or statewide made the choice. Republicans dominated the Senate and House, and Republican senators and congressmen caucused and chose their party's candidates. Since most state legislatures were controlled by Republicans, the party's candidates were certain to be elected. If the procedure called for a popular vote, the result would be the same. Adams had, in 1808, helped end his career as a senator from a Federalist state by taking part in the caucus that nominated Madison. But as a matter of principle, he had not then approved and did not now approve of "King Caucus." He had held major elective office only once, and that office had come to him. He had tried to be above politics, at least to the extent of following principle on the crucial issues: the implementation of the Louisiana Purchase and the Jefferson-Madison embargo policy. That he would allow his name to be put in nomination but not actively seek any office had been his code from the start.

But now, to become president, he would have to run at least a little, a slow jog if not a trot. His obligation, he felt, was to do his duty as secretary of state, to encourage no one to support him for the presidency, to take no part in political activities, and to let the forces that be determine the outcome. But why, he asked himself, given his principles, should he decline to declare that he was not a candidate? Surely others might be asking that question. He had, he wrote in his diary, lived his entire public life on the principle that he would always be at the call of his country. He would serve if asked. Not that he did not want to be president. His election would demonstrate how wrong the Massachusetts Federalists had been to reject him, and it would vindicate his father. Most important, he had a vision for the country that, he believed, would lead to a prosperous future and a powerful nation. But what was he willing to do to persuade doubters and satisfy those looking for some benefit in exchange for their support? As little as possible, he hoped, perhaps nothing at all.

From the start, he made clear that he would not make efforts on his own behalf. In late February 1821, a close friend, Joseph Hopkinson, a Philadelphia Federalist whose father had signed the Declaration of Independence, urged him to let his supporters know that he approved of their desire to work for his candidacy. No doubt Crawford and his minions, Hopkinson emphasized, never let a day go by without working for Crawford's election. Adams responded to a query from Robert Walsh, a Philadelphia writer and editor, "I have no countermining at work to blast the reputation of others and seldom attempt even to defend my own. I make no bargains. I listen to no overtures for coalition. I give no money." Crawford's major talent, in Adams' view, was for intrigue. An incompetent secretary of the treasury totally devoted to advancing his presidential candidacy, Crawford had few scruples, Adams observed. Policy followed out of personal interest, and slyly bad-mouthing the competition with unsupported suspicions seemed the hallmark of the nasty political environment. "Whenever a man resorts to suspicion to account for the conduct of others, his belief is governed more by his wishes than by his judgment." Adams' own trump card, as he told Hopkinson, was to refuse to play the game. Anyway, he believed that the likelihood that he would get the nomination was small to nonexistent. "If there has ever been," he wrote in his diary, "an election of a President of the United States without canvassing and intrigue, there has been none since that of my father. There will probably never be another."

It was clear, though, that the other candidates were seeking support, proposing alliances, and attacking opponents through surrogates. Much of Congress was for sale. "About one half the members," Adams observed, "are seekers for office at the nomination of the President. Of the remainder, at least one half have some appointment or favor to ask for their relatives;

but there are two modes of obtaining their ends; one by subserviency, and the other by opposition." Jackson had little need in 1821 to pursue the office. Others were pursuing it for him. His two years in Florida, where Monroe had appointed him territorial governor, worked to his advantage. The appointment kept his military halo bright, gave him executive experience, and sequestered him from Washington politics. Monroe at first could not imagine that any rational person would support Jackson for president. When the Tennessee legislature nominated Jackson in 1822, and in 1823 elected him to the Senate, it became clear that his rivals would have to take him seriously, as Crawford had always done. He had been maneuvering since 1819 to prevent Jackson from becoming a rival for Southern support. Clay, with Western support and national prominence, seemed equally formidable.

To Crawford and Clay, eliminating Adams was a necessity. It would put New England in play. Crawford's supporters flooded newspapers with so-called facts: like his father, John Quincy was a monarchist and his wife a haughty Englishwoman. The accusation that at Ghent Adams had sold out Western interests by proposing that the British have navigation rights on the Mississippi in exchange for the continuation of New England fishing rights in Newfoundland came, perhaps with a wink by Clay, from a man who had been one of the negotiators at Ghent. In April 1822, Adams was blindsided by Jonathan Russell's charge. The accusation shocked Adams. The facts did not support it, and it was an assault on his honor, patriotism, and service to the country. Opponents jumped at the opportunity to deal a blow to his presidential chances. "I have been too much and too long the servant of the whole Union," he wrote to the editor of the National Gazette, "to be the favorite of any one part of it. The whole course of my public life has been that of crossing political interests whether sectional, political or geographical." But he felt he had to answer Russell's charge. And Russell's claim that the War of 1812 voided the peace treaty of 1783 needed to be denounced once again.

A discontented politician and diplomat, Russell had manipulated the crucial document on which his accusation was based, a letter Russell had written in early 1815 to Secretary of State Monroe. Adams, who suspected the fraud, soon located the original. In a stinging rebuttal titled *The Duplicate Letters, the Fisheries and the Mississippi*, Adams proved that, in fact, the Mississippi proposal had originated with Clay, and Russell had supported it. All the commissioners had believed that the navigation rights were so narrowly defined as to be useless to the British, and it was clearly within their instructions to make the proposal, which the British had rejected. Russell's signature was on all the documents. In 1815, in his private letter to Monroe, Russell had exaggerated his role at Ghent and expressed reservations about the terms of the treaty, almost as if he had not signed it. Monroe had filed the letter away. Then, in 1822, assuming that his letter of 1815 was lost, Russell published a doctored version, with significant changes and additions, as if it were the one written in 1815. The doctored letter denounced Adams and charged that the commission had violated instructions; this charge also implicated the other members of the commission.

With the original in hand, Adams highlighted the discrepancies, exposing Russell's lies. If they had not been exposed, Adams' candidacy might have ended. Actually, the widespread attention paid to his rebuttal helped sustain it while Russell's career was destroyed by the forgery, a form of self-destructive recklessness consistent with his temperament. But the controversy also took its personal toll on Adams. "I am weary and sick at heart at what they call the 'diplomatic controversy,' and have never been much more mortified than proud of a victory over brother Jonathan. I had never any ill will to him and did all but entreat him not to force himself and me before Congress and the nation." Ironically, the politician upon whose behalf Russell believed he was acting was to play the decisive role favorable to Adams in the election of 1824.

Living in Washington had its difficulties. Southern values, visible in the prevalence of duels and the mistreatment of slaves, dominated the city. To the Adamses, the claim that the South sustained a higher level of civilization than the North seemed bitterly laughable. Dueling seemed barbaric. Although illegal, it occurred regularly, almost always among Southerners defending their honor. In February 1819, two well-known cousins, political leaders from a prominent Virginia family, fought a fatal duel. Adams was horrified. In October 1820, he was appalled when a mulatto boy, having found banknotes that had fallen from a shopkeeper's pocket, "was tortured, thumb screwed, and hung by the neck . . . to extort confession from him. . . . This is a sample of the treatment of colored people under criminal charges or suspicions here." Summers in Quincy consequently became even more precious. Although Washington's heat made Adams claustrophobic, as if he could not breathe, his heavy workload made him spend more time in Washington than Quincy, including the entire summers of 1820 and 1822. Swimming in the Potomac became a compulsive necessity. Often he went to the river at dawn, in the darkness, with Antoine Giusta, who had married Mary Newell; both were domestic pillars of the F Street household. Frequently young John swam with them. Adams' strength was sometimes put to the test. Sudden storms, a leaky boat, a stronger than expected tide, and a misjudgment about distance turned innocent swims into frightening challenges. Each time he vowed to be more prudent. Partially stripped, he swam weighted with shirt, stockings, and underwear. A few times, he and his companions pulled off their wet clothes and dried themselves in the sun. Still, he much preferred Black's Creek in Quincy or the wharf in Boston.

Like everything else, his Quincy world had changed. His mother was no longer there. As he approached the Old House in the first summer after her death, his feelings on meeting his father were "inexpressible. The world, and life itself, are no longer the same to me that they were while my mother lived, but here it is that I feel in all its desolateness the privation of the blessing which her presence always diffused around her." His heart ached at the sight of his father's decline. John Adams was becoming feeble and hard of hearing. Even worse, he could not see well enough to read or write. "He is near eighty-four-years of age, and he told me that he felt his constitution both of body and mind was breaking up. . . . I endeavored as well as I could to cheer him." He himself as a father had anxious concerns for the next generation. In 1819, George completed his second year at Harvard, John entered as a freshman, and Charles continued at the Boston Latin School. Each seemed to be doing barely well enough. When John Quincy returned to Washington in the fall of 1819, he took Charles with him, to study under his father's direction. When the eldest boys joined them for the Christmas holiday, George appeared to Louisa to be "as eccentric as ever and John as wild." Twelve-year-old Mary Hellen, attractive and flirtatious, enjoyed the presence of the three young men, and each began to take a special interest in her.

The next November, when his parents learned of a rebellion at Harvard, they worried that John might have participated. "My Children seem to have some very intemperate blood in them, and are certainly not very easy to govern," Louisa wrote in her diary. "John is somewhat like his Mother a little hot headed, and want of timely reflection will I fear often lead him to error, as long as he suffers the first naturally strong impulse to guide him." They worried about all three. Floating in a romantic haze, a lover of poetry and Byron especially, George "magnifies his joys and sorrows," Louisa noted, "until the real world in which he moves vanishes from his sight." He was "likely to suffer through life from his impetuosity and the absurdity of his notions." When the boys were together in Washington, their "jealousies and squabbles" and their insensitivity to their mother made her miserable. Their departure made her even more unhappy. It leaves "a void which cannot be supplied. . . . My sons are coming successively to the age when those feelings are the most intense. I have often parted from them before, but never with so deep an anxiety."

The Adamses observed that their boys were becoming men. In early 1821, Charles, infatuated with Mary Hellen, left for Cambridge to prepare for Harvard. Louisa worried that he was too young. In Boston, George had a "foolish entanglement." When it was decided that he would study law in Washington with William Wirt, Louisa predicted that it would lead to a relationship with Mary. "There are many objections in my eyes to a connection with Mary. . . . She may yet make a very fine showy woman but she will have a great deal to learn and it is nearly impossible to correct a number of vulgar habits . . . which . . . if suffered to grow will make her what I most thoroughly despise a woman of loose conversation of coarse of impure mind." She deplored Mary's materialism, her pettiness, her eagerness for compliments. Every small triumph was an occasion for laughter, every defeat for tears. At Cambridge, Charles had a rocky start. When Adams' successor as Boylston Professor of Rhetoric, whose appointment he had opposed, examined Charles, he made the boy's entry conditional. Adams suspected bias. He resisted President John Kirkland's uncomplimentary assessment of George and John with a touch of bitterness, even paranoia, which George stoked. Adams speculated that George had been discriminated against in his assignment at graduation exercises. When Charles at first did miserably, his father recommended that he withdraw and reapply later. Charles decided to stay. Then he asked to withdraw. But his father had no alternative place for him. Gradually, he began to find his footing. George had been an inconsistent student, and so was John. If only he would apply himself, his father urged. John made promises and efforts. But hopes were kindled only to be dashed. "Cares for the welfare and future prospects of my children; mortification at the discovery how much they have wasted of their time at Cambridge . . . agitated my mind so much that through the night I could not close my eyes. I had hoped that at least one of my sons would have been ambitious to excel. I find then all three, coming to manhood, with indolent minds. . . . It is bitter disappointment."

In early 1823, it seemed to Louisa that John had "very much improved." But that spring brought a devastating disappointment. The previous year John had been reprimanded for misbehavior. When, in March, he was assigned a prime end-of-year recitation, his father took it as a sign of rehabilitation. But, in May, he was one of forty students expelled for disorderly conduct, the incident probably having been fueled by alcohol. His devastated father wrote two letters to President Kirkland requesting that his son's punishment be mitigated. The sentence, though, was final. What would twenty-year-old John do now? When he returned to Washington, his parents were self-controlled enough not to make matters worse by harsh criticism. He was, as his father had already concluded, not suited to academic success. There was, though, still hope for George. It had been arranged that, after two years of apprenticeship in Washington, he would spend a year in the Boston office of the prominent lawyer and Federalist congressman Daniel Webster. George would then take the Suffolk County bar exam. But, in late July 1823, twenty-two-year-old George stunned his father. He asked his consent to become engaged to fifteen-year-old Mary Hellen. How long would the engagement last? his father asked. "He said," Adams wrote in his diary, "perhaps five or six years," until he would be in a position to marry. "I gave my consent." George left for Boston, never again to live with his parents. John took his permanent place in the Adams household.

At Quincy in the late summer of 1823 there was both gloom and resignation. For the first time, Adams declined an invitation to attend the Harvard commencement. He could not bear to be there on the day when "John was to have taken his degree." His own father was now even more feeble, unable to walk across a room without assistance, his failing eyesight reducing this insatiable reader to having to be read to. Still, he could be mounted on a horse, and father and son took short rides together. Worried about his own future, John Quincy tried to put his financial affairs in order, which meant having a clear summary of assets and debts and proper oversight of his Boston and Quincy properties. It seemed likely that his $6,000 government salary would soon come to an end. He had had his fifty-sixth birthday in July. Retirement could not be far off. Two things concerned him: money and occupation. His income from the Boston properties, his investments, and his father's gifts provided a barely sufficient income. He had an additional reason to be worried about money after he returned to the capital. He committed himself to the purchase of the Columbian Mills, on Rock Creek above Georgetown, from Louisa's uncle Roger Johnson.

Built in about 1800, it was an unprofitable, dilapi-
dated facility for grinding wheat and corn, entan-
gled in debt and imminent foreclosure. Roger's son,
George Johnson, proposed that John Quincy purchase
the water-powered mills and employ him to manage
the business, allowing Johnson the option to buy back
half ownership. Adams did no due diligence, trusting
to Johnson's eagerness. He knew there was risk, but he
was optimistic. "This affair involves the comfort of my
future life, and that of all my family. . . . Less than two
years will terminate my political career, and leave me
to the support of my private resources." Perhaps, he
thought, this was "a gracious offer" from providence to
provide him with something useful to do in his remaining years. "Such is the hope I have conceived. But it is
yet beclouded with doubts. . . . It must be . . . a leap in
the dark. But man must trust."

By the end of July 1823, the deal was in place. It
required mortgaging property to raise cash for the
purchase and repairs. "I have made a disposal of a
large portion of my private property," about $20,000.
This "may have great influence upon the remainder
of my days." By the end of 1824, it was clear that he
had made a serious mistake. Matters went from bad
to worse. Rock Creek went dry. The foreign market
plunged. Flour spoiled. "All the labor of the year has

been lost; and instead of a resource for retirement, is likely to prove a heavy clog upon my affairs. Yet I am embarked, and must share the fortunes of the ship. I shall continue the experiment for another year." When, in 1825, Adams asked Johnson for an accounting, it turned out that every penny of the money Adams had given him had been used to pay his own debts. The mill continued to be a financial drain for another ten years.

What also distressed them both was Louisa's health, which sometimes kept her in bed even when the F Street house, to which a sizable addition had been constructed, was crowded with guests. The social duties required of the wife of the secretary of state made her feel resentful, exhausted, and exploited. What would life be like for the wife of the president? "I am so thoroughly unfit for so exalted a situation that I think the wisest thing I can do is to retire to some quiet place where I may at least spend my life in peace." Her absence from social events was noted by the Washington rumor mill that she detested, as if her life were being lived on a public stage. "Very sick all day again," she wrote for her father-in-law's eyes in mid-December 1821. "My old friend the Erisepelas seizes on me with more than usual violence and is I believe excepting yourself the most attached friend I have." There was another complication. At the age of forty-six, she was pregnant again. Louisa feared for her life. Lonely in her confinement, she managed a few visitors and outings, including one to see the president's wife, who was "very desirous of finding out what is the matter with me. . . . Something or other is perpetually occurring to make me the fable of the City." She feared she might never see her sons again. John Quincy had required they stay in Massachusetts for the winter holiday, as punishment for their poor grades. "Mr. A— acts for the best I wish he may not err in his judgment." She could not get him to change his mind. "I am absolutely refused the sight of my children—I must submit because I have no recourse but it grieves me to the Soul and much I fear my unborn Infant will feel the shock I have sustained." John Quincy did change his mind. But when the boys arrived, Louisa remained depressed. She resented the gaiety in the house, as if it were purposeful insensitivity. She thought of her dead infant daughter, believing that God had granted Louisa the gift of her death because, if she had lived into later life, any threat to her would have driven her mother to madness or suicide. On January 10, 1821, she passed "the most dreadful day I ever passed in my life. I must have been raving mad." When the miscarriage occurred can only be guessed at.

Both husband and wife had a sense of their age and burdens. John Quincy had the usual colds and sore throats. His eye infections continued to plague him, especially at times of the heaviest workload and strain. Sustained by love and loyalty, the marriage absorbed all the disappointments inherent in unchangeable patterns. "With the dawn of this morning," the twenty-fifth anniversary of their wedding day, he awoke and blessed "Heaven upon the semi-jubilee of our marriage," he wrote to Louisa. "More than a half of your life and nearly half of mine have we travelled hand in hand in our pilgrimage through this valley, not alone of tears. We have enjoyed together great and manifold blessings and for many of them I have been indebted to you." On the same day, Louisa, in a bleaker mood, wrote privately that on such occasions they had "once celebrated in a different way but alas old age has come and all attraction is flown forever." John Quincy took the opportunity in a poem to Louisa to place the losses in the context of truths he believed eternal. Life and time had taken their toll,

Yet while the ecstasies of youth
  In blunted senses pall
The heart that leans on Love and Truth
  Is not bereft of all

For him when earth shall pass away
  Celestial spheres shall roll.
And every sensual joy's decay
  Yield Rapture of the Soul.

What Adams most feared was that he would lose his sense of purpose, that he would be unable to contribute to what he believed was his primary mission: serving his country and contributing to human progress. The wear and tear of aging, family and financial pressures, the stress of a demanding workload, and his concern that his career might soon be over took their toll. After all, he asked himself, for what other work was he suited? And what would he do if, in his late fifties, he were to return to private life? He would need "some permanent employment, to be substituted for the public duties" that now absorbed his time. "Some employment which may engage my feelings, and excite a constant interest; leaving as few hours as possible for listlessness, repining or discontented broodings." The law was out of the question. Farming might suit, except that his ignorance of it would be "a great, if not an insuperable obstacle to my success." That he might return to teaching or be offered a university presidency does not seem to have occurred to him. Just being literary had its attractions:

writing and translating poetry, reading classical authors, drafting essays on political and philosophical subjects. But he knew a literary life no longer had even minor plausibility. Louisa thought the same. "I know my husbands character. . . . I fear that he could not live long out of an active sphere of publick life and that it is absolutely essential to his existence." What he principally dreaded, Adams confided to his diary, "is a dejection of Spirits, and atrophy of mind. . . . I shall want an object of pursuit." But, John Quincy and Louisa recognized, he had become addicted to the excitement of leadership. He loved being at the fulcrum of the action that defined the nation. He had a vision of the nation's future, to which he had always hoped to devote all his energy, skills, learning, and imagination. The presidency would give him the best opportunity to realize that vision.

Nonetheless, he would not campaign. In his view, since 1800 the contest for the presidency had become a take-no-prisoners killing field inseparable from the gridlocked nastiness of Congress and the divisions that threatened the union. Although he saw no easy way out of this madness, he believed he could best serve his country by providing an example of fairness, moral values, cooperation, and nonpartisanship in government. Still, he was realist enough to know that the odds were against his candidacy coming into play at all. It was unlikely that he would receive the Republican Party nomination. Crawford was at the head of the line, and Clay probably next. Jackson was by now tied with Clay, perhaps edging even with Crawford, a battle between a formidable Washingtonian and a popular war hero. Adams had no reason to believe that he could leapfrog over Crawford or Clay. But he had supporters who, even in the face of his "Macbeth Policy," thought him the best-qualified candidate. And though he kept his pledge to keep his nose to the State Department grindstone, he and Louisa did not curtail her Tuesday evenings at home, their dinners for the Washington elite, and occasional large parties, the appearance of campaigning notwithstanding. Anyway, his famous name made it inevitable that he would be on everyone's short list.

On the evening of January 8, 1824, the Adamses hosted a ball at their expanded F Street house in honor of Andrew Jackson and the Battle of New Orleans. About a thousand guests came, including almost all the members of Congress, to attend the largest, most splendid event of the season. Louisa had opposed the ball, which would require weeks of preparation: the furniture removed, temporary pillars to reinforce the ceilings, an alphabetical list of guests, five hundred invitations printed, days of "running about town" to deliver them, the expense of the military band. But she was "overpowered by John's arguments. . . . The four lower rooms . . . decorated with wreaths and roses . . . were ornamented alike and all the doors removed which afforded as much space as we could make and looked very showy." Still, the enlarged house would barely contain the guests. Probably John had in mind that such a glorious party would throw a bright spotlight on his father, who was widely talked about as the possible next president. Jackson arrived at 8 P.M. Arm in arm with Louisa, he was escorted to the receiving line. Gaunt, leathery, white-haired, showing the wear and tear of duels, illnesses, and battles, the general seemed a model of gracious composure, his best party manner a purposeful demonstration that he was not the rumored uncouth backwoodsman. The portly, balding secretary of state and his slim, elegant wife greeted him. Adams looked like what he was, the epitome of the civil servant, the master of the desk and midnight oil, a man who smiled when the occasion was merry but never when business was at hand or people needed to be wooed. "I took mankind as you must take your wife," he later wrote, "for better or worse, and believed that the secret of dealing with my fellow creatures . . . was to

keep my heart and my judgment always *cool*." When
the guest of honor left, much before the dancing ended
at 1 A.M., neither principal could have guessed how involved their lives would be over the next five years.

Apart from dinners and this grand party, Adams
kept close to his desk. "Whatever talents I possess, that
of intrigues is not among them. And instead of toiling
for a future election . . . my only wisdom is to prepare
myself for voluntary, or for unwilling retirement." It was
not always easy, though, to keep politics at a distance.
Enemies attacked; friends made overtures. Crawford's
and Clay's partisans were unremitting in their assault,
and every dinner party was a potential battlefield.
Louisa worked at creating guest lists of politically compatible people. It became difficult to do as the election
approached. "We had a dinner today of gentlemen,"
she wrote: "Mr Clay Mr Calhoun . . . Mr Webster,"
with about fifteen others, including a Supreme Court
justice, the minister of the First Unitarian Church,
and Robert Hayne, a recently elected South Carolina
senator. Calhoun, now out of the political closet, was
maneuvering for the nomination, with the vice presidency as his fail-safe position. Eager to leave the
House, Webster had in mind an appointment to the
Court of St. James's. "This was what we call a snarling dinner," a tired Louisa observed, "composed of

such opposite materials it was not possible to prevent sharp speeches—Mr Clay as usual assumed a very high tone . . . the oddest compound of vulgarity and courtesy I have ever met with"; he was the master of bluff, playing for high stakes. If this had ever been an era of good feelings, the good feelings were long gone.

At the State Department and White House, governance needed close attention. Adams was peeved when he had to devote many hours to preparing testimony and being deposed in the suit that Levett Harris, the former consul at St. Petersburg, brought to defend himself against charges of corruption. He felt sorry for and detested Harris, "one of those mixed characters . . . with some very good qualities" who "made a princely fortune, by selling his duty and his office, at the most enormous prices. . . . Yet he sold his signature for little as well as for much." Congress, though, was even more annoying and demanding than Harris. Republican Party factions created frustrating stalemates. Foreign affairs had to be addressed. When the Russian government claimed that the waters of the Pacific Northwest up to a huge distance offshore belonged to Russia, Adams rejected the claim in no uncertain terms. While diplomatic discussions with Britain and France continued on matters of trade, little progress was made. Albert Gallatin, in Paris, tried to settle the French demand for special trade privileges. When his approach differed from his instructions, Adams was appalled, especially as it implied that his instructions were "not worth a straw." Adams thought that Gallatin, denied a path to the presidency because of his European birth, disguised and yet revealed "a supercilious prejudice of European superiority of intellect . . . holding principles pliable to circumstances." Stratford Canning, the British minister, seemed to Adams argumentative and arrogant. When Canning raised the British right to control the Columbia River, Adams objected to his tone and rejected the claim. " 'But, I *now* only have to say to you, Sir," Canning replied, "that henceforth, whatever may happen, I shall never forget the respect due from me *to the American Government.*' I made no reply, but bowed, to signify that I considered the conversation as closed and he withdrew." None of these issues, though, threatened an impending crisis.

Great Britain had continued trying to persuade the United States to join it in repressing the slave trade. But Adams repeated and repeated, as did Richard Rush in London, why that was a nonstarter. What was to the British a minor aspect of the larger issue had serious domestic reverberations. When a British ship with black sailors docked in Charleston, local officials arrested them, with the support of the governor and legislature.

The state wanted no free blacks, let alone gainfully employed black sailors, walking its streets. South Carolina had given such arrests legal sanction. With Monroe's agreement, Adams made it clear that the state's legislation conflicted with federal law: the Constitution reserved foreign affairs to the executive. In effect, South Carolina was making and enforcing its own foreign policy. South Carolina replied, loudly, that Adams could affirm as much as he liked that the Constitution made a treaty the supreme law of the land, but no Negro, free or otherwise, would be allowed to walk the streets of Charleston at will. In Georgia, a different but parallel confrontation developed. The Cherokee and other Indian tribes, Georgia had decided, would not be allowed to retain any but token lands in the state's territory. This had been, from the start, the aim of the Washington, Jefferson, and Madison administrations. But removal was to be a matter of negotiation, not force. The Indian tribes were independent nations. Treaties had to be signed between the tribes and the federal government.

Under pressure from Georgia, the United States had signed a new treaty with the Cherokee that required immediate eviction, with minimal compensation. When it became clear that it had been signed under illegal circumstances, including bribes to a self-interested splinter group in disregard of other tribal leaders, Adams and Monroe had the disagreeable challenge of creating another new treaty. When they demanded noncorrupt negotiations and some degree of justice for the Cherokee, the governor of Georgia and its legislature were furious. If Washington did not propose a treaty that satisfied the state's commitment to immediate total eviction, Georgia would take matters into its own hands. Adams "suspected this bursting forth of Georgia upon the Government of the United States was ominous of other Events." Like South Carolina, Georgia created an irresolvable conflict. Would the administration be willing to use federal troops to enforce federal law and the Constitution? If it was willing, would soldiers with Southern affiliations cooperate? In addition, the actual military resources available were modest. Congress had forced severe cutbacks. By late 1824, with the election impending, the conflict was left to simmer. And, in Adams' mind, the overarching question about Indian policy was still open: should the tribes be integrated into white society or sequestered into reservations west of the Mississippi?

Serious as these matters were, the issue that kept Adams and the Monroe administration in a state of tense indecisiveness in late 1822 and throughout 1823 was the latest development in the dissolution of the

Spanish empire. Would Spain's European allies, as was rumored, help bring its former South American colonies back under Spanish control? European autocracy, under assault during the Napoleonic years, had been revived, especially in Prussia and Austria. It had never weakened in Russia. Under the banner of the divine right of kings and Christian orthodoxy, the three countries formed the "Holy Alliance," a group of modern-day crusaders whose mission was to fight republicanism and stamp out any vestiges of the French Revolution. France joined the expanded alliance. When dissident antimonarchical forces in Spain forced Ferdinand VII from his throne, a French army, on behalf of the Holy Alliance, marched across the Pyrenees and soon restored autocratic rule. Why not, the Alliance muttered, send armies to South America and restore the former colonies to their rightful owners? That would be an object lesson to their own colonies and dissidents. Or why not put on the throne of these reconquered colonies newly created kings from the royal families of Europe? Within the Holy Alliance there were both war hawks and skeptics. Great Britain stood aloof, its leading Tories and George IV sharing some values with the alliance but aware that Britain's commercial interests and comparative liberalism required that it encourage free trade with South America and not involve itself in unrealistic and expensive military schemes. It would not be to Britain's advantage for Spain to regain its lost possessions or for the European monarchies to gain new colonies.

Nor was it in the interests of the United States. Guided by Adams, Monroe had moved cautiously. Arguing that recognition should occur at a time least dangerous to the United States, Adams had long believed "that all the Spanish Colonies will be either independent," as he wrote in 1810, "or at least have an existence totally different from that which they have been under from the discovery of Columbus to these times." All the colonial systems seemed rightfully doomed to extinction. Once the Adams-Onís Treaty was signed, there were only minor reasons to delay recognition. The highest priority of the British government, which also believed recognition inevitable, was not to cede or even share its advantage in South American trade. Britain's navy controlled the seas. How, Adams wondered, was it possible for the Holy Alliance, with negligible to no navies, to transport large numbers of troops to South America and support them there? But, as the British and American governments were aware, impracticality was no bar to troublemaking folly. In 1821, when the Greeks rebelled against Turkish rule, many Americans, led by Webster and Clay, championed recognition of Greek independence and financial and military aid. When Adams warned in his July 4, 1821, speech that the greatest danger to the United States was becoming a colonial power, and argued that it should continue its policy of neutrality, he had in mind the South American and Greek situations. America should respect the independence of others, and it should not seek colonies of its own. It should recognize, at the appropriate time, former colonies or subjugated countries, but it should not enter into any alliances. He successfully argued, against Monroe's inclination to do more, that the United States should give only moral support to Greece. Recognition needed to await clear evidence of independence, and to recognize Greece now would give the Holy Alliance an additional reason to be hostile to the United States.

After years of strategic postponement, in 1822 the Monroe administration recognized the independence of Argentina, Chile, Peru, Colombia, and Mexico. Adams reminded his colleagues that there was good reason not to expect the new South American countries to become American-style republics. They had for centuries been church-ridden autocracies with no respect for or experience with democracy, free speech, shared governance, and civil liberties. Now, in 1823, the question was what to say or do if the Holy Alliance attempted to resubjugate them. In late October, Adams

was astounded by news from London that George
Canning, the British minister for foreign affairs, had
proposed to the American minister that the two countries announce together that both believed Spain would
never recover its colonies; that widespread recognition was inevitable; that neither country desired to
take possession of any former colony, and that neither
would look with "indifference" on the transfer of any
of the former colonies to another country. As he sat
in Canning's office in Whitehall, Richard Rush could
hardly believe his ears. It was an extraordinary offer,
among other reasons because it would be, if only in a
modest way, a public alliance between two former enemies whose doubts and resentments about one another
still had force in British ruling circles and American
public opinion. By the time the August 1823 proposal
reached Adams' desk, Canning had turned in another
direction. Britain's politics, its relationship with the
Holy Alliance, and politics within the Holy Alliance
itself were in flux. But in Washington, in the councils
of the Monroe administration, the proposal provoked a
vigorous debate. How to respond? At the same time, as
Monroe began to give thought to his annual message,
he and Adams pondered how to react to pressure from
Russia for the United States to mind its own business
about the Holy Alliance and South America.

Between late October and early December 1823, the Monroe cabinet tried to reach a consensus. A cautious, prudent Monroe was indecisive when Adams urged that he incorporate into his annual message a declaration of the republican principles of the United States as a response to Emperor Alexander's assertion of the principles of the Holy Alliance. Monroe worried that such a direct rebuttal would offend the emperor and provoke the alliance into military action. Adams thought both outcomes unlikely. Calhoun, fearing that the alliance might actually send troops to South America, urged acceptance of Canning's proposal. Monroe had doubts. Adams opposed it. To accept would be a sign of weakness and inconsistent with the policy of neutrality. "It would be more candid as well as more dignified to avow our principles explicitly to Russia and France, than to come in as a Cock-boat in the wake of the British man of War."

But there was still no consensus. The cabinet, under intense pressure, debated every aspect of the issue. Would Great Britain go it alone if the United States declined the proposal? If so, would that not result in South American trade becoming exclusively British? If the United States declined to take any stand at all, would that encourage the Holy Alliance to send troops to Mexico, Texas, and California, which might once again become European colonial possessions?

The United States would then have hostile regimes on its southern and western borders. What Adams had viewed as America's transcontinental destiny, accomplished peacefully, at the initiative and with the consent of the Western inhabitants, would be blocked. And if the United States asserted, on its own or in concert with Britain, that it would look with disfavor on any effort by the European powers to reassert control over South America, would this commit the United States to military action if such an effort should occur? Wirt argued that the country would not support a war to save South American independence. Adams responded that since only Congress could declare war, it would do so only if it believed the country favored war.

Monroe wavered. He seemed to Adams "alarmed far beyond anything that I could have conceived possible." He feared that the Holy Alliance was "about to restore immediately all South-America to Spain. Calhoun stimulates the panic." Adams kept applying pressure. By mid-November, the cabinet had decided against a joint statement with Britain. By the next week, Adams had convinced Monroe that his legacy would best be served by stating in his annual address that the United States opposed any attempt to resubjugate the former Spanish colonies, that there should be no further colonization of South or North America, and that the United States would abstain from any involvement in European affairs. It was not realistic to think, he argued, that the Holy Alliance would go to war in response to the American statement. "My purpose would be in a moderate and conciliatory manner, but with a firm and determined Spirit, to declare our dissent from [its] principles." We should "assert those upon which our own Government is founded; and while disclaiming all intention of attempting to propagate them by force and all interference with the political affairs of Europe, to declare our expectation and hope that the European Powers will equally abstain from the attempt to spread their principles in the American Hemispheres or to subjugate by force any part of these Continents to their will." Would the United States go to war if colonization continued? There was no need to make a statement, Adams maintained, about what the United States would or would not do. That could be determined at the time of the provocation. But if the Holy Alliance breached the American wall, the United States should counterattack.

Monroe was at last convinced, though how to respond to Alexander still worried him. Adams continued to urge that a strong statement of republican principles be included in the message. He wrote a draft. Monroe wavered. One day he agreed. The next he changed his mind. Then he changed his mind again.

Adams made a strong plea for its retention. Finally, "Mr. Brent," the chief State Department clerk, "called me out and gave me a note from the President returning my original draft—expressing the apprehension that the paragraph of principles contained a direct attack upon the Holy allies; by a statement of principles which they had violated—but yet consenting that I should reinsert the paragraph on account of the importance that I attached to it." On December 6, 1823, Monroe sent to Congress his State of the Union message. The foreign policy statement had been composed in exchanges of drafts between Monroe and Adams as Monroe gradually embraced Adams' views. The main point was unmistakable. "We owe it, therefore, to candor and to the amicable relations existing between the United States and those powers to declare that we should consider any attempt on their part to extend their system to any portion of this hemisphere as dangerous to our peace and safety." Much of the language, slightly revised, was Adams'. Decades later, it came to be called the Monroe Doctrine.

But the overriding preoccupation of Washington was less with foreign affairs than with presidential politics. Who would be Monroe's successor—Crawford, Calhoun, Clay, Jackson, or Adams? The next president would benefit from or reap the consequences of the bold assertion of America for the Americans. And each of the competitors could find some advantage in the resolution of the dilemma created by Canning's proposal, though Adams, as secretary of state, probably had the most to gain. That probably contributed to Calhoun's opposition and the line he took during the cabinet debates. For all of them, on New Year's Day 1824, that issue was now off the table. When, in the first six months of the new year, the danger of war with the Holy Alliance proved as insubstantial as Adams had believed it to be, everyone could now devote full attention to the question that most preoccupied the country: Who would be the next president? "Every day brings forth a new rumor," Louisa noted, "not one of which can be believed or relied on." Inevitably, despite his Macbeth Policy, it preoccupied Adams also. By late summer 1824, "the bitterness and violence of Presidential electioneering," he noted, "increase as the time advances. . . . It distracts my attention from public business, and consumes precious time." In August, he took what he thought would probably be his "last bath in the Potomac. . . . For on my return the bathing season will be past, and I shall probably never pass another summer in this city."

He had a moment of distraction in early September on his way to Quincy. As he headed north with Louisa on the narrow Raritan River in New Jersey, their steamboat, *Legislator*, raced the steamboat *Thistle* at full speed at "the rate of ten miles an hour," with the *Thistle*'s bow to the *Legislator*'s stern, to see which would be the first to get to New York. When the river widened, the *Thistle* raced ahead. At New Brunswick, the pier crowded with barrels of New Jersey peaches, "the market-man insisted in my accepting two or three of his peaches as a brother Yankee." In Quincy, he walked in the graveyard where four generations of Adamses were buried. The three living generations, he observed, would in another century "all be moldering in the same dust. . . . Who then of our posterity shall visit this yard? And what shall he read engraved upon the stones?" Would his stone say, he was indirectly asking, that John Quincy Adams had been the sixth president of the United States? If so, how could that happen? The odds were considerably against him, though they had improved slightly. Crawford had suffered what was probably a heart attack. Worse, he had had a near-death response to drugs he had been administered, particularly digitalis, which had left him partly paralyzed and practically blind. Miraculously, over the next six months, he recovered enough to appear at some cabinet meetings. His eyesight had returned, though weakened. His speech was slurred. Rumor had initially reported him dead or about to die. Then he seemed capable again, as if partly resurrected.

Not everyone was sympathetic. The heaviest burden was on Crawford supporters, such as Martin Van Buren, the senator and political heavyweight from New York, who alternated between hope and pessimism about the health of the man to whom he had tied his political fortunes. "Mr. Crawford's friends were beginning to consider the state of his health as desperate," one of Adams' supporters remarked, "and that it would be necessary for them to fix upon another candidate." The collapse of "King Caucus" also deprived Crawford of a political advantage, and he had to deal with Illinois senator Ninian Edwards' charge that he had mishandled government money. In May 1824, he had a relapse. By the fall his support had eroded. Monroe, who became aware that the secretary of the treasury had not been fully loyal to the administration, felt deceived by Crawford's manipulations. When Crawford spoke disrespectfully to the president and brandished his cane, Monroe raised the fireplace tongs to defend himself against what he thought would be a physical attack. Clay, Calhoun, and Jackson looked to be the gainers.

In the fall and winter, rumors circulated that Crawford and Jackson had reconciled. There were other rumors concerning every possible combination of candidates. Alliances were being raised and promoted, initiated by the candidates' supporters, sometimes without the knowledge of the principals. Crawford's supporters proposed a coalition with Adams that would reward Adams with the vice presidency or any cabinet position of his choice. Had not Adams' supporters offered, at his own initiative, he was asked, to support Clay for the vice presidency if Clay would back him for the presidency? No, he responded. "I further said, that although I never had authorized any man to make such a proposal to Clay, yet friends of mine, and friends of Clay too had often suggested it to me as desirable; nor is there anything in it, unconstitutional, illegal or dishonorable. The friends of every one of the candidates have sought to gain strength for their favorite by coalition with the friends of others, and to deny very indignantly an imputation of that which is not wrong in itself is giving the adversary the advantage of fastening upon you a consciousness of wrong where there is none." No doubt the congruence between Adams' and Clay's visions of a union made strong and prosperous by federal leadership gave the combination plausibility.

When Pennsylvania, where Calhoun had initially been strong, turned to Jackson, Calhoun announced that he would run for the vice presidency. The key to the presidency seemed to be New York, where, a supporter told Adams, "everything political . . . is an article of purchase and sale." But could Van Buren keep the legislature and the electors in his control? If not, how many would stay with Crawford? How many would vote for Adams? Van Buren's enemy, the influential and charismatic De Witt Clinton, had his own presidential ambitions. After those proved unrealizable, he threw his influence behind Jackson, hoping to be appointed secretary of state. When Adams was asked to explain his position on the Missouri statehood battle, he sought to soften Southern unease by explaining that, though he had supported the right of Missouri to determine whether it would be a free or slave state, what he had considered unconstitutional was Missouri's attempt to exclude free blacks from residence. He was not opposing Missouri. He was supporting the Constitution. Crawford shared support of the South with Jackson, while Clay and Jackson competed for Western support. New England would vote for Adams, though Massachusetts Federalists needed to be assured that he would not exercise his resentment over how he had been treated in 1807 by refusing to appoint Federalists to government positions. In 1800, many Federalist leaders had hated John Adams enough to contribute to the election of Jefferson. Had that turned out well for them, Adams asked? Did they now really prefer Crawford?

By the fall of 1824, the war hero had been packaged for the public as the Washington outsider unrivaled in courage and integrity, the military superman charging into the fray to disperse the Washington elite and govern on behalf of the people. Pennsylvania seemed to have been preempted by Jackson partisans. If the New York electors favored Jackson, he would, Adams recognized, almost certainly be the next president, despite the objection, particularly Jefferson's and Madison's, that he was a mere military man. Eighteen of the twenty-four states now chose electors by popular vote. But to gain a majority of the electoral votes, Jackson needed some of Crawford's or New York's or both. He got neither. Crawford's supporters, even the ambitious Van Buren, resisted. And New York was split by intrigue, bribery, and rival ambitions. In the popular vote, Crawford held Georgia and Virginia. Clay got Ohio, Kentucky, and Missouri. Jackson won in Pennsylvania, New Jersey, Delaware, Maryland, Indiana, Illinois, Tennessee, North Carolina, South Carolina, Alabama, Mississippi, and Louisiana. Adams got all of New

England and New York. Turnout was low. There was no official popular vote tally, and in a few states the legislatures still chose the electors. Regardless, neither Jackson nor Adams got the 131 electoral votes necessary to win. Jackson had about 41 percent of the popular vote and 99 electoral votes, Adams 31 percent and 84 electoral votes, Crawford 11 percent and 41 electoral votes. Clay had only 13 percent with 37 electoral votes. The three-fifths provision in the Constitution produced Jackson's lead in the electoral college. "Without it," one of the premier modern scholars of the period has noted, Jackson "would have received 77 electoral votes and Adams 83."

It was, in its peculiar constitutional way, a nightmare scenario. Each state would have a single vote in the House of Representatives to determine a winner. Since the House could consider only the three candidates with the largest number of electoral votes, Clay, though he had apparently obtained more popular votes than Crawford, was eliminated. The Clay states might go to Adams or Jackson, and only if Jackson got a majority of the state delegations in the House would the winner of the popular vote be president. If they all went to Crawford and he got some of the Jackson and Adams states, he could become president, but it was an unlikely prospect. If Adams got the votes of the three

Clay states, he might become president. There were twenty-four state delegations, and the magic number was thirteen. Jackson already had eleven. Adams needed an additional six. Each delegation, not bound by any previous votes or commitments, could vote for any one of the three candidates. Informed opinion anticipated that it would take multiple votes and days for the House to choose between Jackson and Adams. Virginia and North Carolina would stick with Crawford. That meant that the three states that had voted for Clay would be decisive if any one of them chose Jackson, as long as Jackson held the eleven states he had won in the general election. If Clay's three states—Kentucky, Ohio, and Missouri—voted for Adams, he would still need three more, assuming he retained the seven states that had voted for him in the general election. Jackson and his supporters had good reason to believe he would win.

For months Adams had wrestled with the burden of his candidacy. He could not help asking himself whether he even "ought to wish for success." To lose would be painful. But so little could be known about the future, especially if he should become president, "that whether I ought to wish for success is among the greatest uncertainties of the election." Victory might prove more adverse than defeat. "Yet a man qualified for the elective Chief Magistracy of ten millions of people should

be a man proof alike to prosperous and to adverse fortune. If I am able to bear success, I must be tempered to
endure defeat." Would his election by the House, without a popular or electoral mandate, allow him to govern?
"He who is equal to the task of serving a nation as her
chief-ruler must possess resources of a power to serve
her even against her own will. . . . This is the principle
that I would impress indelibly upon my own mind."
Such an election would represent the will of the founding fathers embodied in the Constitution. He would
not withdraw. If defeat came, he could bear it. And his
commitment to the Constitution sanctioned his allowing
the House to fulfill its constitutional responsibility. If
chosen, he would serve, even in the potentially difficult
circumstances that would exist for a president who had
not received the most popular or electoral votes. Not to
proceed would be a betrayal of his personal and civic
values. And it would forfeit the opportunity to lead the
nation toward the fulfillment of his vision for its future.

Adams now moved toward the House election in a
mood and with a conviction that allowed him to balance
his Macbeth Policy with the political realities. He would
not himself try by direct persuasion to get individual
members of the House to vote for him in their delegation caucuses. But he would make himself as visible as
possible. He would not campaign, though his surrogates

might, but he would show the attentions and courtesies that would give presence to his candidacy. At home, he and Louisa entertained almost nightly. At home and at the State Department, he had hundreds of visitors throughout December and January. Many wanted something. They got a candidate who gave only his promise that he would be the president of the entire country and his assurance that he would be moderate and compromising on the contentious issues of the day. Visitors were free to write into his general promise whatever particulars suited them. Although this was a little less than the Macbeth Policy, he could live with that.

And the path to victory, he also knew, almost certainly required that the state delegations representing the three states that had voted for Clay vote for him. The Kentucky, Ohio, and Missouri delegations were more loyal to Clay than to instructions from the legislatures of their states. Clay could deliver those three votes to Adams, if he chose. He had no sympathy for Crawford's views, which were not nearly nationalistic enough for him: they were too much narrowed by a commitment to states' rights and a limited federal government. Clay thought little of Jackson. It was widely rumored that a young supporter of Jackson, Congressman James Buchanan, had offered Clay, with or without Jackson's knowledge, the Department of State. But Clay believed that Jackson would act in the presidency in the same high-handed way he had acted as a general. "The elevation of the Hero," Clay told Adams, would be "the greatest calamity which could befall the country." And who knew what Jackson's public policy positions were? He and his handlers had been clever. What they seemed most interested in was power, but to what service they would put it nobody knew.

That left Adams. Early in December, Clay made his view known orally and in writing. Despite differences of temperament and background, he and Adams had worked well together at Ghent. They shared a vision for the future of the country. Each had a good sense of the other's strengths and weaknesses. Both favored building a national infrastructure, promoting a national banking system and currency, supporting manufacturing and commerce, maintaining a strong military, constructing a national educational system, and ensuring that the federal government was the support and shield of the union. Clay's friends assured Adams that Clay had no personal hostility to him. Adams responded in kind. Through surrogates, Clay made clear to Adams, on January 17, 1825, that he had decided that the three state delegations he controlled would vote for Adams.

Actually, Clay had made that decision by early December, when he had told his friend Missouri senator Thomas Hart Benton, who told others, including Jackson's supporters. Would Adams meet with Clay, intermediaries asked? Clay told them he had no need for a meeting. A surrogate put it directly to Adams. He hoped Clay "would be a member of the next administration. . . . I told him that he would not expect me to enter upon details with regard to the formation of an Administration, but that if I should be elected by the suffrages of the West I should naturally look to the West for much of the support I should need." Samuel Southard would be asked to stay on as secretary of the navy, Crawford to remain at Treasury. That left War and State. Since Clay had turned down Monroe's offer of the former in 1817, it would not be sensible to offer that to him again. That left State, which was what Clay wanted and Adams had in mind, an appropriate position for a speaker of the House who had taken an active interest in foreign affairs. At the end of January, they had a long discussion. Clay "spoke to me with the utmost freedom of men and things." Adams listened. Neither spoke explicitly about a cabinet appointment. But it was clear that an alliance was in place. It seemed a practical and honorable arrangement of a sort that was customary in American politics and essential to the machinery of governance.

With "the excitement of electioneering . . . kindling into fury," Adams tried hard to keep his composure and navigate the tumultuous political waters. He provided friends and enemies with whatever assurances he could. The city was swarming with strangers, positioning themselves for appointments, and with rumors, spin, and lies. "Duplicity pervades the conduct of so many men in this country," he wrote in his diary, "that it is scarcely possible to know upon whom any reliance can be placed." But the sensible and natural alliance held. On February 9, 1825, the House convened. Kentucky, Ohio, and Missouri voted for Adams. He needed four more states. Maryland, Louisiana, and Illinois decided that Adams was preferable to Jackson. After some hesitation, the Crawford partisans in the New York delegation switched. Jackson now had only seven votes, Crawford had four, and Adams the necessary thirteen. To the shock and surprise of many, including the president-elect, John Quincy Adams had been elected the sixth president of the United States. "Never did I feel so much solemnity as upon this occasion," his father wrote to him. "The multitude of my thoughts and the intensity of my feelings are too much for a mind like mine, in its ninetieth year." The son, like the father, knew, as he was to write three years later, that his presidency would not be "a bed of roses."

Chapter 13
No Bed of Roses
1825–1829

After two sleepless nights, having spent weeks writing and rewriting his inaugural address, Adams left his F Street house at 11:30 A.M. on March 4, 1825. With an escort of militiamen and a parade of citizens, followed by President Monroe in his own carriage, Adams went to the Senate chamber where John Calhoun, who had been on both Adams' and Jackson's electoral tickets, had just been sworn in as vice president. The senators and the justices of the Supreme Court accompanied Adams to the House of Representatives, where the president-elect was escorted to the speaker's rostrum. Every seat in the hall was occupied, the gallery crowded. John Marshall waited to administer the oath. He had hoped Adams would be elected, partly because, as Rufus King had told Adams, it was a vindication of Adams' father but also because Adams and Marshall shared a vision of the United States as a unified nation-state. Not everyone in the audience shared this vision or wished him well. "Fellow citizens, you are acquainted with the peculiar circumstances of the recent election. . . . Less possessed of your confidence in advance than any of my predecessors, I am deeply conscious of the prospect that I shall stand more and oftener in need of your indulgence."

From the start, Adams pledged himself to an idealistic agenda and a post-partisan tone. He would, he decided, revive and remain true to the nonpartisanship that he associated with Washington, who had attempted to rise above party. Factions were inevitable. Parties were not. "Whatever vices there are, Federalism and Republicanism will cover them all," John Adams had remarked. Some abhorred political parties because they were narrow, selfish, and divisive. Some abhorred only the *other* party because it was always dangerously wrong. Like Jefferson, they preferred no party, which was the same as *one* party, as long as it was theirs. The Constitution, a practical and idealistic document, had not provided for parties. Although their existence had taken on the force of inevitability, it was worthwhile, Adams believed, to take a stand, even if it were an object lesson in the impossible. He would not be a party builder, and he would not create a politically inflected bureaucracy. No one would be fired from or appointed to a government job, Adams announced, because of his political views. Competency was the only requirement. He hoped that his fellow citizens would have the wisdom to embrace virtue, though, from the start, he did not fully expect it. At any rate, he framed the issue as one of principle versus political expediency, and he had no doubt what his own sense of self required. Nonpartisanship in appointments was a moral imperative.

Discussion of philosophical and political differences, Adams believed, was essential to forging a unified public policy. But differences "founded on geographical divisions, adverse interests of soil, and modes of domestic life," by which he meant slavery, "are more permanent, and, therefore, perhaps, more dangerous," Adams noted in his inaugural address. Monroe had made a good start. But there was still a distance to go. And it would be traveled, Adams hoped, by a country unified by internal improvements. Like previous presidents, Monroe had advanced such projects, though they had been, more or less, constrained by his constitutional scruples. The federal government, Adams insisted, was empowered to undertake such projects in order "to provide for the general welfare," and even those who advocated a narrow interpretation of the Constitution had interpreted it broadly when it suited them. He would take as precedent the infrastructure projects of his predecessors. "Nearly twenty years have passed since the construction of the first national road was commenced. . . . To how many thousands of our countrymen has it proved a benefit? To what single individual has it ever proved an injury?" He hoped that "the extent and limitation of the powers of the general government in regard to this transcendently important interest will be settled and acknowledged to the common satisfaction of all, and every speculative scruple will be solved by a practical public blessing." He would continue many of the policies of his five predecessors. But he would, he announced, urge Congress to sponsor many more such projects than his predecessors did.

By the time of his inauguration, his cabinet choices were in place. He chose, as did Monroe, to emphasize continuity and sectional balance. He had to replace the secretary of war. Jackson would have been an appropriate choice, but he made it known that he was not interested, his eye exclusively on the glittering prize. James Barbour, a former Virginia governor and now a U.S. senator, admired for his oratory and his Republican principles, replaced Calhoun. It was a noncontroversial choice intended to give Adams credit with the Virginia elite, though he could still expect them to oppose his nation-building program. Adams had learned to respect Samuel Southard, Monroe's secretary of the navy and a former New Jersey senator; and William Wirt, Monroe's attorney general, a Virginia lawyer and the first biographer of Patrick Henry. Both agreed to stay on. John McLean, a former Ohio judge who had shown administrative skill as postmaster general, agreed to continue managing the postal system. Looking to enlist Crawford and his supporters, Adams offered him the Treasury Department. Crawford declined. Richard Rush, now minister to the Court of St. James's, accepted instead. The two powerhouse states —Virginia and Pennsylvania—now had cabinet seats; the two most prominent cabinet members, the president and secretary of state, brought New England and the Western states to the table. Although there was some opposition, Clay was confirmed by a two-to-one majority. It was to be, as it had been with Monroe, a cabinet government. The executive would lead and, in the end, be decisive. But it would be government by consensus. In all domestic matters, the executive would pay due deference to Congress. After all, that is what the Constitution and Republican principles required.

The presidential bed of roses soon revealed its thorns. Even before his inauguration, the factions and candidates were preparing for the next election. Everyone, including Adams, recognized that he was politically vulnerable. The misrepresentations made during the campaign continued with hardly a pause. That he had been elected by the House became a weapon to wield against him. Some were sincerely and in principle opposed to what the Constitution had required. It was undemocratic. Since the people had spoken, the candidate with the most votes should have become president. That flaw in the Constitution needed to be addressed. But many who took up the cudgels for a more democratic electoral process would not have in the least objected if Crawford or Jackson had been the victor by the same system. When the new Congress elected a pro-administration speaker, it was by the narrowest of margins. Soon after his inauguration, Adams was warned that forces were coalescing against him. His opponents would look for any way to make him a one-term president. In Congress, Jackson, Crawford, and Calhoun partisans together would have more votes than Adams' supporters.

And the president's claim that a federal program of internal improvements was essential for the unity and prosperity of the country immediately met the resistance of those who did not want Adams as president and who viewed national planning as a step toward the subordination of the states to Washington. That, in turn, would threaten slavery. In fact, the South in general believed that Adams was in his heart an abolitionist. Nothing he could say could convince them otherwise. By calling attention in his inaugural address to the dangers "founded on geographical divisions, adverse interests of soil, and modes of domestic life," he demonstrated his opposition to the three-fifths rule. And his call for a broad interpretation of the Constitution raised the question: Might he not find in the Constitution grounds to argue that slavery is incompatible with the Constitution's commitment to "the general welfare"?

Then there was Clay. His appointment as secretary of state had angered his and Adams' enemies. They assumed that Adams had anointed Clay as his successor, though the 1824 election had made moot the assumption that the secretary of state was invariably going to become the next president. Jackson at first declined to make a fuss. He was disappointed. He believed that he had been the people's choice, but he seemed to have accepted the outcome, even to the extent of being cordial to the new president—though not to Clay, who had vigorously condemned Jackson's incursions into Florida and had led the congressional attempt to censure him. Those who had almost prevented Adams' election felt confident they could stop Clay from becoming his successor. The ideological grounds for opposition were the same. Clay's "American System" and Adams' vision overlapped: national union, a protective tariff, the encouragement of manufacturing, internal improvements, and a strong navy. A slaveholder who supported the American Colonization Society, Clay had made himself questionable to some Southerners on the slavery issue, though he was not suspected of being a closet abolitionist. That Clay had delivered the decisive votes that made Adams president was a handy club with which to pummel them. In order to make Jackson or Crawford the next president, Adams and Clay had to be brought down. Each swing of the club would inevitably hit both. Adams had been warned. If he made Clay secretary of state, there would be hell to pay.

In the meantime, the issue of Georgia versus the Creeks and Cherokee had to be addressed. Georgia wanted the Indian land for its white citizens. The federal government had the responsibility to ensure that treaties were honored and new ones fairly negotiated. Expediency and morality clashed. Georgia's governor pressured the administration: either the federal government would get the Indians out or else. "Or else" implied that Georgia would forcibly evict them. Adams, Clay, and Barbour exerted themselves to keep the situation from exploding. When Governor George Troup declared that he had the right to call out the militia to clear off the savages, the cabinet laughed at and lamented his abusive letters. Still, they shared the same mind-set. The Indians had to be disposed of. The cabinet discussed assimilation, which Barbour favored, or eviction, which Clay supported. Adams wanted consensual agreements and fair treatment. When, in January 1821, a group of Missouri Indians sat in the gallery of the House, they occasionally dropped their blankets. It was, a shocked Louisa Adams remarked, "a sight loathsome and disgusting to the Spectators." Clearly, all eyes were drawn to the savage spectacle. How could such people be assimilated? Assimilation, Clay argued, was inconceivable.

Three years later, Adams, with Monroe and Calhoun, met with a Cherokee deputation. "This is the most civilized of all the tribes of North American Indians. They have abandoned altogether the life of hunters, and betaken themselves to tillage." Dressed like white men, two of them spoke English "with grammatical accuracy." Still, Adams had as deep a commitment to the white settlement of Western lands as did his colleagues. "There is not upon this globe of earth," he had written a decade before, "a spectacle exhibited by man so interesting to my mind or so consolatory to my heart as this metamorphosis of howling deserts into cultivated fields and populous villages which is yearly, daily, hourly, going on by the hands chiefly of New England men in our western states and territories."

In Adams' view, Indian dispossession should be accomplished by gradual assimilation. It also required fair treatment. This was not the majority view. On the frontier, Indians were feared and hated. Many whites thought that, as savages, they should have no rights. A superior civilization was on the move. As Adams struggled with the Georgia issue, Clay expressed the reality as most Americans saw it. The Indian tribes were doomed to extinction, sooner or later—better sooner. In January 1826, Adams reluctantly signed a new treaty with the Creeks, ceding much but not all of their land to Georgia but stipulating that they could remain until they left voluntarily. Angry and impatient, Troup rejected the treaty. Georgia citizens were given full range to evict the Creeks and Cherokee. In February 1827 Troup informed the secretary of war that Georgia would fight if federal troops should enter the state. At the end of the year, a third treaty was signed. As usual, the chiefs who signed did not necessarily have the consent of their people or know exactly what they were agreeing to. Although the treaty ceded all Indian land to Georgia, it soon became clear that many Indians would leave their land only if forced to. For the time being, the federal government did nothing. In July 1828 Adams noted that "Col. Brearley . . . has accompanied one party of Creek Indians from Georgia, to the Alabama Territory, and is soon going to muster another detachment of them for that purpose; but he says they will not go till they are starved into another sale of their lands, which must soon happen, as they will exercise no industry; and now live only upon what they receive from the U.S."

When a delegation of Seneca Indians came in March 1828 to "speak with the great father the President" about the cession of their lands to New York state, Adams told the elderly chief that he "was glad to see him and to hear his talk." Red Jacket brought a silver plate he had been given by George Washington, a parchment copy of the 1794 treaty between the Seneca nations and the United States, and a petition protesting the recent nonconsensual sale of their lands. "I told them I had read all their papers, that I had given no permission or consent for the sale of their lands. That I should further consult with the Secretary of War concerning their petitions, and would do anything just and reasonable within my power to gratify their wishes."

Later that year, he spent three hours with a delegation of Winnebago Indians who were also being forced off their land. "They said the white people had swarmed upon their lands, till the game had all disappeared, and there was not a rabbit to be found. . . . I exhorted them to peace and friendship." Although he pardoned two chiefs who had been sentenced to death for murder, the larger issue, the cession of land, Adams believed, could not be prevented. He "told them they had better let us have the land where the land was of no use to them." The view widely held by white Americans was that land was for settlement. The country had no room for hunting and nomadic societies. America was moving westward, and the Indians had to get out of the way.

Two other problems needed attention. There had been no progress in settling the border dispute between Maine and Canada. As secretary of state and now as president, Adams urged the American and British commissioners to resolve all outstanding differences that the Treaty of Ghent had not settled. When Rufus King, whom Adams had persuaded to return to public service as his minister to Great Britain, had to resign as minister and from the commission because of ill health, Adams sent his negotiating partner at Ghent, Albert Gallatin, to negotiate with the British. He was successful only on one point, the settlement of the claim that

Great Britain owed American slave owners compensation for the slaves British officers had taken during the War of 1812. Under pressure from the anti-slavery lobby in Parliament, the British government agreed to provide $1.2 million to be distributed to the claimants. A new issue, which caused serious problems, contributed to British-American tension. In 1824, trade with the British West Indies had been opened to American vessels. In June 1825, the British reversed the policy. Almost all American trade was prohibited. It was a shock to U.S. trading interests and a political embarrassment for the Adams administration. After all, his strong point was foreign affairs. How had he allowed this to happen? Congress, urged by Adams, soon added to the British trade bill retaliatory restrictions on British vessels. At the same time, it fully placed the blame for losing the West Indies trade on President Adams. The opposition to the administration was delighted to have another club to use.

When a number of South American countries, concerned that Spain might attempt to reassert control, decided to send representatives to a conference to be held in Panama in 1826, Adams agreed to an American presence. The agenda was loose: to consult about mutual interests, "to deliberate upon objects important to the welfare of all," as Adams put it to Congress. Any increase in the security of the South American republics, he believed, strengthened the security of the United States. At the same time, the instability of Central and South American regimes, particularly Mexico, was a potential threat. The purchase of Texas, which Adams considered desirable, would best be served by hemispheric solidarity. He thought it sensible to send a delegation. It would not have the authority to enter into agreements, and this restriction should, Adams thought, obviate any concern that the administration was initiating alliances or treaties without consulting Congress. The objective was discussion, information, and good relations. The Constitution, he believed, granted the executive the power to engage in such discussions without Senate approval. It only remained for the House to fund the delegation. Adams had also been made aware of another South American possibility.

In late January 1825, he was brought "a project for opening a passage from the Atlantic to the Pacific Ocean." The Erie Canal, to be completed at the end of 1825, was demonstrating what engineering miracles could be accomplished by American ingenuity. Adams was presented with "a map of the Isthmus of Panama, showing the projected communication between the two seas." Would the navy make available an armed vessel to ascend the San Juan River? In July, he had a "conversation about cutting a canal from Panama to the Pacific-Bay covering a million of square miles." Adams was interested. Nothing came of it, but the vision was there.

Adams presented his first annual message on the state of the union to Congress and the nation on December 6, 1825. It set a visionary agenda, a comprehensive policy by the federal government to use tax dollars to create a national transportation infrastructure. He also called for a national university, a naval academy, a Pacific exploratory expedition, and scientific and humanistic projects, such as an astronomical observatory. For years, he had studied astronomy, his eyes not only on the wonders of space but on the importance of science and engineering to America's future. "Would it not be worthwhile," he had mused in 1819, "among the public institutions of a nation to have a school for the education of a certain number of civil engineers?" In a gracefully phrased and eloquent message, Adams reviewed the state of the nation. Its treasury was in good shape; its finances were sound. Foreign affairs were on an even keel; the administration was attentive to the issues that needed to be addressed, including national and hemispheric security. "The Board of Engineers for Internal Improvement," implementing the act passed by Congress in April 1824, had made estimable progress. Surveys for the Chesapeake Bay to the Ohio River Canal were under way. A national road from Washington to New Orleans; roads in Florida, Arkansas, and Michigan; a road from Missouri to Mexico; and the continuation of the Cumberland Road were all in some state of planning or actual implementation. So too were various lighthouses. And, Adams emphasized, the funding of these and even more ambitious projects would be forthcoming from the sale of public land, "an abundant source of revenue . . . a swelling tide of wealth."

He could not close his message, Adams wrote, without recommending that Congress and the country consider the subject of improvements in a larger framework. Man was here on earth to improve himself, his society, and the earth on which he lived. That was the point of government. It existed for "the improvement of the condition of those who are parties to the social compact." This meant not only material improvement but "moral, political, and intellectual improvement." Governments received power for this purpose, and the Constitution of the United States provides both express and delegated powers. "The exercise of delegated powers is a duty as sacred and indispensable as the usurpation of powers not granted is criminal and odious." Everywhere, "the spirit of improvement is abroad upon the earth," an observation that, thirty-four years later, Lincoln was to extend at length in his essay "Discoveries and Inventions." Europe had led the way toward America. The genius of Europe and the voyages of discovery had made possible the next stage of human progress, the existence of a country in which liberty is power: both are inseparable from and based upon "the improvement of human knowledge."

It was now America's turn to take the lead, Adams continued. Its mission was to cultivate and enlighten its homeland, to make progress toward improvement and union, and to build a better future on the achievements of the past. Individual states were already making efforts. "I shall await with cheering hope and faithful cooperation the result of your deliberations, assured that, without encroaching upon the powers reserved to the authorities of the respective states or to the people, you will, with a due sense of your obligations to your country and of the high responsibilities weighing upon yourselves, give efficacy to the means committed to you for the common good." Inherent in Adams' vision, to be realized over time and in other hands, were land-grant universities, the Panama Canal, the Tennessee Valley Authority, the Federal Reserve, and the Interstate Highway system.

Adams knew, of course, as he composed his eloquent appeal, that he did not have a completely willing audience or electorate. There would be strong opposition. For many it would be a vision too far. The extent to which he hoped to transform previous limited commitments into the creation of a national infrastructure staked out what to many seemed a breathtaking change. His political friends were made nervous by the boldness of his vision. His opponents furiously objected. It was one thing for Congress, case by case, to allocate funds for individual internal improvements. It had been doing that since the creation of the republic. It was another to have a cohesive national policy— the difference between having money occasionally put in one's pocket and an overall vision of the allocation of funds as part of a plan for national unification. In December 1825 and for the next year Adams maintained his hope that principle would triumph over politics, that visionary leadership would be recognized and rewarded.

He had no doubt that the country actually wanted internal improvements. But he knew that it did not want to pay for them, especially if this meant that Western land would not be sold cheaply, that there might be internal taxes, and that the revenue of the federal government would be increased. It also did not want to acknowledge openly, as Adams' grand statements did, its desire for internal improvements, let alone the benefits of federal spending to local communities. Rational planning frightened those for whom big government was the ultimate evil. Many valued individualism and unregulated entrepreneurship more than social community and beneficial regulation. The American spirit, particularly in the West, contained a hefty dose of creative anarchy: the landscape existed to be turned into cash through planting, grazing, logging, mining, and hunting, at whatever cost to the earth and future generations. What the country would in the long run benefit from most, Adams proposed, was some constructive balance between individual enterprise and communal action. Government leadership and rational planning were, he believed, compatible with capitalism and private property. And the divisive issues that threatened the stability of the country could be resolved only by stronger bonds of union. Union provided security and prosperity. The most effective agents of union were public improvements. Better to go down fighting for a stronger future than to serve a second term at the cost of forfeiting the opportunities for leadership that the presidency provided. There was the long-term future to consider, and the leadership that was unsuccessful today might sow the ground for successes tomorrow.

By late 1825, the factions eager to make Adams a one-term president had coalesced. His vision represented everything they detested. It would make government a player in their everyday lives by creating a transportation infrastructure, regulating financial institutions, and supporting education and research. Ultimately, it would be the institutional voice of national values. Adams' decision to send delegates to the Congress of South American countries in Panama provided the opposition with a concrete issue. In December, he told Congress that there were good reasons for the United States to be represented: the Panama Congress would defend South American independence, support the Monroe Doctrine, promote trade, encourage religious liberty, foster a friendly spirit between North and South America, seek cooperation in suppressing the slave trade, and help control the volatile issue of Haiti and its black government. Adams and Clay believed that consultation with the new nations was a natural consequence of the doctrine. In announcing the mission, Adams stated that executive authority in foreign affairs allowed him to accept the invitation without the Senate's advice and consent, though the Senate would have to approve the ministers he would nominate. Many disagreed, in principle and as a tactic. Was acceptance not tantamount to a treaty commitment, and were these representatives not ministers to a foreign entity? Was this not an unconstitutional extension of executive power? No, he responded: the decision to be represented at a conference belonged to the president, not the Senate. It was not a treaty commitment. The furor "was altogether unexpected to me," he later wrote, "for although well aware that there were in both Houses of Congress individuals little disposed to consider the executive measure with favor, I had not anticipated either that they would so speedily coalesce into one mass of opposition or that they would select this measure as their first trial of strength in hostility to the administration." The Senate delegated a select committee, its members to be appointed by the vice president, to review the matter. Four of the five were Southerners and slave owners. They discussed it as a constitutional issue, but in reality it was mostly about ideology, politics, and slavery. The committee unanimously recommended against American participation.

A seemingly innocuous proposal had become a club with which to bash the administration. In the House, the opposition was in the minority. In the Senate, where Calhoun presided and helped plot strategy, it was four votes short. For the first three months of

1826, the issue preoccupied the Senate. Its opponents mounted a relentless attack. Led by Martin Van Buren, Robert Hayne, Thomas Hart Benton, and especially John Randolph, they tortured every aspect of the issue. At first the Senate requested all relevant executive documents. The rules, though, required that executive branch documents on foreign policy be kept confidential. Van Buren, who was moving himself and his New York influence from Crawford to Jackson, requested that the president inform the Senate whether he would object to the publication of the documents. If not, would he not agree to the publication of selections? And if so, which selections? It was not his business to advise the Senate, Adams responded. If he did so, it would resent his advice. All the documents, he replied, should stay confidential unless the Senate had justification for making them public. He declined to identify which documents, if published, would "be prejudicial to existing negotiations" and which would not.

The opposition exploded into outrage and vituperation. The Southern contingent, in defense of its personal honor and the integrity of the Senate, accused Adams of questioning its motivation, as if there were base reasons for the request—though everyone knew there were. Adams, however, had not meant to make an issue of it. The opposition filled the Senate chamber with wounded, angry, and aggressive denials. It focused on the constitutional issue, arguing that the executive was usurping the power granted to the Senate, and on the foreign policy issue: it would be a breach of neutrality because it would ally the United States with the South American countries against Spain. It was "utterly foreign to us, and dangerous to our peace and institutions," Thomas Hart Benton wrote years later. Adams thought the objection thin on constitutional grounds. The executive had the authority to open discussions with any foreign country. "Advise and consent" was relevant only when the executive had a treaty to submit or was nominating ministers to serve abroad. And it must have seemed especially ironic to Adams, who had always advocated neutrality, that he was being accused of courting foreign entanglements. These were exploratory talks, at which the United States would be more observer than participant. Any initiatives that might be discussed would have to be brought to the president and secretary of state. If they then decided that an alliance or treaty was desirable, Congress would be informed, negotiations might begin, and the procedure of advise and consent might be initiated. So why all this angry fuss now? Adams had no doubt that its basis was political.

It was also racial. Neither Adams nor the opposition doubted that the darkest stain on the Senate's consciousness was Haiti. To the Southern-dominated Senate the volatile ex–slave state in the Caribbean was as large as a hostile continent. It represented what white Southerners most feared: slave rebellion. Slaves would become masters, masters slaves. And free blacks were a standing encouragement to black slaves to think about freedom. What about Cuba? John Randolph asked in a pro-slavery speech, ironically for a man who in principle abhorred slavery but who mocked the idea of universal emancipation as an unrealistic fantasy. What if the new South American republics should invade Cuba, with its huge black population? "It is unquestionable that this invasion will be made with this principle— this genius of universal emancipation—this sweeping anathema against the white population . . . and then, sir, what is the position of the Southern States?" If we should accede, "we should deserve to have negroes for our task-masters, and for the husbands of our wives." And was it not the case that the American representatives might encounter at Panama delegates who were black? Would there not be black ministers sent to Washington from South America or Haiti or Cuba? Would they expect to be treated as equals? "The Southern States will look to their safety as States and as individuals," Randolph warned, "whatever the ink and sheepskin may say, whatever Congress may decree."

Although the fulfillment of Randolph's apocalyptic rhetoric was reserved to a later decade, its flame was intended to destroy an immediate target. He had, he claimed, no personal animus against Adams. But Randolph believed that Adams, like his father, favored monarchy and a federal government that reduced the states to dependency. He remembered, he told the Senate, John Adams' coachman whipping his brother away from the vice presidential coach on a New York street in 1793. "The step you are about to take," he warned Adams, is the match to a barrel of powder that "is enough to blow—not the first of the Stuarts—but the last of *another dynasty*—sky high." And the explosion would blow up the entire administration.

In mid-February 1826, the Senate narrowly confirmed Adams' nominees to the conference. But the verbal attacks on the administration continued. Randolph's rhetoric became even more slashing. It was, Adams remarked, "the eloquence of Hogarth's Gin Lane and Beer Alley." "My resentments," Randolph announced, "are entirely political—they are for my country's enemies, not mine," though there had been rancorous personal antagonism between Randolph and Clay since 1812. "Sir, let these unhappy persons retire to the obscurity that becomes their imbecility, and befits their shame, and they shall never hear from me the language of sarcasm or reproach." He had, he claimed, plausible circumstantial certainty that the invitation to attend the Panama conference did not come from South America. He would demonstrate by stylistic analysis that the secretary of state himself had written it. "I will prove" to the satisfaction of the Senate "that the President has Jonathan Russelled himself . . . he has done that which has damned Jonathan Russell to everlasting infamy." And how has he done it? "By the aid and instrumentality of this very new ally," his secretary of state. "I shall not say which is Blifil and which is Black George," two characters in Henry Fielding's *Tom Jones.* The administration, he charged, is based on a coalition of "the puritan with the black leg." Which was Adams? Which was Clay? Most people assumed Adams was being compared to Blifil, a hypocritical, self-serving, and cowardly rival to Tom; and Clay to Black George, a rough, low-minded poacher who leads Tom astray.

Without evidence to support it, the accusation of forgery seemed the extravagance of a man whose sanity his friends had reason to question. Even some of his Senate sympathizers found the comparison outrageous and counterproductive. Randolph's volatile emotions, his gift for language, and his literary enthusiasms all tempted him into jeremiads that had consequences he had no intention of creating. Adams did not respond, except to encourage or at least permit a State Department clerk to place, in newspaper essays, much of the blame for Randolph's remarks on the president of the Senate. Although the rules forbade personal attacks, Calhoun, who used a pseudonym to respond to "Patrick Henry," whom he assumed was Adams, argued that the rules did not give the presiding officer the power to stop a senator from speaking, whatever his tone and language. Only the Senate as a whole could do that. Both Adams and Clay had no doubt that Calhoun had gone entirely over to the enemy, the hypocritical strict constructionists and the antinationalists.

Clay, though, could not allow Randolph's accusation to go unanswered. The unsavory literary comparisons were barely within the acceptable bounds of political nastiness. But an accusation of forgery, even if a qualification could be distantly glimpsed, was beyond the pale, an insult that required satisfaction. He challenged Randolph to a duel. Although neither man was hurt, its potential lethal absurdity could only have depressed Adams' spirits even more. When, in the fall and winter of 1826, the antiadministration coalition became the majority in the House and Senate, it was clear that the administration could expect to pass little of the legislation it favored. The opposition to the Panama conference was symptomatic. An American presence there never materialized. Illness, death, and distance prevented the representatives from ever reaching it. "No president could have commenced his administration under more unfavorable auspices," Benton later wrote. The forces arrayed against it were politically overwhelming.

By New Year's Day 1827, Washington was once again preoccupied with the question: who would be the next president? The betting was against Adams. With the expansion of the electorate, expectations were changing. Backslapping electioneering seemed to Adams to degrade public service. To his opponents, his intellect often appeared as a red flag to the bull of American anti-intellectualism. His competence was never in question, and he fulfilled for four years what his predecessors had defined as the primary mission of a president: to manage effectively, with the help of a capable cabinet, the day-in-day-out administration of the executive departments. Hoping for the best from human nature and believing that it could reach noble heights, he was saddened and sometimes depressed by the corruption of public life. Besieged by appeals and pressured by office seekers, he found that his attempt to award only on the basis of merit made him more enemies than friends. And he was aware from the start that his vision of the United States as an economic powerhouse, which required union and liberty, seemed threatening to those who feared that a national community would undermine local rule and inhibit individual freedom.

Another accusation soon replaced the Panama conference as the predominant issue with which to attack the administration: Adams and Clay had entered into a corrupt bargain in which the office of secretary of state had been sold for votes. What had been considered a normal political alliance with implied compensation in every other case became in this instance a dastardly crime. Repeated with the regularity of a drumbeat, it became increasingly damaging. In the political world, as Adams well knew, the best defense is an attack, no matter how hypocritical and inaccurate. How to answer it? Adams declined to respond, but Clay tried. His denial was supported by written and oral testimony, even from an opponent of the administration like Benton. It did no good. The opposition was in the process of creating a political machine with a powerful propaganda apparatus. And it had waiting in the wings its popular champion, the hero of the Battle of New Orleans. Jackson and his supporters had convinced themselves that they had been nefariously deprived of what was rightfully theirs. It was imperative for the well-being of the soul of America that the corrupt coalition be driven out of the holy temple.

When, in July 1826, fifty-nine-year-old John Quincy walked through the graveyard in Quincy, ruminating on the five generations of Adamses who were now buried there, he regretted that he had arrived too late to see his father one last time. He had traveled as fast as possible from Washington to Quincy. "It is among the rarest ingredients of happiness," he wrote to William Cranch, "to have a father yet living till a son is far advanced in years. This to a certain extent was your good fortune, and it has been much longer mine." Now John Adams was buried next to Abigail, near and around them the graves of his long-departed kinsmen. Even the inscriptions on those tombstones were difficult to read. How little was known about them. They had no stories, only the "short and simple annals of the poor." For his father and himself there was a price to be paid for distinction. The more they were public figures, the more their personal narratives had a public presence, the more vulnerable they were to the hostile judgment of enemies. John Quincy had had a bruising year and a half as president.

But it had been his pleasure in late 1824 and throughout 1825 to have hosted the sixty-seven-year-old Marquis de Lafayette, who made a last visit to the country whose founding was inseparable from his own fame. The United States gave him a triumphal welcome and a sizable gift of cash and land. For a year the heroic figures and flattering memories of the Revolutionary War were vividly reinvoked at public celebrations. In August 1824, soon after his arrival, Lafayette had visited Quincy and paid his respects to John Adams. John Quincy was then secretary of state. When Lafayette stayed at the President's House in August 1825, Louisa worried that the stifling heat would exhaust them all. But it did not prevent Adams and Lafayette from traveling together to Virginia, where, on "a blazing and suffocating day," Adams, Lafayette, and Monroe attended a ceremony in honor of the distinguished visitor. On this "oppressively hot night . . . a chamber with two beds was allotted to Mr. Monroe and me," Adams noted in his diary.

A diary, John Quincy wrote to George, is "the time piece of life and will never fail of keeping time, or of getting out of order with it. A diary if honestly kept is one of the best preservatives of morals. A man who commits to paper from day to day the employment of his time, the places he frequents, the persons with whom he converses, the actions with which he is occupied, will have a perpetual guard over himself. His record is a second conscience, of steady exertion and of composure in disappointment." The diary as catharsis helped him to keep his emotional composure in the face of personal and public blows.

But high-level public service severely reduced the time he could devote to it. Starting in 1819, business and social obligations began to press on his writing time. "I consider every day as lost in which there is no writing done." He "took a firm resolve" to resume "the interrupted thread of this Journal." But less than full daily resumption galled him. "Visitors at home, evening engagements abroad, newspaper reading, inflammations of the eyes, and involuntary drowsiness . . . absorbed many of the evening hours." State Department business preempted the days. But he also blamed himself. "Above all an instinctive idleness, an obstinate do-nothing habit wasting hours of leisure . . . have consumed at least half the time which I devoted by anticipation to this work." And would it not have been better to use his time and energy to better purpose? "Had I spent upon any work of science or literature the time employed upon this diary, it might perhaps have been permanently useful to my children and my country." His writing hand literally faltered. It "will not accomplish the purpose of my mind." Actually, he kept at the diary with astonishing regularity, despite lapses and gaps, between 1819 and 1824, and even during the four years of his presidency. He lamented, "I find it absolutely impossible to keep up the current of my diary." But he did, mostly.

Reading, like writing, required self-creation, self-assessment. But how much time could he devote to reading? "I cannot indulge myself in the luxury of giving two hours a day to these writers, but to live without having a Cicero and a Tacitus at hand, seems to me as if it was a privation of one of my limbs." He was not, though, a literalist reader. "All the facts related of the life and death of Jesus may be disbelieved, and his precepts as a teacher of morals and religion be adopted. . . . The narrative part of the Bible, Old Testament and New, according to all the rules of human evidence, is more fabulous than the metamorphoses of Ovid." He admired people of faith, but he gave his fullest allegiance to reason. After all, how could one have faith in absurdities? "The idea that the execution as a malefactor of one human being should redeem the whole human race from a curse entailed upon them by a single act of disobedience by the first created man is a compound of absurdities, which sets at once all wisdom, all the reasoning faculties at defiance." It was no more literally believable than pagan mythology. "The resurrection of the Spirit disencumbered of the body, I can imagine; but the resurrection of the body—a body too which will no longer be flesh and blood is beyond the compass of my understanding." And how could Jesus literally be the son of God and born to a virgin? "If Christian preachers would take my advice; they would seldom say much about it to those of their hearers who are accustomed to exercise their reasoning faculties. It is enough for us to know that God hath made foolish the wisdom of this world."

Adams did, though, have an amiably disputatious mind. In September 1825, traveling on the steamboat *Fulton* from New York to Providence, he conversed for hours with a young Calvinist minister, Samuel Hanson Cox. The minister was eager to help the president return to his Calvinist origins. "I think God is too good to punish men forever," Adams responded. "Why is it, think you, that they are so punished? What end is to be gained by it? Does God delight in the miseries of his creatures?" Cox admired Adams' simplicity of dress, his manifestation of the First Magistrate as First Citizen. Only an identifying banner flew from the masthead to let the city of arrival prepare in whatever way it chose. He also slept in the common bunkroom, where he and Cox continued the conversation they had begun on deck. Was scripture inspired? Cox asked him. "Yes, I believe the scripture was inspired, and also the Iliad of Homer." Adams asked Cox what he "thought of the inspiration of Milton, the Homer of our language. He then descanted in free and full sway on the grandeur of his master-piece, Paradise Lost," of Milton's "great thought and rich expression. He especially honored his celebrated and richly excellent invocation to light, with which the third book commences." Adams then recited "its opening passages, with comments and praises; seemed enthusiastic and almost absorbed; and when his familiar critique was ended, we disrobed for the night." In the morning, they continued a conversation in which Adams delighted and Cox felt honored, despite his inability to change any of Adams' views. Having taken note of Cox's tight budget, Adams insisted on reimbursing him for his fare. At Providence, escaping as quickly as he could from the waiting crowd, the president had the minister ride with him to Boston.

There were many ways to be a Christian, Adams remarked to Cox. He remained committed to the Congregational church. It is the church "to which I was bred, and in which I will die." But there was no Congregational church in Washington. In 1819, he purchased a pew at St. John's Episcopal Church, partly because of Louisa's Anglican preference. But he alternated between Sunday attendance there, at the Second Presbyterian Church, led by Daniel Baker, and at the Unitarian Church, organized in 1821 and led by Robert Little. Adams had good words and financial support for all three congregations, stretching his resources to make personal loans and sustain building projects. On Sundays, when not overwhelmed by work, he attended one church in the morning, another in the afternoon. "I can frequent without scruple the church of any other sect of Christians," as he had done in Europe, from Anglican and Lutheran to Roman Catholic and Russian Orthodox, "and join with cheerfulness in the social worship of all, without subscribing implicitly to the doctrines of any."

What he most focused on was the minister's sermon, which often prompted comment and analysis. One test was of stylistic effectiveness. He disapproved of ministers of one denomination attacking the theology of another. He noticed that personality was often in conflict with theology. The Unitarian minister "sends forth the goodness of God and the Presbyterian minister . . . holds out his terrors. . . . Yet there is more harshness in the personal character of Little and more affectionate kindness in that of Baker." Calvinist orthodoxy emphasized fear more than hope. They are, Adams recognized, "the two pillars upon which all religious faith is built. . . . Hope is always attended with doubt, and conscious of its own delusions. Fear is therefore always a more efficacious agent than hope, and has far more powerful operation upon religious faith." He put his own emphasis on hope, and he urged his sons to avoid sectarian partisanship, to use their own judgment to evaluate theological claims, and to value guides to conduct more than disputatious arguments. He felt the consoling and restorative power of prayer, and he created opportunities to pray other than in church. "Religious liberty for ourselves, [tolerance for the] religious opinions of others, are the only doctrines which I deem essential to all, and the only creed which I earnestly hope may become universal." Immediately after his father's death, honoring him, his tradition, and his own deepest loyalties, John Quincy took communion for the first time at the Quincy Congregational Church. It apparently gave him great satisfaction—a sense of rootedness not only in his Quincy inheritance but also in the community whose spirit had for centuries pervaded the New England landscape.

When Louisa gave her husband a present of a small seal with the figure of a rooster on it and the motto "WATCH," he sent an impression to London to be engraved on a semiprecious stone and then set on a gold ring. He wore it always. The bird and the motto impressed themselves on his mind and heart. "The cock is a bird of extraordinary properties," he told Charles Francis on the last day of the year 1827. It is "the emblem of vigilance, of generous tenderness, of unconquerable courage." In the gospel of Mark, Jesus admonishes his disciples always to "watch" for the end of days. Judgment day may come at any time. To John Quincy, it meant that he should live his life in the knowledge of the inevitable coming of its final moment. And the image on the ring symbolized Jesus' prediction to Peter that before the third crowing of the cock, Peter would betray his master.

Each morning in every season, as he left the White House for a walk "between the peep of dawn and sunrise," he had in mind both physical and moral exercise. "I rise from bed, dress myself, and sally forth from my door in darkness and solitude. I have not descended from the stoop" when, in the distance, the song of Chanticleer "greets me with a welcome. It is immediately responded to by another, and re-echoed by a third, and from that moment a chorus . . . follows me as I proceed: the song of triumph is wafted along upon every gleam of the brightening twilight."

Early rising, he never tired of repeating, helped form moral character. Maintain watch on yourselves, he advised his sons. That was one of the functions of a diary. And self-betrayal was a constant temptation. Jesus' warning of universal apocalypse had most to offer as moral and psychological guidance. It was about how a human life should be lived. Obstacles and temptations could be overcome; the struggle was not with the world but with oneself. George, Louisa noted, was subject to "depression of spirits and a nervous irritability." So were both his parents. But the sun would rise, John Quincy affirmed. Chanticleer would sing. Depression was "a disease from which I have suffered much at various periods of my life and in which the patient must emphatically minister to himself." Exercise helped. So too did writing poems, the latest of which he named "Watch and Pray, Sonnet to Chanticleer," who is "the minstrel of morn . . . the Bird of the Brave." Depression, he wrote to Charles, "is the disorder of meditative minds and the remedy may be found in the very source of the disease." The best cure was constant work. "The more you have of necessary occupation, the less you will feel of this depressing despondency. I have had less of it for the last ten years than ever before, precisely because I have had no time for admitting its visits."

"Watch and Pray" seemed to Adams an appropriate motto for his public and personal life. As president, he struggled to keep to his usual routines, especially exercise, reading, and writing. His door was always open. That was how he defined his role as a public servant. That was what the public expected, especially the curious who wanted to shake the hand of a president or see if he looked like a great man or a monster. Protocol required that he not accept invitations to dinners or parties at other people's homes. "Having neither lot nor part in the debates of the morning, or the dances of the night, I pass both in my prison." Everyone wanted something, either reflected glory or financial advantage. He found that he had too little time to "watch and pray," to pay attention to the long-term patterns. But when he did look to the future, for himself and his country, the vision often seemed dark.

By late 1826, he began to prepare himself for the likelihood that he would be a one-term president. His vision for the country's future, he was aware, would not be achieved under his leadership, except in small gestures. One gave him great satisfaction: the groundbreaking on July 4, 1828, for the Chesapeake and Ohio Canal, an infrastructure project built partly with federal government funds. He impressed the crowd by taking off his coat to resume pushing his spade through a resistant tree root. He met the challenge. But the distant future would have to be the testing ground of his vision of internal improvements. And he had forebodings of another challenge, the equivalent of the apocalypse for which Jesus exhorts his disciples to watch and that Adams predicted would be a war between the North and South. There would be, he believed, no other way to end slavery. While the future script would be written by the providential hand, he also assumed that chance would play some part in the events. Still, the true agency of prayer was hope, not luck. And it was based on his belief that prayer would be heard, that an inscrutable but loving God controlled human destiny. Resigned and stoic submission was the wisest, most prayerful response to events he could not control. In politics, that meant sticking to his principles. In his personal life, it meant alert attention to whatever could be done and then hopeful prayer. In November 1825, he pasted into his diary a lithograph of "the President's House." Beneath it, he attached a card with his translation of four lines from a poem by Horace, which expressed Adams' belief that the human condition was the same for everyone, even presidents.

Oh! in the hour of sharp distress
Preserve thy spirit calm and high
And in the transport of success
Remember Dellius thou shalt die.

He needed to keep his spirits as high as possible regarding Louisa and his sons. She continued to suffer erysipelas eruptions, sick headaches, and fainting spells. Every month, with occasional exceptions, she was "violently sick," some of the symptoms perhaps resulting from menopause. She now also suffered from "inflammatory rheumatism." Her illnesses frightened both of them. The doctors, who did their best, offered mostly pain relief, leeches, and general advice. In 1823, as she had considered the possibility that she might become first lady, she feared that "the exchange to a more elevated station must put me in a prison." She did her best to fulfill the duties of the post, her illnesses and insecurity notwithstanding. The Washington world usually found her at her appropriate station, gracefully social and elegantly dressed, entertaining notables. The foreign ministers especially admired her French. Her guests enjoyed her social evenings with little or no sense of the private reality. She consulted in Philadelphia with the most famous doctor of the day and took restorative waters in Maryland and Lebanon, New York. She managed some summer visits to Quincy, though fewer than her husband. He traveled as quickly as possible; she made rest stops along the way. She also wrote poems, mostly, it would seem, from her sickbed. It was something to do, a way to pass the time, but it was also the expression of her literary sensibility and emotional preoccupations.

In two poems about the biblical Judith's slaying of Holofernes, she focused on Judith's preoccupation with her vision of saving Israel. Much of the poetry emphasized loss and Christian hope. As Louisa's first Christmas in the White House approached, she sent a poem to her sons, a verbal accompaniment to a recent portrait:

> Go flattered image tell the tale
> Of years long past away;
> Of faded youth, of sorrow's wail
> Of times too sure decay. . . .

She often brooded about the daughter she had lost.

> Thro every pleasure, every pain
> Thy smile my babe I see,
> Oft in yon blue ethereal plain
> My spirit flies to thee. . . .

And about her own death:

> Lay me in consecrated ground. . . .
> My name my only eulogy. . . .

Here where my Mother lies I seek
A shady spot to lay my head
Near to the humble Rocky Creek
I fain would make my lowly bed
On solemn calm repose to sleep
Till Angels wake me to adore
That God whose mercy vast and deep
Shall raise my Soul to die no more.

Charles Francis' bouts of ill health and low spirits concerned his parents. He kept out of trouble at Harvard, though he failed to achieve any major distinction. With his father's approval, he joined his mother and brother John in Washington. In 1825 and 1826, he seemed to be settling on a future consistent with the sobriety of his personality. His father assured him that his choice of profession was entirely his own, though he hoped all his children would settle in the Quincy and Boston area. That is where he expected, after "being discharged from the public service," to spend "the remainder of my days. . . . My wish would be that all my children should prefer this to any other spot in the world, for their abode in life and their repose in death." Charles decided on the law. It suited his temperament, his intellect, and the family tradition. In Washington, he studied under his father's direction.

In Boston, he was welcomed into Daniel Webster's office, a mutual accommodation between two of the political elite, though Webster paid little to no attention to his apprentices.

Early in 1827, twenty-year-old Charles asked Peter Chardon Brooks for the hand of his nineteen-year-old daughter, Abigail. The families on both sides had been in Massachusetts since the founding. Adams and Brooks had known one another at least since 1804, when Brooks had employed Adams to argue a Supreme Court case. Edward Everett, Adams' strong supporter and a family friend, now a member of the House from Massachusetts, had married Abigail's sister Charlotte. The stars seemed aligned. The Adamses had prestige, the Brooks family wealth. The only impediments were Charles' age and dependence. But, they all recognized, this was a marriage worth waiting for. Charles Francis rose in his mother's estimation. "I do not hesitate to say that you suit me better than either of your brothers as your manners are more like my own, and in consequence of having been so much with me your sense of the proprieties of life is more strongly defined." Just as his father had embraced Louisa, John Quincy embraced Abigail. The name itself must have warmed his heart. And she was soon to bear the future of the Adams family.

He was also not about to let Charles move forward without the benefit of advice. Between 1827 and late 1829, they engaged in a mutual tutorial, focusing on John Quincy's passion for the life and works of Cicero. "I have not in seven years read so much of classical literature," he wrote to Charles in November 1827, "as since I began these letters to you. And I might add I have not in seven years enjoyed so much luxurious entertainment." The advice was insistent and repetitious. But he was sometimes self-aware enough to cut himself short. "Do not think me a tedious correspondent. I shall not inflict upon you many letters so long as this." His advice, though, sometimes promised too much. "Early rising is so indissolubly connected with many of the most active virtues that it may be laid down as an axiom of almost universal application— give me an early riser and I will give you a virtuous man." He believed Charles had talent as a writer and encouraged him to make a name for himself with articles and essays for newspapers and magazines. He urged him, especially when Charles seemed insufficiently impressed with Cicero, to read Burke, "the master mind who . . . when century after century shall have passed away . . . will remain a teacher of wisdom, virtue and taste. . . . I think of [Cicero] as a father; and indulgent to his foibles and frailties feel uneasy when they are remembered by others." He hoped his son had more genius than he had. But "genius is the child of toil. . . . All my success in the world has been the blessing of heaven upon drudgery, the reward of untiring, of unmitigated labors. May yours be more brilliant and more durable." The transformation of his relationship with Charles began to provide satisfactions that had eluded him with George and John.

In Washington, John acted as his father's secretary, short-term employment for a young man without a degree or career. When, in November 1827, he asked his father's consent to become engaged to Mary Hellen, who had been engaged to George, John Quincy and Louisa could not have been completely shocked. They had noticed that something was brewing. Her engagement to George had ended by the summer of 1825. His feelings, he wrote in his diary, had been "the sport of idle coquetry." John Quincy had reason not to be enthusiastic. "My mother is half inclined to the marriage and half opposed," Charles Francis wrote in his own diary, "my father is tacitly opposed." Charles envisioned Mary as a sexual temptress whom no young man could resist. Mary and John lived in the same house; they dined at the same table; they frequented the same parties; they went on excursions with friends and family. John Quincy took notice in August 1824 that they had gone together, apparently just the two of them, "to the Methodist camp meeting about six miles from the city." "Without self control nothing difficult can be achieved," he wrote to Charles Francis in December 1827, as if he were writing about John, "and the first victory to be won is over sensual indulgence." By late 1827, Louisa felt she had reason to urge the couple to marry soon. In mid-January, John asked his father to consent that they be married in February. Adams "took it for consideration." He had not favored the engagement. And how would George take it? Would it affect his already delicate health? On February 25, 1828, John Adams and Mary Catherine Hellen were married at the White House. It was not a turn of events that called for public celebration. It was far less than the magnificent White House wedding about which Mary Hellen had fantasized. Neither of John's brothers attended. "After the ceremony we had a supper," John Quincy noted in his diary, "the company retired about midnight. May the blessing of God almighty rest upon this Union!"

His parents had even more reason to worry about George. He was fanciful, irresolute, difficult to have rational conversation with, neglectful of serious study or business, drinking immoderately, and gambling. His state of mind fluctuated in response to stimuli that were difficult to identify, let alone track. Alcohol made matters worse. His father, aware of his "licentious life," certainly his drinking and gambling, perhaps his sexual activities, urged him to "return to the laws of unsullied temperance . . . the laws of eternal truth," including church attendance and religious belief. "I have seen in a late Boston newspaper, the account of the meeting of the Society for suppression of intemperance, and of some of their proceedings. I ask of you as a favor to become a member of that Society," an early version of Alcoholics Anonymous. George, who had little insight into and deeply feared his father, wanted nothing more than to please him. But he also wanted nothing more than not to. Constantly promising better performance, he seemed to do almost all he could to disappoint his father, even to give him pain. For some time, George had been spinning out of control, at first alternating between short-lived efforts at reform and slothful paralysis, at other times escaping into compulsive gambling, sleeping, drunkenness, romantic desire, especially about his cousin Abigail, Thomas' daughter, and a sexual relationship with a working-class mistress, which he hid from almost everyone.

In August 1825 George made another brief attempt to keep a diary, which included a plan for a book on the history and practice of poetry. With a few friends, he took restorative excursions. Occasionally he seemed to

have happy days and be back on track. "I rejoice heartily therefore in the observation that you have already
put your shoulder to the wheel," his father responded.
Elected, by the power of his name and the influence of
family friends, to the state legislature, he was not re-
elected. His appointment as an officer in a Boston militia company more often led to carousing than discipline.
The relationship between father and son became characterized less by the time they spent together than by
their painful correspondence when apart. John Quincy
paid his eldest son $1,000 a year to handle his Quincy
and Boston business. But the quarterly accounts came
very late, often with serious inaccuracies, sometimes
not at all. "Irregular accounts are the direct road to
ruin. I have told you this so often that I am ashamed
now to repeat it." If George could not correspond with
John Quincy as his father, then do so as his client, he
exhorted. To pay off George's $2,000 worth of debts,
John Quincy manufactured a phantom sale of his son's
library. George would keep the books and get the money.
But he would need to make a list of them. George never
sent the list. Praise, advice, admonition, and incentives
had no effect, except perhaps to make him feel worse, if
that were possible. To please his father he made promises, which he fulfilled partly or not at all. "Believe me
when I say," John Quincy wrote to him, "that there is

nothing wanting but the *will* to perform them. All the rest is self delusion." He could not accept that what had worked for him would not work for his son, that all he could do was "watch and pray." His eldest son was disappearing into silence. John Quincy alternated between anger, depression, and despair.

March 2, 1828, was clear and cold. It was almost to the day the third anniversary of his inauguration as president. It commenced, he wrote in his diary, "the last year of my public service." In the meantime, he had public business to attend to. At his request, Congress, retaliating against Great Britain, had prohibited American trade with its West Indies colonies and increased tariff duties on select British products. When the British responded by forbidding American trade with any of its colonies except Canada, American business interests were damaged, and that in turn damaged Adams, whose congressional enemies were happy not to help repair the situation. The defiant stalemate between Georgia and the federal government also frustrated the administration, which put energy and effort into finding an acceptable resolution. All parties to these issues knew what the outcomes would be, though the pro-Jackson forces found it politically expedient to create delay: eventually West

Indian trade would be resumed and the Cherokee and Creeks moved beyond the Mississippi. Although criticism of the administration's inability to resolve these problems became prominent in the pro-Jackson press by early 1828, it was mainly because there was little else of substance to use against Adams.

Still, defending himself and his record was not to Adams' taste, and in fact the administration's lack of spectacular achievements characterized its success. There were no wars or threats of war. If there had been, Adams would have been a formidable commander in chief. His calmness, rationality, analytic skills, and administrative competence would have served the country well. If there had been serious internal threats, his devotion to the union and the Constitution would have held the country together. Fortunately, he had inherited from Monroe a stable, prosperous, and growing country. And, unlike Jefferson and Madison, he had the advantage of a peaceful Europe. There were difficulties in the relationships with Britain, France, and Russia about trade, reparations, the Oregon territory, and the northeastern boundary. But there was also every reason to think they would be resolved peacefully. The fuss about the Panama conference had no foreign policy consequences. And Adams' vision of an America extending its reach to the Pacific was

widely shared; the differences were in regard to means, not ends—for Adams, anything other than peaceful and democratic means would betray American ideals. The economy was prospering. Land sales and a tariff provided the income that paid for the infrastructure projects Congress approved without regard to constitutional issues. The cost of running the government was prudently managed. Where there was corruption, it was individual, especially in the collection of import duties and awarding of contracts in the land and post offices. In the daily management of the executive workload, Adams and his cabinet excelled.

Consequently, with the country prosperous and at peace, the Jackson party focused almost exclusively on personalities and vague generalities. It had the advantage of a military hero to market to an already impressed electorate. Adams had the private support of those who judged Jackson unsuited to serve as president. The two living former presidents feared, as had Jefferson, Jackson's philistinism and his recklessness, and deplored his disregard of lawful authority and constitutional restraint. They knew and respected Adams. Jackson seemed the embodiment of prejudices and opinions totally independent of ideas. His military temperament, threatening the triumph of power over liberty, seemed unsuited for civilian leadership. And since

Jackson's national popularity was based exclusively on the hero worship of a military victor, it inevitably brought to the elite Republican mind the long-standing fear that one day an American president would become an American dictator, a tyrant like Caesar or Napoléon.

For the general public, Jackson was a political blank slate, which made him the perfect anti-Washington candidate. He was defined exclusively by his military victories. Almost nothing was known about his views, or even whether he had any. It was to be, as Adams knew by early 1826, a campaign much like that of 1800—about personalities, slander, and slogans, and more about power and money than good governance. To the extent that there were ideological differences between the two wings of the Republican Party, they were expressed not in discussions of public policy but in personal attacks, particularly the accusations that Adams was a monarchist; that he favored big government, the wealthy, the Northeast, and manufacturing; that he held the West, the South, and agriculture in low regard; that he preferred raising tariff duties over the interests of the South; that he advocated abolition; and that he had presided over a corrupt administration, which had insulted Republican virtue and honesty and which was especially exemplified by the corrupt bargain between Adams and Clay.

The lack of evidence to support the accusations did not matter. A powerful coalition, with little or nothing in the way of binding principles, wanted him out of office. It was doing what no American political group had done before, at least to this extent: organizing a political bureaucracy, led by pro-Jackson editors, senators, congressmen, and assorted political operatives; raising money; sponsoring newspapers; forming local clubs; and promising rewards to its supporters. Its sole purpose was to make Jackson president. For the first time in American history, large sums of money were spent in a political campaign. The Jackson money machine and party apparatus represented to Adams the blatant corruption of American political life. And even if he attempted to compete with the Jackson machine, the political calculus was still unfavorable. No slaveholder would vote for him. No Old Republican would. Nor would anyone who had supported Crawford or Calhoun. Now Jackson's running mate, Calhoun had allied himself with Jackson, as had Van Buren, forging an alliance between Old Republicans, especially in Virginia, and disparate Northern Republicans. Jackson would neutralize Clay's influence in the West. And Pennsylvania, despite Adams' selection of Rush as his running mate, would again support Jackson, partly because it assumed that Jackson would be in favor of strengthening the protective tariff, a subject about which he said nothing since his Southern and Northern supporters differed on the issue. Most of all, Adams, unlike Jackson, seemed untrustworthy about slavery, on which the South and much of the North, especially New York traders and merchants, depended for their bank balances. He would probably not get a single electoral vote in the South.

He had two other disadvantages. One was the widespread cultural bias against New England, the home of the Jews of the North. In cahoots with the Bank of the United States, Boston bankers controlled the money supply, so Jackson and his followers believed, exploiting farmers and settlers. Also, unlike Jackson, Adams had a record as a foreign minister, secretary of state, and president, all of which could be attacked. Not that he took the bait, except to refute lies about his Louisiana votes and his private character, such as the charge that he had urged Martha Godfrey, Louisa's American servant in St. Petersburg, into the bed of Alexander in order "to seduce the passions of the Emperor . . . and sway him to political purposes . . . one of the thousand malicious lies which out-venom all the worms of Nile, and are circulated in every part of the country in newspapers and pamphlets." When anti-Masonry became an issue in New York, pro-Jackson operatives claimed

Adams was a Mason, which he was not. Jackson was, though that seemed to make no difference. By mid-1828, the campaign had become slanderously personal. Adams' supporters publicized Jackson's legally questionable marriage and depicted him with hands bloody from duels and peremptory executions. Smarting from the personal attacks, Adams, who had little to no control over his supporters, may not have wanted to stop the barrage. After all, he and Clay were constantly being called corrupt. The slanderous and simplistic nature of the Jackson campaign, which emphasized no-holds-barred political warfare; the changes in demography and the electorate; the rise of a new political structure dominated by a well-developed party apparatus; and the difficulty of overcoming the conditions under which he had been elected in 1824 made it certain, Adams recognized, that he would have to find the strength to bear the burden of rejection.

On medical advice, John Quincy spent eleven weeks in Quincy in the summer of 1827. This was mainly precautionary, as his constitution was strong. But he was tired and dispirited. As often, he suffered from chest and throat colds that were difficult to shake off. Under stress, his eyes were regularly inflamed or at a minimum watery. It was an unusually long period to be away from Washington responsibilities. But he had cabinet surrogates in place and postal transmission that took advantage of improved roads and speedy steamboats. Anyway, the presidency had been defined by his predecessors, influenced by Washington's summer heat and mosquito-breeding marshes, as a nine-month job. Jefferson, Madison, and Monroe had decamped to the Virginia mountains. Adams, with or without Louisa, depending on her health, went north to Quincy, where he stayed with his brother Thomas, whose residence in the Old House he subsidized. He provided assistance and love to three of Thomas' children, one of whom, Elizabeth, had become almost a member of his household. His brother was no longer managing John Quincy's Boston and Quincy income. He was having difficulty managing himself. Bouts of drinking and depression made him unstable. Paying for Thomas' eldest son to attend West Point, John Quincy worried that his brother's family would become his permanent financial responsibility.

The family was shocked and depressed in April 1828 when John, entering the Capitol to deliver an official document, was assaulted by a pro-Jackson newspaperman over a perceived insult. An anti-Adams congressional committee refused to take any action. "Slander and assassination are working hand in hand against us,"

John Quincy wrote to Charles. Adams' assessment of his enemies, as always, was inflected by a slight touch of paranoia. But in 1828, personal and political conduct had the potential to combine in an explosive mixture. Insulting as the assault was, it was characteristic of the time and place. It depressed the mood of the Adams household, among other reasons because he had no recourse other than to condemn it, and because it fed into the image that the Jackson campaign emphasized of Jackson as a man of action and Adams as a bookish scholar. His reputation as an intellectual was a red flag to the bull of American anti-intellectualism. The action-worshipping American voters, especially in the South and West, knew whom they preferred.

Also on Adams' mind was how to support his family when he no longer had a government salary. His $25,000 salary had to cover all the expenses of White House life except building maintenance. His Boston and Washington properties were his only source of income other than his salary. And Adams was in effect also supporting his sons. There was a rationale in each case. John's salary was payment for services rendered, though the source of funds would soon disappear. Perhaps, it was proposed, John, who might or might not have business skills, could redeem the Columbian Mills fiasco. When John Quincy, in the spring of 1828, relinquished his guardianship of Mary Hellen's inheritance, it provided the newly married couple with some money. But the marriage would need financial support. George also was on the family payroll. Charles, engaged for an indefinite time to Abigail Brooks, candidly asked his father how much support he could expect and for how long. "It has been my endeavor to educate my children," John Quincy responded, "to the class of society to which I belong by indulging them in every expense for education which it was in my power to procure and which could possibly be useful to them." He would not deny Charles. In addition to the support of his sons and his own household, he also assisted his sister Nabby's and his brother Thomas' children. He was paying the bequests from his father's will, which required that he raise cash. He supported charities with gifts and loans. Most onerous of all was a huge debt from the Columbian Mills. It was now "about thirty thousand dollars upon which I am paying interest, and of which I must pay off six thousand dollars during the present year." Most of his Boston and Washington properties were mortgaged. Although he would deny himself to help his sons, he asked them "to remember that I am retiring from the public service with scanty resources, no light embarrassments, and a family unavoidably and heavily expensive." He would leave the White House poorer than when he had entered.

Although the summer and fall of 1828 were a dismal time for Adams, he found distraction and pleasure in gardening. He was not especially interested in flowers and shrubs, though he enjoyed their flowering. He took an interest in the White House vegetable garden and nursery, which he urged the White House gardener, John Ousley, to expand. But what he became even more preoccupied with than before were trees. He began an ambitious planting program at the White House. "I have laid the foundation of a fine grove of forest trees around this house," he noted, "and of a fruit garden as of all Washington cannot now exhibit." Gardening might, he thought, fill a place in his retirement activities. The mighty oaks and other hardwoods were an important part of the national wealth: timber for homes, furniture, and firewood, and the wood to build America's maritime fleet. Why should the federal government not, he asked, sponsor lumber farms and reforestation?

By executive order, he funded a plantation of live oaks in Florida, the earliest instance of a government-sponsored agricultural station. He regretted that "the cultivation of forest trees has scarcely received any attention in this country. Yet it is one of the most important branches of political economy. . . . My purpose is . . . to draw the attention of my countrymen to it."

He tried "without much success" to influence Congress to support such projects. What he planted at the White House could not be fully separated from his condemnation of the narrow vision of his political opponents. "I intend [these groves of trees] as a memento for them which I hope they will have the means of cherishing when the recommendation of an astronomical observatory will no longer be denounced as a deep conspiracy to subvert the Constitution. . . . I leave them as a bequest to my successor." He did not think it likely that his successor would see that the union had a vital interest in the planting of trees and preservation of forests—a moral, material, and aesthetic commitment to the future.

In Washington and especially in Quincy, Adams had some success in raising saplings. He had failures also as he became a knowledgeable amateur horticulturist, learning from books and garden experience, planting over two hundred trees of more than twenty varieties at the White House, and planting hundreds of acres of orchards and hardwood trees on his properties in Quincy: oak and maple of various types, chestnut, walnut, shagbark, cherry, peach, plum, and apple. He preferred to start from seeds or acorns. He requested them from American consuls in various parts of the world, including olive tree seeds, which of course he had no success with. He did some of the planting himself, though workmen did the heavy labor. George and Charles also planted seeds and acorns he sent from Washington. That most such efforts failed did not discourage him. He filled his diary with planting notations, a record and schedule of each stage in the efforts. "Dendrology has become to me precisely what Uncle Toby's fortifications were to him," he wrote to George, referring to the eccentric, all-consuming hobby of a character in *Tristram Shandy*. "By devoting to it all my leisure moment, I think I am answering [one of the] great ends of my existence." It was the expression of a generational vision. He recalled that he had planted his first tree in Quincy at the age of thirty-seven, in 1804. "If I had begun to plant seventeen years before . . . what a noble forest of oaks, elms, walnuts and maples, I might have had this day. . . . Had I commenced the attempt at your age I might even now have walked under the shade of oak, chestnut and maple forests of my own raising. I can now plant only for posterity." But his sons could plant for their sons. As he prepared to leave office, it was a lesson and a legacy.

Gilbert Stuart's portrait of
Abigail Adams, started in
1800 and finished in 1812,
depicts Abigail at the end of
John Adams' presidency. At
her death in 1818, John Quincy
wrote that "the world feels to
me like a solitude."

Jane Stuart's version of her
father's 1824 portrait of John
Adams two years before
his death. It is the John
Adams that President John
Quincy Adams, when he left
Washington in early July 1826,
hoped to see one more time
before his death.

John Quincy and Louisa's eldest son, George Washington Adams, depicted by Charles Bird King in the 1820s. Erratic and unstable, George was to commit suicide in 1829.

John Adams II, depicted by an unknown artist in 1820, served as his father's secretary during his presidency and died of alcoholism-related illnesses at the age of thirty-one.

Charles Bird King's 1827 portrait of twenty-year-old Charles Francis, a recent Harvard graduate and now engaged to Abigail Brooks. "Your letters are becoming a necessity of life to me," his father wrote to him. "I have not in seven years read so much of classical literature as since I began these letters to you."

On the inside front cover of Diary 37, 1825–1828, President John Quincy Adams pasted an engraving of the President's House and his own translation of four lines from the Roman poet Horace's "Ode to Delius." The theme is stoic courage and acceptance of adversity and death.

Gilbert Stuart's 1821–1826 portrait of Louisa "speaks too much of inward suffering and a half broken heart," Louisa wrote. She was surely looking into the mirror of her feelings rather than at the engaging, delicate, and attractive portrait.

Edward Dalton Marchant painted in 1843 the only portrait of Adams that his son Charles Francis liked. Like Copley's 1796 portrait, it is a softened idealization, a peaceful, benign, and content Adams, hardly a realistic representation of his inner life or outward appearance in 1843.

An 1846 view of the "Old House," John Adams' and then John Quincy's residence in Quincy, much as the house looked two years before John Quincy's death. It is now part of the Adams National Historical Park.

Adams' home in Washington at 1333 F Street in a photograph taken by Mathew Brady during the Civil War, when it was the central office of the Sanitary Commission, which raised money to support wounded soldiers. It was destroyed in the twentieth century.

Washington, Monday 29. March 1841.   Five

29. IV:30. Monday   Rain the greater part of the last night, and of this
Munroe E   day, with a chilling east wind requiring a small
                     fire in my chamber, just enough to be kept burning

Mr Munroe was a stranger from Boston, who brought a parcel for Elizabeth. I completed the assortment and filing of my Letters received since the beginning of this year, and find myself with a task before me perfectly appalling—I am yet to revise for publication my argument in the case of the Amistad Africans, and in merely glancing over the Parliamentary Slave-trade papers lent me by Mr Fox, I find impulses of duty upon my own conscience which I cannot resist, while on the other hand, the magnitude, the danger, the insurmountable burden of labour to be encountered in the undertaking to touch upon the Slave-trade. No one else will undertake it. No one but a Spirit unconquerable by Man Woman or Fiend, can undertake it, but with the heart of Martyrdom—The world, the flesh, and all the devils in hell are arrayed against any man, who now, in this North-American Union, shall dare to join the standard of Almighty God, to put down the African Slave-trade—and what can I, upon the verge of my seventy-fourth birth-day, with a shaking hand, a darkening eye, a drowsy brain, and with all my faculties dropping from me, one by one, as the teeth are dropping from my head, what can I do for the cause of God and Man? for the progress of human emancipation? for the suppression of the African Slave trade?—Yet my conscience presses me on—let me but die upon the breach——I walked about half an hour for exercise before dinner and called at the house of Mr S. H. Fox the British Minister to have some conversation with him. It was 2. o'Clock P.M. The servant at the door told me that he was not up, and that he was unwell. I enquired at what time he was usually visible—he said between 3. and 4. I had heard that his usual hour of rising was 3. In my second walk after dinner I met Mr Jesse D. Miller, first Auditor of the Treasury; from which Office it is said he is to retire at the close of the present month and quarter. This evening I answered an old and repeated invitation to deliver a Lecture at Richmond Virginia; and postponed answering the Letters received last Evening from the Amistad Committee, and from Lewis Tappan. I read judge Betts's opinion upon the 14th Section of the Tariff Act of July 1832. and the reversal of his decision by judge Thompson. And I made several minutes from the Parliamentary Slave trade papers Class A. 1839-40. shewing the enormous extent to which that trade was in those and the two preceding years carried on in American vessels under the patronage of N. P. Trist.

In his diary entry for March 29, 1841, Adams, who had just recently argued the *Amistad* case before the Supreme Court, writes that he finds "impulses of duty upon my own conscience, which I cannot resist" to attack the slave trade. "No one else will undertake it. No one but a Spirit unconquerable . . . can undertake it, but with the heart of Martyrdom. The world, the flesh, and all the devils in hell are arrayed against any man, who now, in this North-American Union, shall dare to join the standard of Almighty God, to put down the African Slave-trade."

A rare extant photograph of Adams, this riveting 1843 daguerreotype by Philip Hass captures the way Congressman Adams looked to the camera and to his contemporaries.

"Sculptor!" John Quincy responded to Hiram Powers' 1837 marble bust: "Thy hand has molded into form / The haggard features of a toil worn face: / And whosoever views thy work shall trace / An age of sorrow and a life of storm."

1848.20 Feby.

Fair Lady, thou of human life
Hast yet but little seen.
Thy days of sorrow and of strife
Are few and far between.

The last entry in Adams' diary, three days before his death, a brief poem to a female admirer about innocence, experience, and time.

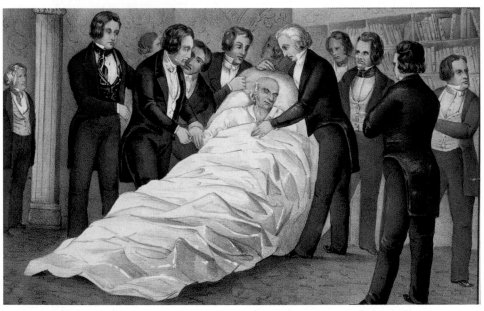

Nathaniel Currier's popular 1848 lithograph, based on newspaper reports, depicts Congressman Adams' deathbed, surrounded by colleagues, in the speaker's office in the Capitol building. Currier took as accurate the report that Adams' last words were "This is the end of earth—I am content." They may have been "This is the last of earth, but I am composed." Louisa wrote: "Thy fiat has gone forth O Lord my God: and I am left a helpless Widow to morn his loss which nothing on this dreary earth can supply."

Chapter 14
The Uses of Adversity
1829–1833

I n the early morning of the first day of 1829, as Adams began to write in his diary, the light of his shaded lamp flickered and went out, "self-extinguished. . . . It was only for lack of oil; and the notice of so trivial an incident may serve but to mark the present temper of my mind." That afternoon the largest crowd of his four years in office attended the president's New Year's Day reception. There is no record of who came, but a large number of congressmen and citizens, whether or not they had voted for him, availed themselves of this public entertainment, the ritual start of the new year in Washington. It went off quietly, in a mood of resignation, farewells, post-mortems, anxiety, and the gleeful expectation of Jackson partisans who looked forward to the advantages of office, soon to be happily referred to as "the spoils." Government employees who had not supported Jackson had good reason to be worried. His political machine had campaigned almost exclusively on the need to reform a corrupt administration. No one knew exactly what that meant. Louisa, who struggled to act as hostess, soon went to bed. Having begun the new year in the same poor health that she had endured through much of 1828, she had been awake most of the night, painfully sick. Late on the day of his New Year's reception, Adams "witnessed the cloudless setting of the sun, with a feeling of hope and consolation."

December 1828 had been an exhausting month. The results of the election had turned out to be what he had expected. He had won the popular and most of the electoral vote in eight states, Jackson in sixteen. In three Jackson had won—New York, Ohio, and Maryland—it had been close. In New York, which divided its electoral votes proportionally, Adams got sixteen out of thirty-six electoral votes. He did not win in any Southern or Western state. He was not on the ballot in Georgia, where animus against the Indian-loving Adams united otherwise hostile factions to exclude him entirely. Jackson received 647,276 popular and 178 electoral votes, Adams 508,074 popular and 83 electoral. A switch of a small number of votes in Ohio and Maryland would have given him those states and made the race closer. If New York had not divided its votes proportionally, one of three states to do so, and Adams had gotten 6,000 more votes, the result would have narrowed substantially. If the slave states had not had their electoral numbers swollen by the three-fifths rule, it would have been even closer still. In fact, Jackson had won barely half the votes in the free states, which gave him 73 electoral votes, but over 70 percent of the votes in the slave states, which earned him 105 out of his 178. The less-populated South was the tail that wagged the country. But even if these counter-factuals had been the reality, Adams still would have lost. The alliance between the slave-holding South and the free land–loving West controlled too many votes. It was a "change election." And the change, while in many ways itself corrupt, had at its ideological core the fear and hatred of a central government of the sort represented by Adams' vision. At least half the country claimed it wanted as little government as possible. "In looking back," Adams concluded, "I see nothing that I could have avoided, nothing that I ought to repent."

He had another two months in which to tidy up whatever he could. He did not look as far back as the tariff mess that had contributed to his defeat. Although he had opposed it, he had signed in May 1828 what was widely called the "Tariff of Abominations," the result of a series of maneuvers concocted by Van Buren to keep the South and Pennsylvania solid for Jackson. The Jackson forces blamed Adams for it. Since the Constitution allowed, in Adams' view, a veto only if in the judgment of the president the act itself was unconstitutional, he did not veto the tariff. To his dismay, Adams discovered that his postmaster general—this was not yet a cabinet position—had been making patronage appointments, playing a double game in which he worked for Adams but supported Jackson. At first Adams refused to believe the charge. But when the evidence defeated denial, he declined to fire John McLean. He had not been corrupt or criminal. And he had managed the regularization and expansion of the post office brilliantly. With Clay, Adams had developed a relationship that put the tension of their Ghent days far behind. Clay, who had feared Adams would be peremptory with his cabinet, testified to Adams' commitment to open, fair, and consultative decision making. Every member of the cabinet experienced the president's respect. Decisions were made by consensus. Despite differences of personality, their shared vision of the American future, which had initially created their partnership, kept them together during four stressful years.

Inevitably the barrage of accusations of a corrupt bargain baffled, pained, and infuriated Adams and Clay. As president, Adams felt it beneath the dignity of his office to respond. From the start, the likelihood of a second term was doubtful, even without the charge, and Adams assumed he would retire from public service after his presidency, whether it was one term or two. But Clay, fifty-one in 1828, had no doubt that he would make another run for the presidency, almost certainly in 1832. He had reason to assume that Jackson, who had given people to believe he would limit himself to one term because of age, health, and predisposition, would not run again. When, in late 1828, Adams offered Clay a Supreme Court nomination, he declined. Aware of the damaging accusation that he was "the Judas of the West," Clay took every opportunity to disprove it. Suffering in silence would be counterproductive, though Adams pointed out that counterargument would only give the accusations more publicity. Those who chose to believe them would not be swayed, no matter what the evidence. Clay, though, persisted. The wound was deep. "For near two years and a half I have been assailed with a rancor and bitterness which have few examples." He had "freely examined, and . . . justly condemned the conduct of General Jackson in some of our Indian wars. I believed . . . him to have trampled upon the

Constitution of his country, and to have violated the principles of humanity." In Lexington in July 1827, he analyzed the sequence of events, Jackson's duplicity, and his own motivations. "Had I voted for him, I must have gone counter to every fixed principle of my public life. I believed him incompetent, and his election fraught with danger." He still believed that. So did Adams.

When the new administration announced its cabinet, Adams was appalled by its lack of experience and of expertise. The secretary of state, Martin Van Buren, had no credentials in foreign affairs. When John McLean was asked to stay on, as his reward for betraying Adams, he declined to run the post office, now raised to a cabinet position, as a political arm of the administration, though Adams predicted that it will "be brought under the control of the Executive, and all the post offices are to be subject to its patronage," an "overt market for political partisans. . . . Memorable reform!" Jackson appointed his friend from Tennessee, John Eaton, secretary of war. He established an informal "kitchen cabinet," the earliest form of a White House political staff, reducing the influence of the actual cabinet. Pennsylvania was rewarded with the appointment to Treasury of an undistinguished businessman-politician. Two of Jackson's cabinet appointees were from the South, one from the West. "President, Vice President, Secretary of

War, Secretary of the Navy, and Attorney General, all within the compass of four adjoining Southern States," Adams remarked, "two of them," Georgia and South Carolina, "in all but insurrection against the Union." New England had no presence at all. A few days after Jackson's inauguration, Adams nevertheless wrote to his supporters that he was "disposed to hope and pray for the best, to extend to the Administration every reasonable indulgence, which they may need; and to give them credit for every good deed they may perform, for the promotion of the general welfare."

Four years of a beleaguered presidency had been relentlessly demanding. Adams' depleted energy was put to the test in late 1828 and early 1829 in a struggle that started with a mistaken account in 1825 by the elderly Jefferson of an 1808 interview with Adams. The private letter from Jefferson responded to a query from Virginia senator William Giles, who had been Adams' Senate colleague and now supported Jackson. Jefferson's letter stated that Adams had urged him to repeal the embargo, warning that there was a conspiracy to separate New England from the union, that some Massachusetts Federalists were secretly negotiating with British agents. In the letter, Jefferson mistakenly attributed to the 1808 discussion information he could have learned only at a later date. The inference that Giles promoted was that Adams had slandered his Massachusetts constituents in order to curry favor with Jefferson. His appointment as minister to St. Petersburg in 1809 was payment for services rendered. In October 1828, Giles published the now-deceased Jefferson's 1825 letter.

Adams soon published a correction in the *National Intelligencer*. It was a single long paragraph, phrased in the third person to maintain the fiction that a president in office was above self-defense in the press. Giles, Adams alleged, had misled an already confused Jefferson. In fact, as Adams' diary entry for March 15, 1808, revealed, he had told Jefferson nothing more than that he had seen a letter, written by the governor of Nova Scotia in 1807, that supported the Federalist claim that Jefferson was in league with France against Great Britain. Jefferson had assured him that "there had never been any negotiation with France on the subject." In late 1808 and early 1809, Adams explained, he had been queried by Giles about the extent of New England's objection to the embargo. As a private citizen, he had "urged that a continuance of the embargo . . . would certainly be met by forcible resistance, supported by the legislature, and probably by the judiciary, of the State." Its object would be, "and had been for several years, a dissolution of the Union, and the establishment of a separate confederation," which "he knew from unequivocal evidence, although not provable in a court of law."

To his dismay, he soon had in hand a letter, published in the *Boston Advocate* at the end of November 1828, censuring him for having accused the Federalists of 1804, 1808, and 1814 of treason. It was signed by twelve prominent Massachusetts Federalists, some of whom had once been his friends, led by Harrison Gray Otis. Theophilus Parsons' son had signed. Others were names poignantly remembered from friendships past, including Daniel Sargent. Adams had unintentionally provoked a blistering response by post–Essex Junto Federalists. Although Charles Francis had alerted him in late November that some people had taken offense, his statement "was rendered indispensable," he explained, "by the publication of Mr. Jefferson's letter in which I was named . . . as having given him information of a treasonable correspondence during the war, which induced him to consent to the substitution of the non-intercourse for the embargo in 1809," though, by the very dates, this was impossible. "My troubles thicken upon me." He made no specific accusations now about treasonous Federalist plans in 1804, but only corrected the record about his March 1808 discussion with Jefferson. He believed his statement to be factually accurate: there had indeed been discussions between 1804 and 1814 among some New England Federalists about the possibility of secession. The response of his antagonists now seemed to him hostile and manipulative, and some of their errors of fact willful, others accidental. But he had no doubt he had inherited and further incensed his father's enemies. "They persecuted my father as unrelentingly as they now will me, and they persecuted me from the time of the affair of the Chesapeake and Leopard in 1807, till I returned from Europe as Secretary of State in 1817. . . . Whatever success I have had in the world has been without their aid and in despite of them. They have been equally ready to crush me in adversity and to fawn upon me in prosperity."

As always, he needed to respond. Even "retirement will not shield me from persecution. I was born for a controversial world, and cannot escape my destiny. But what of that? I am inured to it. . . . As for the wish expressed in your late letter," he wrote to Charles, "that your father may be left to enjoy peace in this old age, do not expect it." Urged to desist, he agreed that it was "an unprofitable contest. . . . But my character has been brought into question."

He took to the pen with willful energy, publishing a lengthy response. He had no doubt that the Essex Junto had the right to petition Congress for redress and for constitutional change. For "the right of petition is well known and understood: it is a sacred right," one he would defend at almost all costs. But Rufus King, he explained, had told him in 1804 that some Federalists were planning secession, not petition. Although his report of King's information would not be admissible in a court of law, he could cite evidence from public statements and activities, culminating in the Hartford Convention: the Essex Junto had publicly claimed that the Constitution allowed any state to withdraw from the union if it should be deemed in the state's best interests. Jefferson and Madison had made the same claim in the Kentucky and Virginia Resolutions in 1799. South Carolina was now making it again, proclaiming that its legislature would nullify any congressional act it deemed unconstitutional. Georgia was nullifying congressional legislation and Supreme Court decisions it considered unconstitutional. Why was it not reasonable to assume that between 1804 and 1814, the principle of secession had led to plans for secession? "Such is the spirit not of federalism, but of the section of the party once known by the name of Essex Junto. Almost all the signers of this last summons to me, are of the remnant of that party. . . . My principles," Adams wrote, "do not admit the right even of the people, still less of the legislature of any one state in the Union, to secede at pleasure from the Union. . . . To say of men holding these principles [that] they were intending a dissolution of the Union, and the formation of a new Confederacy, is charging them with nothing more than with acting up to their principles." Adams' review of the politics of the American republic from early Federalism to the end of the War of 1812 is brilliantly written, acerbic yet accurate, at least from the perspective of the Adams family.

On an afternoon in the middle of June 1829, a young boy, standing on the shore of City Island near New York City, saw a body drifting with the tide on Long Island Sound. He watched as it slowly brought the corpse to land. The coroner's verdict was "death by drowning." The identity of the corpse was quickly established from the contents of its pockets: a watch, chain, and seal; an account book; a penknife; a comb; a silver pencil case; and eight $5 bills. On April 29, twenty-eight-year-old George Washington Adams had boarded the steamer *Benjamin Franklin* at Providence on his way to Washington to visit his parents. George's mother had urged him to come. By the end of February, most of the family and some of the servants had moved to Meridian Hill, their rented house in Columbia Heights, about a mile and a half from the

White House. It would be, John Quincy thought, "a temporary shelter till the mildness of the season may admit my wife's journeying northward to Quincy."

Until the day before the new president's inauguration, Adams had stayed on, taking care of public business, which pressed him until the last moment. At first Jackson had invited the Adamses to stay in the White House beyond the inauguration, if they needed more time. Adams accepted the offer. Then Jackson requested that the house be made available for the Inauguration Day reception. It seemed an abrupt and inconsiderate reversal. Adams complied but made it a point not to attend the inauguration, as his father had not attended Jefferson's. Why should he inconvenience himself, given the circumstances? And he probably sensed that his participation would be awkward for others. Soon he found that he was almost as busy at Meridian Hill as he had been at the White House, but more pleasantly. He took long horseback rides, read Cicero's *Philippics*, and translated some of La Fontaine's fables. "My retirement is as complete as if I was a thousand miles from the Capitol, although I hear that the Court Newspaper administers its regular daily portion of abuse upon me, and lies as lustily about me as it did when I was President." But he was far from happy. His retirement was enmeshed in old and new miseries. Late in April he noted that "my wife, my son's wife and child are all ill, and I am sick at heart." Since he had known in advance what the outcome of the election would be, he had "not been disappointed. . . . But I had indulged a hope that after burying myself in total retirement, as was my design to do . . . I should be suffered to repose in quiet, and left to the pursuit of agriculture, letters, and history. . . . I now see I was mistaken; that I shall be hunted down in privacy as I have been in public; and that my life must be *militant to its close.*" If there was to be a new beginning, what and where was it to be? If it were to be in Quincy, he would have at least two of his sons nearby, perhaps all three. If not, Washington would suit Louisa, but could he reside there in a non-political post-presidency?

Throughout the second half of April, they were expecting George's arrival. He would accompany them to Quincy, probably to reside there, partly to save money but also to seek out a rudder for his drifting life. As usual, he was late in responding to letters. At the end of the month they were on watch, expecting him at any moment. On the afternoon of May 2, Louisa's brother-in-law came to the house. Had Adams received any letters that day? "I had not. He asked if I had heard anything of my son George. No. He said he had seen a short paragraph in the *Baltimore American* of this morning; that George had been lost from the steamboat *Benjamin Franklin*, between Providence and New York on Thursday morning before daylight." William Cranch arrived with the dreaded confirmation. Louisa's condition was indescribable. "Have compassion upon the partner of my soul," John Quincy prayed, "and bear her up with Thine everlasting love. Deep have been her afflictions heretofore. But this! Oh This! Stay thy hand, God of Mercy. Let her not say, my God! My God! Why hast thou forsaken me? Teach her and me; to bear thy holy will; and to bless thy name." They both struggled to keep their crippled souls from despair. "There is a pressure upon my heart and upon my spirits, inexpressible and which I never knew before. As it subsides, it gives way to dejection and despondency."

A fellow passenger came to Meridian Hill. His riveting story became a lengthy entry in John Quincy's diary. George had been unwell from the start, he told the grieving parents. A sick headache made him nauseated. He had become hallucinatory and thought that someone was trying to break into his cabin. When he searched, he found no one. He had "the impression that the birds were speaking to him; and that the machinery of the steam-boat seemed also, as if it were speaking." He asked another passenger if he had been "circulating reports" about him and then went with a candle to examine the berths of other passengers. At 3 A.M., he asked the captain to put him ashore. "Captain Bunker asked him why he wished to be gotten on shore. He said because there was a combination of all the passengers against him, and he had heard them talking and laughing against him." The captain was for the moment distracted by some pressing business. George, who was then noticed "near the end of the upper deck of the boat," disappeared from sight. Had anyone seen him in the last few minutes? The boat was searched. He was not to be found. From the situation of his hat and cloak on the deck, "it was inferred that in the wanderings of his mind, he had fallen overboard. It was too late for human help." His death was mysterious, his father concluded. But he recognized that even a rational man in George's situation might have concluded that life was not worth living. "Oh! My unhappy son! What a paradise of earthly enjoyment I had figured to myself as awaiting thee and me." He had never given up hope for George's reformation. Now he could hope only for his resurrection. Through George's death, he asked, was he not being punished for his every flaw of character? "I believe that special Providences enter into the general purposes of the Creator, and that [it] was in his design to chastise me in the immature and lamentable fate of my son." If he was being tested, trial could not be separated from blame. He kept trying to stop imagining George falling or jumping to his death. "It is inexpressibly painful for me to think; and every day more and more difficult to write."

Where, in the deep and wide ocean, he wondered, was the body? He had no belief in bodily resurrection, but he and Louisa hoped that George's body would be found, prayed over, and buried next to his ancestors in Quincy. Commodore Isaac Chauncey, the commander of the New York Navy Yard, conducted a search of the shorelines. Adams sent the thanks of "a grateful heart . . . to all the men who have lent their assistance in these repeated searches." A month later, the body drifted ashore. He went with John to East Chester, where the temporary tomb was opened and "the burial service . . . performed at the entrance. . . . The verdict is death by drowning," he wrote to Louisa. "I cannot describe to the most affectionate of mothers, the mingled sensations of distress and of consolation [with] which I went through the day." Louisa had two healing pleasures, her loving attention to her two-year-old granddaughter Fanny and the poems she wrote during the next year about her lost son, one addressed "To Him Who Is Lost to Me Forever":

So long in memory shalt thou live
In that fond heart enshrined;
And God in pity will forgive
A weak and erring mind.

John Quincy turned to prose, immersing himself in what most revivified him. Work, he believed, was a form of prayer, and work made men whole, the best antidote for depression. In mid-May 1829, two weeks after George's death, he began to write "some remarks on parties in the United States," a perceptive account of the rise of the party system and the conflicting ideologies of the founding generation. What he intended as a short piece became an extended account of how the divisions at the time of the country's founding produced two distinct political ideologies. "The remnants of the Tories of the Revolution generally sided with the Federalists," he wrote, "and produced an effect doubly disadvantageous to them." They infused into Federalism "opinions adverse to the Revolution and to republican government." All of Federalism was tainted with the Toryism of its most extreme wing. "This mixture of Tory doctrines with the principles of Federalism was the primary cause of their disasters and of all their subsequent errors, till their ostensible dissolution as a party." It seemed to him richly ironic that the circumstances that gave rise to the Republican Party produced in each of the two parties public policies that would have had comprehensive merit if combined. In the end, the total victory of either was a defeat for the country. But, he concluded, given the historical forces at work, the cleverness and ruthlessness of Jefferson, and the suicidal folly of the Essex Junto, the triumph of Republicanism was inevitable. Its worst flaw, even up to the present moment, was its "deeply rooted sentiment that commerce is in its nature an inferior and ancillary interest, not a primary and equal interest in society, and to be encouraged only as it is subservient to the higher and master-interest of agriculture." Agriculture, the Jeffersonians believed, was to commerce "in the relative position and importance of master and slave." That view "could have originated," Adams sarcastically remarked, "only on a tobacco plantation."

He had no doubt that political parties and sectional partisanship were here to stay. Both seemed more evil than good. During the embargo period, they had almost led to a national disaster. Jefferson had been horribly wrong. "The prejudice against naval force would be overcome by nothing but bitter experience." In the end, the best part of Federalism had been destroyed by the worst. What would happen now that a highly partisan slave owner of unstable temperament, who had begun to shape his administration as if he were a dictator and the country his personal kingdom, had been elected president? Jackson reminded Adams of his family's bitterest enemy, Timothy Pickering, the epitome of the self-interested destructiveness of the Essex Junto leaders. "If a man should look at a person having but one eye, exclusively in profile, on the dark side, till he convinced himself that the person was blind, and then should go before a magistrate and make oath that he was so, it would be precisely what Mr. Pickering was constantly doing in his political controversies. The effect was the same as if this falsehood were wilful, but he always began by imposing upon himself." That had also been the case with Jefferson. John Adams had been no more of a monarchist than Jefferson. He had been no more of a Republican than John Adams. But Jefferson had been ruthless and partisan enough to convince himself that his opponents were blind. John Adams' great flaw in the political arena was his total and compulsive honesty. Jackson's flaw was his character. He believed what he saw, without regard to the limits of his perception.

Urged on by Charles Francis, John Quincy began the major literary project he had promised would be the first work of his retirement: a biography of his father designed to "redeem his fame." When it did not go as well as he had hoped, he put it aside, to be resumed the next summer. At the beginning of September 1829, he attended the wedding of Charles Francis to Abigail Brooks. Louisa, in Washington, attempted to reach Quincy for the ceremony but, after collapsing in New York, returned to Washington, sick and depressed. She had a new daughter-in-law, but George was gone forever. John Quincy may or may not have known that George had seduced a servant and fathered a child. Once Charles Francis got over his initial shock, he concluded that George's death had not been untimely. "He would have lived probably to give much misery to his friends and more to himself, and he died when his fate was not so evident as not to admit of doubt, and a favorable construction, so that his memory will be cherished by his friends, and his end lamented." In late November, his casket was placed in the Adams family vault next to John Quincy's sister.

In the summer and fall of 1829, Adams looked to the benefits of no longer being a public man. He had a sense of his age and the limitation of his days. At Quincy he had visits from local friends. He missed Louisa, but by late summer, having determined to stay until George's remains were laid to rest, he felt physically better than he had for years, more limber and less arthritic. His eyes seem not to have troubled him. His writing hand mostly held steady, though he did not put it nearly to the test he had hoped for. He got done only part of a first chapter of the biography of his father. He kept getting sidetracked by research and reading, spending whole days absorbed in books that he had opened only to check a fact or remind himself of some event. The improvement in his health he attributed to exercise, particularly long rides and hours surveying woodlands he had inherited. He supervised renovations at the Old House to make it ready for his and Louisa's permanent residence. He wanted to retrieve all his books and give them the home he had never before been able to provide. Without an income other than from rents and investments, he watched expenditures closely. "But with a large and necessarily expensive family to maintain, with debtors who cannot and will not pay principal or interest, and with creditors to whom I *must* punctually pay both, my means and resources are so absorbed that I am . . . living upon the marrow of my property." John's health and finances, the latter sustained by Mary's inheritance, most of which went into building a house in Washington, continued to worry his father. "My anxieties for both my sons are deeper than I can express," he wrote to Louisa, "and the double experiment they are about to commence, one at Washington and the other here, weighs upon my mind more, or at least in a different manner, than upon themselves."

From the summer of 1829 to the fall of 1830, he spent more time reading and writing than he had for more than a decade. Literature and the humanities were a protection and a pleasure. In late September 1829, Charles Francis had an evening of intense recognition. "The conversation was literary. . . . It astonishes me more and more to perceive the extent and reach of the acquisitions of my father. There is no subject on which he does not know a great deal and explain it with the greatest beauty of language. . . . He is a wonderful though a singular man, and now displays more of his real character than I have ever before seen." From Washington, John Quincy assured Charles Francis that he felt the deepest sympathy "with everything that may contribute to your happiness," encouraging him to find a public voice as a newspaper writer on political subjects. He opened a correspondence with his new daughter-in-law, his first letter a humorous account of having been tongue-tied in the presence of one of his traveling companions, enclosing a poem he had written in praise of the unnamed "fair damsel." Since he would probably never see the lady again, he told Abby, she should consider the compliments as addressed to her. Forced to vacate Meridian Hill, he and Louisa moved into John and Mary's new house, expecting that they would move permanently to Quincy the next summer. It seemed likely that they would be leaving Washington forever.

By the start of 1830, some of his worst fears about Jackson were being realized. The administration vigorously redistributed "the spoils of office," which appalled Adams. He knew many of the government clerks and low-level officeholders. "There is beating of hearts and wringing of hands every night at what the next morning may bring forth." An unsettled Washington discovered that the volatile temperament they had elected had a peculiarity that touched on the moral stature of his administration and led to a political mess. Jackson's close friend John Eaton, the secretary of war, had had an affair with Mrs. Peggy Timberlake, who ran the boardinghouse in which he lived. Later, after her husband committed suicide, they married. Jackson insisted that Peggy's virtue was as white as snow. He would take it as an insult to him and to the memory of his late saintly wife if Mrs. Eaton were not given all the courtesies due to the wife of a cabinet member. Neither evidence not rational analysis mattered. By mid-1831, after a year and a half dominated by what Washington considered the "Eaton scandal," Jackson replaced four of his five cabinet members in the service of his insistence that Peggy Eaton's morals were unimpeachable. It was, Adams commented, "an administration one half of which holds the other half too infamous to admit of social intercourse with it." Since Calhoun's wife, Washington's social arbiter, refused to recognize Mrs. Eaton, it began to seem that the vice president had lost Jackson's support for the succession. The president, though, found an even better reason to end their relationship: Calhoun's heretofore unrevealed condemnation of Jackson's military activities in the Seminole War. Calhoun had brought down this misery on himself, Adams noted, by acceding to Jackson's petty and unpresidential demand that he respond to the charge that he had not supported Jackson in 1818 and 1819. The only conclusion to be drawn "is that the affairs of a great nation have got into the hands of very small men."

On Easter day 1829 Adams was experiencing his own version of the passion. His Boston and Washington enemies, from Otis to Jackson, weighed heavily on him as he read Cicero's "second Philippic." It was, he wrote to Charles Francis, "a holy and ever burning altar to liberty. I cannot tell you how my soul is affected by the contemplation of the condition in which Cicero existed from the death of Julius Caesar to his own. It is a feeling akin to that which many Christians feel on this anniversary." So intense was his identification with Cicero that "sometimes I feel as if I wished that Antony were alive, and I had the power to torture him. But he is in good hands." His hope was that now, in retirement, he could stay entirely away from politics. He was successful from the spring of 1829 to the spring of 1830, restricting his observations to private letters and conversations in which he surveyed the main flaws of the Jackson administration: it had come to power by perverting the democratic process through its propaganda machine; it elevated party spirit above principle; it corrupted government by making employment dependent on party affiliation; it attacked the Bank of the United States as if the bank were the enemy rather than the engine of economic growth; it opposed a protective tariff to support American manufacturing; it stained America's honor and founding principles by evicting Indian tribes from their homes regardless of legal treaties; it was at the service of a president whose narcissistic willfulness, uncontrollable temper, and deep prejudices determined administration policy; and it was increasingly caught in an internal contradiction that it could straddle but not solve. It was pro-Southern, pro-slavery, pro-states' rights, and in favor of selling cheaply or giving to the states the federally owned Western lands that belonged to all the people. But since its strongest support was in the South, how would it respond to the increasingly widespread Southern doctrine that a state legislature had the constitutional right to void any law passed by Congress if the state deemed the law unconstitutional?

That view, Adams wrote in late summer 1830, "now rages in the South. . . . Georgia has effected it so far as respects the Indians. So have Alabama and Mississippi. South Carolina is attempting it with regard to the Tariff, and I think will succeed. In both these cases the federal *Administration* is leagued with the nullifying states; and the laws of the Union are sinking under the weight of the combination." But it was not clear in 1830 and 1831 how Jackson would respond if South Carolina formally enacted legislation that gave it the power to nullify federal laws. That would put his presidency and the union at stake. In Adams' view, the stitching that bound the nation together, woven by his father's generation, was being unraveled. The modern nation-state, set in place by Washington, Hamilton, and John Adams, had been tampered with by Jefferson, but it had been adapted, even reinforced, by Madison and Monroe. And by himself. But just as the policy of his administration had been to "strengthen the ties of the

Union, that of the present . . . is to loosen and weaken them." Ultimately, "all remedy for political errors must in this country come from the people; and until they show signs of reconsideration, it is of little use to speak to them, though it were with the tongues of angels."

His own, he recognized, was not an angelic tongue. Anything he could say would be taken amiss, even dismissed as sour grapes. He was also restrained by distrust of his "own prejudices and resentments. If I ever review the acts of this administration it shall be not only with coolness but impartiality." But he was not capable of such coolness now. Behind nullification was slavery. And though he granted that each of the issues to which the South claimed nullification applied had a political rationale, at the center of its concern was the protection of its most vital institution. Nullification was the ultimate Southern protection against possible federal legislation about slavery. "It is the odious nature of the question," Adams concluded, "that it can be settled only at the cannon's mouth."

Early in 1830 he was reminded of how much his appearance had changed since his youth. At a casual gathering John Quincy encountered a man he at first believed he had never met before. When told that they had met in London in 1796, John Quincy observed that the man had undergone as great a change as he himself had. "No one knows me from the portrait painted of me by Copley that year, though it was then a very exact likeness. . . . We are always present to ourselves." He had once written to George that when we meet people we have not seen for decades, "memory returns to us the images of men and things as we *last* saw them; and it is a strange sight to meet a man in the decrepitude of old age, whom your memory presents to you as no other than in the sprightly vigor of twenty five." That was frequently Adams' experience in Quincy. Acquaintances from youth shocked him into awareness that he appeared to them as changed as they did to him. It also reminded him that his generation was passing away. In March 1832 he became the last of his parents' children to survive. "I have lost an always kind and affectionate brother," he wrote to Thomas' wife, "reflecting as I do how dear . . . he has been to me through a long life of which in the course of nature he should have been my survivor." But a lifetime of losses had taught him that nature makes many exceptions. His brother had been the closest companion of his early years. Now he provided for Thomas' children and widow from the estates that Thomas had mortgaged to him.

In the summer of 1830, John Quincy worked at creating a meaningful retirement. He had no reason to believe that he would ever be in the public arena again except as a writer on historical topics, such as a series of articles on Greece and the Russian-Turkish war he published in the *Annual Register.* When Alexander Everett, the editor of the *North American Review*, urged him to write a review-essay of Jefferson's recently published letters, he declined. The letters, though, renewed his consideration of a deeply flawed great man with whom his and his father's lives had been deeply enmeshed. "Mr. Jefferson's *infidelity*, his *antijudiciality*, and his *nullification* were three great and portentous errors." The harm that Jefferson had done outweighed the good. Jackson and the Southern slave owners were his heirs. Adams' one remaining mission was to rebalance the historical record. Controlled by Jefferson partisans, it had deprived John Adams of the credit he deserved, as if Jefferson, not Adams, had been the first, boldest, and most outspoken voice for independence.

In mid-September, working in the garden, he was called into the house to receive two visitors. He had the previous day attended the celebration of the two-hundredth anniversary of the founding of Boston, where he had been approached by Joseph Richardson, the Plymouth district congressman, and his colleague, John B. Davis. Richardson had declined to be nominated for re-election. They asked if they might call on him. Ten days before, the *Boston Courier* had proposed that Plymouth should elect Adams to represent it in the next Congress. He had assumed the tone was ironic, but there was a possible pro-Adams coalition. The moderate Republican center or National Republicans, many of whom had voted for Adams in 1828; a few Jackson Republicans, now called Democrats; the Federalist remnant in the Plymouth district; and the anti-Masons, a movement in reaction to the murder of a Mason in upstate New York to prevent his making public secret oaths and rituals, all might support Adams. Richardson "came purposely to enquire of me if I would serve, if elected. . . . I had not thought [the newspaper paragraph] serious, or that any person had a thought of holding me up for the election." If he declined, Richardson observed, the district might not be able to agree on any candidate at all. The Plymouth district's pride in two native sons' having gone to the White House might also influence electoral decisions. Would not a man of Adams' experience, knowledge, talent, and fame better represent the district than any alternative?

With almost whirlwind speed, the moderate Republicans, after overcoming opposition and a walkout, unanimously voted to support him. The National

Republicans soon did the same. The anti-Masons approved of his strong condemnation of Masonic secret oaths, and prudence, local loyalties, and historical ideologies made it better to nominate Adams than risk no representation at all. On November 6, 1830, he felt unalloyed pleasure at the result of the election. He had gotten 1,817 of the 2,565 ballots, with the difference divided between a hard-line Jacksonian and a right-wing Federalist candidate. "And so I am launched again," he noted, "upon the faithless wave of politics." The question for himself, his family, and his friends was not what but why. Had he not committed himself to retirement? Did he not have an important mission to fulfill on behalf of his family and family history? Was he physically and mentally up to the demands of a congressional seat? Was it appropriate, even seemly, for someone who had held the highest office in the land to afterward serve in a lower office? And why had he allowed these questions to arise? He could have said no to Richardson, and that would have ended it. Later he explained to Samuel Southard, "I saw no warrantable ground upon which I could withhold my services if demanded. This was strictly the principle by which I was governed. Had I perceived any sound reason upon which refusal could stand I should have refused." Others did see warrantable grounds. To those who said it was a "derogatory descent," he responded that he was not so arrogant as to think he was above serving his neighbors in the People's House. And future former presidents might be asked to use their talents and experience in the service of their country. His decision would provide a useful precedent.

The most serious challenge came from those to whom he was closest. Charles Francis expressed, with blistering politeness, his and his mother's disappointment that his father had again chosen to subject himself to the abuse of public life. Equally deplorable was the need to postpone his obligation to write the biography that John Adams deserved. Charles might have been less critical if he had known that in 1796, his grandfather had considered becoming a candidate for the House of Representatives, if Jefferson had won the presidential election, in order to drive out some of the demons that haunted him. John Quincy made clear to himself and Charles that there were demons *he* needed to confront. He would go into the House as a voice of dissent against policies he believed were destructive to the nation.

He also had been as disappointed as Charles in how little he had accomplished in the last two years. This paralysis "has hung upon me like a spell. . . . I pray God that I may recover from it while life and health may remain and enable me to tell the tale of my father's wrongs and my own. But if I do it must be to the men of other times." He felt his fortunes to have much in common with Cicero's. They had both been "deserted by all mankind . . . attacked by enemies, expelled from power, and rendered miserable by the loss of a much-loved child. . . . This unexpected stroke of divine displeasure" shook "the steadiness of my mind and unnerved the vigor of my arm. It has blunted the sensation of all earthly pleasure, and prompted indifference, and apathy to everything that the present age can enjoy or the future can inherit." Retirement had not provided healing. It had left his emotional equilibrium dangerously fragile. His self-worth was inseparable from work, from assiduous application to the task at hand, the hallmark of his life. Now the people of his district had called on him to represent them in Congress. "My election as President of the United States was not half so gratifying to my inmost soul. No election or appointment ever gave me so much pleasure. I say this to record my sentiments," for "the dearest of my friends have no sympathy with my sensations." This was an unexpected path that had opened for him. He had little doubt from the moment it had been proposed that he would walk down it.

Late in November 1831 he was awakened at half past four in the morning. So much brightness streamed into his Washington bedroom that he thought it was

THE USES OF ADVERSITY · 755

daylight. Snow, which had fallen overnight, covered the ground and the nearby roofs. There was no
more talk of selling the F Street house. In March, he
took the oath of office as a member of the House of
Representatives while the lame-duck Congress finished its business. His two-year term would start in
December. What would Washington and Boston do
with a former president turned congressman at the
age of sixty-four, and what would he do with them? In
Boston and Washington, naysayers mumbled about his
return to public life. It was an odd retirement from the
presidency. Some of the hostile press was nasty. "I am
suffering some of the penalties of Job, for daring to be
alive after suffering political death."

His energy was tested in the summer of 1831. On
Independence Day, he delivered the annual oration in
Quincy. It took almost a full month to prepare and an
hour and twenty-five minutes to present, though he
omitted about a third of the written text. "Eloquence
herself," he began, "perhaps best performs her appropriate office by silence upon exhausted topics." But his
had been renewed by recent events. His subject was
once again the narrative of the founding of the country, emphasizing the union as a compact among the
people of the states, enacted in special conventions, an
unbreakable national covenant. No state had the right
to nullify national legislation, he claimed, or withdraw.

Ironically, the administrations that had most stretched federal power, when it was in their interest to do so, were those that strongly advocated states' rights and nullification. The doctrine was elastic; the needs and use of power seemed always to trump abstract principle. In fact, Adams argued, "our collisions of principle have been . . . very little more than conflicts for place; and in the meantime the nation has been advancing, with gigantic strides, in population, wealth and power." Nullification and secession were threats to break the compact. Nullification "substitutes physical force in the place of deliberate legislation. . . . It would not the less be levying war against the Union if conducted under the auspices of state sovereignty." The federal government would have no choice but to resort to force. Otherwise rebellion would put an end to the vision of the founding fathers, the ideals of the Declaration of Independence would be betrayed, and the legal structure that the Constitution embodied would be destroyed. It would be a "calamity to all." He concluded with a distant echo of Webster's words in his widely lauded Senate speech in January 1830. "Were the breath which now gives utterance to my feelings the last vital air I should draw, my expiring words to you and your children should be, independence and union forever!"

On a rainy afternoon in late August, he met his second test, a eulogy of James Monroe, who had died on July 4, 1831, delivered to an overflow crowd in the Old South Church. The text that Adams had created was twice as long as the three-hour oration. The printed version, divided into *Life of James Monroe* and *Monroe's Administration*, had the distinction of being the first biographical study of the fifth president. Like the authors of the Constitution, Adams wrote, Monroe believed that government could transcend party. Faction could be made minor or nonexistent. His successor shared the wish but not the expectation. Motivated by loyalty to the president who had nominated him to be secretary of state, "it was a tribute of affection and gratitude to the memory of my venerated friend" whom he had served for eight years. Their work together had produced the second diplomatic triumph of John Quincy's career, the Adams-Onís Treaty. Their major disagreement had been about degree, not direction: both thought a national infrastructure program desirable, but Monroe had concluded that it required a constitutional amendment. Adams believed the federal government already had that power.

Reviewing Monroe's career, Adams combined affection and respect with a gracious but impartial overview. Monroe, he emphasized, had been true to the tradition of Old Virginia Republicanism but also to the changing conditions of America. He had accurately assessed the desirable relationship between federalism and states' rights, between Federalists and Republicans, and recognized, as Adams had, that the center of American political sentiment had shifted. The extremes had been pushed further away. The dominant center, consisting of most Jeffersonians and most Federalists, had gradually repositioned itself in what had formerly been only Republican territory.

In the heavy rainstorm, the lamps at the side of the pulpit in the Old South Church had to be lit in order for him to see his manuscript. Adams' family worried: would he falter, would he lose his place, would his voice or his eyes give out? And were these two speeches a prelude to work as a member of the House that would test him beyond his capacity? But the sight of a man who looked at least his age—in an age when a man of sixty-four was considered old—who had runny eyes and a shaking hand, speaking for three hours under difficult circumstances, presenting an artfully written and sweepingly well-informed overview of American ideas about government since the revolution, had to be impressive. He had obviously decided to go out on his feet in the midst of the life that suited him best. "Your beau ideal of life for a retired President may have its charms for minds and hearts better suited to it than mine," he wrote to a friend. In December 1832, as the Twenty-Second Congress began, Henry Clay, now a senator again, teased him. "Mr. Clay asked me how I felt upon turning boy again to go into the House of Representatives. . . . He repeated several times that I should find my situation extremely laborious; and that I knew right well before. Labor I shall not refuse, so long as my hand, my eyes, and my brain do not desert me; but what shall I do for that which I cannot give?"

He still had time and energy for writing poetry. In 1828 he had taken special note of David Hume's account in his *History of England* of Henry II's conquest of Ireland in the twelfth century. Adams took a self-referential interest in Dermot MacMurrough, the king of Leinster in Ireland. In order to retain his throne in a struggle against other Irish kings, Dermot had entered into an alliance with Henry that eventually resulted in the subjection of Ireland to England. It was a story of ambition, treason, assassination, and the assertion of brute power in family life, in the struggle for national consolidation, and in international affairs. "Courage and force, though exercised in the commission of crimes, were more honored than any pacific virtues," he wrote to Charles Francis during the presidential campaign, drawing a parallel between the politically motivated attack on his son John, the assassination threats mailed to him during his presidency, Andrew Jackson's appeal to the American public, and the widespread use of violence as a political tool in the twelfth century. He granted that the last "is not exactly congenial with the manners of our own age and country." But "it is with the incipient practice" and "the portentous doctrines of the leaders in the present canvass." Who is "our Dermot MacMorrogh . . . as impatient of the life of his *predecessor* as he has proved himself to be of the good name of his rival?" It was Andrew Jackson, an intemperate, willful, and violent man whose reputation had been forged in blood and who had unleashed his minions to slaughter the good name of John Quincy Adams.

Between mid-1829 and April 1831, he escaped his demons and entertained himself with the composition of a narrative poem based on Dermot MacMurrough's life. He brought it to an end after ninety rhymed stanzas of eight lines each, modeled after Byron's use of rhyme royal in his satirical epic *Don Juan*, a poem whose facility, wit, sweep, and political vision Adams admired but whose moral tenor he detested. "If I could take [Byron's] confession for an earnest of repentance and amendment, I should have hopes of him; but I expect it is only a pretence to more cantos of *Don Juan*." Over the years, he had read more and more Byron, both because his son George had admired him and because Byron had become the most highly lauded poet of the times. He was not, though, a wisdom poet of the sort Adams usually admired, like Milton, Pope, and Shakespeare; or, if he were, it was not in regard to morality and the good life. Adams, though, recognized and tussled with Byron's genius, and in his own partly satirical, mostly serious small-scale epic, *Dermot MacMorrogh; or, The Conquest of Ireland: An Historical Tale of the Twelfth Century in Four Cantos*, he borrowed from Byron the rhyme scheme, the theme of an epic without a hero, and the flexibility that allowed him to use narrative to dramatize his views about honor, morality, marital relations, and national independence.

He finished the poem in April 1831, some of it having been written in his head while he was walking or riding. The narrator, in imitation of *Don Juan*, is the author himself. The story is mediated through the voice and values of its author; his view of himself as a poet and his thoughts on other topics are incorporated into the narrative at any convenient and appropriate place. In some sequences, Adams manages to be charming, revealing, self-expressive, and aesthetically capable in a minor but impressive key:

XVII: But further now before my tale proceeds,
It seems a fit occasion to display
Before the man, or boy my work who reads—
(For ladies long ere this, I fear, will lay
My book aside, for Arthur's noble deeds;
For Marmion, Lalla Rookh, the minstrel's Lay—
Don Juan, or the Waverleys by scores:
Bright offspring of the Byrons, Scotts, or Moores.)

XVIII. To please the ladies is my dear delight;
For I have had a sister and a mother—
I have a wife, and if I could but write
The bliss they beam on husband, son and brother,
Scarce Heaven itself could purer joys unite,
Life to embellish, or bestow another.
I had a daughter—darling of my love—
She is an angel in the realms above.

XIX. No mortal on this earth then, better knows
The charms that women scatter o'er our lives;
Or more intensely feels the bliss that flows
From them, as sisters, mothers, daughters, wives.
But then I must admit, in verse or prose,
The dull and tedious seldom with them thrives:
They cannot bear a wearisome composer,
And from their very souls despise a proser.

They must already, Adams writes at the start of the next stanza, in a mixture of sincerity and mock humility, have tossed aside his book. "My style is the mock-heroic; but it wants vivacity, humor, poetical invention, and a large command of language. . . . I want a facility of inventing . . . character, of naturalizing familiar dialogue, and of spicing my treat with keen and cutting satire. I want the faculty of picturesque description, of penetrating into the inmost recesses of human nature, of moralizing in harmonious verse, of passing from grave to gay, from lively to severe; of touching the chords of sympathy with the tender and sublime, and to consecrate the whole by a perpetual tendency to a pure and elevated morality."

After its publication in October 1832, he noted that the critics tended to focus more on him than the poem. Since its pervasive theme is the conflict between foreign tyranny and national independence, there must have been readers in the United States, Great Britain, and Ireland who learned with interest that former president Adams, the epitome of Protestant New England, identified with modern Ireland's struggle for independence and drew a parallel between the liberation of the American colonies from British tyranny and the current relationship between Ireland and Britain. *Dermot* is many things but, more than anything else, like *Don*

Juan, though hardly on the same level as poetry, it conveys a political and moral vision. On that ground, Adams' and Byron's values are much the same, their views indistinguishable in regard to man, society, and the primacy of economic and political freedom.

A man who writes so well in prose perhaps should have no need to be a poet. But Adams did. As *Dermot* affirms, he was an excellent versifier and rhymer, with an abundance of learned skill and a touch of estimable talent. He was, though, aware that his poetry suffered from his overreliance on what he did best. It left it short of what the greatest poetry has: imaginative genius that transforms the mastery of technical devices into a totality in which the devices are in service of something intelligently beautiful and aesthetically profound. The poetry that he loved meant a great deal to him. Byron's offered too many irresolvable difficulties, and Byron's life was not, for Adams, an advertisement for his poetry. But he was always keenly aware of the gap between his highest models and his own poetic performances.

Still, the act of writing poetry never lost its power or its meaningfulness for him. It brought him closer to and gave him even more respect for the genius of the great poets. When he took satisfaction in his own best works, it was because they were good, not great, and good in a way that served a valuable purpose in his life.

They gave him pleasure; they provided balance to the prosaic affairs of life and state; they encouraged him to express emotions as well as ideas, from lyric poetry to poetic prayer to humor to satirical attack and even to mock epic. Some of these poems were published in his lifetime and numbers of them posthumously in the 1848 volume *Poems of Religion and Society.* Some have never been published. But there are many stanzas here and there and more than a few poems worth reading and remembering as one of the tonal registers of a rich life.

His expectation that he would be given a light workload and the privilege of following his own interests was shattered when Congress organized itself in December 1831. To his shock, the speaker assigned him the chairmanship of the Committee on Manufactures. There were other members of Congress who had taken a special interest in legislating on the desirable balance between the tariff as a means of raising revenue and as protection for American industry. Adams knew the arguments and the issues. He had had ample experience with the conflict between the Southern claim that the tariff was a penalty leveled on its agriculture, raising the cost of manufactured goods from abroad, and the claim by Northern manufacturers that a protective tariff was essential to

America's future as a manufacturing nation and to its national security. The conflict was between clashing visions of America: the Old Republican belief that the agricultural life exclusively nurtured democratic values, the good life rooted in the soil, and the belief that commerce and manufacturing, combining free workers, business entrepreneurship, and competitive capitalism, were equal partners in the national economy. On a practical level, the Southern lobby increasingly blamed the protective tariff for the South's financial difficulties, as if the additional costs that the tariff imposed on imported goods were the main, even the sole, reason for lackluster economic performance. It was as if Jefferson's impoverishment had had nothing to do with poor management, reckless soil depletion, extravagant indebtedness for a lavish lifestyle, and the inefficiency of slave labor.

Although it was not his special interest, Adams knew as much about the tariff as anyone. But the chairmanship would immerse him in the kind of partisan issue he believed hopeless. Every Jacksonian, including the speaker, opposed a protective tariff. Andrew Stevenson had appointed five Old Republicans to the seven-person committee. Every New Englander, by conviction or sectional interest, was bound to support the tariff. South Carolina's promise that its legislature would pass a law nullifying the federal tariff made the chairman-ship a frontline position in a national conflict. Adams had assumed he would be appointed to the Committee on Foreign Affairs. "I have petitioned almost upon my knees to the Speaker," he wrote to Charles Francis, "to be released from the performance of the duties of Chairman of the Committee of Manufactures. . . . He says he has no power over the arrangement of the Committee when once made and in point of *form* he is right." When Adams proposed to Edward Everett, the chair of the Committee on Foreign Affairs, that they switch positions, his Massachusetts colleague agreed. Stevenson did not.

In fact, it was not the form but the politics of the tariff that resulted in Adams' being appointed to the position. The Jackson majority had no desire to make Adams' life comfortable. The chairmanship of the Committee on Manufactures would keep him tied down, and he could do little damage, it was assumed, since he would always be outvoted. And the speaker could take credit for courtesy to a former president. Five of the seven members had voted against him in the election of 1828, and he was about to become a member of a body of 240 legislators who by a majority of two to one had opposed his presidency. They could not be certain he would not run for that office again, unlikely as that might seem. The national party was now including him in its consideration of candidates it might support in 1832. Adams had been slightly tempted by the prospect. In mid-September, William Seward, a young "zealous anti-Mason" and member of the New York state senate, an immense admirer of Adams, called on him before the start of the Anti-Masonic Party convention in Baltimore. They had a three-hour conversation. "I repeated to him that I had not the slightest desire for a nomination, nor for the office of President of the U.S itself." But "I would not reject the nomination . . . if made." He could, though, he told Seward, be "more useful to their cause" if he remained in Congress.

By August 1832, he was groaning under the burden. "I had voluntarily put myself [in Congress], against the opinion of some of my warmest friends, and the advice of more than one. . . . I had flattered myself that I should be at liberty to pursue my own inclination . . . with regard to the subjects of legislation to which I should devote my attention. . . . I had formed to myself the anticipation of a winter of little responsibility. . . . I was soon disabused from this day-dream." The South Carolina legislature had announced that if the tariff was not modified, it would make good its nullification threat. "I clutch like a drowning man at straws, at anything that may relieve me from the enormous responsibility under which they have affected to place me on this subject, attributing to me an influence which they well know I do not possess either over themselves or others." On the committee, the vote on every aspect of tariff modification rejected significant protection of Northern industries, though Adams had no difficulty keeping its meetings collegial. As a diplomat, he had learned the art of keeping one's enemies close. And it was possible to like some of the House members whose public policies he deplored, especially George McDuffie, the chairman of the House Ways and Means Committee, soon to become the governor of South Carolina. The art of persuasion, in Congress and in public opinion, would triumph, Adams hoped, in the long run. What did frighten him, though, was the threat by the South to pack up and go home, to become a separate country, if it did not get its way. And while that fear did not intimidate Adams into submission, it was clear that on the issue of the tariff, the South would, in the end, get its way.

That was also the case on the subject of the bank. He had been appointed to a special committee to look into charges against the Bank of the United States. Adams himself was a strong advocate for it. But the president held the bank in contempt, obsessively hostile to it, though it was not yet clear how far he would take his opposition and what might substitute for a national bank. In July 1832, there was no longer much doubt. Congress passed legislation to renew the bank's charter. Jackson vetoed it. In November 1831, Adams had been the guest of Nicholas Biddle, the head of the bank, whom he respected and liked, at a dinner party where he renewed his acquaintance with twenty-five-year-old Alexis de Tocqueville, who was gathering impressions about America for a book. They had already met at Edward Everett's home in Boston, and the French visitor was to be Adams' dinner guest in Washington i. February 1832.

In Boston, Adams had sat for an interview, an early contribution to oral history in which they quickly got to the heart of what he saw as the dominating influences in the country and its major problem. "There are two facts which have had a great influence on our character. In the North the political and religious doctrines of the founders of New England; in the South, slavery. Q. Do you look on slavery as a great plague for the United States? A. Yes, certainly. That is the root of almost all the troubles of the present and fears for the future." De Tocqueville asked him about "the immediate dangers to the Union and the causes which might lead to its dissolution. Mr. Adams did not answer at all, but it was easy to see that in this matter he felt no more confidence in the future than I did." If the South would threaten nullification and even secession over the tariff, what would it do if slavery was attacked in Congress? And what could the chairman of the Committee on Manufactures do other than write a minority report laying out as vividly and energetically as possible the web of connection from the tariff to slavery and its danger to the future of America? Jackson's veto of the bank bill seemed another facet of the attempt by the South to keep America's economy as slave-based and agricultural as possible.

Adams had much to brood about during the second half of 1832. Late in the summer, to his surprise, Jackson signed a tariff bill drafted by the pro-Southern majority of the Committee on Manufactures. It contained just enough protection for Jackson's Northern supporters to vote for it and so little that half the Southern members also did. It might, Adams hoped, defuse the nullification crisis. When, in November, South Carolina declared the new tariff law unconstitutional, Jackson decided that he preferred union to secession. South Carolina would stay in the Union, he made clear in his State of the Union speech in December, and obey federal law even if his will had to be imposed by force. Adams did not undervalue Jackson's opposition to nullification. But he understood that the doctrine transcended Jackson. It defined differences between the North and South that had slavery as their most visible manifestation. To Adams, the destructiveness of nullification was inseparably connected to a view of the union and the future of America exclusively entrenched in the South. The nullifiers and the slave masters of the 1830s had left behind Jefferson's ambivalence about slavery and Union. As the 1830s began, South Carolina and its allies proclaimed, as Jefferson's generation had not, that slavery was a superb institution, the Southern way of life was superior, and every state had the right to nullify federal law.

As chairman of the Committee on Manufactures, Adams seized an opportunity to make his voice heard. Jackson had given him his opening, asserting in his State of the Union address that agriculture was and should be the dominant industry in the country. Like Jefferson, though less elegantly, he claimed that "the best part" of the country's population are the cultivators of the soil. "Independent farmers are, everywhere, the basis of society, and true friends of liberty." Adams had no doubt what that meant. Commerce, industry, and Northern free labor should be subordinated to agriculture. Farmers of influence and wealth, most of them Southerners and slave owners, knew best what liberty is and were best suited to run the country. A small minority of a minority, Adams responded in a report that he submitted to the House in late February 1833 and which was soon published, ruled the nation.

In a long essay of about thirty thousand words, analyzing the philosophical and political underpinnings of the conflict, Adams surveyed the full range and implications of the tariff, the nullification controversy, and other administration policies: the end of a federal role in internal improvements; the elimination of the public lands as a source of revenue; the termination of the national bank; the refusal of fair protection for industry; the twisting and evasion of the words of the Constitution and Declaration of Independence; the preference for slave rather than free labor; and the privileging of those engaged in agriculture as an expression of the belief that the country was divided into superior and inferior people by occupation, geography, and birth. This "is the fundamental axiom of all landed aristocracies . . . holding in oppressive servitude the real cultivators of the soil, and ruling, with a hand of iron, over all the other occupations and professions of men. . . . The assumption of such a principle . . . for the future government of these United States, is an occurrence of the most dangerous and alarming tendency; as threatening . . . not only the prosperity but the peace of the country, and as directly leading to the most fatal of catastrophes—the dissolution of the Union by a complicated, civil, and servile war."

The Southern propaganda machine, built on an edifice of widely propagated lies, particularly appalled him:

Falsified logic—falsified history—falsified constitutional law, falsified morality, falsified statistics, and falsified and slanderous imputations upon the majorities of both Houses of Congress for a long series of years. All—all is false and hollow. And for what is this enormous edifice of fraud and falsehood erected? To rob the free workingman of the North of the wages of his labor—to take money from his pocket and to put it into that of the Southern [plantation owner]. . . . It has been said that there is no philosophic falsehood so absurd but it has been maintained by some sublime philosopher. Surely there is no invention so senseless, no fiction so baseless or so base, but it has been maintained by some learned, intelligent, amiable and virtuous, but exasperated and bewildered statesman. Nor was there ever in the annals of mankind an example of a community fretted into madness and goaded into rebellion, by a concerted and persevering clamor of grievances so totally destitute of foundation, and pretences so preposterously fictitious, as that which has found its consummation in the nullifying ordinances of the South Carolina convention.

With his command of language, logic, literary references, and historical evidence, Adams sustained his voice and argument in an authoritative and devastating rebuttal of Jackson, nullification, and Southern principles. Reprinted and widely distributed, the essay is one of Adams' most brilliant achievements as a writer on political and cultural topics. Officially written with the one committee colleague who voted with him, Lewis Condict, a congressman from New Jersey, it is entirely Adams' in style and substance. It deserves a place in the canon of American historical documents and American literary essays. In April 1833, the *Washington Globe* printed a comment on Adams as a writer and his two most recent achievements that, despite some hyperbole, is accurate enough to be perceptive. "*Dermot MacMorrough* is all prose," and the *Report of the Minority of the Committee on Manufactures* is "all poetry. The *Dermot* is all Ice and the *Report* all Fire."

Chapter 15
A Suitable Sphere of Action
1834–1839

On a Friday in early November 1833, Adams had a close brush with death. He left New York City, the halfway point on his trip from Quincy to Washington. Steamboat and railroad had accelerated it to a head-spinning three days. Each of the two engine cars of the crowded train from Perth Amboy, New Jersey, to Philadelphia pulled an open platform with railings and benches on which sat about fifty people; they were followed by four or five closed cars, like large stagecoaches, each with three compartments that had doors on both sides and benches running parallel to the length of the train. Then came an open baggage car covered by a tarp. Each bench in the closed cars could hold four passengers, facing the opposite bench.

The train stopped to have its wooden wheels oiled, a regular procedure to prevent friction from causing fire. Adams was soon to offer heartfelt thanks to heaven that none of his family was with him. Mary and John's two daughters lived with their grandparents. Abby and Charles now had two children; his first grandson, John Quincy Adams II, was the repository of John Quincy's hope that he would carry on the family name. Louisa, with one granddaughter, was a few days ahead on their return from Quincy to Washington; Charles and his family were in their Boston home.

As the train regained its full speed of thirty miles an hour, Adams smelled something burning. "It was so strong that all the passengers . . . noticed it." Moments after examining their own clothing and the seat cushions, as smoke seeped into the compartment, they realized that the fire was outside and beneath them. About to shout to have the train stopped, he felt the carriage lurch sharply to the left, the left front wheel burning. It slipped off the track, pulling the car behind, to which it was linked, violently to its left. The train kept moving "about sixty feet in a second but the wheel of our car . . . dragging on the ground operated as a lever on the other side of the car hoisted up from the rail . . . when by its link of connection with the car immediately behind it raised the left side of that from the rail, and that car went completely over drawing back as it went over our car to stand upon its four wheels, so that it did not overset."

Miraculously, no one in his car received the slightest injury. But the ground on the raised railroad bed and twenty feet below was littered with broken and dead bodies from the carriages behind his. "Men, women, and a child scattered along the road, bleeding, mangled, groaning, writhing in torture, and dying." It was "a trial of feeling to which I had never before been called. . . . The sight was heart-rending." A few more "yards of pressure on the car in which I was would have laid me a prostrate corpse for life; and, more unsupportable still, what if my wife and grandchild had been in the car behind me!" Helping the victims, he "felt a strange and mingled sensation of terror, superstition and awful gratitude to God." But he would not allow fear to determine his actions. Adversity was a test of character. He continued to Washington by train, as he would have done "if no accident had happened."

Having quietly noted his sixty-sixth birthday that July, Adams had good reason to be preoccupied with mutability. How quickly time passes and how soon his end was likely to be. His religion and the wisdom literature he cherished taught the importance of preparation. At the Harvard commencement he had recently attended, he noticed that half his graduating class was no longer alive. In a contemplative mood, he provided two months later a self-assessment. He had once been, he wrote, a warmhearted young man. Eventually, "I brought myself to the conclusion that Justice, and Benevolence, and humanity, and kind-offices, were virtues due from man to man . . . but that warm heartedness was not of the number. . . . I took mankind as you must take your wife, for better [and] for worse, and believed that the secret of dealing with my fellow creatures of my own times was to keep my heart and my judgment always cool. I have not always succeeded." Still, he was a man of feeling, capable of angry and tearful flare-ups, and also of deep reserve.

In 1831, he had written a prose soliloquy on the passage of time. Shakespeare's seven ages of man and two lines from a poem of James Thomson, whose poetry he had learned as a child, came frequently to mind: "As those we love decay we die in part. / String after string is severed from the heart." That he himself was soon to be in the seventh age of man elicited a variation on François Villon's fifteenth-century poem "Ballade des Dames du Temps Jadis," with its refrain, "Where are the snows of yesteryear?" Having spent much of his career among the most renowned rulers of Europe, Adams started his soliloquy at the point when, sixteen years before, he had had an audience with Charlotte, the wife of George III:

Where is that Queen of the British Islands? Where is her Imperial Consort, once our Sovereign and long our oppressor . . . ? Where is his daughter then the blossom of the land, presumptive heiress to the throne? Where his brother next in succession, the soldier of the family and once the leader of their armies? Where his son . . . then Regent, since crowned King of the Imperial Isles? All are gone. And where are the statesmen their Ministers who seemed seated as on a rock of adamant to sway the destinies of their country and of mankind? Where are Castlereagh, Liverpool, Canning . . . ? And where is Alexander the Autocrat founder of the Holy Alliance? And where with him is the Holy Alliance itself? Where is Napoleon . . . ? To come nearer home I ask myself where are my own father and mother then in full life, and forming links of the tenderest attachment between this world and me? . . . In the compass of these sixteen years all has changed around me, and yet I am here scarcely conscious how much I have changed myself. . . . I have neither room nor heart to speak of anything else.

His debts weighed heavily on him, as if they were a blot on his character. His obligations, aggravated by his insistence on paying down debts quickly rather than extending loans even on favorable terms, continued to exceed his income. He wavered between agreement with and rejection of Charles Francis' financial advice. Charles now managed his Boston properties, all of which were mortgaged. His youngest son thought his father imprudent, incapable of economy, and a poor money manager. John Quincy did not entirely disagree. With a constant cash flow problem, he regularly asked Charles to raise money from his Boston assets to cover living expenses, and he borrowed thousands of dollars from Antoine Giusta, his former servant, who now owned a profitable tavern in Washington. His congressional salary of $8 a day hardly made a difference. And he had an extended family to support: Louisa; John, Mary, and their two children; and nephews and nieces. His account balance benefited from renting out the F Street house; the entire family lived in John's house, though John had no income other than what his father paid him to manage the Columbian Mills. When John Quincy was in Massachusetts, they exchanged letter after letter about his "imaginary gold mine at Rock Creek."

He did his best to get solid information and give sound advice, to no avail.

Between 1832 and 1836, he felt chained to his desk in the House, overburdened with the tariff and bank issues. He felt compelled to speak out, which made him anxious. His views, he feared, would alienate friends and make enemies hate him even more. He also worried about re-election. In the summer of 1834, he thought it likely that the fall election would make him a one-term congressman. Rivalries and divisiveness among Jackson Republicans, now called Democrats; National Republicans, now called Whigs; and the Anti-Masonic Party seemed likely to diminish his chance for re-election, and the possibility of election by the Massachusetts legislature to the U.S. Senate. The Whigs were split between Clay's and Webster's factions. The Democrats were enough of a presence to cause trouble but not to elect their own candidate. Antagonistic to Clay and Jackson, who were Masons, the anti-Masons were united only by their anti-Masonry, which Adams shared. If supported by the large number of anti-Masons in his Plymouth district and enough pro-Webster Whigs, he had a chance of being elected senator. He gradually learned, however, that Webster, behind the scenes, opposed him. It did not help that some of Adams' supporters hoped and his opponents feared that he might become a candidate for the presidency in 1836. When the anti-Masons nominated him for governor of Massachusetts in the summer of 1833, he was inserted into a three-way race that he was unlikely to win for a position he did not really want. Eventually, when the election went to the legislature, he withdrew. But the Senate appealed to him. It would be vindication, and the six-year cycle would be preferable to the two-year House cycle. But in the summer of 1834, it was only his House seat that was at issue. In November, he would receive almost 90 percent of the 3,720 votes.

Louisa wished his political life did not exist. "Why will he waste all the energies of his fine mind upon a people who do not either understand or appreciate his talents?" If he would retire "from all these ill requited troubles, how much of bitter strife, of endless toil, of mortified vanity, and of disappointed ambition, would be saved to himself and to his family!!" She knew, though, what was at stake. He could not retire "without risking a total extinction of life; or perhaps of those powers even more valuable than life, for the want of a suitable sphere of action." It was an anxiety-ridden dilemma. In order to live, he needed the life he had chosen. But "the extreme uneasiness I feel at the state of Mr Adams health is beyond language to express!"

His work ethic required that he be at his post regardless of weather or health. He suffered from colds and influenza, throat and chest coughs, arthritis, rheumatism, and essential tremor. But his cough was not tubercular. He had no chronic disease. His insistence on exercise in the summer and winter did him good. If he sometimes felt that his home was an infirmary, it was partly because infections were passed around for months on end. An excursion to the White Mountains in the late summer of 1833 helped. When he came "through the narrow pass between the two stupendous ridges of the White Mountains," he was "reminded of the mountains and valleys and roads and people of Silesia and Bohemia at every step of our way."

At the start of the summer of 1834, he agreed to the request of the Harvard Overseers that he chair a committee to investigate a recent student rebellion. Was President Josiah Quincy's leadership at fault? And to what extent should the destruction of university property be subject to legal prosecution? "This is a subject the importance of which grows upon me as I proceed, and the earnestness in writing grows in proportion." He interviewed the principals, presided over meetings, and labored over a report that in the end vindicated the president. Privately Adams, who shared Quincy's abhorrence of unlawful student dissent, believed that his personality limited his effectiveness. Adams recognized, though, that Quincy had in 1829 taken the helm of a poorly governed university. It was, Adams had noted ten years before, "more of a caucus club . . . bigoted to religious liberality, and illiberal in political principle. When they have a place to fill, their question is not who is fit for the place, but who is to be provided for. And their whole range of candidates is a parson or a partisan, or both." Ten years later it was still controlled by a "political coterie," and "the offices and revenues of the University" had been turned into a "foundling hospital for the dunces and bankrupt of the Junto. I hope and trust it has come to its end just in time to save the University from total bankruptcy." Quincy's economic stewardship had been successful. He was also, in Adams' view, freeing Harvard from the control of exclusivist Unitarians and ultra-right-wing Federalists. Still, Adams privately concluded, the president "can never regain his popularity with the students and the public treat him as they treat all old men with cold neglect and insulting compassion. He is not made of stuff to struggle long against this."

When, at the end of July 1834, he had dinner at the home of Dr. George Parkman, who owned most of Beacon Hill, he was anxiously awaiting Louisa's arrival from Washington. With the other guests, including

786 · JOHN QUINCY ADAMS

Everett and Webster, he went to the roof of the building to watch a balloon ascension. "We followed it until it had diminished to a speck and vanished." But he felt no elation as he had in his youth in Paris when he had been excited by the miracle of flight. He was in a worrying mood, tired, headachy, and feverish. He had enough excess anxiety to worry even about the danger to the man in the balloon. He was relieved when he learned that the balloonist had come down about five miles off Cape Ann and been rescued by a passing boat. His main concern was Louisa, who had been too ill to accompany him to Quincy, but who at last had gotten sufficiently better to be able to travel. He also worried about John, not only about the fiasco of the mills but about his health. "I have lost my appetite. I am losing my sleep. I feel as if heavy calamities were impending over me."

On July 4, he had noted in his diary: "Independence Day. My son John's birthday. He is 31. Day of my father's death." John had been "in a declining and drooping state of health more than three years . . . several times afflicted with severe and acute disease," a long-lasting eye infection and scarlet fever. The family made no extant comment about the degree to which John may have also developed a drinking problem. The ghosts of alcoholism in the family may have induced paralysis and a sense of unavoidable doom. John may have become a heavy drinker, perhaps by genetic fiat, perhaps also to escape his disappointment with himself. Later, Charles Francis, commenting on John's "moral ruin" and "extravagant tastes," partly blamed Washington, whose "climate and habits" he believed unfavorable to New Englanders. In June and July 1834, John Quincy urged his Washington family to join him. "The thoughts of the future haunt me in my dreams; of which I had a cruel one last night." He could not shake off fear that one or more of them would not survive the capital's unhealthy summer.

Before leaving Washington, he had agreed to a Senate and House resolution that he deliver the official eulogy for Lafayette, who had died in May. But he felt too anxious to make headway with writing it. Whether re-elected or not, he had decided it would be best to sell his Washington properties. "It is my undoubting conviction," he wrote to John, "that nothing but ruin, utter and irretrievable ruin can await you and your wife and children" if they made Washington their permanent home. "I have here a large landed estate. . . . A large portion of it, I intend shall pass to you or your children. . . . This . . . is the only expedient now left to save your children and yourself, and to give quiet to my last hours." Overjoyed when he received word that

Louisa was in transit, he was disappointed that John and Mary were not with her. But Louisa's arrival relieved him on her account at least, and she had brought his granddaughter Mary Louisa, with whom he began to spend a few hours each day teaching her to read.

At the end of the month, he heard Ralph Waldo Emerson, the son of his deceased friend, deliver the annual Phi Beta Kappa poem at Harvard, "a slovenly performance, the only merit of which was its brevity." Adams may have given the poem even less than its due since it was an evocation in praise of Daniel Webster. That same evening he dined at Webster's. "The dinner was excellent, the wines various, the conversation cold, cautious, and under constant restraint," almost as if he could see the opportunistic Webster silently calculating his thoughts and words in regard to his chances for the presidency. Adams walked around the border of the Commons, passing the house on Boylston Street in which his son Charles had been born, noting that he had left from that house for St. Petersburg almost exactly twenty-five years before. "What an age of hope and disappointment, of good and evil fortune I have since gone through. Prosperous on the whole beyond the ordinary chances of human life—but with how many deep afflictions, with how many cruel humiliations!"

Early in September, Louisa became sick with chest pain, a burning fever, a racking cough, and fainting fits: how much was influenza, how much erysipelas, how much nervous exhaustion? The erysipelas dominated, "the most violent and long protracted attack of her old disorder." Meanwhile, Mary Louisa, teething and with a bad rash, cried constantly. John Quincy had a severe cough and runny eyes; on some days he was "half blind." When news came that Mary and John were still detained by "a bilious fever," Louisa advised that they not come to Quincy so late in the summer. Early in October, Mary Louisa came down with a severe earache; Louisa had a relapse. Worse, she "was much distressed by a letter this day received from her sister Mrs. Frye, mentioning that both our son John and his wife are very ill, confined with chills and fevers." On October 18, a letter came from Walter Hellen "saying that . . . John was extremely ill and urging that if his mother could not come on, Charles and I should come immediately." Louisa, whose distress at the news was agonizing but who could hardly walk, insisted that she would leave immediately, despite "excessive sickness, faintings and cramps." The doctor convinced her that it would be at the risk of her life.

In Boston, there were letters with further news. John's condition, though varying from day to day, was critical. As he headed toward Providence in a heavy rain, John Quincy felt the "difficulty of resignation to the will of God." But he still cherished "hope where hope is without rational support." On the steamer to New York, he took part in a lively discussion with other passengers, lasting until after midnight, on "the controversial character and divinity of Christ, and much upon the slave question, the prospects of universal emancipation and the intellectual capacity of the African race." Apparently it did not seem to him in the least odd to take an active part in public conversation as he traveled from a house of sickness to a house of death. Perhaps it made the voyage easier.

Less than three days later, he was at the door of John's house, where he was greeted by the mournful faces of Louisa's sister Carolina and her husband, Nathaniel Frye. John was "very low." Carolina advised that he wait to see John until morning. A few hours later, she changed her mind. "I sent for Dr. Hewitt who advised me to go see him. I went to his bedside twice, and saw and heard him. He had no consciousness of anything on earth. His wife was in bed in the upper chamber, very sick herself. On seeing me she burst into tears. I promised her that I would be a father to her and her children." Exhausted and unwell, he went to bed at two in the morning. "At half past four I rose again and going in to the bed-chamber where my son lay found Mr. Frye closing his eyes. He had just ceased to breathe. May god in his infinite mercy have received him to the joys of heaven!" Exhausted and numb, he made funeral arrangements: pallbearers, clergy, newspaper notices. He wrote to Quincy. When he asked the doctor what illness he thought had killed John, the doctor responded that "he did not know," probably an evasion that Adams did not make any further attempt to resolve. Mary was now a widow with two daughters to raise. Her father-in-law reaffirmed that he would care for her and the children; they were to live permanently with him.

The next day, taking a last look at the shrouded body of his son, he found solace in the speculative thought that "his death was tranquil, undisturbed by any conscious pang of voluntary wrong." He memorialized "all the previous services that my darling child had rendered me—his discharge of all his filial duties, his kindness, his affection, his devotion to all my interests, and all my desires, the uncomplaining patience with which he has endured misfortune, sickness and disappointment the hardness of his fate and the meekness with which he has submitted to it—even his infirmities." It was not a time for realism about John's disservice and the penalties paid for meekness in a competitive world.

Overwhelmed by his loss, Adams exerted the fragile willfulness of his Christian faith. "Let me believe, that for suffering upon earth there is some compensation in heaven, and that there the tears of sorrow are wiped away, and that every virtue shall be blessed with its reward. My child, my child." In a funereal October and a miserable November, he found it beyond his capacity to return to a normal schedule. He could not work or sleep. "The resources of wretchedness are stoic fortitude which is insensibility, and Christian resignation to the will of God. The first is not in my nature. The last I know to be my duty, but is not always in my power. Reason is no salve for pain." Exercise might help. But each morning's walk seemed a "melancholy pilgrimage in which to divert my thoughts from the bitterness of my misfortune. . . . The extirpation of my own and my father's race is the issue over haunting my imagination."

At a little past noon on the last day of 1834, Congressman and former president Adams mounted the speaker's platform in the House to deliver the nation's final tribute to the Marquis de Lafayette. Adams had conceived of it as a narrative of the general's life and an evocation of his political beliefs. The president, the executive officers, the judiciary, and both houses of Congress had marched in procession to their assigned seats. The French delegation led the diplomatic corps, bestowing posthumous honor on a man whose country had not always treated him reverently. With unhesitating emphasis, Adams explained why Lafayette had found a second and better home in America. He had been born into a Europe that believed in hereditary succession. No two people more unfit to occupy the thrones of Great Britain and France, Adams began, could have been found than the German George II, who was ignorant of the customs and language of the English, and the infant Louis XV. "Yet, strange as it may sound to the ear of unsophisticated reason, the British nation were wedded to the belief that . . . fixing their Crown upon the heads of this succession of total strangers, was the brightest and most glorious exemplification of their national freedom; and not less strange . . . was that deep conviction of the French people . . . that their chief glory and happiness consisted in the vehemence of their affection for their King, because he was descended in an unbroken male line of genealogy from Saint Louis."

The French diplomatic corps could not have been pleased with Adams' theme. But American ears found it both gratifying and true. As usual, his two-hour presentation was a shorter version, reduced as he spoke, of a longer text. Keenly aware that he did not have the time or energy to write less, he found "it much easier to enlarge than to abridge." By mid-December, he had regained enough focus to finish it, and on the last day of the month, his distinguished audience was riveted. Lafayette's principles, Adams remarked, "were in advance of the age and hemisphere in which he lived. . . . The principle of hereditary power was, in his opinion, the bane of all republican liberty in Europe." Lafayette's genius was his love of liberty, and where it came from, given the family and country into which he had been born, could be understood only as an expression of the inalienable right to life, liberty, and the pursuit of happiness, the birthright of every human being. The time will come, Adams concluded, when France will once again be a Republic, when hereditary privilege will be extinguished, when its people will themselves be the repository of national authority. "Then will be the time for contemplating the character of Lafayette," a disinterested benefactor of mankind, "not merely in the events of his life, but in the full development of his intellectual conceptions, of his fervent aspirations, of the labors and perils and sacrifices of his long and eventful career upon earth." America had more right to take possession of Lafayette than the country into which he had been born.

No sooner had Adams taken his congressional seat than he was appalled by the flood of petitions "for the plunder by whole and retail of the Public Lands. I am more and more convinced it will be impossible to save them from the grasp of the western people, aided by the abandonment of the South, and the treachery of the northern traffickers for the Presidency." In April 1834, he had been prevented from delivering a speech condemning the removal of all federal deposits from the Bank of the United States. Although President Jackson had vetoed congressional recharter, the existing charter was still in force. Yet he had ordered the deposits removed and transferred to the state banks of his choosing. Outraged, the Massachusetts legislature passed resolutions opposing Jackson's removal of the deposits and his hard currency policy. With difficulty and only by the courtesy of his colleague George McDuffie, Adams was able to place the resolutions on the record.

The battle among various political factions, the bank, and the administration intensified. Hoping to ride the issue to the Whig nomination, Clay sided with the bank. Van Buren, now Jackson's heir, favored removal. The result, Adams predicted, would be "a great catastrophe" for the economy. As he attempted to get the floor, the call for the previous question by a supporter of the administration stopped any further debate, and this lesson in parliamentary maneuvering reminded Adams he would do well to make himself a master of the rules of the House. In Boston, bank supporters published his speech under the title "SUPPRESSED BY THE PREVIOUS QUESTION." It was a phrase with which he was to become intimate over the next ten years.

At the beginning of 1835, what was most on Adams' mind was the possibility that the administration would force the country into a war with France. Long-standing claims for indemnity for the confiscation of American ships had been resolved by a treaty in which France agreed to pay $4.6 million in six equal install-ments. When, in April 1834, the French Chamber of Deputies rejected the treaty, Jackson condemned the French government, threatening to confiscate its ships in retaliation, his language belligerent, his tone inflam-matory. The country thought it would soon be at war. Those who thought Jackson unsuited to be president believed he was again justifying their predictions. Indeed, the administration and its allies had little expe-rience and less finesse in foreign affairs. In the House, the semi-dysfunctional Committee on Foreign Affairs dithered about the degree to which it should support the president. Although a pro-Jackson committee, it was not entirely pro-war. How could it be both loyal to and more temperate than the president? The business community panicked. War would damage commerce; revenue from the tariff would be diminished; the cost of defense might require internal taxes. Adams led the passage of a resolution requiring that the administration send all relevant documents to the House. In April 1835, when the Committee on Foreign Affairs rejected further negotiations, Adams had ready three tactful, effectively written substitute resolutions making it clear that if the matter was not settled through negotiations, there would be consequences. Suddenly Adams was in the thick of things, and his colleagues listened to him attentively. After all, he was not only a former president; he had been a professional diplomat for most of his life. No one in the House had even a tiny fraction of his foreign policy experience and skill.

After thirteen hours of debate, the House accepted two of his three substitute resolutions, which were then consolidated into one, the language of negotiation replacing the language of threat. But the velvet glove contained an iron fist. The indemnity would have to be paid. Would there be war first? Although the Jackson forces, the nullification Southerners, and the Clay and Webster contingents spoke against the resolution, in the end they begrudgingly acknowledged that it was the best alternative. "I breasted them all," Adams boasted to his son, "and . . . closed with a vote of two hundred and ten ayes for my resolution and not one solitary *nay*. . . . I will not attempt to describe my feelings. . . . It was one of those moments which compensate for a life of sorrows, but I well knew what would follow. Nemesis was upon the watch." Much of Whig New England focused on what it considered Adams' prowar speech. And was it not also a pro-Jackson speech? Led by Webster, maneuvering to prevent Adams from obtaining a Senate seat, many New England Whigs were furious. His speech and resolution were pilloried as pro-war and pro-Jackson. Maneuvered by pro-Webster Whigs, the Massachusetts legislature chose Webster's friend John Davis to be senator and Edward Everett to be governor. In the end, Adams concluded that the price he had paid for his independence was not too high. He had had his first great triumph in the House, though it might be, he speculated, "the last trophy of my public life."

Much to his disgust, the question of who would succeed Jackson dominated Congress. Every policy debate was inflected by the competition for office. It now seemed inevitable that every president's last two years would be dominated by the race to succeed him. As much as Adams had tried to prevent that in 1827 and 1828, forces beyond his control had made it impossible. Despite pressure to support his longtime colleague, he could not back Clay. Webster, in Adams' view, divided his time between scheming and orating. He still respected Calhoun. "There is more of elementary good in his composition than there is in that of any of his competitors, or in that of all of them put together." But Calhoun's embrace of an extreme version of states' rights separated them irrevocably. He abhorred Calhoun's principles but admired him for openly sticking to them. "Every other leading statesman of the day is a trimmer, a man of compromise." Anyway, Jackson's feud with Calhoun eliminated him as a viable successor.

No matter how damaging or immoral Jackson's bank and Indian removal bills had been, his popularity remained great enough that it was widely assumed his chosen heir would be elected—"the battle of New Orleans [was] . . . a victory more complete over the people of the U.S. than over the soldiers of Great Britain." Democracy, Adams believed, required a respect for diversity in view and honesty in the evaluation of character. Jackson failed both standards. "If I were the Grand Elector," Adams observed, "I would be much puzzled to make my selection. The tendency of our institutions is to promote men of capacity just sufficient for deep intrigue and nothing more . . . and consequently all the candidates now before the public are intriguers. Which of them is deepest, is not yet discernible." It turned out to be Van Buren. It would be best, Adams decided, not to support any candidate.

By 1835, the daily rhythms of the House had become as much a part of him as the circulation of his blood. He took energy and life from having a forum in which to exercise his natural combativeness. His reason, he confessed to Charles, told him to withdraw. But "a sense of duty; perhaps a lingering tenacity to existence as a public man, controlling my better judgment, keeps me yet in the field of politics. I have no rational prospect before me, but of disappointment upon all three of the great measures of policy, frowning in my face." But he would not, could not, desert his post. "I must stay." And he had become a formidable presence, partly because of his distinction as a former president. Even those most opposed to his views felt they had to tread carefully, though there were exceptions in the heat of debate. Adams had, from the start of his service, felt reverence for the House's rituals and procedures. The House epitomized one of the three branches of an ideal form of government, the triumph of his father's vision of what the revolutionary generation had created.

On a rainy, gloomy Monday in February 1832, soon after his first election to Congress, he made the prose of his diary dance with poetic energy as he observed one of the sacred rites of democracy, the call of the roll of states for the presentation of petitions. When divided and lightly edited into poetic lines, it has a more literary presence than some of his explicit poems.

> *Being Monday, the States were*
> *successively called*
> *for presentation of petitions;*
> *a most tedious operation*
> *in the practice, though to*
> *a reflecting mind*
> *a very striking exemplification*
> *of the magnificent grandeur of this nation*
> *and of the sublime principles upon which*
> *our Government is founded.*
> *The forms and proceedings of the House,*
> *this calling over of States for petitions,*
> *the colossal emblem of the Union*
> *over the Speaker's Chair,*
> *the historic Muse at the clock,*
> *the echoing pillars of the Hall,*
> *the tripping Mercuries*
> *who bear the resolutions and amendments*

> *between the members and the chair,*
> *the calls of ayes and noes,*
> *with the different intonations*
> *of the answers from the different voices,*
> *the tone of the Speaker*
> *in announcing the vote,*
> *and the varied shades of pleasure and pain*
> *in the countenances of the members*
> *on hearing it, would form a fine subject*
> *for a descriptive poem.*

Starting in 1835, the Monday ritual and call for petitions was to become the focus of his conviction that Americans had been guaranteed the right to address their government about the one issue whose moral and practical complications seemed to him, more and more, to define the destiny of the United States.

Some of Adams' most cogent thoughts about slavery were provoked by his reading of Shakespeare's *Othello*. Adams "read Shakespeare as a *teacher of morals,* as a student of human nature, as a painter of life and manners, as an anatomical dissector of the passions, as an artificer of imaginary worlds, and at once the sublimest and most philosophical of poets." In November 1835 he became obsessed with *Othello*, especially Desdemona, whose "sensual passions" were so "over-ardent . . . as to reconcile her to a passion for a black man. . . . I thought the poet had painted her as a lady of rather easy virtue." Desdemona had, in Adams' view, a fatal flaw as a dramatic character, for "the pleasure that we take in witnessing a performance upon the stage depends much upon the sympathy that we feel with the sufferings and enjoyments of the good characters represented, and upon the punishment of the bad. We never can sympathize much with Desdemona or with Lear, because we never can separate them from the estimate that the lady is little less than a wanton, and the old king nothing less than a dotard." As a student of Aristotle, Adams required that successful tragedy represent "human action and passion, to purify the heart of the spectator through the instrumentality of terror and pity." *Othello* fell short of that standard, he told George Parkman and the British actress Fanny Kemble. Desdemona's passion for Othello was, Adams believed, "unnatural." Her physical intimacy with Othello had to offend and "disgust. . . . She not only violates her duties to her father, her family, her sex, and her country, but she makes the first advances." Her disloyalty, her elevation of passion over filial duty, was morally reprehensible. In Adams' view, the play was widely misunderstood. "If the color of Othello is not as vital to the whole tragedy as the age of Juliet is to her character and destiny, then I have read Shakespeare in vain."

In fact, as he argued in two essays, "Misconceptions of Shakespeare upon the Stage" and "The Character of Desdemona," and as he also wrote to Parkman, "the moral of the tragedy is that the marrying of black and white blood is a violation of the law of Nature. That is the lesson to be learnt from the play." That "lesson" embodied one of the complications, even for many pro-abolition Northerners, of the politics of slavery and race. Like many of his anti-slavery contemporaries, Adams had no difficulty believing both that slavery was immoral and that sexual relations between whites and blacks were "unnatural." The former belief affirmed the inherent right of personal sovereignty granted by natural law; the latter breached the natural law prohibiting the mixing of races. It was a short distance from the relationship of Desdemona and Othello to the relationships between white masters and female slaves. Abigail Adams had railed against the white Southern male propagation of mulattoes; her target was the self-righteous Southern elite who dominated the national government. So too had Louisa and John Quincy. And, in Washington, Southern congressmen made no effort to conceal their black mistresses and mulatto children. What the Adams family believed morally reprehensible and an assault on the dignity of all women seemed inseparable from the existence of slavery itself. After all, the slaveocracy flourished by breeding new slaves, some of them the children of white masters.

Adams had made the connection between Othello and breeding slaves in an ironic comment to Charles Francis in 1830, prompted by their discussion of a case argued by Cicero in which the issue was partial or full ownership rights in a slave. With his characteristic mixture of irony and outrage, Adams proposed that the "managers of our theatres might take the hint" and "might breed many a slave to make excellent Othellos," though that would require that the slave raised to perform the role learn to read, a felony in certain jurisdictions. But race and slavery could be distinguished from one another. Race was an unalterable fact, with certain givens. Adams gave great latitude to the equality before God and man of all human beings, regardless of race. Natural law and its embodiment in America's sacred texts required that. Miscegenation, though, was a step too far. Slavery also was a fact. But slavery was a step not far enough. It fell below the requirements of natural law. And it was not unalterable.

When Adams returned, in late 1835, to his seat in the House, slavery had once again become an explosively divisive issue. It had been comparatively quiescent for the twelve years following the Missouri Compromise though still a source of tension. Various factions in the South, disagreeing about gradual emancipation, had coalesced into a widespread agreement that slavery would be forever essential to the Southern way of life. In the North, the small number of abolitionists, divided between those who favored immediate and those who favored gradual abolition, expressed themselves vigorously, pricking the Northern conscience where there was one and attacking widespread indifference to a disease that could not be cured except at a cost most Northerners were unwilling to pay. Abolitionists from Philadelphia to Boston, many of them members of the Anti-Slavery Society, regularly sent petitions to Congress urging the end of slavery, especially in the District of Columbia. Would Adams cooperate? Yes, he would submit to the House any petition, whether he agreed with it or not, if it was a sincere statement of grievances to be redressed and couched in respectable language. Every citizen had a constitutional right to petition his government for redress. Congress had the obligation to receive the petition.

But he did not wish to foment discussion that "would lead to ill will, to heart-burnings, to mutual hatred, where the first of wants was harmony; and without accomplishing anything else." What, he had asked a Quaker visitor, would he think if the people of the District of Columbia "should petition the Legislature of Pennsylvania to enact a law to compel all citizens of that state to bear arms in defence of their country? He said he should think he was meddling with what did not concern them." He would always, Adams emphasized, present such petitions to the House. Their reception, though, would not be hospitable in a Congress dominated by Southerners and their Northern allies, some even willing to threaten violence and the dissolution of the union. He could expect the petitions, once tabled or sent to committee, never to be heard of again. And he could make no friends for himself or his abolitionist associates by antagonizing his colleagues. But what, his visitor asked, were his own views about slavery? His Southern and many of his Northern colleagues assumed that he favored abolition. Did he? In private and in his diary, Adams was unhesitant. "I abhorred slavery," he told his visitor, "did not suffer it in my family, and felt proud of belonging to the only State in the Union which at the very first census of population in 1790 had returned in the column of slaves none." Behind the conflict over nullification, he told a supporter of Van Buren and Jackson, "was whether a population spread over an immense territory," one half of which contained only freemen and the other half mostly "masters and slaves, could exist permanently together as members of one community or not; that, to go a step further back, the question at issue was slavery." It was the issue convulsing the nation.

But how could slavery be eliminated? The Constitution had inscribed it as a permanent legacy in the states in which it existed and that did not reject it voluntarily. Those who opposed it should concentrate, as a practical matter, Adams maintained, on preventing its spread. The Colonization Society was a farce, "the day-dream of some if its members," impractical nonsense diverting attention, sometimes naively, sometimes purposely, from what might actually be done. But what could be done? Very little, he thought in the early 1830s, and with great danger. "Slavery is, in all probability, the wedge which will ultimately split up this Union. It is the source of all the disaffection to it in both parts of the country." The obstacles to a resolution were immense. After all, both sides increasingly based their justification on natural law, though the Northern and the Southern views on this issue were incompatible. Was the slave a person or possession? Had the

Negro race been created by God and nature inferior and forever subordinate to the white race? And, as a legal matter, had not the Constitution guaranteed by omission that the federal government would have no say in the matter in the states in which slavery was allowed?

Adams saw no grounds for a peaceful resolution. There was tactical wisdom, then, in keeping his mouth shut on the issue. The less antagonism, the better the chance for some comity in the House. But how could one discuss Indian policy, nullification, the tariff on manufactures, the role of the federal government in public improvements, the price of government-owned Western lands, the attempt to purchase Texas from Mexico, and the transformation of territories into states without noticing that they all were inseparable from views about slavery, especially the conflict between free soil and slave soil, free labor and slave labor?

By 1835 it was clear to abolition leaders that Adams was an ideological colleague, though a prudent one. He was to be visited by and soon to work with numbers of radical abolitionists like the Grimké sisters, and especially Benjamin Lundy, Arthur Tappan, Roger Baldwin, Theodore Weld, and James Mott. When Adams attended a large Quaker gathering in Philadelphia, he had much to say about the main topic of the evening, slavery and abolition. He read *The Emancipator* and William Lloyd Garrison's *Liberator*, though always reserving judgment on what was politically and practically sound. He had no doubt that the federal government, dominated by the Jackson–Van Buren forces and the Calhoun Southerners, was controlled by slaveholders, as it had always been. As the election of 1836 came closer, he lamented that Clay, his former ideological ally, was so eager for the Whig nomination that he had compromised with Southern Whigs on the tariff and made an alliance with the slaveocracy. Adams sought out the most intellectually respected attempts to defend slavery, to engage with the best salvos of the enemy, including Thomas Dew's *Review of the Debate in the Virginia Legislature of 1831 and 1832*, "a monument of the intellectual perversion produced by the existence of slavery in a free community. . . . To the mind of Mr. Dew," soon to become president of William and Mary College, "slavery is the source of all virtue in the heart of the master." Since the Declaration of Independence, many of the best minds and hearts of the South had been defensive about slavery. Some had hoped it would be eliminated. They had now gone entirely on the offensive: slavery was a positive virtue.

In Quincy in the summer of 1835, he had additional reasons to find the situation increasingly threatening.

Washington mobs rioted against blacks, provoked by an attempt by a slave to murder her mistress. There had been anti-abolitionist riots in New York and Baltimore; a number of abolitionists had been hanged in Mississippi for circulating anti-slavery literature; and when, in July, an anti-abolition mob in Charleston made a bonfire out of abolitionist literature seized from the United States mail, the governor, the legislature, and the Jackson administration decided to allow local nullification to deny mail delivery. Ironically, in Adams' view, the Democratic Party had elevated populism, race prejudice, and Southerners' fear of slave insurrection into the sanction of mob violence, the demos at work not at the ballot box but in the streets. At the same time, aggressive abolitionism, between the lines of its most strident literature, advocated slave rebellion in the name of natural law. Adams' heart, principles, and moral imagination embraced emancipation. But a "servile insurrection" would come at heavy cost in blood. Resistance even to gradual emancipation would narrow the freedoms Americans otherwise took for granted, such as freedom of the mails. Congressmen who spoke out against slavery had already been told, in no uncertain terms, that the introduction of the topic into debate in the House would be an insult that Southern honor would not tolerate.

In the mid-Atlantic states and New England, the threat of disunion struck horror into peaceful and commercial hearts. Everywhere abolitionism was blamed for the violence against abolitionists. In Boston, the leaders of amelioration, with an eye on the 1836 election, organized an anti-abolition meeting at Faneuil Hall in late August. Busy with Quincy civic business, Adams did not attend. But he probably read and would have identified with the abolitionist poet John Greenleaf Whittier's response. Must we be told our

> *Freedom stands*
> *On Slavery's dark foundations strong—*
> *On breaking hearts and fettered hands,*
> *On robbery, and crime, and wrong? . . .*
> *That Freedom's emblem is the chain?—*
> *Its life, its soul, from Slavery drawn? . . .*
> *Rail on, then, "brethren of the South"—*
> *Ye shall not hear the truth the less—*
> *No seal is on the Yankee's mouth,*
> *No fetter on the Yankee's press!*
> *From our Green Mountains to the sea,*
> *One voice shall thunder—WE ARE FREE!*

Whittier agreed with Adams' analysis: "The slave drivers will forever under all circumstances rule the union by that same engine of slave representation," the three-fifths rule of the Constitution. But how could the analysis drive a solution? "If I am now finally dismissed from public life," as Adams thought he might be when facing re-election in 1834, "I shall perhaps undertake to expose the whole system and its operation to the freemen of the north who cannot be thus made the tools and victims of the Southern machinery but by their own acquiescence and their own cooperation. It is the system of the South colleaguing with the devil; for the domination of the union. . . . If the freemen of the North will submit to it and bow their necks to be trampled upon, be the responsibility upon themselves." Gradual emancipation, Adams was convinced, depended on changing public opinion. But that was impossible, given the vested interests, the fear of slave rebellion, and the inability of anyone to conceive of a practical transition between slavery and freedom. He feared the likelihood that tensions would increase, the crisis grow worse. Ultimately, the soul of the country would be washed clean in a bath of blood. Slavery would be ended only if a "servile insurrection" in the South or secession gave an anti-slavery president the legal power under the implied military necessity provision of the Constitution to impose emancipation by force. How to avoid that and still eliminate slavery seemed the impossible challenge.

What to do in the meantime? He had no desire to initiate legislation that would create turmoil in a legislature dominated by Democrats eager to smooth the path to Van Buren's election. Many Northern Whigs, especially the business community, thought cotton more important than conscience. Since Southern Whigs usually supported the Democrats on issues critical to the South, he could expect to be not only voted down but also denied recognition to speak. The speaker, James K. Polk of Tennessee, who appointed committees and controlled access to the floor, had total commitment to Southern dominance. Unexpectedly, in the middle of December 1835, early in the first session of the Twenty-Fourth Congress, an opportunity Adams had not anticipated arose.

A Whig from Maine, John Fairfield, presented a petition from his constituents to end slavery in the District of Columbia. There had already been many such petitions. Adams had determined that he would support them only if they came from citizens of the District. The procedure that the House had followed was either to table them, acknowledging them for the record but leaving them forever unaddressed, or to send them to the appropriate committee, where they would probably also remain forever unaddressed. Sometimes the House would be requested to print the petition, which meant to place the text in the official journal. Adams had, between 1832 and 1834, submitted numerous petitions from his constituents for the abolition of slavery in the District, though he had publicly stated that he did not concur with all their views. In each case, the petitions had been received and sent to the Committee on the District of Columbia, where they went, as Adams phrased it, "to sleep the sleep of death." The House rules required only a brief oral statement of the contents. Since discussion would be useless, even counterproductive, this seemed satisfactory. It acknowledged the obligation to receive petitions, regardless of their content.

But the House now had a change of view about how to handle anti-slavery petitions. Fed up with attacks on an institution they considered constitutionally protected, Southern congressmen argued that the House should never be a forum for the subject of slavery, including petitions. South Carolina's James Henry Hammond moved that it not receive Fairfield's petition, which differed from either tabling it or sending it to a committee. No brief oral statement of the contents would be permissible since, if the speaker was careless or inattentive, a crafty congressman might expand the statement into a speech. Still, how could the subject of the petition be determined and any mention of its subject also avoided? And if the petition was about slavery, then the forbidden subject had already been introduced, and how could Southern congressmen, once the subject had been raised, resist condemning its introduction, which would then open the floor to further debate?

That is exactly what Hammond did at the beginning of February 1836. As matters stood, he and his fellow Southerners could not prevent anti-slavery remarks from accompanying petitions, and Hammond could not resist responding. "Slavery can never be abolished. . . . I believe it to be the greatest of all the great blessings which Providence has bestowed upon our glorious region." His reference to the possibility that, if abolitionists had their way, there might one day be in the presidency "some Othello . . . gifted with genius and inspired by ambition" who would "wield the destinies of this great republic," may have reminded Adams of his own view of Othello and Desdemona. "From such a picture," Hammond continued, "I turn with irrepressible disgust." The possibility of a black president may have been beyond Adams' ken. Natural law, he believed, condemned miscegenation. But it approved of political equality. Eventually slaves would have the same rights of citizenship as free blacks, and he was prescient enough to know that, over time, in a future he would not live to see, emancipation would radically change American public life.

In the meantime, his path was clear. Every attempt to deny the right of petition needed to be opposed with whatever tools were available. In 1829, in a long essay about New England federalism that he decided not to publish but which his grandson Henry edited and published in 1877 as *Documents Relating to New England Federalism*, John Quincy had quoted Otis' question, justifying the Hartford Convention, about the right of petition: "Who shall dare to set limits to its exercise, or to prescribe to us the manner in which it shall be exerted?" While Otis' logic was flawed and his inconsistency self-serving, the principle was correct. "The right of petition is well known and understood: it is a sacred right," Adams wrote. In 1836, Hammond and his colleagues responded that when two sacred rights are in conflict, the one that did most to make the country functional and without which the South could not prosper took priority. Like his anti-slavery colleagues, Adams did not agree that slavery was a sacred right. And, he sensibly insisted, whatever right it was, the South was not in any significant way deprived of it by Congress' receiving petitions about it. He hoped that the Southern firebrands would restrain themselves. In 1835 and early 1836, he urged the House to sustain the right to petition. Why not handle petitions in the same way they had always been handled? Surely common sense and the Constitution could prevail.

But, he worried, was he the right messenger for this message? "Men of proud and haughty minds," he had written many years before, "often have connected with those qualities a real fund of generosity in their tempers, and much more may be done with them by conciliating than by exasperating means." He did have friends among Southern congressmen. Opposition did not always mean enmity, though there was enmity enough against him, he believed, from every section, including New England. But when he composed in his head and his diary a list of enemies, his bitterness had a touch of paranoia. "Jean Jacques Rousseau," he told Charles, "was tortured by an imagination that the whole human race without a single exception were leagued together in a conspiracy for his destruction. I am not yet quite come to that." But he was close. "Jackson, Van Buren, Webster, White, and Harrison, by themselves, or by their ambassadors are parties to this Holy Alliance. They hunt me like a partridge upon the mountains." To a degree, he was right. There were some who found him difficult to like, and he had rivals who would do anything in their power to damage him. Some of the antipathy had deep historical roots in the conflict among John Adams and Jefferson and Hamilton. Some arose from the ignorance and intemperateness of a few House members, especially when interests such as slavery, the tariff, and Western lands were at issue. But there was also widespread respect for Adams, in deference to his age and accomplishments, to his verbal skill, persistence, and evocative presence.

It soon became clear that the South would accept nothing short of prohibiting any mention of slavery. In early February 1836, Henry Laurens Pinckney moved that any petition mentioning slavery be automatically tabled, though it was still unclear how the House would know that a petition did mention this. In May, a special committee made that a formal recommendation. "Congress possesses," it stated, "no constitutional authority to interfere, in any way, with the institution of slavery in any State. . . . Secondly . . . Congress ought not to interfere in any way with slavery in the District of Columbia. And, for the purpose of arresting agitation, and restoring tranquillity to the public mind . . . all petitions, memorials, resolutions, propositions, or papers relating in any way to the subject of slavery, or the abolition of slavery, shall, without either being printed or referred, be laid upon the table; and that no further action whatever be had upon them." An outraged Adams felt "under the necessity of making some remarks offensive to the slaveholders," he wrote to Charles, "and in no way pleasing to the northern men with southern principles." When the speaker cut off debate, Adams exploded in anger. "Am I being gagged or am I not?" The resolution, he shouted above cries for order, was "in direct violation of the Constitution of the United States, of the rules of the House, and of the rights of my constituents."

Southern and Northern Democrats, by a vote of 117 to 68, supported Speaker Polk's enforcement of the so-called gag rule. "Throughout the South," Adams noted, "the Northern man rolls the stone of Sisyphus. If he can purchase the southern votes it will be at the price of the public lands, or irredeemable negro slavery." An outraged Adams and his colleagues took every possible opportunity to speak their minds, refusing to be "gagged," an evocative word suggesting violent physical restraint. Although not all were anti-slavery in any serious way, most Whigs supported Adams.

If the House would refer all anti-slavery petitions to the relevant committee, Adams told his colleagues, he would say nothing more on the subject. But if not, he would speak at every possible opportunity about the gag itself, the "sacred right" of petition, and the right of free speech. Since the gag rule, the right of petition, and now also Texas and the admission of new states to the Union were inextricably linked by the subject of slavery, to speak about one was to speak, directly or indirectly, about the others. When, in June 1836, the House debated for twenty-five consecutive hours a statehood resolution for the pro-slavery territory of Arkansas, Adams spoke vigorously against its admission because its constitution prohibited its legislature from ever "giving freedom to the slave." How could the House attempt to prevent him from speaking about slavery when the issue was not a petition against slavery but the admission of a new state? Congress, he argued, just as it did not have the power to abolish slavery, did not have the power to admit a territory whose constitution violated the U.S. Constitution. When he finally yielded the floor, the House voted by an overwhelming margin to admit Arkansas to the union, providing the Democrats with two new senators and additional House members. Constitutional niceties, Adams realized, would get him nowhere.

In the Senate, Calhoun repeatedly failed to pass a gag resolution. Northern senators of both parties would not universally vote for it; Southern Whigs would not always support Southern Democrats. Just as Adams was in the minority in the House, Calhoun was in the Senate. Pro-Union Southerners rejected nullification, which "has the effect," James Madison wrote to Henry Clay in 1833, "of putting powder under the constitution and the union . . . and a match in the hand of every party to blow them up at pleasure. . . . The first and most obvious step is nullification, the next secession, and the last a farewell separation. . . . What *madness* in the South, to look for greater safety in disunion. It would be worse than jumping out of the Frying-pan into the fire: it would be jumping into the fire for fear of the Frying-pan." That was Adams' view as well. And nullification, the expansion of slavery, and the gag rule seemed to him inseparable threats to the Union.

When Adams got the floor in a discussion of how to help victims of Seminole attacks, he broached the topic of Congress' war powers and slavery, which he extended into an evocation of the likely wars that would result from the annexation of Texas, raising the level of his attack on Southern anti-Union recklessness to the height of a caustic Ciceronian Philippic. Mexico had abolished slavery in Texas. Could we not expect slavery to be reinstituted and Texas divided into half a dozen slave states? And would that not lead to war with Great Britain or France or another Indian war or "a servile insurrection"? Would not the federal government, then, under the war powers provision, have the military necessity and legal right to interfere with slavery in the Southern states? "The right of interference, in every way, in the case of war, appears to me so clear [in the Constitution] that I know not how it can be contested." Did his colleagues not realize that prohibiting petitions about slavery worked against their professed aim? In the drama of his rhetoric, Adams felt himself in the modern role of his Roman hero. He would speak truth to power, to the Southern Caesars who would destroy the republic. By the spring of 1837, he knew the battle would be a long one. That it would last until 1865 was beyond his or anyone's predictive powers. The first stage was the gag rule and Texas, of that he was sure. He worried, as he approached his seventieth birthday, that his strength would not hold out, that he would collapse attempting to fulfill what he believed to be his sacred duty.

When the Twenty-Fourth Congress had assembled for its second session in December 1836, the speaker was compelled to declare that the gag rule of the previous session had expired. There would have to be a new one. Without a gag in place, Adams went directly at it again. In January and February 1837, the House was in turmoil. All they asked, Adams and his colleagues persistently argued, was that the House receive all petitions and dispose of them, as it previously had done, by sending them to the appropriate committee.

In response, the Jackson–Van Buren Democrats put a new total gag in place. On February 6, 1837, Adams introduced a petition from nine Virginia women asking for the abolition of slavery in the District. The speaker ruled that it be tabled without discussion. Adams asked whether another petition he had in hand was admissible. Signed with scrawls, it purported to be from slaves. Angry screams of outraged objection came from every section of the House. Legislative hysteria erupted. Did Adams believe that slaves had the right to petition Congress? He had not, Adams responded, proposed that slaves be acknowledged to have the right to petition. He had only asked a question of the chair. It called for a ruling, not a condemnation.

He was, in fact, sticking it to the anti-petition extremists. How could the House even think of denying slaves the right to have their petitions also tabled? A frothing Waddy Thompson of Virginia responded that Adams had not only insulted the House but committed a crime. He "threatened me," Adams later wrote to his constituents, "with indictment by a grand jury of the District, as a felon and an incendiary, for words spoken in this House!" Having mastered the art of inciting his opponents to act in ways damaging to themselves, he could not have been surprised when Thompson moved that he be "called to the bar of the

House, and censured by the Speaker." Thompson had not considered that the speaker would have to rule that Adams would then have the right to defend himself. If the right was denied, a precedent would be set that might be used against any congressman. But once the accused had the floor, he could respond to every aspect of the charge against him. What Thompson and his colleagues had instituted the gag rule to prevent, they had now made certain.

On February 9, 1837, Adams held the floor for a full day. "Let that gentleman, let every member of this house, ask his own heart, with what confidence, with what boldness, with what freedom, with what firmness, he would give utterance to his opinions on this floor, if, for every word, for a mere question asked of the speaker, involving a question belonging to human freedom, to the rights of man, he was liable to be tried as a felon or an incendiary, and sent to the penitentiary!" With both a sharp scalpel and a blunt hammer, he eviscerated and bludgeoned the pro-slavery faction. He even dared to refer to the sexual exploitation of female slaves. "If it is the law of South Carolina, that the members of her legislature are held amenable to petit and grand juries, for words spoken in debate, God almighty receive my thanks that I am not a citizen of South Carolina!" The resolution of censure failed by a vote of 106 to 93. It had been a step too far. But it had also been too close for Adams not to realize in what dangerous waters he was navigating. The next day the House resolved, 162 to 18, that slaves did not have the right of petition. At his F Street home, his mailbox was flooded with more than the usual number of death threats. "Dark Terror round my spirit cling," Louisa wrote. "Protect us 'gainst the murderer's hand / . . . O hear our cry in pity Lord / For blood! for blood they lust!"

It was a liberating moment for Adams. He no longer felt restrained by comity. On February 25, 1839, he proposed a constitutional amendment "abolishing hereditary slavery in the United States, prohibiting admission of new slave states, and abolishing slavery and the slave trade in the District of Columbia." He had moved from discreet tactfulness to outspoken radicalism. On January 28, 1840, the House passed, 114 to 108, a continuing gag rule. It would no longer have to be voted on at the start of each session. It was a nasty defeat for Adams, and for William Slade and Joshua Giddings, his most outspoken allies. "The difference between the resolution[s] of the four preceding sessions of Congress and the new rule of the House is the difference between petty larceny and highway robbery." It would make it more difficult for them to submit petitions about slavery and Texas. But the vote was suggestively close.

As he walked through Mount Auburn cemetery in Cambridge in November 1838, the ground "covered with dry and scruff-colored leaves," John Quincy was startled by a moment of recognition. In front of him was a gravestone with a name he recognized: Maria Osborne Sargent, born 1804, died in 1835. His tears flowed. It was the grave of the daughter of Mary Frazier, whom he had loved forty years before, "the most beautiful and beloved" of women, he wrote in his diary. At the grave site he had felt "a mingled emotion of tenderness, of melancholy, and yet of gratitude to Heaven. . . . I imagined to myself what would have been her fate and mine, had our union been accomplished." What would his life have been like if he and Mary had married, if he had been her husband and the father of their child? Would he now be a widower grieving for a dead daughter? As he walked through the autumnal leaves, he imagined how heartbroken he would have felt if her daughter had been his own. It was a path not taken, which he visited now as both loss and gain. The moment had to have resonated with the death of two sons and the loss of his infant daughter in St. Petersburg. The cemetery path led him to a moment of thankfulness for a turn in life that had brought him "to the formation of other and more propitious ties, by which I am yet happily bound."

Inevitable deaths sometimes touched him with special force. In 1818, he had said good-bye to his mother, in 1826 to his father. But those losses had the ameliorating feeling of generational change. Their pain could be dissipated by respect for the natural order and the hope, if not promise, of meeting again in heaven. He did not feel that way about George and John II. Those deaths seemed out of the course of nature. Neither he nor Louisa could carry that weight of grief without bowing under it. Louisa bent more than he did. He had the advantage of a public mission, the hard work that helped him keep private grief under control. He felt only pain at the news in December 1837 of the death of Lieutenant Thomas Boylston Adams Jr., "the most distressing event to me which has occurred since that of my own dear son John. He was almost a son to me." And, as he told Charles, the pain included the loss of one more possible perpetuator of the Adams name. To his pleasure, Charles Francis added another heir when Henry Adams was born in February 1838, a grandson who got to know his grandfather well enough to remember him with awe and love. At the death in 1831 of Dr. Thomas Welsh, with whom Adams had lived during four of the most difficult years of his life, he was overcome with melancholy. What most saddened him was how heavily weighed down with afflictions his friend's last decades had been. In July 1834, in New York, he noted that it was "the fiftieth year" since he had been there "and what a world of life I had since then gone through." The city had had "20,000 inhabitants. It has now ten times that number but not one tenth now of those it had then—perhaps not one hundredth. Not a single one that I know." How to make sense of all this, of nature, time, and change? Would religion help? Would philosophy? "I must wait for light and hearken to human reason upon more comprehensible things."

In March 1837 he looked at the plaster cast Hiram Powers had created of him for a marble bust. He saw a representation of what he felt daily. Time and events had shaped him. He responded to the clay cast with a sonnet:

Sculptor! Thy hand has molded into form
The haggard features of a toil worn face:
And whosoever views thy work shall trace
An age of sorrow and a life of storm. . . .
Oh! Snatch the fire from Heaven, Prometheus stole,
And give a breathless block a living soul.

As he observed America changing, he also worried about its "living soul." In September 1836 he had delivered, at the request of the citizens of Boston, a eulogy of another of the great figures of the revolutionary generation, James Madison. Every such departure reminded him that almost everyone who had signed the country's founding documents, starting with the Continental Congress' address to George III in 1774, was dead or soon to be. With these deaths, the country had a sense of being on its own. Adams had reasons to think well of Madison, especially in contrast to Jefferson, whose newly published letters, he noted in his diary as he began to write the eulogy, revealed "his craft and duplicity in very glaring colors. I incline to the opinion that he was not altogether conscious of his own insincerity, and deceived himself as well as others. His success . . . seems to my imperfect vision, a slur upon the moral government of the world." It was Madison who had "moderated some of [Jefferson's] excesses" and was "in truth a greater and a far more estimable man."

Starting in July 1836, and all through August and most of September, Adams found himself writing a full-length study of the *Life and Character of James Madison*. He found justification for giving the main demerits to Jefferson, not Madison, for the doctrine of nullification in the Virginia and Kentucky Resolutions, which Madison, in his later years, rejected. "Jefferson was the father of South Carolina Nullification," Adams privately wrote, "which points directly to the dissolution of the Union. Madison shrunk from his conclusions, but I think admitted rather too many of his premises." Madison's stumbles as president and his mistakes as commander in chief Adams touched on gently. After all, this was a eulogy, and there was much positive to emphasize, especially Madison's recent pro-Union exhortations to his fellow Southerners.

Although Adams had promised himself that he would undertake no more public speeches and declined almost every one of the many invitations he received, he had become the obvious first choice in the Northeast for patriotic celebratory occasions, a trial by public demand that he thought would kill him if he accepted more than a few. But some could not be turned down, partly because he felt he still had something important to say, and also because he recognized he could bring to some occasions, like eulogizing Madison, knowledge and experience that no one else alive had. At the same time, he could speak to his father's place in American history. In reviewing the lives of Monroe and Madison, he had the opportunity to evaluate the achievements of the great men of the founding generation. To some extent, his eulogy of Madison was a defense of his father's leadership. And if Jefferson, as usual, was cast as the villain, Adams could readily emphasize all that was positive about Madison, especially his ultimate rejection of nullification. And it had been Madison who had appointed John Quincy minister to Russia and then Great Britain. Without those appointments, he would not have become secretary of state and then president. His current role in the House of Representatives might not have been possible without the name and fame of his previous service.

When he stepped to the lectern before a packed audience in Boston, he was already exhausted. Nervous, concerned that his strength would not hold out, he feared he could not get through his text, even the shortened version that still took two and a half hours to read. As on the day he had eulogized Monroe, the sky was dark with rain clouds. He had marched in the procession from the statehouse up and down State Street. While he spoke, a heavy shower fell. He could hardly see his papers. At times he had to rely on memory. The delivery, he later confessed to his diary, was "accordingly bad." The eulogy itself was a reminder to his countrymen that they had to live up to a high standard and needed to continue to look to the founders for inspiration:

What then is our duty? Is it not to preserve, to cherish, to improve the inheritance which they have left us—won by their toils—watered by their tears—saddened but fertilized by their blood? . . . You too have the solemn duty to perform, of improving the condition of your species, by improving your own. Not in the great and strong wind of a revolution, which rent the mountains and brake in pieces the rocks before the Lord—for the Lord is not in the wind—not in the earthquake of a revolutionary war, marching to the onset between the battle field and the scaffold—for the Lord is not in the earthquake—Not in the fire of civil dissension—In war between the members and the head—In nullification of the laws of the Union by the forcible resistance of one refractory State—for the Lord is not in the fire; and that fire was never kindled by your fathers! No! it is in the still small voice that succeeded the whirlwind, the earthquake and the fire. The voice that stills the raging of the waves and the tumults of the people—that spoke the words of peace—of harmony—of union. And for that voice, may you and your children's children "to the last syllable of recorded time," fix your eyes upon the memory, and listen with your ears to the life of James Madison.

The eulogy was also a poetic evocation of a generation that had passed away by a member of a generation that was now passing away. It was a testimony to time and change. For where are they now, he asked.

We look around in vain. To them this crowded theatre, full of human life, in all its stages of existence, full of the glowing exultation of youth, of the steady maturity of manhood, the sparkling eyes of beauty and the grey hairs of reverend age—all this to them is as the solitude of the sepulchre. We think of this and say, how short is human life! But then, then, we turn back our thoughts again, to the scene over which the falling curtain has but now closed upon the drama of the day. From the saddening thought that they are no more, we call for comfort upon the memory of what they were, and our hearts leap for joy, that they were our fathers.

In Quincy in late October 1840, Adams received two visitors on a mission: Lewis Tappan and Ellis Gray Loring. Neither was a stranger, and Adams could not have been surprised by the subject of the visit. A successful New York businessman, Tappan had become an abolitionist, partly driven by his religious moralism and influenced by the British abolitionist

William Wilberforce. With his brother Arthur, with whom he had founded the American Anti-Slavery Society in New York, Tappan opened his deep pockets to abolitionist causes. The brothers were radical activists in every sense, envisioning an America that blended black and white into a mulatto nation that, in its freedom from either extreme, triumphed not only over slavery but also over racism. For Adams, it was Othello and Desdemona with a vengeance, a view of what natural law required quite different from what he believed. But though he did not share Tappan's view about miscegenation, he shared Tappan's abhorrence of slavery and respected his abolitionist commitment. What Adams did not support were the kinds of anti-slavery agitation he believed did more harm than good. Neither did his other visitor, a well-known lawyer from a long-established Boston family. A close friend of Emerson, Loring had rebelled against upper-class conformity, had contrived his own expulsion from Harvard, and in 1832 had become a founding member of the New England Anti-Slavery Society. His anti-slavery venue was mainly the law and the courts. Like Adams, he believed that the natural right of the individual to personal freedom overrode man-made laws. But was there a legal venue in which the argument could be made? And if so, was it not likely that claims of natural law, despite their moral, religious, and historical foundations, would have less purchase than specific legislative acts and judicial statutes? And would there not be a tangled web of conflicting interests, even within the judiciary, that made the law a poor forum for abolitionist efforts?

What Adams and his two visitors had most on their minds was a jail in New Haven, Connecticut. In August 1839, a two-masted black schooner, the *Amistad*, had come to the end of its long voyage off the shore of northern Long Island near Montauk Point. It had started two months before in Havana, its cargo fifty-three people of African origin, designated slaves, who had mutinied against their putative owners, two Spanish slave traders, as they were being transported from slave pens in Havana to another coastal Cuban port for resale. With the complicity of the Spanish government, Spanish Cuba was infamous for a flourishing illegal trade in slaves captured in Africa and brought to Havana. On the *Amistad*, blood was shed. The captain and cook were killed. The two Spanish slave traders, Ruiz and Montes, were kept alive in order to navigate the ship toward Africa. The insurrectionists knew no Spanish, and Ruiz and Montes did not speak the language of their captors. During the day, the Spaniards sailed eastward, in circles and often into the wind, since the blacks knew nothing about navigation; at night they went westward, hoping to be intercepted by American or European vessels. Under a hot sun, with little to no food or water, the *Amistad* drifted and sailed about 1,500 miles northwestward up the coast, to the prolonged suffering of everyone on board and the increasing suspicion of the blacks. The *Amistad* was spotted by various vessels, and reports reached shore: a mysterious ship, most likely manned by pirates, was making its nefarious way along the coast. It came to the end of its voyage when Cinque, the leader of the Africans, went ashore with a small band, desperate for water and food. A few local seamen notified the authorities and claimed salvage rights. The next day, the USS *Washington*, under the command of Lieutenant Thomas Gedney, took possession of the *Amistad*. With salvage rights in mind, he tugged it to New London, informed enough to know that slavery was illegal in New York but not in Connecticut.

It soon became clear to the authorities that they had a problem on their hands, though exactly whose hands was unclear. There were status issues, property issues, legal issues, political issues, and diplomatic issues. Ruiz and Montes claimed they owned the blacks. The blacks claimed they were free Africans who had been kidnapped. Two separate parties claimed the right of salvage of the ship and men, on the assumption that the men were slaves. It would be up to the district court in New Haven to hear the case, at least for a start. Were these men murderers and pirates? Should they be treated as criminals? Were they in fact legally slaves? If they were, by Spanish law, which was also Cuban law, they could not have been brought legally to Cuba after 1820. But there were three girls among them who were too young for that. If the black men had been in Cuba since 1820, why did they not speak Spanish? Why did they speak a non-European language, which was soon identified as African? If they were Africans newly, and consequently against Spanish and international law, imported into Cuba, would not their mutiny, with its consequences, be an attempt to free themselves from kidnapping and enslavement? If so, natural law made them free men. When word of what had happened and what issues were at stake became national news, Tappan and Loring realized they had a blessed opportunity. Would Adams help them?

Chapter 16
Adhering to the World
1838–1843

On Sunday, December 2, 1838, Adams, brooding and anxious, sat in his pew in St. John's Church, a short walk from his F Street home and not far from the House of Representatives. He was aware that no matter how forceful the preacher or how hallowed the place, he could hardly hear the sermon or pay his respects to the Sabbath. The things of this world were very much in his thoughts. He was preparing to speak out in Congress again, and he feared the consequences of what he would say. He feared almost as much that he would fail in his own eyes to perform well as he feared the defeat of every political position he advocated. He had no doubt that nothing he proposed would pass, and that much that would pass this Southern-dominated

Congress would, over time, direct the nation toward disunion and ruin. The direction seemed long taken and at every session more certain. But "more than sixty years of incessant active intercourse with the world has made political movement to me as much a necessary of life as atmospheric air. This is the weakness of my nature, which I have intellect enough left to perceive, but not energy to control. . . . The world will retire from me before I shall retire from the world." If he did not continue, he doubted there was anything he could do that would interest him enough to keep him active. He tried but could not by any exercise of his will convince himself that he had "nothing better or more urgent to do than to pack up and make ready for my voyage." He insisted on urgency, on mission.

Not that everyone applauded his decision. His constituents, though, kept re-electing him, mostly with healthy margins. In March 1841, he received a letter from a stranger advising him "to retire from the world." The reason he did not, he explained, was that he could not afford it, a rationale that had an element of truth as well as of sardonic humor. His finances were being stretched by the nationwide depression that began in the first year of Van Buren's presidency. He no longer had to support Charles Francis, whose marriage, law practice, and newspaper articles made him independent.

But Adams' real estate hardly carried itself. Since his debts kept him cash poor, his congressional salary had its place in his pressured budget. He had not expected to escape his share of the "universal . . . calamity; and if my creditors must suffer by my delinquency it is some consolation to know that it is more the fault of my debtors," particularly his tenants, "than my own." He was even more worried about the depressed condition of the national economy. He had left the presidency with the nation's finances sound. Like Monroe, he had favored a balanced budget. Like every previous president, he had also favored what Republican presidents did not advocate but mostly practiced, a reasonable national debt if incurred for special expenses, such as the Louisiana Purchase and the War of 1812, and an international line of credit. As Madison and Monroe had recognized, this required a national bank and a sound banking system.

Jackson had given the highest priority to eliminating government debt entirely. As a hater of banks and paper currency, he had eliminated most financial leverage from the system, returning the United States to what seemed to many a semi-primitive economic condition in which cash was king. But starting in 1837, very few people had cash. Adams, who had many reasons to detest Jackson, blamed him for the depression, though it had probably been largely caused by international economic realities. "A divorce of Bank and State! Why a divorce of Trade and Shipping would have been as wise to carry on the business of a merchant! A divorce of Army and Fire Arms, in the face of an invading enemy. A divorce of Laws, and a bench of Judges to carry into execution the statutes of the land would be as reasonable!" Van Buren had decided to sink or swim on the anti-bank policy. "Multitudes of [state] banks must break," Adams wrote to Charles Francis, "and then the ruin will fall upon the poor and hungry, and what will be their remedy?" It would be as painful as the disease.

Although Adams recognized that Nicholas Biddle had made errors that contributed to Jackson's veto of the bank bill, Adams placed most of the blame on Jackson's paranoia. "Biddle broods," he noticed when they dined together in November 1840, "with smiling face and stifled groans over the wreck of . . . ruined hopes." He also blamed those who would benefit from government funds being deposited elsewhere, particularly the collaboration of state banks and self-interested politicians, which he thought more dangerous to the country than the minor corruption that had characterized some of the national bank's activities. In addition, Jackson's interference in the prerogative

of the House to make law about revenue seemed blatantly unconstitutional, another instance of his tyrannical tendencies. Congress voted twice to recharter
the bank, though not by large enough margins to
overcome Jackson's and then Van Buren's vetoes. State
banks suspended specie payments. "We are now in
the midst of a national bankruptcy," Adams remarked
in 1837. What would happen to the federal revenue?
Where would it be deposited and for whose benefit?
What had caused problems was the misuse of banks,
not their existence.

When, in 1838, Boston banks, short of specie, halted
payment, as if they had the right to decline to repay
depositors, Adams was outraged. In any strict sense,
the State Street banks were bankrupt. But with the
support of the state legislature, they could continue to
hold their past and present profits, relieved from financial or legal responsibility for having leveraged a small
amount of specie into large loans. "Oh! Your Boston
Banks! How I blush to think what exposures are made
from day to day, and the worst and basest of all, in my
own, my native land! . . . Alas! There is something
worse than that. It is the coldness and indifference
with which these disclosures are received! It is to see
these insolent banks demanding to be absolved from
the penalties of their own delinquencies." The country

844 · JOHN QUINCY ADAMS

needed banks, Adams argued, but the banks needed to be regulated.

Although he continued to believe that he would be happiest if he could die doing what he had always done, his political opponents did not share this view. He was an honored member of the House, up to a point. But he exceeded that regard on the one issue that counted more than any other. There was a price to pay, and it was especially high because, despite his reputation as a tough fighter who "must have sulphuric acid in his tea," as Emerson described him in 1843, he placed high value on comity and fellowship. "When they talk about his age and venerableness and nearness to the grave," Emerson remarked, "he knows better, he is like one of those old cardinals who as quick as he is chosen pope, throws away his crutches and his crookedness, and is straight as a boy." But Emerson did not see Adams whole. The aggression was less in Adams than in the situation. He believed a collegial legislature to be in the best interests of the country. Almost always a brawling ground, the House had been made even more combustible by the South's altered approach to slavery and the increased opposition to slavery by a vocal minority in the North. Adams' role in the House was additionally difficult since, behind his self-confident, argumentative exterior, he suffered from nerves, anxiety, and occasional doubt about whether his strategy was constructive. His serious troubles began, he wrote to Charles Francis, when he had asked whether slaves had the right to petition Congress. "How they have weighed upon me night and day. How they still weigh upon me, you may imagine but cannot conceive." He was a warrior, but rarely a happy one.

In February 1837, he had defended himself against the censure motion.

> Censure would be the heaviest misfortune of a long life, checkered as it has been by many and severe vicissitudes. Yes, sir, I avow, that, if a vote of censure should pass upon my name, for any act of mine in this House, it would be the heaviest of all calamities that have ever befallen me. Sir, am I guilty, have I ever been guilty, of contempt to this House? Have I not guarded the honor of this House as a cherished sentiment of my heart? Have I not respected this House as the representatives of the whole people of the whole union? Have I ever been regardless of the great representative principle of the people, here exhibited? Have I ever been wanting, as a member of this body, in a proper esprit

de corps . . . and am I now to be censured for doing
what I have not done, or for not doing what I did
not do, under pretence of a contempt of this House,
in an act which was done from motives of the highest possible respect to this House?

He also felt that he needed to defend himself to his
constituents. He was anathema to the Jacksonians. He
had only soft and localized Whig support. The anti-
Masons were still a factor, but a minor one. In the
spring of 1837, he composed a lengthy letter to the
voters of his district, printed as a pamphlet, in which
he included his February speech to Congress. "Petition
was prayer," he eloquently argued. "It was the cry of
the suffering for relief; of the oppressed for mercy. It
was what God did not disdain to receive from man,
whom he had created; and to listen to prayer—to hearken to the groan of wretchedness—was not merely a
duty; it was a privilege; it was enjoyment; it was the
exercise of a godlike attribute, indulged to the kindly
sympathies of man. I would therefore not deny the right
of petition to slaves—I would not deny it to a horse or
a dog, if they could articulate their sufferings, and I
could relieve them."

But he had not, he emphasized, presented a petition
from slaves. That issue and the right of petition almost

always, though, led to the larger issue of slavery itself. "We are told that the national government has no right to interfere with the institution of domestic slavery in the states, in any manner. What right, then, has domestic slavery to interfere in any manner with the national government? What right has slavery to interfere in the free states with the dearest institutions of their freedom? with the right of habeas corpus? with the right of trial by jury? with the freedom of the press? with the freedom of speech? with the sacred privacy of correspondence by the mail?" Slavery in the South imposed slavery's values on the North. "Would the freemen of Massachusetts, the descendents of the Pilgrims, allow themselves to be manacled? Children of Carver, and Bradford, and Winslow and Alden! the pen drops from my hand!"

But he continued to hold it firmly, to address the public and shape opinion, which he believed to be the key to the eventual implementation of his vision of a union strengthened by public works and advances in education, science, and the arts. In the House, much of the drama inhered in attempts to thwart the introduction of anti-slavery petitions. Outside, he could speak freely about the issue that underlay all others. From 1837 to 1844, he continued to oppose the gag rule, nullification, the extension of slavery to new states, and the annexation of Texas as a slave state. Nullification and territorial expansion endangered national cohesion. Behind it all was "the pestilence of exotic slavery," he told his audience in Newburyport on July 4, 1837. Those claiming that slavery was a blessing, not a curse, contaminated the national atmosphere.

> We are now told, indeed, by the learned doctors of the nullification school, that color operates as a forfeiture of the rights of human nature; that a dark skin turns a man into a chattel; that crispy hair transforms a human being into a four-footed beast. The master-priest informs you, that slavery is consecrated and sanctified by the Holy Scriptures of the old and new Testament. . . . My countrymen! These are the tenets of the modern nullification school. Can you wonder that they shrink from the light of free discussion? That they skulk from the grasp of freedom and truth?

With a gift for the melodrama of high patriotism, he merged the power of who he was with the power of his message, comparing the material and cultural progress of the human species with its moral progress, especially as manifested in the message of Christianity and the Declaration of Independence. The lessons of

Christianity were "lessons of peace, of benevolence, of meekness, of brotherly love, of charity . . . utterly incompatible with the ferocious spirit of slavery. . . . The first words uttered by the Genius of our country, in announcing his existence to the world of mankind, was,—Freedom to the slave! Liberty to the captives! Redemption! Redemption forever to the race of man, from the yoke of oppression! It is not the work of a day; it is not the labor of an age; it is not the consummation of a century, that we are assembled to commemorate. It is the emancipation of our race. It is the emancipation of man from the thralldom of man!"

In June 1838, having been mostly silenced since the censure resolution, Adams seized an opportunity to get and hold the House floor. In March, the controversial resolution to annex Texas had been referred to the Committee on Foreign Affairs. At the same time, it was clear to the leadership that it did not have the votes to annex Texas. When the committee recommended that all resolutions about Texas be laid on the table, Adams had his chance. It was a debatable motion. He could hold the floor indefinitely, as long as his voice and stamina held out, though he might be and was rebuked by the House speaker for irrelevancy. But it was impossible to deprive him of the floor if he chose to hold it, which he did from the middle of June to the end of the first week of July, when the end of the congressional session put a stop to his series of speeches on Texas, territorial expansion, slavery, and the right of petition.

He held the floor to an even larger audience when he published the *Speech of John Quincy Adams . . . upon the Right of the People, Men and Women, to Petition; On the Freedom of Speech and of Debate in the House of Representatives of the United States on the Resolutions of Seven Legislatures, and the Petitions of More Than One Hundred Thousand Petitioners Relating to the Annexation of Texas to This Union.* "My intent, my sole intent," he assured his readers, "is, by the power of truth, of justice, and of ripening public opinion," to bring Congress back "to the path of honor, of honesty, and of peace." When he had mentioned slavery to the House, he had been called to order for irrelevancy; this, he told his readers, had placed him "much in the circumstances of a company of strollers, who advertised to perform the tragedy of Hamlet, the part of Hamlet being, for this evening, omitted." Public opinion saw the absurdity of that. "I entertain the hope that another Congress is at no great distance, when we shall hear nothing of the gag, or of a resolution first to receive, and then not to consider, but to lay on the table, memorials or petitions from the People on any subject whatever. There is a reform for the People of the United States to effect. I thank Heaven there is reason to conclude they are seriously thinking of it at this time."

The annexation of Texas, he argued, almost certainly meant war with Mexico. It would mean the extension of slavery, a major increase in the power of the slave states in Congress, and an assault on Northern freedom that would rock the union to its foundations. "The People of this Union will not go to war with Mexico on the false pretence of petty spoliations, and the real impulse of a craving for Texas, and the Paradise restored of SLAVERY. If the lion roar of Jackson could not rouse them to battle for an unrighteous cause, the sucking-dove roar of his successor will scarcely serve even to frighten the ladies. War then is out of the question; negotiation must be renewed, formally—fully renewed; and it must be by diplomatic agents having neither personal interests of speculation in Texian lands, nor nullification sympathies with Texian slavery."

There were other issues on his mind. One was temperance, a nationwide movement dominated by Northern Whigs. Adams had personal reasons to counsel abstention. Still, he declined to join any temperance society, though he "heartily approved of their object, and wished them success in their cause." But he feared that his support would result in a misunderstanding. He was not against moderate drinking. All his life he had been "in the habit of drinking wine, generally with moderation, sometimes freely." The Bible did not prohibit it, and he "believed it conducive to health." The remedy was moderation, not abstinence. In 1842, he accepted an invitation to address the Norfolk County Temperance Society. With typical rigor, he provided a historical review of the subject, particularly emphasizing that the Bible had not counseled prohibition. But alcoholism, he observed, had infected all levels of society. Some congressmen stumbled to their seats barely able to rise or speak. Although he himself was not and never had been in danger, he was not "lukewarm to the cause—but I had no need of a pledge for my own person, and other duties so absorbed all my time and engrossed all my attention, that I did not feel myself called upon" when so many others were active. "To the general cause of temperance throughout the world, I would say, as I would say of all moral reform, God speed you." But please do not forget, "in the ardor of your zeal for moral reform . . . the rights of personal freedom." Legislation was not the answer. True reform came to the alcoholic only "by the dictate of his own conscience, and the energy of his own will."

Whether women should have the right to vote had not been a major concern for Adams. The exclusion of his mother from the suffrage did not seem to him in any way to diminish her civic presence. Her private influence on her son had been so powerful that it would be natural for him to wonder why any woman should seek a forum beyond the family. Louisa sometimes thought female suffrage a good idea, sometimes not. But it had less hold on the public imagination than temperance. In 1838, ten years before the Seneca Falls women's rights convention, Adams was caught off guard as he was defending his right to introduce petitions about slavery by a secondary attack: many of the petitions about the forbidden subject were from women. That insulted the purity of the female sex, outraged congressmen claimed, and the proper role of women.

But, Adams responded, the right of petition was guaranteed to all citizens. In fact, since petition was a form of prayer, was it not even more natural for women than men to bring their moral views to the issues under discussion and the attention of the body politic? Such petitions were not "discreditable" to their authors. "It is to the wives and to the daughters of my constituents that he applies this language. . . . He says that women have no right to petition Congress on political subjects. . . . Everything in which this House has an agency—everything which relates to peace and relates to war, or to any other of the great interests of society, is a political subject. Are women to have no opinions or action on subjects relating to the general welfare?" And had the gentleman forgotten "the role of patriotic women in the Revolutionary War? . . . They loved their country . . . they felt the impulse of patriotism, and manifested it in action; they entered into the hottest political controversies of the time." And why was it, he asked, that those who most vociferously defended the purity of women and the sanctity of motherhood limited their concern to white women?

In Washington, he had witnessed with special sympathy the separation of slave mothers from their children. In 1837, he recounted in his diary and then to the House an instance close to home that exemplified "the tyrannical, the hard-hearted master . . . the profligate villain who procreates children from his slaves, and then sells his own children as slaves." These were the true destroyers of families and women, not the women of Massachusetts who sent petitions to Congress. A black slave woman and her four children had been taken from her husband, a free Negro, and imprisoned in Alexandria by order of a slave dealer who claimed to have purchased them. Death seemed to her a better fate than slavery. She killed two of her children. Trying her for murder, the jury decided she was insane. When the slave dealer, claiming he owned the two remaining children, tried to take possession of them, Adams intervened. "Sir," he later told the House, "it was the verdict of an honest jury. The act was not murder. I have seen the woman and her surviving children. She attempted to kill the other two, but they were saved from her hands, and I hope are now free. I say the jury was an honest jury. They did not dare to convict her of murder, though the fact that she killed her children with her own hand was clearly demonstrated before them. The woman was asked how she could perpetrate such an act. . . . She replied, that wrong had been done to her and to them; that she was entitled to her freedom, though she had been sold to go to Georgia; and that she had sent her children to a better world."

This kind of predation had a stronger hold on his conscience than female suffrage, and, beyond any question of priority, suffrage seemed to him not only a step too far but an undesirable one—though he sometimes softened this view, particularly in his awareness of exceptional women. Still, the argument against suffrage was, in Adams' mind, strong. It had its basis in natural law and in the origin of the social compact. In 1842, he put his thoughts on the subject into a lecture and pamphlet on the social compact titled *Social Compact*

Exemplified in the Constitution of the Commonwealth of Massachusetts, with Remarks on the Theories of Divine Right of Hobbes and of Filmer, and the Counter Theories of Sidney, Locke, Montesquieu, and Rousseau, Concerning the Origin and Nature of Government. A Lecture Delivered Before the Franklin Lyceum at Providence, R.I., November 25, 1842. He based his argument on the origins of society, with its foundation in "the laws of Nature and of Nature's God." It "presupposes a permanent family compact formed by the will of the man, and the consent of the woman, and that by the same laws of Nature, and of God, in the formation of the Social Compact, the will or vote of every family must be given by its head, the husband and father." That no member of the family other than the husband and father could enter into a contract seemed to Adams incontestable. The constitutions of Massachusetts and the United States were, he observed, compatible social compacts, of an exemplary sort. For motives founded in natural law and God's will, both denied women the right to vote, and any argument in favor of changing the role of women in public affairs would need to deal with these weighty givens. For Adams, there seemed no likelihood of that and no pressing reason to pursue the subject. The preservation of the values of the Declaration and the Constitution did not require it.

But the Constitution, he believed, did need amendment. It was not, as the radical abolitionists maintained, a document of death. Its major flaws, the three-fifths rule and the fugitive slave provision, required correction, while the right to petition and other civil liberties needed protection and reaffirmation. In 1804, as a senator, he had proposed an amendment to eliminate the three-fifths provision. In 1839, he pushed much further, proposing a constitutional amendment abolishing hereditary slavery, prohibiting the admission to the union of new slave states, and ending slavery and the slave trade in the District of Columbia. In March 1839, he accepted an invitation from the New-York Historical Society to address it on the fiftieth anniversary of Washington's inauguration and the jubilee of the Constitution. Each spring he promised himself and Charles Francis to devote the summer to the long-postponed continuation of his biography of his father. "My principal apprehension is of being beset with invitations to preach politics, and literature and abolition, and temperance, and history and education, and whatever other fancy the multitudinous humors of the people may start for the fashion of the day. I must if possible be impenetrable to all such demands upon my time." But he was not. On the contrary, he felt compelled to make his voice heard. Between 1832 and 1842, he gave on average two speeches a year, every word of which he wrote and then rewrote for publication, in addition to his work in Congress. Although there are no statistics available, he probably reached more eyes and ears, either in the full text or in extracts, than any other contemporary public figure.

In Manhattan in April 1839, flanked by Peter Gerard Stuyvesant, the president of the New-York Historical Society, and by Philip Hone, the former mayor, he had the energy to address his elite audience for two hours. All eyes were on Adams, impressive for the gravitas of his appearance and his unique position in American public life. The values that should shape the present, he argued, as he had done many times before, were those inherent in the history of the nation and in its foundational documents. His major target was the doctrine of nullification, its arbitrary and willful insult to the facts and values of American history. The history he summarized was, of course, Adams' version, with its emphasis on the rationale for the revolution in specific events and natural law. Freedom was everything. Liberty and morality were the highest values. The Declaration of Independence provided the principles, the Constitution the implementation. The brilliance of the Constitution was that it was both stable and elastic—it was a living document. The Bill of Rights, he observed, had corrected a flaw. More corrections were necessary. The

Constitution mandated that the federal government "promote the general welfare." That meant new initiatives, such as public improvements, that the authors could not have anticipated but had made provision for in general terms. And its most precious gift was union. It was a sacred contract. "And now the future is all before us, and Providence our guide." In private, he was less optimistic.

An exhausted John Quincy had spent a sleepless night early in May 1840 composing a poem. It had been a difficult, draining week. He had labored day and night on the tedious details of an appropriations bill. Then debate had started on a revenue bill he had written as chairman of the Committee on Manufactures. It attempted to correct and reform the corrupt practices at the federal customhouses, especially in New York, Philadelphia, and Baltimore. Fraud was rampant. Money was being stolen. Naturally, parts of the bill faced strong opposition. In his nervousness about whether the bill would be gutted or passed at all he turned to writing verse, "the only mode of relieving myself from the continual pressure upon the brain of thought upon one subject through a sleepless night." It took his mind to another place, and the search for rhymes and rhythms soothed him, as if the meter calmed his body and mind. The subject of the twenty-five-stanza poem was "The Wants of Man," a phrase taken from a poem by Oliver Goldsmith. It had been a theme of self-examination for much of Adams' life, a subject that was both practical and philosophical. What did man want and what did he need to be happy?

> "Man wants but little here below,
> Nor wants that little long."
> 'Tis not with me exactly so;
> But 'tis so in the song.
> My wants are many and, if told,
> Would muster many a score;
> And were each wish a mint of gold,
> I still should long for more.
>   What first I want is daily bread—
> And canvas-backs,—and wine—
> And all the realms of nature spread
> Before me, when I dine.
> Four courses scarcely can provide
> My appetite to quell;
> With four choice cooks from France beside,
> To dress my dinner well. . . .
>   I want (who does not want?) a wife,—
> Affectionate and fair;
> To solace all the woes of life,

And all its joys to share.
Of temper sweet, of yielding will,
Of firm, yet placid mind,—
With all my faults to love me still
With sentiment refined. . . .

I want a warm and faithful friend,
To cheer the adverse hour,
Who ne'er to flatter will descend,
Nor bend the knee to power,—
A friend to chide me when I'm wrong,
My inmost soul to see;
And that my friendship prove as strong
For him as his for me. . . .
       I want the seals of power and place,
The ensigns of command;
Charged by the People's unbought grace
To rule my native land.
Nor crown nor sceptre would I ask
But from my country's will,
By day, by night, to ply the task
Her cup of bliss to fill. . . .

These are the Wants of mortal Man,—
I cannot want them long,
For life itself is but a span,

And earthly bliss—a song.
My last great Want—absorbing all—
Is, when beneath the sod,
And summoned to my final call,
The Mercy of my God.

Charmingly autobiographical, the poem asks the same questions Adams had asked many times before. Even when he had not asked them directly, he had provided the answers in the ongoing text of his diary, which had taught him that his mission was self-improvement, the welfare of his fellow human beings, and the development of a moral compass flexible enough to provide for human frailty and strong enough to affirm his commitment to a moral order; and that he should aim for a satisfying continuum between the attractions of a material life—wine and food, wife and family, exercise and friendship, art and literature—and the moral life. From the start, he had believed in the value of knowledge for itself and as a tool for virtuous action. He had learned that they both required hard work, as if work combined both prayer and advancement. Some of his questions were about religious belief. As a rationalist, he worked his way through or around theological disputes to the ethical core that he believed sustained religious values and transcended sectarian commitments, though he could not imagine a moral order not based on "a responsible hereafter." Without that, "right and wrong have no meaning." By temperament and, he would argue, by experience, his life consisted mostly of disappointments, the existence of "another world" "indispensable . . . to reprieve the injustice of this." For "how little is the fruit of upright intentions and unremitting toil?" Believing that he had been more sinned against than sinning, he needed to believe in a justice that transcended the limitations of this world. It was, he generalized, one of the wants of man. It sustained his own belief "in the existence of a Supreme Creator . . . of an immortal principle within myself, responsible to that Creator for my conduct upon earth, and of the divine mission of the crucified Savior, proclaiming immortal life and preaching peace on earth, good will to men, the natural equality of all mankind, and the law to love thy neighbor as oneself." But he also had moments of "involuntary and agonizing doubts," which he could "neither silence nor expel."

When Tappan and Loring visited Adams in Quincy in October 1840, they had good reason to think he would agree to help defend the *Amistad* prisoners before the Supreme Court. The case was scheduled for January 1841. As soon as the ship and its passengers had been taken into custody in Connecticut in September 1839, Loring had written for Adams' opinion about the complicated case. Ruiz and Montes had been set free. The so-called insurrectionary slaves had been imprisoned in New Haven for what was turning out to be a protracted judicial process to determine what should be done with them and the ship. The two Spaniards wanted their slaves back. Lieutenant Gedney wanted the right of salvage, which meant the ship and its merchandise, a word loosely used to include the alleged slaves. So did the Long Island sailors who had first come upon Cinque and his men. The Spanish minister in Washington demanded that the Spanish-American Treaty of 1795 be applied: the law of nations required that the "pirates and murderers" be handed over to their rightful owners or, at a minimum, returned to Havana for a Spanish court to determine their status. It was not a matter for the U.S. courts to decide but an international issue, to be adjudicated between the representatives of the two nations on the basis of existing treaties and previous practice. It was, though, too late, if it ever had been possible, for the Van Buren administration to comply with the Spanish demand. The president, who feared alienating Southern supporters if slaves were represented in the court system, faced re-election in 1840. He was eager to settle the matter quickly and quietly. The prisoners, though, were already in the hands of the courts. Since it seemed essentially a case about property rights, the federal government could not play a role, except to assure the Spanish that it would use its influence to make the result mutually satisfactory.

From the start, the abolitionist committee formed to help the *Amistad* Africans had thought it desirable to consult Adams, though not necessarily to have him join the legal defense. He had not tried a case for over thirty years. But when Tappan and Loring approached him, they had already been turned down by two well-known lawyers, Daniel Webster and Rufus Choate. Webster had no sympathy with abolitionists. He did not want to alienate Southern Whigs whose support he needed for his presidential hopes. An elite Boston lawyer, Choate claimed to be too busy. When, in October 1839, Loring had written to Adams for advice, he had good reason to assume that Adams would respond. He probably knew that Adams, on learning of the capture of the *Amistad*, had written to a prominent anti-slavery advocate, the son of John Jay, that he could not get out of his mind the "unfortunate Africans." They were on trial before the U.S. circuit court in Hartford for having seized "their own right of liberty by executing the justice of Heaven upon one pirate murderer. . . . I have seen with horror a statement by the organ of the Executive Government at Washington, that if demanded by the Spanish Government they *must* be delivered [to] slave-trading justice and mercy," which would return them to Havana for a mock trial. Adams was outraged. He could not believe that an American jury would deny them the freedom that was rightfully theirs. Justice, liberty, and natural law required that. Still, he recognized, courts did not always give these values priority.

The defense immediately requested that the circuit court issue a writ for their release. The lead attorney, Roger Baldwin, granted that if they were slaves, they had indeed committed crimes. But since, he argued, they were not, they should be freed. When the grand jury pressured Associate Supreme Court Justice Smith Thompson, who sat on the circuit court in Connecticut, a duty of U.S. Supreme Court justices until 1891, to charge or free them, he ruled that whether these were slaves or free Africans illegally imported into Cuba would have to be decided first by the district court.

Having received Loring's letter soon after Thompson's ruling, Adams confided to his diary that almost all of his time and his feelings were absorbed by the *Amistad* case. First, he sketched out a detailed narrative and chronology. Next, he highlighted the relevant facts and issues needing clarification. Baldwin had argued almost exclusively the salvage issue. But, Adams advised, the crucial issue was "the right of capture." He suggested several strategies. "If they were guilty of piracy and murder, it was the duty of the Circuit Court . . . to try and convict them. If they were not guilty it was the duty of the Court to try and discharge them, or to discharge them without trial. As persons found within both our territorial and maritime jurisdiction, accused of piracy and murder, their right to a trial was a right to which they were entitled by the universal law of nations, and the refusal to try them was the greatest wrong that could be done to them." Adams and Loring recognized, as did Baldwin, that Thompson was trying to thread a needle through complex difficulties, including pressure from the secretary of state and the U.S. attorney to rule that the *Amistad* prisoners were slaves to be returned immediately to their rightful owners or to Havana. Adams was outraged when Thompson decided that American law did not allow trying "piracies committed in foreign vessels. . . . He therefore cannot try them for piracy or murder but . . . the District Court may try [if] they are slaves or not." Since "it is doubtful whether this trial shall be held in Connecticut or in NY and it must take time to ascertain in which, they shall in the meantime be held as slaves to abide the issue. Is this compassion? is it *sympathy*? is it *justice*? But here the case now stands." The district court would meet in November. The Africans would have to remain in jail.

Racism, pro-slavery sentiment, and national politics were a toxic mix. It seemed evident to the anti-slavery movement that the circuit court would have granted a writ of release if the prisoners had been white. The movement, it was decided, should do everything it could to elicit sympathy for the *Amistad* victims; every legal maneuver to prolong the opportunity for pro-abolition publicity should be pursued; and Adams should request that the president give Congress the letters exchanged between the administration and the Spanish minister. If it were, as the Spanish claimed, a matter to be adjudicated between nations, oversight was the responsibility of Congress. When Tappan, in New York, initiated a civil suit on behalf of the Africans against Ruiz and Montes, charging them with assault and false imprisonment, they fled to Cuba. In Hartford, the defense had little difficulty proving conclusively that the prisoners were Africans who could not possibly have been brought to Cuba before 1820. In January 1840, Judge Andrew Judson ruled that since they were not slaves they should be returned to Africa. First, though, the case had to be returned to the circuit court in New Haven, the court of jurisdiction for a final review. In April, when Thompson affirmed the decision of the district court, the U.S. district attorney filed an appeal. The case would go to the Supreme Court.

In October 1840, Adams pondered the request that he change from a friend of the defense to a member of the legal team. Baldwin would argue the details. Adams would provide the broader view and closing argument. "I endeavored to excuse myself upon the plea of my age and inefficiency, of the oppressive burden of my duties . . . and of my inexperience, after a lapse of more than thirty years. . . . They urged me so much, and represented the case of those unfortunate men [as] a case of life and death, that I yielded." In November, en route to Washington, he stopped in New Haven, where he consulted with Baldwin and visited the prisoners. Studying the documents the House had forced the executive to hand over to Congress, he was stunned and appalled by a translation from the Spanish so blatantly false that he had no doubt it was a purposeful alteration by the State Department to favor the Spanish argument that the Africans were slaves. He erupted into furious but logically powerful tirades against the government's duplicity. Although he could never conclusively prove that the mistranslation was purposeful, he remained convinced that it was, and there was also now incontrovertible evidence that the administration had placed a naval vessel, the *Grampus*, in dangerously icy waters off New Haven to spirit away the *Amistad* prisoners to Cuba if the case were to be decided in its favor and before the defense could file an appeal. It seemed a blatant abuse of executive power.

By the end of the year, Van Buren was a lame-duck president, defeated in a campaign in which the *Amistad* case played no role. For the first time, the Whigs had a president, though Van Buren's administration still declined to have the case dismissed, despite or even perhaps because Adams tried to persuade them to do so. It was necessary, the administration countered, to give the Spanish claims their day in court, though they had already been adjudicated in two courts in Connecticut, in one of which a U.S. Supreme Court justice had decided against the government.

Adams prepared himself for the trial. The lives of the prisoners were at stake. If he failed, they would be returned to Cuba, most likely to be executed as murderers, at best to be sentenced to a lifetime of slave labor. How could he best defend them? And how awful the prospect of failure. One precedent particularly concerned him, the 1825 *Antelope* case, in which the ruling was that the possession of a slave on board a ship was evidence that the slave was property and which many thought supported the government's brief, though Adams was certain that on close inspection it did not. He could rely on Baldwin, who would open the argument for the defense. Then he would have as much time as he needed to argue before the court, which had recently been expanded from seven to nine members. One of them was Thompson, another Joseph Story, but five were Southerners and slave owners, including Chief Justice Roger Taney. There was no certainty, other than about Thompson, as to how the non-Southerners would rule on the legal issues. Six of the judges had been appointed by Jackson and Van Buren. And even Story, who opposed slavery, detested abolitionists. He thought them a threat more to peace than to slavery. Adams, who had a severe inflammation of his left eye and who had not yet gotten through the entire *Antelope* case record, was relieved when, on the first day of the trial in January 1841, the chief justice postponed it to February 20. Two days before the rescheduled trial was to start, Adams' coachman, his longtime employee Jeremy Leary, was fatally injured as he was leaving the Capitol courtyard when his coach horses were frightened by shots from a nearby demonstration of the newly invented Colt repeating rifle. "As soon as I heard of the disaster, I found him, in excruciating torture." He died shortly thereafter. The next day Adams, "with a heart melted in sorrow, and a mind agitated and confused," asked the court for a one-day postponement so that he could attend the funeral. The chief justice answered, "Certainly."

As he walked to the Capitol for the first day of the trial, Adams felt "bewildered." He did not yet have a frame for his argument. Attorney General Henry Gilpin would open for the government. Then Baldwin would use the rest of that day and all of the next to present the case for the defense. His argument would be essentially what had already been argued at length in the lower courts. The justices, all of whom probably had read the transcripts, could not expect new facts. Adams would need to pay close attention to Baldwin's presentation and devise his arguments to supplement it. Adams had prepared an index of the documents so that the order of his presentation would have the coherence of a chronological sequence, which would allow him to "follow some order in extemporizing on them." Although he had written out his opening remarks, most of his phrasing would have to be devised spontaneously. Entering the courtroom, he felt that his only resource was a "fervent prayer that presence of mind may not utterly fail me at the trial I am about to go through." He thought Baldwin's argument "sound and eloquent," particularly tailored to the strategy to avoid as much as possible offending Southern sentiments. While Baldwin made what seemed a "powerful and perhaps conclusive" argument on the facts, Adams fixed on aspects that would reward additional discussion. He most worried that the court would be influenced by executive pressure. Would it have the strength of mind and heart to make an independent decision? That night and the next day, he continued to prepare, anxious, "little short of agony." The day before he was scheduled to start, he confided to his diary, "the very skeleton of my argument is not yet put together." On February 24, before a crowded courtroom, the chief justice announced "that the Court was ready to hear me. . . . I had been deeply distressed and agitated till the moment when I rose and then my spirit did not sink within me."

The structure of his argument fell into place as he spoke, "perfectly simple and comprehensive," in the service of his overarching theme—that American law, embodied in the Declaration, the Constitution, and common law, placed the highest value on justice. Legal quibbles had no place in a system in which it was the "inalienable right" of each individual to enjoy the liberty that natural law and nature's God had bestowed upon him. He argued, referencing the copy of the Declaration of Independence that hung in the courtroom, that the mission of the Supreme Court, which is "a Court of JUSTICE," was to ensure that the

Amistad prisoners received the justice that American standards required. Not that the court need worry that the facts and the law were inconsistent with the liberty they deserved, Adams assured them. For four and a half hours, he reviewed and analyzed the evidence and the law, much as Baldwin had done. But he also provided a passionate voice and a focused theme. The New Haven lawyer had been restrained and tactful. Adams was forcefully outspoken, regardless and in the face of Southern and Northern sensitivity. After all, he had already been forced into that role in Congress. He embraced it before the court. For one side, the issue was slavery and politics, a conjunction to be harmonized. For Adams, the issue was slavery as a subcategory of Justice and Liberty, always in capitals, values already in harmony. Slashing through the evidence with outrage, sarcasm, and logical precision, and also with eloquent appeals to the overarching framework, he asked the question, "What is Justice and was this Justice?" As he reviewed the documents, particularly the letters between the secretary of state and the Spanish minister, he castigated the administration's political machinations and the minister's arrogance. At the end of the day, having presented about half his argument, Adams felt he had done at least well enough. But he was exhausted.

After a restless night, he awakened "encouraged and cheerful."

Unexpectedly, he was not able to conclude until five days later. On the night of February 24, 1841, Associate Justice Philip Barbour died in his sleep. The chief justice announced that the court would be in recess. On Monday, March 1, Adams resumed. For another four and a half hours he reviewed and analyzed the facts. The complicated *Antelope* case needed to be disposed of. With the clock running out, he improvised a condensed version of his argument: the case so differed from that of the *Amistad* that it could not be a governing precedent, especially since Chief Justice Marshall had declared that "no principle" had been settled by the decision. It was possible, Adams feared, that the court would now misapply the *Antelope* decision to reach a verdict favorable to the government.

Having almost exhausted himself in his two daylong efforts, he still had the energy to conclude with a powerful and personal appeal. He spoke to a hushed courtroom words of memorable eloquence about himself and America. The issues were slavery, freedom, the gag rule, the founding documents, the revolutionary past, the clotted present, and the uncertain future. Justice was the issue and the cause. Powerfully present in this summarizing moment was the theme of how quickly life passes, how short our days are even when extended to the ripeness of old age, and how fully we stand in the footsteps of those before us who sought to do their duty in the light of the American vision. Hovering over his shoulders were the ghosts of the great men in his own pantheon, Washington, Madison, Marshall, and especially John Adams, that first generation of revolutionary patriots who had created the new nation. Less visible but nevertheless there in principle were future presences, including Abraham Lincoln, who seven years later was to become Adams' colleague in the House of Representatives and twenty years later would take the oath of office from the same presiding chief justice.

May it please your Honors: On the 7th of February, 1804, now more than thirty-seven years past, my name was entered, and yet stands recorded . . . as one of the Attorneys and Counselors of this Court. Five years later . . . I appeared for the last time before this Court, in defence of the cause of justice, and of important rights. . . . Very shortly afterwards, I was called to the discharge of other duties—first in distant lands, and in later years, within our own country. . . . Little did I imagine that I should ever again be required to claim the right of appearing in the capacity of an officer of this Court; yet such has been the dictate of my destiny—and I appear again to plead the cause of justice, and now of liberty and life, in behalf of many of my fellow men. . . . I stand again, I trust for the last time, before the same court. . . . I stand before the same Court, but not before the same judges. . . . As I cast my eyes along those seats of honor and of public trust, now occupied by you, they seek in vain for one of those honored and honorable persons whose indulgence listened then to my voice. Marshall—Cushing—Chase—Washington—Johnson—Livingston—Todd—Where are they? Where is that eloquent statesman and learned lawyer who was my associate counsel in the management of that cause, Robert Goodloe Harper? Where is that brilliant luminary, so long the pride of Maryland and of the American Bar, then my opposing counsel, Luther Martin?

Where is the excellent clerk of that day, whose name has been inscribed on the shores of Africa, as a monument of his abhorrence of the African slave-trade, Elias B. Caldwell? Where is the marshal—where are the criers of the Court? Alas! Where is one of the very judges of the Court, arbiters of life and death, before whom I commenced this anxious argument, even now prematurely closed? Where

are they all? Gone! Gone! All gone! . . . I humbly
hope, and fondly trust, that they have gone to receive the rewards of blessedness on high. In taking,
then, my final leave of this Bar, and of this Honorable Court, I can only ejaculate a fervent petition to
Heaven, that every member of it may go to his final
account with as little of earthly frailty to answer for
as those illustrious dead, and that you may, every
one, after the close of a long and virtuous career in
this world, be received at the portals of the next
with the approving sentence—"Well done, good
and faithful servant; enter thou into the joy of thy
Lord."

On "tenterhooks" about what the court would decide,
Adams and his colleagues soon entered an earthly house
not of joy but of relief. On March 9, Story read the
court's decision. Speaking for the seven-to-one majority, he announced that "upon the merits of the case . . .
there does not seem to us to be any ground for doubt,
that these negroes ought to be deemed free." The documents produced by Ruiz and Montes were fraudulent;
the *Amistad* blacks had been brought illegally to Cuba;
the 1795 treaty with Spain was inapplicable since fraud
had been committed; the prisoners had never been and
were not slaves; and the *Antelope* case, Story implied,

was irrelevant. Chief Justice Taney signed the decision, as did every justice except Pennsylvania's Henry Baldwin, whose mumbled words about his grounds for dissent were not recorded. The decision was made as narrowly as possible and was based on the facts that, once established, required the court to free the prisoners. It made no comment, let alone judgment, about slavery in broad terms, and it made no reference to Adams' argument except to the extent that it dealt with the evidence about whether the prisoners were slaves. If they were, nothing else mattered. If they were not, they were free men.

Abolitionists could rejoice that they had saved the Africans from re-enslavement or death. Beyond that, though, they had at best gained moderately in the court of public opinion. Southerners, like Taney, could satisfy their distaste at the very nature of the case by segregating it from the key challenges to slavery itself. The Constitution remained their inviolable fire wall, the *Amistad* decision an outlier. Like Tappan, Loring, and Baldwin, Adams felt relief or at best tempered joy. He had not failed. Justice had been done to the *Amistad* Africans; their lives had been saved; their freedom had been sustained. But slavery as a legally protected American institution remained untouchable; the iron force of the gag rule still controlled free speech in the

House; and public opinion, even in the North, favored containment, not emancipation. What more could he do, Adams pondered, to change this? Although the result of the *Amistad* trial gave him some limited satisfaction, slavery still seemed the rock against which the ship of state would be split apart.

Peaceful change would come, Adams believed, only through education. Who was to be educated, in what, to what degree, and at whose expense were questions to be addressed. In 1835, he had an exchange with George Bancroft, the historian and founder of the naval academy at Annapolis with whom he shared a dedication to scholarship and cultural progress. A committed Democrat in both theory and political practice, Bancroft was a Jacksonian who entered politics when Van Buren appointed him customs collector of the Port of Boston, a lucrative patronage position that allowed Bancroft to employ Nathaniel Hawthorne to measure coal and salt. "When you speak of Democracy, the government of *the People*, whom do you mean by the People?" Adams asked. "In the word People, do you include women and children? In the word People South of the Mason Dickson's line, do you include the slaves? do you include the colored free?" While Bancroft did in fact favor female suffrage, at this stage in his career he supported the pro-slavery policies of Jackson and Van Buren, including new slave states and the annexation of Texas. "I heard Mr. Calhoun give as a toast *universal education*," a sharp-tongued Adams continued, "and I had it on my lips to ask him to add *skin deep* but I thought he would not understand me, nor be likely to relish my explanation, if he should call for it. I said nothing, but mused on the probable consequences of universal education extended to the People of N.C."

What, if any, were the limits to be placed on universal education? Calhoun shared the Southern assumption that slaves should not be taught to read and that it was preposterous to educate free blacks. The very phrase was a contradiction in terms. No black deserved to be free. And who deserved to be educated? And to what purpose? Adams had no doubt that education was as much a human birthright as freedom, for females as well as males, for slaves as well as free blacks. Freedom and education were inseparable. All hope for the fulfillment of his vision of America's future depended on an educated electorate whose power to shape the future would be expressed through enlightened public opinion. America and its democratic ethos depended on universal literacy, a view shared by most Whigs and many Democrats. It was the responsibility of government, in Adams' view, to sponsor education at every level; knowledge was the fundamental building block of the country's future greatness. An educated public opinion, he believed, would eventually act to implement policies of public improvement that would bind the union together. It would help to create widespread economic prosperity. Over time, it would vindicate the policies he had proposed as president, his vision of a government that supported the search for knowledge of the earth below and the sky above, an engine of material and moral progress. He felt proud that Massachusetts was leading the way, from free mandated elementary education to Harvard University. And, he argued, in a striking reinterpretation of the Puritan tradition, it had done so from the time of the first settlement.

In October 1839, Adams delivered a lecture, *A Discourse on Education*, in fulfillment of a promise he had made to his native town. If he was tired of orating, as he sometimes moaned, he could not resist appeals to his loyalty or his desire to speak out on subjects about which he had something to contribute. In an age that encouraged secular preaching of every kind, his was the gospel of education, the pursuit of knowledge, and the role of education in advancing the greatness of America. His theme in 1839 was that the governing priority of the New England settlement had been education, not religion. The religion of the Puritans depended on the literacy that allowed them to become, like the ancient Hebrews, the people of the Book. Education came first. "May I be permitted to inquire, whether Religion is not herself the child of *Education*, and whether it would not be more proper to say, that *Education* was from its first origin the governing principle of the settlement of New England, or in other words that education was the mother of New England?"

The engine, Adams argued, that powered Protestant Christianity, inseparable from the message of the gospel, was the learning process. For "he came to teach, and not to compel. His Law was a Law of Liberty. He left the human mind and human action free," conveying a message of free and active inquiry, of individual and national regeneration through liberty and personal choice, that was the inward and outward infrastructure of the American vision. When Catholic Europe failed and Protestant America faltered, "the radical cause of this deplorable inconsistency, and of all this melancholy depravity in the history of mankind [was] defective education," the education that requires obedience rather than freedom. An educated people will be a free people. "In the laws enacted by the first settlers," he told his Braintree audience, "is a complete system of instruction, based upon the principle that human life, from the cradle to the grave, is a school—That at every period of his existence, man wants a teacher, and that his pilgrimage upon earth is but a term of childhood, in which he is to be educated for the manhood of a brighter world. . . . Let us impress it indelibly on our own minds; let us impress it to the extent of our ability upon others, that education is the business of human life." And it was, consequently, the business of all levels of government.

In December 1835, Adams had begun a ten-year effort to guide the progress of an unexpected opportunity to advance knowledge in the United States. In his State of the Union message, Jackson told a startled and suspicious nation that an Englishman who had died in 1826 and of whom no one in America had ever heard had left the equivalent of $500,000 to the United States for the purpose of establishing "the Smithsonian Institution for the increase and diffusion of knowledge among men." It seemed implausible that an aristocratic Englishman with no connection to the United States would choose a nation with which his own country had fought two wars and with which it still had unsettled grievances as his residual legatee. When the primary legatee died, the United States became his heir. Congress should handle the matter, Jackson decided. This put the decision into the hands of a legislature controlled by states' rights Democrats dedicated to keeping government as small as possible. Would not a Smithsonian Institution in Washington expand the activities of the federal government? What were James Smithson's motives? Was he an idealistic but deluded republican? Was he insane? Some congressmen hoped to benefit from the disposition of such a large sum, and very few had an interest in "the increase and diffusion of knowledge."

Adams immediately realized that, if he was to have any influence, it would probably be to require that Congress keep to the terms of the bequest. The danger was that, in a practical and business-dominated country, it would be assumed that the money should be used to support a vocational teaching institution, such as an agricultural or engineering school. American interest in research and scholarship was minimal to nonexistent. Obtaining the money, which required action in London, transfer to Washington, and a congressional bill stipulating the structure and governance of a Smithsonian Institution, would take effort. Still, $500,000 was a huge sum, and the speaker, whatever he thought of Adams on other issues, found it sensible to appoint him, given his international experience and scholarly credentials, chair of a special committee for the Smithsonian bequest. In early 1836, Adams composed a report that sketched out all the information available, which he expanded into a characteristic evocation: "The attainment of knowledge is the high and exclusive privilege of man." God had bestowed upon human beings exclusively "the power and capacity of acquiring knowledge." It is the essence of human nature. It enables man "to improve his condition upon earth. . . . To furnish the means of acquiring knowledge is, therefore, the greatest benefit that can be conferred upon mankind. It prolongs life itself, and enlarges the sphere of existence." It is the engine of human happiness and prosperity. The committee unanimously approved the report.

So did both houses of Congress, which enacted legislation pledging that the money would be used according to the stipulation of Smithson's will. When $10,000 was appropriated to cover the costs required to attain the money, Richard Rush, previously the American minister in London, was assigned to obtain literal possession. At the beginning of September 1838, having returned with the glittering gift in hand, he deposited $500,000 in gold in the Philadelphia mint. Adams recommended that the money be used to establish an astronomical observatory in Washington, the equivalent of the Greenwich observatory, to conduct research and provide calculations whose diffusion would be to the benefit of the nation. Some of his political antagonists made distorted reference to his use of the phrase "lighthouses of the skies" in his 1825 inaugural address, as if an astronomical observatory were some egghead boondoggle or government power grab. Other interested parties weighed in with proposals, most advocating that the money be used to create an institution of vocational or higher education or to provide funds to rescue and strengthen existing schools. Each proposal, whether from congressmen or educational entrepreneurs, included a well-paid position for the proposer. "The private interests and sordid passions into which the fund has already fallen fill me with anxiety and apprehensions that it will be squandered upon cormorants or wasted in electioneering bribery." Van Buren gave Adams lip service but seemed indifferent. The Calhoun-led Southern contingent preferred that the money be absorbed into the general treasury or even returned.

To Adams' shock, the Van Buren–dominated Congress voted to invest almost all the Smithson money in Arkansas state bonds to make certain that the new state would support Van Buren in the election of 1840, especially if the election should go to the House. To add insult to injury, the interest rate was less than what was readily available on the open market.

Also, since the bonds provided money for infrastructure projects, which Adams favored in general but the Democrats opposed in principle, it seemed an act of supreme political hypocrisy. And was it fair, Adams asked, to provide funds for one state and nothing for all the others? Amid the hurly-burly of political fund grabbing, a dispirited Adams, with the subject weighing heavily on him, took his case to the public. He had already been to the podium enough times to feel the weight of such activities, and he had additional promises to keep. As the dates for his lectures in Quincy and Boston about the Smithson bequest came closer, he was preoccupied with the illness of his nine-year-old granddaughter Fanny, who had become one of the delights of his life. He gave the first part of a lecture on the bequest to the Quincy Lyceum on November 13, 1839, and at the Mechanic Apprentices' Library Association on November 14. He was forced to have the second part, scheduled for November 20, read by a friend. The night before, he had stayed up all night near Fanny's bedside. Her death the next day was another excruciating blow to Louisa and John Quincy. Louisa appealed to the mercy of Jesus and a heavenly reunion. John Quincy carried on.

Probably the published two-part lecture had little to no effect on the competition for the Smithson spoils.

It did, though, conveniently summarize the history of the bequest and reprint Adams' report to Congress, emphasizing his plan for a disinterested board of trustees, his preference for an astronomical observatory, and the obligation to fulfill the specific terms of the bequest. He again invoked the high mission that the bequest would fulfill if it should be devoted to creating an astronomical observatory. He was so honored by the invitation to address them, he told the mechanics' apprentices, and so respected what they represented that "can it be wondered that with the consciousness of blunted instruments and organs all decayed I should have exclaimed in the fullness of my heart—nothing shall deter me! I will address them, were it to cost me my life." The possibility that the funds would be misused drove him into an anguished passion. "Oh! my countrymen—can you think of seeing this fund wasted upon the rapacity of favorite partizans, squandered upon frivolous and visionary mountebanks, or embezzled in political electioneering, without mortification and disgust?"

For the next five years, he argued, cajoled, and maneuvered, in committee, on the floor of the House, and in private conversations, to preserve the money for an institution that would increase and diffuse knowledge. When he failed to have the funds used to create an observatory, he had a compensatory satisfaction in the successful creation of an astronomical observatory at Harvard and a small-scale government-sponsored naval observatory in Washington. He did successfully fight back efforts to use the bequest for educational institutions, which would benefit individuals rather than the public as a whole. Gradually, in the first half of the 1840s, a compromise developed. An independent board of trustees was given legal authority and a director put in place for a museum that would serve the public and the national interest. It would be devoted to collecting materials for research and the dissemination of knowledge. It was not what Adams had hoped for, but he had faith that it would turn into something estimable.

What was a promising start to the fulfillment of his vision for American science was the decision by the Cincinnati Astronomical Society to create an observatory, the first in the United States. Adams did not, in August 1843, hesitate to accept an invitation to speak at the laying of the cornerstone in November. He had had a restful and restorative summer in Quincy, his awareness of old age mitigated by how much energy he still had for reading and writing, and also by the pleasure he felt on July 4 when Charles Francis delivered the Independence Day oration in Boston, "an incident of the most intense interest to me, it being this day fifty years since I performed the same service to the town." It struck him that the last time he had been in Boston on Independence Day was in 1809, shortly before he departed for Russia. He remembered that he had walked to the highest point of Beacon Hill to watch the fireworks on the Common. Fifty years later to the day, when he had delivered an Independence Day oration in Boston, he watched the sunset from his son's Beacon Hill house. A cannonade saluted the close of the celebration. As he "saw the smoke ascending from the side of the [Bunker Hill] pyramid, the top of which was full in view, there came in forcible impulse to my memory the cannonade and the smoke, and the fire, of the 17th of June, 1775." He recognized that he was well into the time of life when much in the present resonated with the emotions of the past, and the coordinates that he could map in his memory were distinctive.

At the beginning of August, in Utica, New York, on a traveling holiday with his daughter-in-law, her father, and his ten-year-old grandson to Quebec, Buffalo, and Niagara Falls, he was the honored guest at a convocation of the students and faculty of the recently founded Utica Female Seminary. The chairman of the trustees put him into tears when he read extracts, written between 1774 and 1778, from his mother's letters to his father. "Oh, my mother! Is there anything so affecting to me as thy name? . . . My heart was too full for my head to think, and my presence of mind was gone." At Niagara, he was approached by Ormsby Mitchel, the professor of astronomy at Cincinnati College, with an invitation from the Astronomical Society. He could not suppress what he thought of as "a rash promise." His age, he felt, was against him. Every day was precious and might be his last. Any day in which he did not create an entry in his diary seemed like a day lost to the passage of time. Like "that irresistible current which hurries and dashes over the cataract of Niagara . . . so am I hurried down the stream of time, and day after day turns over the precipice and is lost." But he still had it in his power to make up arrears.

The New York holiday tour turned out to be triumphal. A private vacation became, without his anticipation, a celebratory public event. Millard Fillmore and crowds welcomed him in Buffalo. He was escorted by a torchlight parade to his overnight stay at Governor William Seward's home and was enticed to give short speeches in a dozen cities. Near Albany he visited the home of Stephen Van Rensselaer, now deceased, the man whose vote in the House had perhaps been decisive in Adams' election to the presidency. When Adams left Quincy for Cincinnati on October 25, 1843, he had the sense that he was on the last journey of his life, this time as far westward as he had ever been. Publicity had spread the word throughout the Ohio River Valley that Adams would be visiting. As he had worked through the latter part of the summer on his speech, he had declined almost all of the many invitations from Ohio, Kentucky, and Pennsylvania. His age, his health, and the pressure of time made attending impossible. He turned down Henry Clay's invitation that he visit Lexington. Loyalty and personal affection for an old but troublesome colleague pulled him toward Ashland. But his concern about time and health was genuine, and he probably did not want to create the impression that he was endorsing Clay, or anyone else, for the 1844 Whig nomination. His plan was to take the northern route, via Buffalo and Lake Erie to Cleveland, and then southward across Ohio by canal boat and stagecoach. He would return by the Ohio River into Pennsylvania, spending four days in Pittsburgh, his destination Washington, where a new congressional session would start in December.

By the time he crossed New York state, early winter covered much of the landscape. Winds howled on Lake Erie. His steamer had to take refuge from a storm on the Canadian side, though he enlivened himself and fellow passengers with conversation and a scrapbook poem in honor of the birthday of one of the young passengers. The descent from Cleveland to Columbus by canal was slow and tedious. He caught a cold along the way, with a persistent cough, fever, and chills. In Akron, as everywhere, he was greeted by dignitaries and crowds. He took advantage of being kissed on the cheek by a lovely young lady, returning the kiss on the mouth. Mischievously, he did the same to the line of ladies that followed. "Some made faces," he noted, "but none refused." On the canal boat, "I write amidst perpetual interruptions, in the presence of half a dozen strangers, who seem to think me a strange, sulky person, to spend so much time in writing." In Columbus, just after dawn, a mulatto came to see him, "the only time, he said, when he could expect to obtain access to me, to return the thanks of the colored people of this city for my exertions in defence of their rights." To many blacks he was, in the North, a heroic figure.

In Cincinnati, the night before the laying of the cornerstone, he sat up late, still writing his speech. Exhausted, anxious, with a heavy chest cold, he went to sleep with it unfinished. He was up at 4 A.M. On the way to the site, it began to rain heavily on the procession. Beneath him was a sea of mud. From the hilltop he could see "the whole plain . . . covered with an auditory of umbrellas, instead of faces." His manuscript became streaked with rain. His oration was postponed to the next day.

At 10 A.M. on November 10, indoors in the Wesleyan Methodist Chapel, he spoke for two hours. The address synthesized the message about education, knowledge, and national progress that had been one of his lifelong preoccupations. Inherent within the ideals upon which the country was founded, he told his Cincinnati audience, was the vision of an ever-progressing America. Education was the key to progress, the door that would be opened to a better future, that would make all men free. The founders "spoke of the laws of Nature, and in the name of Nature's God; and by that sacred adjuration, they pledged us, their children, to labor with united and concerted energy from the cradle to the grave, to rid the earth of all slavery."

The chains that held the mind also needed to be broken. Education was the trigger of freedom. Knowledge promoted liberty. Science, represented by this astronomical observatory, with literature and the arts, would triumph over ignorance, superstition, fear, and enslavement. "The whole soul of every citizen . . . must be devoted to improving the condition of his country and of mankind. . . . Education multiplies and sharpens all these faculties. . . . Man is a curious and inquisitive being, and the exercise of his reason, the immortal part of his nature, consists of inquiries into the relations between the effects which fall within the sphere of his observation, and their causes, which are unseen." For "among the modes of self-improvement, and social happiness, there is none so well suited to the nature of man, as the assiduous cultivation of the arts and sciences." The scientific method provided facts and tested theories. Language and the arts promoted moral and emotional well-being and universal truths. Their pursuit was the human mission.

Chapter 17
The Summit of My Ambition
1844–1848

Strong images of happy childhood moments surged within John Quincy when, in May 1843 in Weymouth, he visited, with Charles Francis and Abby, a "dangerously ill" elderly remnant whom he had known from his childhood. They were memories so strong that he wondered where they came from and why. Almost none of the people remained from the days that he had spent there as a child with his mother's family. New structures, new people. "Some of the earliest of my recollections are there, and the localities that remain unchanged bring them back with a pungency for which I scarcely know how to account . . . and a return to them seems to sweep away the interval of time and to make me again a child."

The next month he was much less the child when he inwardly writhed with anger and anguish at what seemed to him a desecration of Bunker Hill. A huge throng, many of them strangers to Boston, led by President John Tyler and inspired by Daniel Webster's oratory, gathered for the nationally publicized ritual marking the completion of the Bunker Hill monument, whose granite had been quarried in Quincy. Seventy-six-year-old John Quincy Adams declined to attend. The memory of himself as a boy, "the thundering cannon, which I heard, and the smoke of burning Charlestown, which I saw on the awful day," had become one of the touchstones of his existence. It had been powerfully formative of his view of what he would dedicate his life to. He had not fought in the war, like Joseph Warren, one of his family's heroes who had died at Bunker Hill. But he had spent his life devoted to public service in continuation and advancement of what John Adams' generation had struggled to bring into existence. Now a slave-holding president, who wanted Texas for his fellow slaveholders, a man who prevaricated and held Congress in contempt, and his Massachusetts handyman, a brilliant orator whose ambition had no moral restraint, were preening themselves on Bunker Hill, disgracing the flag of patriotism and the cause of liberty before an audience that they despised and most of whom despised them. Adams conspicuously absented himself from that evening's celebratory dinner. "What have these to do with a dinner in Faneuil Hall, but to swill like swine, and grunt about the rights of man?"

Adams had enemies who attacked him vigorously, especially volatile pro-slavery advocates in the House, whose personal invective against him occasionally rose almost to the level of violence. Neither Tyler nor Adams had any doubt about what each thought of the other's politics and values. Adams' contempt for Tyler, though, was not personal. Like most Washington functionaries, they performed the rituals of political life with civilized politeness. Defending himself in the House, Adams played offsetting dramatic roles, the cool sophisticate choosing his opportunities strategically and the outraged polemicist irrepressibly angry at injustice. Tyler's, though, was a unique presidency. It dealt a gut-wrenching blow to what had been in 1840 Adams' and the Whigs' great expectations when, benefiting from the Van Buren depression, the country elected the first Whig president, sixty-eight-year-old William Henry Harrison, and Whig majorities in both houses of Congress. To Clay's bitter disappointment, the party, having decided to take a lesson from the Jacksonians, nominated a military hero, the victor of the Battle of

Tippecanoe, a phony triumph over hostile Indians in the War of 1812. The Virginia-born, pro-slavery former governor of Indiana was marketed as a cider-drinking, log cabin populist. Clay-worshippers, even the thirty-one-year-old Illinois lawyer and Whig partisan Abraham Lincoln, worked maniacally for Harrison. Along with "Tippecanoe" came "Tyler too," a maverick Democrat nominated to attract Southern votes on the assumption that, like all vice presidents, he would be mostly invisible.

In March 1841, the same month when the Supreme Court ruled for the *Amistad* Africans, Harrison was inaugurated the ninth president. One month later he was dead, a victim of pneumonia and septicemia. Tyler was notified that he was now president. But was he? Adams and others believed that the Constitution made him the acting president, and he was soon to be dubbed "His Accidency" by an increasingly alienated Whig Party. By late 1841, it disowned the president, who had turned out to be no Whig at all. A states' rights, pro-slavery, pro-Texas, and anti-bank Southerner, he could hardly be distinguished from most Jackson Democrats. Eager to come into power, the careless Whigs had elected an ideological enemy. Twice Tyler vetoed bills to reestablish a national bank, the key plank of the Whig platform.

By the late summer of 1841, the Whigs knew what they were up against. "I see the hand of God in the blow which laid the long cherished hopes of this people in the dust," Adams wrote. "We are battling in the dark." The Whig majority, which lasted until December 1843, learned that a recalcitrant president could balk almost any legislation. Tyler learned that, with the support of Democrats, he could advance his most precious object, the annexation of Texas. He also learned that he would never be trusted by enough Democrats or Whigs to be elected president. "The most extraordinary feature in the phenomenon of Mr. Tyler's career," Adams wrote to Clay, "is not the capacity of his ambition for a genuine election to that office but the means upon which he has fallen to obtain it. Like the Kingdom of Heaven he seems to think that it can be taken only by violence. There is an absurdity in his pretensions . . . because he is without a party, which savors more of Bedlam than of the White House. . . . There is a consolation in the reflection that his term of service is drawing to a close."

In mid-September 1842, Adams addressed his constituents on the state of the nation, with the Plymouth meetinghouse crowded "almost to suffocation." These were, he noted warmly, the people he had served for twelve years, as if a special bond had been forged among them, though he worried that each new election might be his last. He was too forcefully anti-slavery for Whig moderates. His opposition to immediate emancipation alienated radical abolitionists. Although he had so far had handsome majorities, he felt he needed to report regularly to his constituents to counter misrepresentations. A famous native son, he could always command a large audience but not always its votes. His themes were what they had been for over twenty years. The state of the union was dismal, not only because it was bad but because it threatened to become worse. The president's "double dealing," he told his audience, was "urging him at once to his own ruin and to that of his country." The political and ideological structure that Tyler represented championed nullification, slavery, and the cheap sale of public lands. It controlled the balance of power in the executive, the Congress, and the Supreme Court. Tyler "is a Virginian slaveholder. All the affections of his soul are bound up in the system of supporting, spreading, and perpetuating the peculiar institutions of the South. . . . Democracy and Slavery!"

Most of his audience would have been aware of Adams' prediction that, at some point, the slaveholders would overplay their hand. In their arrogant belief in their personal and cultural superiority, they would create an extra-constitutional situation, by either a slave rebellion or secession, that would give the federal government the justification to end slavery. That, though, would require an anti-slavery or at least a pro-Union president and Congress. Harrison had barely fitted the description. But his election had been a substantial movement in the right direction. With Tyler, the course had been fully reversed.

Still, with a Whig majority in place, Adams had not given up hope of victory on an issue dear to his constituents, the right of petition. He wanted his Plymouth neighbors to know that he had come through another harrowing effort to reestablish that right in Congress. It had made headlines throughout the country. "Twice in the space of five years, I have for the single offence of persisting to assert the right of the people to petition, and the freedom of speech, and of the press, been dragged before the House in which I was your representative, as a culprit, to be censured, or expelled." If he had been abrasively militant, if his counterattacks had been seen by some as unnecessarily vigorous, he asked his constituents to keep in mind who and what he was fighting against. Between 1835 and 1839, the pro–gag rule forces at the start of the two sessions of each Congress had easily mustered the majority to rule that any petition dealing with slavery be automatically tabled. Adams had picked away at the gag-rule scab. In some circumstances, its protective shell was thin.

When, in early 1840, the House passed Rule 21, a permanent gag, the vote was close, 114 to 108 in a House composed of 126 Democrats and 116 Whigs. Public opinion in the North favored the right of petition, easily distinguishable from advocating abolition, a distinction that Adams reinforced. When, in November 1840, 144 Whigs and 102 Democrats were elected to Congress, the Whigs' hopes were high. For the first time they had a majority in both houses and a Whig president. Adams knew better. Although public opinion favored rescinding Rule 21, only a portion of the Whig majority agreed. Two attempts in 1841 to have the rule rescinded failed. Then, in January 1841, inflamed by the tactical provocations of their Massachusetts nemesis, the pro–gag rule Southerners repeated the mistake they had made in 1839. The irrepressible Thomas Gilmer of Virginia and Thomas Marshall of Kentucky, soon joined in debate by the intemperate, even unstable Henry Wise of Virginia, introduced a new resolution to censure Adams.

With the Whigs in the majority, Adams had at last been appointed to the Committee on Foreign Affairs, as chairman, though some Whigs, especially the Webster faction, had tried to maneuver to prevent it. In January 1842, Adams introduced a petition purportedly from a Georgia resident requesting that he be removed from the chairmanship because he so much favored people with dark skin that it would be impossible for him to handle impartially the contention between Mexico and the United States. Whether the petition was a practical joke, perhaps by an intemperate Southerner, or Adams' own invention, he used it to taunt, scold, and defy his pro-slavery opponents. When he introduced a petition to dissolve the union, the House erupted. Defying calls for order, he continued to speak, with the House in an uncontrollable uproar. Gilmer and Marshall introduced a resolution to censure Adams. By introducing a petition for the dissolution of the union, their resolution charged, he had committed treason and called on Congress to commit treason. Some Southerners immediately recognized that the resolution was a tactical mistake. James Underwood, a slave-holding congressman from Kentucky, urged the "gentlemen from the South to pause and reflect. . . . Beware, I pray you, how you sacrifice and make a martyr of the gentleman from Massachusetts. . . . Beware how you make a martyr to the right of petition!" And "the whole proceeding is unconstitutional . . . and will, if, carried out, be productive of bitter fruits." The abolitionist Theodore Weld, a regular visitor at the Adams home, where he helped Adams prepare for the gag rule debate, watched his performance with awe and admiration. Now, since he had the right to defend himself, Adams could hold the floor at length, which he did for five days, from February 2 to 7. Having made his point, he then suggested that the House allow a motion to table the resolution. It passed 106 to 93. An embittered Henry Wise later remarked that Adams was "the acutest, the astutest, the archest enemy of Southern slavery that ever existed." By the end of the year, the battle lines had noticeably shifted. Three times Adams offered a resolution to rescind Rule 21. By a close vote each time it was neither tabled nor put to a vote until, finally, it was tabled by a vote of 106 to 102.

Partly because of the unpopularity of Tyler, a nominal Whig, but also because of redistricting and new voters, the Democrats were back in control, 142 to 82, by December 1843. The Whigs were in shock. "Your prediction," Adams had written to Charles Francis, "that we shall never again approach so near to the restoration of the right of petition as we have done at this session, I fear will prove prophetic, at least during the short remnant of my life." But Northern congressmen of both parties were feeling increasing heat on the issue. And some Southern congressmen, who had for years voted in solidarity with their sectional colleagues, had concluded, especially during the attempt to censure Adams, that continued opposition was a losing game. With the Texas issue to the fore, it was simply not possible by any stretch of the rules or tyranny of the majority to keep slavery out of the debates. When, in the first session of the new Congress, Adams moved that a committee be appointed to consider revision of the rules, Wise declared that he would no longer participate in any debate over the gag rule and would vote for the committee. Adams assumed it was a trick. Indeed, it was, but not entirely. Adams had worn down some of the opposition. More important, the opposition now had doubts about where to take its stand, especially because Northern Democrats worried about re-election. Furthermore, there had always been some House members from both parties who had constitutional reservations, drawing a distinction between requiring petitions to be tabled and not allowing them to be received at all. In early January 1844, the committee recommended a revision that omitted Rule 21, a game-changing decision. Over the next months, attempts to reintroduce the rule failed. The House declined to adopt the committee report, but the vote was so close—88 to 87—that Adams knew the battle had reached a turning point. It was not over, but only a matter of time.

Under pressure, the House had also agreed to a committee, chaired by Adams, to report on resolutions from the Massachusetts legislature asking for a constitutional amendment to eliminate the three-fifths provision. It seemed clear that the support of Northern Democrats for this committee was a tactical maneuver. The speaker had made certain, John Quincy wrote to Charles Francis, that "the Slave power with its Whig ally" and "its auxiliary Northern democrats" were in the majority. In April, he informed Charles of a major but expected defeat. "The Treaty for the annexation of Texas" was "signed and is to be laid before the Senate." When the Whig Senate rejected it, Adams felt passionate relief. In May, he wrote to his admiring friend, William Seward, the governor of New York, that "the conflict is between *Freedom* and *Slavery*. No sophistry can disguise it. No browbeating can cowardize it. Slavery and Freedom are in the grapples of death. Where shall NY be found? The Governor has answered for her." In Congress, it was a different matter. All he had managed to get out of the committee on the Massachusetts resolutions was a request to the House that it determine the exact number of slave owners and the value of all slave property.

In May 1844, in anticipation of the end of the gag rule, he received a gift of "a beautiful ivory cane of spotless white encircled with a golden ring bearing an inscription which it would be the summit of my ambition to deserve, and surmounted with the proud Eagle of our country holding a motto, the date of which is yet to be supplied." The motto memorialized the end of the gag rule. The date was left blank. "My time on earth is short," Adams wrote to the creator of the cane, Julius Pratt. "I live in nearly hourly expectation of that summons which will call me to other scenes and other responsibilities than of this world. I accept therefore your beautiful specimen of exquisite American workmanship" to hold in trust at Pratt's request until "the date when your eagle's motto will be realized by the fact, and then to deposit the cane itself in the Patent Office, as a memorial at once of the skill of American Artificers; and the spirit of American Freedom." On December 2, 1844, Adams again submitted a resolution to rescind the gag rule. The next day, a motion to table the resolution was defeated, 81 to 104. The gag rule was gone forever. "Blessed, forever blessed, be the name of God!" he wrote in his diary. The ten-year struggle was over.

Day after day Adams was at his desk on the House floor. Weather did not deter him. Health almost never did. He had "a daily-deepening consciousness of decay in body and mind, an unquenchable thirst for repose, yet a motive for clinging to public life till the last of

my political friends shall cast me off—all this constitutes my present condition. These are my cares and
sorrows." His persistence became a part of his image.
Sitting to the left of and not far from the speaker's
chair, he held the same seat, partly by the courtesy of
his colleagues on both sides of the aisle, an object of
either grudging respect or respectful idolatry. In debate, the venom that came his way seemed to make
him stronger. He kept to his seat, often reading or
writing, for most of the hours the House was in session, one of the few congressmen who seemed always
to be there, an elderly, small, bald, somewhat fragile-
looking legislator who, when he rose to speak, transformed himself into a sharp-tongued, gesticulating
dynamo of moral passion and legislative cunning.

With the Democrats again in control of the House,
he had more reason to feel a persistent undertone of
dark anticipation as the Texas annexation became a
fact. To his distress, Clay and the Whigs also lost the
1844 election. "It has been on many accounts painful to me," he told Clay, "but on none more so as on
the dark shade which it has cast upon our prospects
of futurity," especially regarding Texas, the tariff, the
containment of slavery, the Western lands, and public
improvements. "From the day of the peace of Ghent
and the final provision made for the extinction of the

national debt, *internal improvement* was at once my conscience and my treasure. It was at once the divine law of our nature, and the inexhaustible mine of our wealth." All seemed, for the time being, lost causes. When former speaker of the House James Polk, a slave-owning Tennessean with a passion for Texas and territorial expansion, unexpectedly became president in 1845, he seamlessly advanced Tyler's Texas strategy into annexation by legislative resolution. Adams had no doubt that it was unconstitutional, particularly since it imposed American citizenship and law on all residents of Texas without their approval by ballot. And it soon became clear that Polk had his relentless eye on New Mexico, Arizona, and California, all of which Adams would happily see American if they could be obtained by consent, not conquest.

About the whole Pacific arena, he thought it essential that the United States pursue trade opportunities. He encouraged negotiations, though he discouraged the possibility of himself representing the United States in Peking. When Great Britain used force against rabid Chinese nationalism, he spoke in its defense. It was an unpopular position. The war, he maintained, was not about opium but about trade on honorable terms. Long antagonistic toward Britain, he did not find it ironic that he now defended it against unreasonable attempts by the state of Maine to prevent a final resolution of the long-standing northern boundary dispute. Although the Webster-Ashburton Treaty brought that to a satisfactory conclusion, many Americans still feared British power. Others, for self-serving reasons, exaggerated or even invented British plots against American security. When a number of skirmishes between hotheaded Americans and Canadians on the New York–Canadian border, including the sinking of a ship, threatened war or at least retaliation, Adams spoke strongly against the use of force. Eventually, cooler heads refused to be pressured into military action.

For much of the early 1840s, the Tyler administration, led by its sixth secretary of state, John Calhoun, fabricated the claim that Britain would do anything to prevent the United States from acquiring Texas. Consequently, annexation needed to be fast-tracked. The slaveocracy, Adams concluded, encouraged British-American tension for its own purposes. "All this shall not breed a war, if I can help it, and my Lecture on the China question, and my speech on the McLeod resolution [regarding tensions with Canada] have but one and the same bearing on the policy of the country." He now saw Britain, which in 1834 had abolished slavery in the West Indies, as a moral exemplar. The real hypocrites, Adams maintained, were the

American slave owners preaching the rights of man, by which they meant their right to own slaves. Still, at moments, he worried that Britain's policy regarding Texas and slavery in Texas was only tactical, perhaps at the disposal of some advantage it might gain in the ongoing negotiations over the boundary of the Oregon territory.

The foreign nations about which Adams had considerable moral remorse were the American Indian tribes who had in good faith signed treaties. To his shock and dismay, he had been appointed chairman of the Committee on Indian Affairs in 1841. He immediately begged off what would be, he believed, "a perpetual harrow upon my feelings, with a total impotence to render any useful service." The government, Adams summarized, had promised that if the Cherokee and Creeks gave up hunting for farming they could retain their lands. "Their success was their misfortune. . . . It is among the heinous sins of this nation, for which I believe God will one day bring them to judgment—but at His own time and by His own means."

In spring 1843, he addressed the Massachusetts Historical Society on the two-hundredth anniversary of the New England Confederacy. After all, the original settlers of Massachusetts had, with the founders of the Virginia colony, been the first white Americans to establish a policy for dealing with native tribes. The Virginia colony had set a dismal precedent. The New England colony, Adams had argued before, had done comparatively well, given the complex circumstances. "The whole territory of New England was thus purchased," he told his audience, "for valuable consideration by the new-comers, and the Indian title was extinguished by compact fulfilling the law of justice between man and man." At the same time, he acknowledged, there had been an "exterminating conflict of the races." Gradually, the New England tribes were reduced in size and power, their lands purchased or expropriated. By the eighteenth century, historical memory and the record had been subordinated to the desire for a more favorable view of the treatment of the Indian tribes. Adams limited his correction of the record. The degree to which he chose to disregard or was ignorant of historical fact is unclear. But he had no doubt that the moral sincerity of the settlers of New England, in their effort to establish a just accommodation with the native tribes, was in marked contrast with the recent treatment of the Creeks and Cherokee. His mind was as much on the present as the past, a point he made explicit to his Boston auditors. "We ourselves, assembled here, are yet witnessing, in silent acquiescence, a treatment of the Indian tribes cursed with our protection by the government of our national Union,—a treatment marked with perfidy as faithless, with oppression as grievous, with tyranny as inexorable, as ever presided over the conquests of Cortes or Pizarro."

To prepare his essay, he left Washington alone, eager for the solitude in Quincy that would allow him to make himself knowledgeable about the history of the New England Confederacy. Immersed in historical reading, he found himself fascinated by Roger Williams, a "polemical porcupine," an intemperate holy fool who well deserved his expulsion from the Plymouth colony. "Yet his inflexible and finally triumphant principle of universal toleration makes him a name and a praise for all future time." Williams had argued that "the King had no power to grant the lands, as they belonged to the Indians," a message that the Massachusetts settlers rejected. It was not, he argued, that the colonists did not have a legal right to the land they had been granted by royal charter. But it was also the case that the Indians had a just claim, an instance of the age-old situation of two nationalities each with a degree of legitimate claim to the same territory.

Still, there was an important difference, Adams argued, as he had done in his Plymouth speech in 1804. The white settlers as agriculturists had a more valid claim to the land than the Indians had as hunters. Why? Because God had created the earth for man to till and sow, to make settlements and civilizations, to build cities and advance knowledge. "It is not for us, therefore, to charge with injustice or cruelty towards the original inhabitants of this continent the Puritan English colonists of the seventeenth century. The transition of an extensive region of the globe from a land of hunters to a land of planters is the metamorphosis of a wilderness into a garden." Those who hunt must give way to those who plant. But since the Cherokee and the Creeks had become cultivators of the land, since they had accepted the embrace of our civilization and its values, how can we justify, Adams asked in an extended riff that has the eloquence of a prose poem,

> the expulsion of the Southern tribes, not only from their hunting-grounds, but from their own domain; from the possession of the soil acquired by their conversion, at our instance and under our persuasion, from the hunter to the agricultural state. From their planted lands, from their comfortable dwellings, from their domestic hearths, and the sepulchers of their fathers, pledged by solemn treaties to their perpetual possession, they have been expelled by the rude hand of violence, and driven, like herds

of cattle. . . . The tenant of the wilderness must be dispossessed or withdraw; the game, which furnishes at once his subsistence and the occupation of his life, must be exterminated; flocks and herds of tame animals must take the place of the beaver, the buffalo, and the deer; and the tassels of the maize, the waving grass, the bean-pole and the pea-vine, must open their ripening fruits to the sun, on ground hidden even from the face of the hunter by tangled thickets, and gnarled oaks, and enormous hemlocks in thick array, standing as if in defiance of the genial influence of the sky.

He was called from his desk to the lobby of the House in March 1842 to meet the celebrated thirty-year-old literary lion visiting from England, Charles Dickens. With modern novelists, Adams had drawn the line at Walter Scott, whose romantic temperament and characterizations he found inconsistent with reality but whom he had read with some pleasure. He preferred to stick with Sterne and Fielding, partly because they represented the world in which he had grown up, especially their insistence that the mission of art was moral. For Adams, that mission suffused and transcended all literary genres. "I believe that moral principle should be the alpha and omega of all literary composition, poetry or prose, scientific or literary, written or spoken, and emphatically of every discourse. . . . Pen should never be put to paper but for the discharge of some duty to God or man." Everything he wrote, including his diary, discharged that obligation, on the whole more to man than God, though in Adams' case God was never neglected. Although he declined to find time to read any of what Dickens had published so far (*Sketches by Boz*, *Pickwick Papers*, *Oliver Twist*, *Nicholas Nickleby*, *The Old Curiosity Shop*, and *Barnaby Rudge*), he probably would have recognized, especially if he had read *Oliver Twist*, that Dickens was motivated by a duty to man and a moral passion that was as self-conscious and intense as his own.

Dickens later wrote in *American Notes* that, a few weeks before, Adams, "an aged, grey-haired man, a lasting honour to the land that gave him birth," had defended himself in the House against a censure resolution, "having dared to assert the infamy of that traffic, which has for its accursed merchandise men and women, and their unborn children." If Adams was among those most hated in the South, his reputation in Europe, and especially in Great Britain, had risen much above the mixed view about what was still seen as a dubious or at least fragile American experiment in republican government. The British commitment to total abolition condemned the United States as an international anomaly. Adams' distinction as a diplomat and his presidency retained a presence in the European memory. But the younger generation of English liberals knew him best as the defender of the right of petition and the advocate of the *Amistad* Africans. Adams immediately discovered that Dickens shared his hatred of slavery.

At a dinner party, Dickens noted that Adams "ate, and drunk, and made a speech, and talked, and perfectly astonished me by his . . . freshness, vigor, and intellect." Louisa and John Quincy entertained the novelist and his wife at a private lunch. Louisa had the advantage of having read some of Dickens' fiction. The men had political and literary subjects in common, from slavery to Shakespeare, whom both thought the epitome of literary greatness. Before the young couple left Washington, Adams wrote an album book verse for Catherine Dickens, at her request. "Receive Lady this tardy performance of my promise, with the most cordial good wishes that your visit to this country may prove . . . a source of recollections hereafter as vivid as the pleasures which his and your presence have conferred upon me, in common with the thousands upon thousands of my countrymen, who have had the good fortune of meeting you in this hemisphere." As Adams' countrymen soon found out, Dickens had many reservations about America, especially the contradiction between the principles of its Declaration and slavery. But he had none about Adams. Adams, though, had almost nothing to say about the visiting novelist in his diary. His mind and focus were on what he feared was a gathering storm, the likelihood of a war with Mexico over Texas, and the fight against the gag rule.

At the same time, and especially over the next three years, Adams had small mishaps, occasional health concerns, and an increasing awareness of the limits of his energy. Whenever he previously had had arrears in his diary, he had followed his practice of making brief notes for later expansion. He now began to have longer periods in which he did not expand his notes into full entries and interludes as long as a month in which he made no entries at all. Not that he did not have energy for the main issues: his opposition to the gag rule and to the annexation of Texas. In fact, between 1842 and 1845, from his seventy-fifth to his seventy-eighth year, he wrote and delivered a large number of essays, speeches, and committee reports: his July 1842 *Report on the Apportionment Bill*, the August *Report of the Committee on the President's Veto of the Tariff Bill*, the September *Address of John Quincy Adams to His*

Constituents of the Twelfth Congressional District and his *Address to the Norfolk County Temperance Society,* and his November essay on the *Social Compact*; in May 1843, he addressed the Massachusetts Historical Society on *The New England Confederacy,* then in November the *Cincinnati Historical Society, on the Occasion of Laying the Cornerstone of an Astronomical Observatory*; in February 1844, he addressed the American Bible Society, and in October he addressed the Young Men's Whig Club of Boston on the likelihood that the annexation of Texas would result in a civil war; in July 1845, he published an essay on *Society and Civilization.* As always, he also wrote out every speech he delivered in the House, except when impromptu debate forced him to be spontaneous. Each of these, whatever the forum, had been composed in his trembling hand; the cramped, wavering script was an expression of his decreasing ability to put pen to paper. Sometimes Louisa, more often Mary or Mary Louisa or, in Quincy, his daughter-in-law Abby, took dictation, especially of business and courtesy letters. When he was ill in the summer of 1847, he could not write anything other than his name.

It was hard going, for emotional as well as physical reasons. In 1840, after he tripped and fell to the floor of the House—an accident that disabled him from writing—his doctor asked whether his shoulder had been dislocated before. He had had other accidents that affected his writing hand, leaving him "always unable to write fast, and for the last twenty-five years unable to write at all, as other men do, with the forefinger and thumb." But he had always overcome this disability, "considering it as the business and duty of my life to write." Ill in March 1843 and struggling with a period of religious doubt, he felt disabled for writing not so much by his hand as by his mind. He would resume the diary when "unclouded reason shall return to me." When Judge William Cranch, "my boy companion school and college mate and friend came to compare notes of decrepitude between us," John Quincy noted that William suffered even more from "the infirmity of years" than he did. But they were both, he acknowledged, fading old men. If the day arrived when he could no longer write, he would, he felt, no longer be himself.

The diary itself, the genre he had most fully embraced, was the record of a writing act that had contributed to making him who he was, the main expression of his literary self. "There has perhaps not been another individual of the human race," he wrote in October 1846, "of whose daily existence from early childhood to fourscore years has been noted down with his own hand so minutely as mine." The entry has the tone of a final summary, the imminent end of the diary. As always, he did himself less than justice. "If my intellectual powers had been such as have been sometimes committed by the Creator of man to single individuals . . . my diary would have been, next to the Holy Scriptures, the most precious and valuable book ever written by human hands, and I should have been one of the greatest benefactors of my country and of mankind." In youth, he had wanted to be another Shakespeare or Pope, a poet of the highest distinction. That had not been possible. Parental guidance, personality, and circumstance had directed him to the diary. It was not a form of supreme literary distinction. He had not been granted, he concluded, the intellectual power to transcend the limitations of the genre, to create a secular gospel about liberty and progress in an age of enlightenment, a textbook for America's future, a book that advanced justice and peace.

But he still hoped that future generations of his family and country would benefit from it, not only from a historical record but also from the moral exemplar embodied in the life of his text. The diary contained lessons that provided moral guidance, as did the best of wisdom literature. It embodied his belief "that moral principle should be the alpha and omega of all literary composition." His example, its struggles and values, contained a vision of the good life for the individual and the country, no matter how different the details of the American future would be, and a vision of the dangers inherent in the unresolved contradictions of the American situation. He wished that he could have been able to do more to solve them. "I would, by the irresistible power of genius and the irrepressible energy of will and the favor of Almighty God, have banished war and slavery from the face of the earth forever. But the conceptive power of mind was not conferred upon me by my Maker, and I have not improved the scanty portion of His gifts as I might and ought to have done." He did, though, begrudgingly recognize that he had done more than he found it easy to give himself credit for.

When, in late 1844, James Polk was elected the eleventh president, Adams was dismayed. A Southern slave owner who had served in the House, Polk had supported every retrograde policy about the gag rule, the national bank, the Western lands, territorial expansion, Texas, and slavery. A dark-horse candidate, he had become the choice of a deadlocked convention and narrowly defeated Clay, to Adams' regret. At least Clay could have been counted on for enlightened leadership on the issues about which Adams cared most.

Adams watched, with increasing foreboding, the maneuvers of the pro-Texas constituency as it tried to find a legislative or constitutional path to annexation. Was a constitutional amendment required? Adams thought so, partly because the Constitution had not provided for the purchase of territory, but mostly because annexation included the purchase of people without their consent, the forced imposition of American citizenship on foreigners. If a constitutional amendment was not forthcoming, would a majority vote of both houses be sufficient to bring Texas into the union? How could various scruples and objections, particularly the further unsettling of the slavery issue, be overcome? And how to resolve these issues in a way that sustained the faction in Texas that preferred annexation against those who wanted Texas to remain an independent republic? After Texas, in 1836, had established its independence, with arms and encouragement from the United States, it only remained to find a way to navigate through the legal complications and policy divisions for it to become the first acquisition to go directly from foreign country to statehood, without having territorial status first. At the same time, as it became clear during Tyler's administration that Texas would become a state sooner rather than later, it was also an article of conviction to the pro-Texas forces that the southwestern boundary of Texas should be the Rio Grande, not the Sabine River. Some assumed that the American reach eventually would go beyond the Rio Grande. "Texas will be only a stepping stone to all Mexico," Adams wrote to his daughter-in-law. "Canada will follow, and the whole continent of North America bids fair to become the theatre of one confederation."

As early as 1820, Clay had told Adams that if the union should break up, it would divide into three units. The territory between the Sabine and the Rio Grande "would become indispensably necessary for the Western Confederacy" because it provided a seaport and was suitable for growing coffee. That made slavery, Clay implied, inevitable in Texas. But, Adams had responded, there was no crop that necessarily needed slave labor. The need was "not in the lands but in their inhabitants. Slavery had become in the South and Southwestern Country a condition of existence," not an economic necessity. It was a state of mind. But for Jackson, Tyler, and Polk, Texas without slavery was an incomprehensible impossibility. Texas and slavery *were* an inseparable state of mind. And, for the expansionists, mostly Democrats but also some Whigs, Texas was destined to become part of the United States, whether as a slave state or not. The Jackson and Van Buren administrations had been frustrated in their efforts.

By 1840, conditions and attitudes having changed, Tyler made the acquisition of Texas his highest priority, an attitude Polk shared. Although Adams believed that eventually the United States would extend from sea to sea, he argued that territorial expansion required constitutional procedures, the consent of the new citizens, and purchase, not conquest.

Stung by his opposition and eager to strengthen their argument, pro-Texas forces revived the charge that Adams had sold out Texas in 1819 when negotiating the Adams-Onís Treaty. Some expansionists had argued that Texas was already a part of the United States, included in the Louisiana Purchase. Since there was no evidence to support that claim, the Monroe cabinet had decided that the Sabine River boundary was a reasonable compromise in the context of the larger agreement. General Jackson, in private conversations with Adams, had agreed. In the late 1830s and the first half of the 1840s, the claim, conjured out of political thin air, became an article of faith to those who desired to believe it, including Jackson. The charge did not hold in any rational forum. Adams marshaled his evidence, including relevant letters and entries in his diary. He concluded that Jackson either had forgotten or was lying, though he suspected the latter and was not surprised when documents that purportedly sustained

Jackson's views turned out to be missing. "The memory of violent men is always the slave of their passions." But Tyler's and Polk's talking point was relentlessly advanced: Texas had been, should be, and will once again be part of the United States. Despite opposition and occasional setbacks, the steamroller kept moving forward.

In 1843, Adams, immersing himself in heretofore unseen State Department documents, learned for a fact what he had before only suspected: that Jackson had been plotting all along to obtain Texas, and Tyler and then Polk, with the same passion for new slave territory, would do almost anything to make that happen. By late 1844, Adams saw clearly the handwriting on the wall. The accusations against Great Britain, he realized, were motivated by resentment of its anti-slavery policies and the fear that it would resist American expansion. Calhoun led the anti-British charge with a defiant pledge that would result, Adams feared, in a war in defense of slavery. "Shall we respond affirmatively to that pledge? No! by the God of Justice and of Mercy! No! My heart is full to overflowing; but I have no more room for words. Proceed then to celebrate and solemnize the emancipation of eight hundred thousand British slaves whose bonds have been loosened by British hands." Hope for the day "when the soil of

Texas herself shall be as free as our own." With the gag rule rescinded, Texas became his major preoccupation. Annexation "is written in the Book of Fate," he noted in February 1845. "Mexico . . . is falling to pieces, and if Texas were restored to her she could not hold it. The opposition is now confined to the mere mode of making the acquisition." He was revolted at the degradation of the Constitution. The metaphor that came spontaneously to his pen combined blood and sex. The Constitution is, he wrote, "a mere menstruous rag, and the Union is sinking into a military monarchy." In late February, Congress, urged by Tyler, passed a joint resolution admitting Texas to the union. In the Senate it passed by two votes; the House adopted the Senate resolution, 132 to 76. "I regard it as the apoplexy of the Constitution," Adams wrote. It is "the heaviest calamity that ever befell myself and my country."

For Adams, with Texas formally accepting annexation at the end of 1845, there remained two connected issues: the Oregon boundary dispute and the possibility of war with Mexico. Polk had encouraged those who had opposed annexation to believe that Texas would be balanced by the acquisition of Oregon, from which would be created several free states. In January 1845, Adams listened to Stephen Douglas rave "an hour about democracy and Anglophobia and universal empire," advocating immediate occupation of all of Oregon. To Adams, who argued that the United States had an obligation to notify Great Britain that it would no longer respect the agreement about the boundary rather than arbitrarily break it without formal notice, Polk's and Douglas' approach seemed dishonorable. The commercial convention with Britain, Adams reminded the House, was in effect a treaty. He did not believe Britain would go to war about Oregon. True, he granted, Britain's appetite for new territory was insatiable. "She is heaping conquest upon conquest. . . . She is with one hand reading a *homily* and with the other brandishing her sword." But the United States should annul the convention first. War preparations, if necessary, should follow. Since the facts, he believed, supported American ownership to the fifty-fourth parallel, he would support a war if it was necessary in order to assert American rights.

When, in February 1846, members of the House noticed that Adams was in deep conversation about Oregon with one of his bitterest enemies, a colleague remarked "that it was the meeting of Pilate and Herod." "I would wish my enemies to be transient," Adams responded, "and my friendships to be eternal"—a paraphrase from Cicero after the death of Caesar. When, in June, Polk agreed that the Oregon territory would be divided between Britain and the United States at the forty-ninth parallel, he disappointed many, including Adams, and infuriated others. Most Northerners, both Whigs and Democrats, felt betrayed. Northern Whigs assumed that the president had intended all along to settle for the lower boundary. A war with Britain over Oregon was not in his plans. A war with Mexico was. Polk cared more about Texas than Oregon, more about his Southern than his Northern constituency.

It became clear by 1846 that Polk was determined to obtain the disputed territory of about 150 miles between the Nueces River and the Rio Grande. Although he preferred purchase to conquest, he had good reason to think that a proud, anguished, and paralyzed Mexican government would never agree to sell any of its territory. In Adams' view, if war was originated by or even provoked by the United States, it would be unconstitutional. If it originated with the president, it would be a breach of the separation of powers, a precedent for presidents to start wars first and get congressional approval afterward. Executive initiative, though, was a two-sided sword. At the same time as he feared the misuse of presidential power, Adams foresaw that slavery was likely to be eliminated only by a similar use of presidential power, albeit with an important difference. If Polk provoked or started a war, he was acting unconstitutionally. If a future president used force to respond to invasion or rebellion, he would be acting legally and constitutionally. In the case of rebellion, the Constitution, Adams argued, gave the president the authority to institute military law within rebellious areas. If a state seceded from the union, the commander in chief had the legal power to render all civil law within that state void. Since slavery existed only by virtue of state law, slaves within areas occupied by the union military could and would be declared free. Military necessity trumped civil law. Now, as the possibility of a war with Mexico increased, Adams thought it desirable to make the point even more firmly and loudly, though he had little expectation, let alone optimism, that he could prevent the national apocalypse.

In 1842, tension about the border between Maine and Canada, amplified by American war hawks, had brought Great Britain and the United States to the edge of war, only to be pulled back by the sensible Webster-Ashburton treaty. Adams had spoken at length to the House in response to the pro-war fulminations of Henry Wise and Charles Ingersoll, a Democratic congressman from Pennsylvania with whom Adams frequently clashed. Be careful about what you ask for, Adams had warned. He did not think war with Britain likely on this issue, but he worried about miscalculations and irrational passions. Wise and Ingersoll, among others, had frequently threatened secession, repetitive verbal shots across the bow of the union, if the rights of the South were not respected, including the gag rule. Adams soon segued away from their bellicose fulminations about the Canadian border to the consequences of war to the South. What would happen if the South should foment an external war that might lead to a slave rebellion? Would soldiers from Massachusetts be expected to help suppress it? "Let me be told, let . . . the people of my State be told . . . that they are bound by the Constitution to a long and toilsome march under burning summer suns and a deadly Southern clime for the suppression of a servile war; that they are bound to leave their bodies to rot upon the sands of Carolina, to leave their wives widows and their children orphans," that the federal government cannot "emancipate the slaves. . . . I put it forth not as a dictate of feeling, but as a settled maxim of the laws of nations, that, in such a case, the military supersedes the civil power," that "not only the President of the United States, but the Commander of the Army, has power to order . . . universal emancipation." A war with Mexico would offer the same opportunities.

Although the Oregon boundary issue was peacefully settled in 1846, the tension with Mexico was a different matter. When, in late April, Mexican cavalry attacked an American patrol in the disputed territory, the president and the congressional war hawks had their justification to declare a state of hostility between the two countries. After two days of debate, the House appropriated funds to raise troops. Adams, who thought it a crime against Mexico, a stain on the United States, an aggressive war for territory, and the first stage in the transformation of America into an empire, was one of only fourteen members to vote against a declaration of war. It also exacerbated the battle between the North and the South for national political power. How many slave states would be created out of the new possessions? Adams asked. How many free states? To Adams' dismay, at its start the war was popular. The small cohort of Whigs who spoke out against it risked electoral defeat. "The war has never to this day," Adams wrote two years later to Albert Gallatin, "been declared by the Congress of the US, according to the Constitution. It has been recognized as existing by the Act of Mexico, in direct and notorious violation of the truth."

By the early summer of 1846, volunteer regiments were in training and on their way to the Southwest. As the country mobilized and marched into Mexico, Adams dejectedly brooded about a dishonorable war. "It is now established as an irreversible precedent that the President of the US has but to declare that war exists, with any nation upon earth, by the act of that nation's government, and the war is substantially declared. The most remarkable circumstance of these transactions is that the war thus made has been sanctioned by an overwhelming majority of both Houses of Congress. . . . It is not difficult to foresee what its ultimate issue will be to the people of Mexico, but what it will be to the people of the US is beyond my foresight, and I turn my eyes away from it."

His eyes were averted in revulsion. What was the point of self-torment? His physical health, though, including, as always, his eyesight, was manageable; his trembling right hand was activated by his own efforts and supported often by the efforts of others. Feeble, he was still active. Aware that he was not as mentally sharp as he had been, he felt sharp enough to continue to serve, though he guessed that it would not be for long. He spoke less frequently in the House, often in defense against hostile charges, especially from Southern representatives who ignorantly or with malice misrepresented what he had said in the past in order to attack him on a current issue. Over forty years of participation in diplomacy and politics gave his enemies ample opportunity to cherry-pick the record.

Each time he made a speech on any major issue, he knew he would have to undergo, once again, a trial by fire. Each time he dreaded it. And he worried that he would not be physically capable. Each time he felt exhausted, aware that it took him longer and longer to renew his energy for the next battle.

And he had only a few sources of refreshment. In the summer of 1846, in the Washington heat, two days after he had entered his eightieth year, he resumed for the last time the pleasure of bathing in the Potomac. As always, he rose with the dawn or before, "drawn by an irresistible impulse" to his familiar bathing spot on the bank of the river, protected from passersby by a high bluff. Three young men were already in the water, their clothes on the rock on which he had been used to leaving his. He laid his down on another rock, a short distance higher toward the Potomac Bridge. The atmosphere was "calm, and the sun clear." He dipped his elderly body, naked or perhaps clothed in an undergarment, into the warm water; the tide low, the bath refreshing. He heard one of the young men say, " 'There is John Quincy Adams.' " What did the young men think of the elderly former president just a few feet from them, unaccompanied and unself-conscious, as if it were the most natural thing in the world, refreshing himself in the same river in which they bathed?

In Washington, he sometimes veered toward self-torment about the situation of the country, the wrong paths and the warpath on which the nation had been set by the compromises of the Constitution. He saw no way out but bloodshed. The South would never agree to the elimination of the three-fifths rule, let alone the abolition of slavery. At Quincy, during the long summers, his mind was equally if not more often on his local world. He wandered into the past, particularly the changes from his youth to now, and into his concern about his family's future, inseparable from his sense of location. In the autumn of 1844, he had admitted that "the desolation of the season casts a gloom on my spirits." But the season was a metaphor for mutability. Walking around his garden, with the fruit trees mostly bare, the "sere red and yellow leaves" gathering in wet clods on the ground, he climbed at sunset to a height from which he could see all of Quincy, Boston in the distance, and the "shaft of the Bunker Hill monument." He remembered that this was the same day on which, eighty years before, his father and mother had married.

As he had for decades, he worried about the perpetuation of his family name, aware that he had not written the biography of his father that had been an obligation entrusted to him. What had life been like, he asked, for the people of Quincy and Braintree when his parents had married? What was it like now? What would it be like in eighty years? "The recollection of the past is pleasing and melancholy; the prospect of the future— oh, how gloomy it is! Not a soul now living who was then in the bloom of life. Not a soul now living will be here in 1924. My own term—how soon will it close!" And to whom, he asked, as he looked at the property that he and his Adams ancestors had gathered into a family patrimony, will all this belong? "Will prayer to God preserve the branches and shoots from my father's stock? What a phantasmagoria is human life!"

But the sad and the pleasurable often followed one another closely. Soon after the summer solstice in 1845, he made one of many walks to the top of the hill in Quincy, near Charles Francis' summer home, to watch the sunset. "The pleasure I take in witnessing these magnificent phenomena of physical nature never tires; it is a part of my own nature." Sunrise and sunset embodied the magnificence of "this wonderful universe." Every day he felt thankful that the new day was renewed, his own and the world's. Every evening the departure of daylight certified that behind all this was a Creator, always with a capital "C," a beneficent God responsible for eternal renewal. His own renewal was at issue, and he maintained, with fewer doubts than he had had in earlier days, that the universe was governed by a just and merciful God who, in the end, would forgive his "errors and delinquincies." His had not been an easy life, despite its blessings, so he felt, and he wanted now most of all to be patient and submissive, which did not come easily. He prayed for the well-being of those he would leave behind, particularly Louisa, Charles Francis, and his five grandchildren. At the end of 1845, he wrote "A Prayer Composed in the Sleepless Hours of Last Christmas Night to Close the Year."

Oh Lord my God! of boundless might professed;
In mercy soothe the troubles of my breast.
For all the trials I am doomed to bear
My Soul submissive to thy will prepare.
Through the long night of balmy sleep bereft
How 'ere distressed let patience still be left.
Patience with calm composure to endure
Woes which no human toil or skill can cure.
My Wife, my offspring, all whose fates depend
On man. Oh may they find in thee a friend.
What'ere of blessing is to me denied
For them O gracious God! Thyself provide
And when thy wisdom shall arrest my breath
Fit me to meet serene thy face in death.

He was not serene enough, though, to forgive his enemies. On the same day in July 1845 as he celebrated the forty-eighth anniversary of his marriage, he made a retrospective but abbreviated enemies list: "Jefferson, a hollow and treacherous friend; Jackson . . . Ingersoll, George W. Irving, Jonathan Russell, base, malignant, and lying enemies—a list to which I might, but will not, add other names." If he had, he would have included those of some of the Essex Junto, especially Pickering and perhaps Otis, and the names of Southern members of the House who had attacked him personally and viciously. But his mind, on that same day, was mostly on his marriage, on how many years had passed and how few remained. "We have enjoyed much. We have suffered not a little. Good and evil have followed us alternatively." He had "met with bitter disappointments," particularly the deaths of children, the pain of which Louisa gave words to on each yearly anniversary. "Pardon! pardon! the Sin of thy Servant," she prayed on the eighteenth anniversary of George's death, "for deserting the Children of my tenderest love . . . for mere worldly purposes; at that tender age when they most required a Mothers watchful care. . . . It was thy Will to take both my Cherished Sons from me."

The deaths of his parents, his brothers, and his sister had left John Quincy "the only member of the family of the past and the present generation surviving on this earth." The list of those who had meant most to him had already for some time seemed "a mere necrology." But, about his life in general, he had been, he believed, more successful than he deserved. He had had friends and benefactors in Washington, Madison, and Monroe. He had been given the opportunity to fulfill his and his parents' ambition that he serve his country at the highest level. It was, he confided to his diary, a life that he did not regret and about which he had no remorse. Although he could not refrain from dwelling on those who had damaged and those who had tried to damage him, he checked himself with the admonition, "But I am wandering from my wedding day."

He and Louisa had not wandered away from one another. She had reconciled herself to the reality that he would continue in public life as long as he was physically capable. His district re-elected him in November 1846, his name prominent on a Whig ticket that swept the state. That summer he had the company of Louisa and Mary in Quincy. With Charles Francis busy with his family, his law practice, and Whig politics, his father saw less of him than he desired. When, the next year, troublemakers conjured up a political rift between father and son, he wrote a testimonial that he urged Charles to publish. Privately, he brooded about how much he had hoped to accomplish and how little he had achieved. "My faculties are now declining from day to day into mere helpless impotence."

But not to the extent that he could not join his son, Josiah Quincy, and Boston officials for the ground-breaking of an aqueduct. Although tempted, he resisted the invitation to preside at a meeting to pass resolutions in support of an escaped Negro from New Orleans who had been captured in Boston and shipped back to the South, an "outrage upon the laws of the Commonwealth and upon the rights of human nature." That night he slept badly, with his throat sore, his voice almost gone, and "a redoubled agitation of nerves, and a tremor" that almost made it impossible to write. But, soon feeling better, he attended the meeting, where he was asked to be the chair by acclamation. At church, he thanked God for the blessings of his life. His preference was for the Congregational church of his ancestors. "But there is no Christian church with which I could not join in social worship." Christianity had, all in all, he believed, been a civilizing force, "checking and controlling the anti-social passions of man." The future, he predicted at an optimistic moment, would bring even more progress. "A religious principle that man has no right to take the life of man will soon accomplish the abolition of all capital punishments, and the principles of liberty are daily rendering the life of man more and more precious."

In late November 1846, his life changed drastically. Louisa, Mary, and Mary Louisa left from Quincy for Washington on the newly created Old Colony Railroad, Charles Francis accompanying them as far as Philadelphia. If they traveled straight through, they could reach the capital in about two days, transportation having sped up so rapidly that a Cunard line ship, Adams noted, had recently crossed the Atlantic in just fourteen days. That afternoon he went to Boston, intending to spend the rest of the month as a guest at his son's Mount Vernon Street home, an annual urban cap to his Quincy summers that gave him the pleasure of Abby's and his grandchildren's company. Three days later, he took a walk with his friend Dr. George Parkman to visit the new building of the Harvard University medical college. Suddenly he found himself unable to walk. His knees sank under him. With Parkman's help, he staggered back to his son's house. Although he was in no pain, he was partly paralyzed, "with a suspension of bodily powers . . . and little exercise of intellect." Over the next weeks, against all likelihood, he began to show small signs of recovery, so that by New Year's Day 1847 he was able to take a carriage ride with his daughter-in-law. After a ten-week stay in his son's home, he was well enough to return to Washington. When, in mid-February 1847, he took his seat in the House, his colleagues gave him a standing ovation. If his voice were "more powerful," he told them, he would thank them for their warm reception. "But enfeebled as I am by disease, I beg that you will excuse me."

That he had partly recovered from the stroke did not fool either him or Louisa. When, in March, he was again able to write in his diary, aware that there would be few new entries, he headed the paragraph, "Posthumous Memoir." It was a brief account of his illness, which he referred to as "a paralytic complaint." "From that hour I date my decease, and consider myself, for every useful purpose to myself or to my fellow-creatures, dead." That he could serve no further purpose at all was an exaggeration. Although feeble, he was totally competent. Although he hated not being fully himself, he was enough himself to attend to limited congressional business and some of his life's loose ends. He had Louisa bring him his will, the contents and even the whereabouts of which had been unknown to her. He desired "that I should *remember* where it was deposited. . . . I here solemnly declare that I . . . firmly and faithfully believe that my beloved husband has remembered with kindness and perfect justice all his Family; measuring by his love! not their deserts or *mine*," Louisa wrote. To Charles Francis, who had taken dictation for the will, he gave the management of all business matters. "I have been disabled, both in body and mind, to fulfill the obligations, connected with the duties of social life. Upon you therefore rests many of my responsibilities, for the remnant of my days and which will continue after my decease." His brother-in-law continued to manage the F Street house and other Washington properties.

Louisa watched, waited, and worried. "Spare O Lord," she prayed, "that great mind thy bounteous gift; and enable him to endure with patience and submission the trial which in thy wisdom thou hast imposed on him. . . . Strengthen him to obedience; fortify his spirit in the faith, and graciously encourage him to struggle against the worldly passions, which war with his Soul." He may have confided to Louisa what those worldly passions were. But he could no longer confide them to his diary. Most likely the "struggle" that Louisa posited was a generic one, the challenge to go through the process of dying with courage, patience, and serenity. He had a great deal of courage. He had less of the other two virtues, as he and Louisa well knew.

He had enough strength and willpower to remain active during the spring of 1847. He attended the laying of the cornerstone of the Smithsonian building and walked part of the way home, where he received visitors, including the newly appointed Smithsonian secretary, with whom he had a long conversation, his interest in the future of that institution as alert as ever. He attended church services regularly. When, in May, the world received news of the discovery of the planet Neptune, he felt his usual excitement about all things astronomical. That summer and fall he was well enough to visit the new Harvard observatory. Still, "I am unable to put on my own clothes," he tersely noted, the partial paralysis limiting his arm movements. But he was able to walk, sit, lie down, and speak. He may have seemed to observers no more than a weak elderly man, but he was one who still could perform most of the activities of daily life, which included taking his seat in the House. Unable to project his voice, he now expressed himself almost exclusively by his vote. He remained, though, a presence, especially in regard to Polk and the Mexican War. "I have got beyond the age, the power, or the desire of controversy," he wrote. But that was not entirely true. He was still furiously, though selectively, outspoken about the crimes of the Polk administration and its war for new slave territory.

In the summer of 1847 in Quincy his nights alternated between agitated restlessness and restorative sleep, sometimes with the help of paregoric. When his doctor tried a new therapy, a series of galvanic battery shocks, probably to stimulate muscles or calm his nerves, he had no faith in the treatment. With the help of Mary Louisa, now almost twenty, he resumed his diary, making brief entries in his wavering script. He was able to take short walks and longer carriage rides, to receive visitors, and to attend services at the Stone Temple, the now twenty-year-old church building that had been his father's gift to the town. This "can hardly be called life but 'tis the destiny ordained for me, and at which I ought not to repine. . . . That I shall ever be better, I have scarce any reason to expect. . . . It disqualifies me for all business." In July, he celebrated his eightieth birthday. While he and Louisa had a due appreciation of the benefits they had received, they had only a restrained hope that he could ever be useful again. "O God in thy great mercy," she wrote in her birthday prayer, "raise him up from the sickness which now so heavily oppresses him; and grant him strength to enable him to resist the encroachments of desease, that he may yet exert himself for the benefit of mankind and be the protector of his helpless family." What life would be like for her after his death concerned them. "I shudder to think what is yet in store for me," she confessed, "when I watch the sad change in my loved companion."

Late in July, they celebrated the fiftieth anniversary of their marriage, a few days before Louisa's seventy-second birthday. It "passed most quietly according to our own desire." He gave her "an elegant Bracelet as a memento of the occasion." On that day he left another page blank in his diary, a form of subtraction. "I am almost totally physically disabled from writing," he managed to write, "and find it an insupportable task to be called upon for an autograph." By late summer, he had almost given up his morning walks; his main pleasure was the reading to which he gave most of his time, his mind still capable of this life-long passion. He read Herodotus and the *Vestiges of the Natural History of Creation*, ancient history and modern science. Swedenborg, a biography of whom he read, also interested him. This was the kind of mysticism that he had always found perplexing. He thought Swedenborg's spirits indebted to the celestial voices in Milton's *Paradise Lost*. "The existence of intelligent spiritual beings I cannot deny or affirm as part of the Christian religion. It is a subject of belief but not of proof. . . . I never have heard 'celestial voices singing to the midnight air' nor do I believe that any other human being has heard them." He held to his conviction that the purpose of religion was ethics, not ethereality.

He was also thinking about writers who had given him pleasure, particularly Samuel Rogers, whose poems he had enjoyed more than those of Robert Burns and Robert Southey. But Rogers was almost completely forgotten, "cast into the shade by Scott, Byron, and Shelley," an object lesson in transience and changing tastes. "Coleridge and Wordsworth," he noted, "are hardly to be included in the catalogue." Hovering behind the comment may have been his concern about how he and his father would be remembered. At the beginning of November, as he was starting for Washington, it seemed to him "on leaving home as if it were upon my last great journey."

In Washington, Louisa took "a bad fall" and hurt her face. It "almost crazed me," John Quincy wrote. She was soon deep into the misery of her old illnesses. She feared she might die soon. His main complaint was weakness. He could barely walk the short distance from the Capitol to his F Street home. On New Year's Day 1848 he was well enough to receive the usual visitors but not to brave inclement weather to visit, as was his annual custom, Dolley Madison. An invitation to speak at the laying of the cornerstone of the Washington Monument was an honor he regretfully declined. The spirit was willing but the body was not. On December 7, 1847, when the Whig-dominated House had met to organize itself, he had been given the honor of swearing in the speaker. In the drawing of lots for House seats, he drew a low number. But every member who chose before him declined to select the prominent seat he had occupied for the last ten years. In deference to his health, he was appointed chairman of the Joint Library Committee, a light assignment. But his heart and mind, if not his voice, were still insistent upon doing his duty. No one suggested resignation. He was in his seat every day, his presence noted by his votes but also by his venerability, that word used so frequently that it expressed astonishment as well as respect, the sense that the House had in it a living embodiment of history. Since he was not capable of being active in debate, those who had been abrasively hostile before subsided into respectful neutrality, as if his presence were a lesson in mortality rather than political differences.

Given the Whig majority, Adams' views were less singular than they had been before. There was now a majority, or close to it, expressing what he would have said if he could speak, especially on the two issues that dominated: internal improvements and the Mexican War. When Congress voted for internal improvements, he hoped it would herald the triumph of his vision of a unified nation. Polk vetoed the bill, claiming it required a constitutional amendment. Most of the slave-holding class wanted no improvements at all if they were to be sponsored by the federal government. With the repeal of the gag rule, some of the explosive edge had been taken off the slavery issue, though abolitionists in and out of Congress continued their efforts.

When and how to end the war dominated national discussion. Public opinion had shifted, now favoring a speedy end, including a reasonable settlement with Mexico, though vigorous disagreements persisted as to how much territory, if any, the United States would gain, and how much indemnity, if any, Mexico would be forced to pay. Polk favored a heavy indemnity, though he wanted territory more than cash. Since Mexico had no means to pay any indemnity at all, it seemed to Adams an excuse to keep the war going for other purposes. When, in mid-January 1848, the House debated Polk's claim that this was a just war that the United States had not started, Adams denounced the president's refusal to provide documents about the start of the war that the House had requested. He was in his

House seat when, in January 1848, Abraham Lincoln gave a blistering speech blasting Polk, and both men voted the same way on every resolution on the war, slavery, and internal improvements. There is no record that proves that they spoke to one another. But Lincoln certainly took notice of a man he could not but deeply admire, and Adams noticed everything.

In his shaky handwriting, legible but tremulous, he wrote a brief poem in his diary on February 20, 1848, his last entry. Every page had been blank since early January, as if the diary and the life were separating. A verse of four lines, it was as inconsequential as most of the poems he inscribed in keepsake or autograph books. But its sentiment and timing give it an autobiographical resonance:

> Fair Lady, thou of human life
> Hast yet but little seen.
> Thy days of sorrow and of strife
> Are few and far between.

He had seen a great deal. And his days of strife and sorrow had been many. But the strife had been on behalf of deeply held ideals about his own and his nation's moral life, about justice and the American future. For decades he had been anticipating a better future, a unified nation without slavery, though he believed it would come only after a dark and bloody passage. All he asked for himself is that when he came, in his mind or in some future state, before his nation and his God, he would be judged by the values by which he had lived. Ten years before, in a firmer handwriting, he had written,

And when thy summons calls me to thy bar
Be this my plea thy gracious smile to draw
That all my ways to justice were inclined
And all my aims the good of human kind.

It was not justice nor for "the good of human kind," he decided the next day, to vote for a motion requesting unanimous consent to suspend the rules to pass a resolution thanking the American officers for their "gallantry and military skill" at the battles of Vera Cruz and Cerro Gordo. Eight gold medals were to be struck and presented to the generals. The motion passed, 110 to 54, the resolution 98 to 86, Adams and Lincoln voting against this triumphal declaration, maneuvered by the Polk administration. It boasted of American military prowess without any recognition that the victories had cost the lives of thousands of Americans and Mexicans in an immoral and unconstitutional war. As the speaker began the formality of engrossing the resolution, there was a cry and commotion at Adams' desk. "A sudden cry was heard," the *National Intelligencer* reported. "Mr. ADAMS is dying!"

Everyone's eyes went immediately to the sight of John Quincy, "falling over the left arm of his chair, while his right arm was extended, grasping his desk for support." As he slipped toward the floor, he was "caught in the arms of the member sitting next to him. . . . Members rushed from their seats and gathered around. He was carried to the front of the House, and placed on a sofa, which was carried out into the rotunda." In the Senate, Thomas Benton, who had become an admirer, made the "painful announcement . . . that Mr. Adams has sunk down in his chair, and has been carried into an enjoining room, and may be at this moment passing from the earth, under the roof that covers us, and almost in our presence."

He was still breathing. He seemed aware, but he could not move. Or speak. When a crowd circled the sofa, he was carried on it to the door of the east portico, where there was a fresh breeze but also winter chill and fog. The doctor had him moved into the speaker's office. Some members of the House ran to tell Louisa at the F Street house to come quickly. Adams became articulate for one last time. "This is the end of earth," he said to the tearful colleagues gathered around him. "I am composed." They were his only words. Rushing to the Capitol, Louisa was too late to hear them. A group of doctors made every effort to revive him. "But neither the skill of his physicians, nor the kindness of his friends, nor the prayers and tears of his afflicted family, could avert the stroke of death," the *National Intelligencer* solemnly pronounced.

At a little past seven in the evening the next day, February 23, 1848, he died where he had been taken, still in the People's House. There, he lay in state. Washington soon provided its highest honors. Everyone of consequence attended the funeral. "The whole ceremony was inconceivably impressive," Benton later wrote. "The two Houses of Congress were filled to their utmost capacity . . . the President, his cabinet, foreign ministers, judges of the Supreme Court, senators and representatives, citizens and visitors." Lincoln was appointed to the arrangements committee. Calhoun served as a pallbearer. "Thy fiat has gone forth O Lord my God," Louisa wrote, "and I am left a helpless Widow to morn his loss which nothing on this dreary earth can supply." In March, his remains were brought to Quincy. They were placed in the weathered vault filled with the bones and dust of his ancestors across the street from the Stone Church. Four years later, Louisa joined him.

Acknowledgments

I am deeply indebted to the Massachusetts Historical Society, an institution that, since the early nineteenth century, has handsomely fulfilled its commitment to cherish, protect, and disseminate the documents of its particular area of American history. John Quincy himself lectured to the society on New England history in 1843, the two-hundredth anniversary of the founding of the Massachusetts Bay colony. His descendants have made it the repository of the Adams family papers (my bibliographical essay highlights the resources, printed and online, of the Adams Papers project). Its former managing editor, Margaret Hogan, cooperated with my efforts in every desirable way, providing guidance and assistance, both on-site and at a distance. She and her colleagues shared with me their ongoing transcriptions and electronic versions of documents relevant to this book. HarperCollins and I have also been fortunate in having been able to attain her services as copyeditor. My thanks also to C. James Taylor, the Adams Papers editor in chief, the guiding hand overseeing the entire project; and to Elaine Grublin, the former reference librarian, who introduced me to the society's reference resources. Sara Martin, *Adams Family Correspondence* series editor; Sara Sikes, associate editor; and Sara Georgini, assistant editor, have answered my questions promptly, graciously, and helpfully.

I have benefited immensely from the gift by the twentieth-century Adams family to a large number of research libraries of a microfilm version of the entire collection. Imagine the advantage to the Adams scholar of having the complete archive (608 microfilm reels) available reasonably close to home, in my case at the Bowdoin College Library, with the technological advantage of printing and electronic transmission that modern microfilm machines provide. Without this, a six-year project might have taken ten more. My thanks to Bowdoin College and the Bowdoin College Library for its hospitality, extended by the now retired head librarian, Sherrie Bergman, and especially to Phyllis McQuaide, now also retired, the librarian who made recalcitrant microfilms behave properly and facilitated my work with the Adams archive.

The primary site in which John Quincy Adams and his parents spent much of their lives has been handsomely preserved and maintained by the National Park Service as the Adams National Historical Park. It includes, among other treasures, the house in which John Quincy was born and the house and gardens to which his father retired and which, in turn, later became John Quincy's residence. The National Park Service has admirably met the challenge of creating an urban/suburban national park, weaving together its various strands into a historically and aesthetically satisfying unity. We are all indebted to it. I am particularly indebted to Caroline Keinath, the deputy superintendent of the Adams National Historical Park, for taking a full day out of her busy schedule to guide me through the park to help me understand the places Adams cherished and to feel his presence there.

Books such as this one have a collaborative dimension. My agent, Georges Borchardt, continues to be a source of publishing wisdom, experience, and skill that have helped make this book possible. In the end, a book is always what the author makes it, but he usually needs help. At HarperCollins, I am again indebted to my editor, Tim Duggan, who has provided practical guidance in the writing and publication process. He is a keen reader, and his detailed attention to style and substance has been immensely helpful. Improvements that he and his exemplary associate editor, Emily Cunningham, have contributed make an important difference. My thanks also to John Jusino, the very capable director of the production editorial department, and to Susan Gamer, a superlative proofreader. I'm never surprised and always surprised that no matter how many times I and others have reviewed a manuscript, a pair of fresh eyes always finds things that have been missed. Katherine Beitner, the director of publicity at HarperCollins, achieved superb results for *Lincoln: The Biography of a Writer*. She has my warm thanks for that. I put *John Quincy Adams* into her hands with great respect and confidence.

Fred Kaplan
Boothbay, Maine
2014

Bibliographical Essay

The eighteenth- and nineteenth-century Adamses were obsessive writers. John Quincy Adams worshipped the written word as self-expression, as direct communication, and as the building block of the historical record. The multigenerational *Adams Papers* project at the Massachusetts Historical Society is the fountain from which flows the steady stream of almost every word written by a member of the Adams family or words about them, from selected letters such as *My Dearest Friend: Letters of Abigail and John Adams*, edited by Margaret A. Hogan and C. James Taylor (Cambridge, Mass., 2007), to David McCullough's widely read biography *John Adams* (New York, 2001). Almost everything is there—the letters, diaries, poems, essays, speeches, and public papers—much of it unpublished. The MHS has created a microfilm edition (608 reels), which is widely available in public and university research libraries. It includes an online electronic edition of John Quincy Adams' diary, about half of which is available in a twelve-volume selected edition, *Memoirs of John Quincy Adams* (Philadelphia, 1874–1877), edited by Charles Francis Adams. It omits most of the personal entries of special interest to a biographer. Selections from Charles Francis Adams' twelve-volume edition are available in a one-volume edition (New York, 1928) edited by Allan Nevins. The Massachusetts Historical Society and Harvard University Press continue to publish printed and electronic versions of volumes that will, eventually, provide readers with annotated editions of everything of importance. The currently available volumes in *The Adams Papers* series essential to any biography of John Quincy are *Adams Family Correspondence* (volumes 1–11, 1761–1797), *Diary of John Quincy Adams* (volumes 1–2, 1779–1788), *Diary and Autobiography of John Adams* (volumes 1–4, 1755–1804), *Diary of Charles Francis Adams* (volumes 1–8, 1820–1840), *Diary and Autobiographical Writings of Louisa Catherine Adams* (volumes 1–2, 1778–1849), and Andrew Oliver's *Portraits of John Quincy Adams and His Wife* (1970).

What *The Adams Papers* is in the process of doing for John Adams—a complete edition of his public writings (*Papers of John Adams,* volumes 1–16, 1755–1785)—it has not yet initiated for his son. Worthington C. Ford published in 1913 a seven-volume edition of John Quincy's letters from 1779 to 1823, which includes public as well as personal letters. But John Quincy's speeches, essays, poems, and state papers are available mainly in their first publications and in photographic reprints by modern publishing entrepreneurs. Many are available online in digital libraries and can be downloaded. A substantial one-volume John Quincy Adams anthology for the general reader would be desirable. The Library of America has made a small down payment with the inclusion of three of his poems in volume 1 of *American Poetry: The Nineteenth Century* (New York, 1993). Lynn H. Parsons, *John Quincy Adams, A Bibliography* (Westport, Conn., 1993), provides a helpful listing and description of works by and about Adams. Kenneth V. Jones, *John Quincy Adams, 1767–1848: Chronology, Documents, Bibliographical Aids* (Dobbs Ferry, N.Y., 1970), has an excellent chronology and a small selection of Adams' most important state documents, mostly from his presidency. *A Companion to John Adams and John Quincy Adams* (Malden, Mass., 2013), edited by David Waldstreicher in the *Wiley-Blackwell Companions to American History* series, contains many valuable essays, particularly Waldstreicher's "John Quincy Adams: The Life, the Diary, and the Biographers" and Catherine Allgor and Margery M. Heffron, "A Monarch in a Republic: Louisa Catherine Johnson Adams and Court Culture in Early Washington City."

A year after Adams' death, William H. Seward wrote the first full-length biography, *Life and Public Services of John Quincy Adams, with the Eulogy Delivered Before the Legislature of New York* (Auburn, N.Y., 1849). Much of it was written by John M. Austen, albeit in close collaboration with Seward. Although mostly hagiography, it serves as a conduit between Adams and Lincoln. In its latter portions, it is a strong portrait of Adams as a heroic figure to the anti-slavery forces of the next generation. There is little to notice in Adams biography between Seward and Samuel Flag Bemis, a specialist in diplomatic history who won well-deserved accolades for his two-volume biography, *John Quincy Adams and the Foundations of American Foreign Policy* (New York, 1949) and *John Quincy Adams and the Union* (New York, 1956). Seward's is a partisan biography by a contemporary co-combatant. Bemis attempts objectivity, though the biographer shares between the lines an early-twentieth-century version of Adams' New England values. His volumes focus on Adams' public life and government service; they represent a stage in the slow transition from Victorian conventions about biography to modern attitudes. Bemis' main interest is in policy, not personality.

There have been few noteworthy full-length biographies of Adams since Bemis until Paul C. Nagel, *John Quincy Adams: A Public Life, A Private Life* (Cambridge, Mass., 1997). Marie B. Hecht's *John Quincy Adams: A Personal History of an Independent Man* (New York, 1972) provides a well-researched surface narrative. Jack Shepherd's *Cannibals of the Heart: A Personal Biography of Louisa Catherine and John Quincy Adams* (New York, 1980) is a book whose heart and feminist sympathies are so devoted to Louisa that the depiction of John Quincy is reductive. Although it neglects historical context, it is reader friendly, well paced, and indiscriminately feminist about its main concern. Two short books serve well as introductions to Adams' public life: Lynn Hudson Parsons' *John Quincy Adams* (New York, 1998) and Robert V. Remini's *John Quincy Adams* (New York, 2002). There is also Harlow Giles Unger's recent popular biography, *John Quincy Adams* (Boston, 2012). Nagel's remains the modern biography of record, readable and analytic but also overly compressed and preoccupied with its claim that Adams suffered all his life from severe depression. The public context is subordinated to the personal life rather than held in balance, an understandable attempt to counterbalance Bemis' emphasis on Adams' career. James E. Lewis' short, pithy biography of Adams' public career, *John Quincy Adams: Policymaker for the Union* (Wilmington, Del., 2001), has an excellent bibliographical essay. There have been numbers of partial biographies of John Quincy, one of which, Leonard L. Richards' *The Life and Times of Congressman John Quincy Adams* (New York, 1986), is excellent.

Nagel has made numbers of readable, well-researched contributions to what has become almost a genre of its own, Adams family multigenerational biography, starting with *Descent from Glory: Four Generations of the John Adams Family* (New York, 1983), and *The Adams Women: Abigail and Louisa Adams, Their Sisters and Daughters* (New York, 1987). Both have interesting material about John Quincy. Francis Russell's *Adams: An American Dynasty* (New York, 1976) and Richard Brookhiser's *America's First Dynasty: The Adamses, 1735–1918* (New York, 2002) are in the same genre, though "dynasty" seems an exaggeration. Jack Shepherd's *The Adams Chronicles* (New York, 1976) is the basis for the 1976 PBS miniseries, four episodes of which are devoted to John Quincy. Joseph Ellis'

First Family: Abigail and John Adams (New York, 2010) has the merit of giving Abigail, whose lack of a public life makes her less amenable to full-scale biography, equal billing with her husband. The challenge to the biographer writing about any of the major figures in the Adams family is the importance of the family context, the difficulty of balancing the private and the public life, without undervaluing either sphere, and the huge amount of primary material, much of it unpublished. Woody Holton's well-informed *Abigail Adams* (New York, 2009) is preoccupied with Abigail as a proto-feminist. Phyllis Lee Levin's *Abigail Adams: A Biography* (New York, 1987) is a substantial, informative, and very readable general biography. Joseph J. Ellis' *Passionate Sage: The Character and Legacy of John Adams* (New York, 1993) is the single best book on any member of the Adams family.

Three books, among others, from which I have most benefited in the overview are Stanley Elkins and Eric McKitrick's classic *The Age of Federalism: The Early American Republic, 1788–1800* (New York, 1993), Daniel Walker Howe's *What Hath God Wrought: The Transformation of America, 1815–1848* (New York, 2007), and Gordon S. Wood's *Empire of Liberty: A History of the Early Republic* (New York, 2009). Daniel Feller's *The Jacksonian Promise: America,*

1815–1840 (Baltimore, 1995) provides an excellent thematic overview. A useful specialized study on the subject of the American domestic empire is Thomas R. Hietala's *Manifest Design: American Exceptionalism and Empire* (Ithaca, N.Y., 1985). William Earl Weeks' *John Quincy Adams and American Global Empire* (Lexington, Ky., 1992) effectively delineates Adams' view of how far the American empire should extend. And James E. Lewis Jr., *The American Union and the Problem of Neighborhood: The United States and the Collapse of the Spanish Empire, 1783–1829* (Chapel Hill, N.C., 1998), focuses effectively on the fascinating story of the United States, Spain, and Adams' role in its twists and turns. Bradford Perkins provides an excellent overview, with a starring role for Adams, in *The Creation of a Republican Empire, 1776–1865*, volume 1 of *The Cambridge History of American Foreign Relations* (New York, 1993).

Daniel Walker Howe's contributions in all matters Whig and early republic have been considerable. In addition to *What Hath God Wrought*, they include *Making the American Self: Jonathan Edwards to Abraham Lincoln* (New York, 2009), *The Political Culture of the American Whigs* (Chicago, 1979), and *The Unitarian Conscience: Harvard Moral Philosophy, 1805–1861* (Cambridge, Mass., 1970). The chapters about the political leaders of the early republic in Richard Hofstadter's *The American Political Tradition and the Men Who Made It* (New York, 1948) exemplify how intellectual sharpness can give intelligence narrative power. Hofstadter's ideas about the paranoid style and anti-intellectualism in American history, past and present, seem to me keys to understanding John Quincy's public career. Two valuable books by George Dangerfield have not lost their freshness: *The Era of Good Feelings* (New York, 1952) and *The Awakening of American Nationalism, 1815–1828* (New York, 1965). The city of Washington is itself a character in Adams' life, and Constance McLaughlin Green's *Washington: Village and Capital, 1800–1878* (Princeton, N.J., 1962) has helped transport me to the public spaces of that city in the early nineteenth century. James Sterling Young's *The Washington Community, 1800–1828* (New York, 1966) perceptively maps the intersections between Washington's political and social spaces. Stephen Skowronek, in *The Politics Presidents Make: Leadership from John Adams to Bill Clinton* (Cambridge, Mass., 1993), provides a theoretical and historical overview that contributes to an understanding of the problems of Adams' presidency. Its existence highlights the lack of synthesis or even of dialogue between historians and political scientists, though there are exceptions, such as Richard Franklin Bensel's *Yankee Leviathan: The Origins of Central State Authority in America, 1859–1877* (New York, 1990).

I have benefited from some excellent recent studies of Adams' public life and its dominant issues, particularly William Lee Miller's comprehensive *Arguing About Slavery* (New York, 1997) and Leonard L. Richards' *The Slave Power: The Free North and Southern Domination, 1780–1860* (Baton Rouge, La., 2000). Howard Jones' *Mutiny on the* Amistad: *The Saga of a Slave Revolt and Its Impact on American Abolition, Law, and Diplomacy* (New York, 1987), the inspiration for the 1997 movie, provides an authoritative account of this important episode; and Robin L. Einhorn, *American Taxation, American Slavery* (Chicago, 2006), broadens the subject in an important way. Matthew Mason's *Slavery and Politics in the Early American Republic* (Chapel Hill, N.C., 2006) provides a helpful overview of the first two decades of the nineteenth century. Robert Pierce Forbes' *The Missouri Compromise and Its Aftermath: Slavery and the Meaning of America* (Chapel Hill, N.C., 2007) helps to anchor Adams in one of the key moments in the attempt to hold the Union together. There is no shortage of books about slavery between the Constitution and the Civil War, the context for Adams' preoccupation with slavery, the war powers act, and the Civil War to come, from William W. Freehling's *The Road to Disunion* (New York, 1991) to James Oakes' *Freedom National* (New York, 2013). They are essential to an understanding of Adams' apocalyptic imagination.

John Lauritz Larson, "Liberty by Design: Freedom, Planning, and John Quincy Adams's American System," in *The State and Economic Knowledge: The American and British Experiences*, edited by Mary O. Furner and Barry Supple (Cambridge, Mass., 1990), pp. 73–102, is an excellent article on a neglected aspect of economic history that Adams exemplifies: the relationship between central planning and personal liberty. George A. Lipsky's *John Quincy Adams: His Theory and Ideas* (New York, 1950) provides an analysis of interest but makes Adams overly abstract. Gary V. Wood's *Heir to the Fathers: John Quincy Adams and the Spirit of Constitutional Government* (Lanham, Md., 2004) focuses on the important subject of Adams, natural rights, and constitutional interpretation, past and present. Greg Russell's *John Quincy Adams and the Public Virtues of Diplomacy* (Columbia, Mo., 1995) explores the significant topic of politics and morality, especially Adams' principles about the projection of American power abroad. Money and real estate are unavoidable subjects in dealing with the first and second American generations. Daniel M. Friedenberg, *Life, Liberty, and the Plunder of Early America: Pursuit of Land* (Buffalo, N.Y., 1992), provides a good introduction to a neglected subject. John Lauritz Larson's *Internal Improvements: National Public Works and the Promise of Popular Government in the Early United States* (Chapel Hill, N.C., 2001) is a fine survey of a topic that preoccupied Adams and was central to attempts at national self-definition. It could be argued that Adams' presidency would look much different to modern eyes if he had been successful in signing into law the comprehensive infrastructure projects he believed essential to national unity.

The essays and books about Adams' political career, his presidency, and his successor that have been the most useful to me are Richard R. John's "Affairs of Office: The Executive Departments, the Election of 1828, and the Making of the Democratic Party," in *The Democratic Experiment: New Directions in American Political History*, edited by Meg Jacobs, William J. Novak, and Julian E. Zelizer (Princeton, N.J., 2003), pp. 50–84; Lynn Hudson Parsons' *The Birth of Modern Politics: Andrew Jackson, John Quincy Adams, and the Election of 1828* (New York, 2009); Mary W. M. Hargreaves' *The Presidency of John Quincy Adams* (Lawrence, Kans., 1985); and

Robert V. Rimini's *The Election of Andrew Jackson* (New York, 1963). James M. Banner, *To the Hartford Convention: The Federalists and the Origins of Party Politics in Massachusetts, 1789–1815* (New York, 1970), provides the background for Adams' complicated and painful relationship with the right-wing Boston Federalists. Since politics and war are usually inseparable, the three books on the British-American conflicts that dominated much of Adams' life that I have found especially helpful are by Bradford Perkins: *The First Rapprochement: England and the United States, 1795–1805* (Berkeley, Calif., 1955); *Prologue to War: England and the United States, 1805–1812* (Berkeley, Calif., 1963); and *Castlereagh and Adams: England and the United States, 1812–1823* (Berkeley, Calif., 1964). J. C. A. Stagg's *Mr. Madison's War: Politics, Diplomacy, and Warfare in the Early American Republic, 1783–1830* (Princeton, N.J., 1983) is equally formidable, though two recent revisionist and intellectually lively books are required reading: Jon Latimer's *1812: War with America* (Cambridge, Mass., 2007) and Alan Taylor's *The Civil War of 1812: American Citizens, British Subjects, Irish Rebels, and Indian Allies* (New York, 2010). The modern literature about the Monroe Doctrine, from Ernest R. May's *The Making of the Monroe Doctrine* (Cambridge, Mass., 1975) to Jay

Sexton's *The Monroe Doctrine: Empire and Nation in Nineteenth-Century America* (New York, 2012), is more helpful about the issues and context than about Adams as its author.

In the case of John Quincy Adams, the answer to the question, "Why another biography?" is readily answerable: the existing list is thin, it does not see Adams whole, and key elements of his personality, talent, and vision rarely make an appearance. This book emphasizes, within a total portrait, two undervalued or entirely neglected aspects of Adams—his great talent as a writer, which determined much of his relationship with himself and the world; and his vision of the American future: the war about slavery that would come and his hope for a prosperous country acting as a single entity devoted to improving the life of its people through national projects. The opposition that he met, his successes, and his failures are part of the torn fabric of public discourse in early-twenty-first-century America.

Notes

ABBREVIATIONS

AFC *Adams Family Correspondence*, 1761–1797, ed. L. H. Butterfield, Marc Friedlaender, Richard Alan Ryerson, Margaret A. Hogan, et al. (Cambridge, Mass., 1963–2013).

AP-DJQA *Diary of John Quincy Adams*, 1779–1788, ed. David Grayson Allen, Robert J. Taylor, and Marc Friedlaender (Cambridge, Mass., 1981).

APM Adams Papers Microfilms, Massachusetts Historical Society.

D&A *Diary and Autobiography of John Adams*, ed. L. H. Butterfield (Cambridge, Mass., 1961).

D&AW *Diary and Autobiographical Writings of Louisa Catherine Adams*, ed. Judith S. Graham, Beth Luey, Margaret A. Hogan, and C. James Taylor (Cambridge, Mass., 2013).

DCFA *Diary of Charles Francis Adams*, volumes 1–2, ed. Aida DiPace Donald and David Donald; volumes

3–4, ed. Marc Friedlaender and L. H. Butterfield (Cambridge, Mass., 1964, 1968).

DJQA.APM Diary of John Quincy Adams, Adams Papers Microfilm, Massachusetts Historical Society.

M *Memoirs of John Quincy Adams, Comprising Portions of His Diary from 1795 to 1848*, ed. Charles Francis Adams (Philadelphia, 1874).

MHS Massachusetts Historical Society.

WJQA *Writings of John Quincy Adams*, ed. Worthington Chauncey Ford (New York, 1913).

JA   John Adams
AA   Abigail Adams
MSC   Mary Smith Cranch
ESS   Elizabeth Smith Shaw
AA2   Abigail (Nabby) Adams 2nd
JQA   John Quincy Adams
WC   William Cranch
PJM   Peter Jay Munro
LCA   Louisa Catherine Adams
CA   Charles Adams
TBA   Thomas Boylston Adams
GWA   George Washington Adams
JA2   John Adams, son of John Quincy Adams
CFA   Charles Francis Adams

INTRODUCTION

x *"Literature has been"*: DJQA, 12/15/1820. APM.

xiv *"Every one of the letters"*: JQA/CFA, 2/21/1830. APM.

CHAPTER 1: HOOKS OF STEEL, 1767–1778

1 *"in anxiety and apprehension"*: DJQA, 7/8/1826. M7, 124.

1 *"would survive"*: DJQA, 7/8/1826. M7, 125.

6 *"peculiar circumstances"*: DJQA, 7/11/1826. M7, 128.

7 *"It is repugnant"*: DJQA, 7/14/1826. M7, 130.

7 *"Though it will"*: JQA/LCA, 7/14/1826. APM.

7 *"Everything about"*: DJQA, 7/12/1826. M7, 129.

9 *"that freedom is"*: "Day of My Father's Birth," 10/30/1836. Miscellanies. APM.

9 *"I have at no time"*: DJQA, 7/16/1826. M7, 131.

10 *"to the memory of"*: DJQA, 7/16/1826. M7, 132.

10 *"These papers"*: DJQA, 7/22/1826. M7, 134.

11 *"fourteenth"*: JQA, unfinished biography of JA, summer 1829. APM.

12 *"the stock"*: DJQA, 10/1/1823. APM.

13 *"the Calvinistic"*: JQA, unfinished biography of JA, summer 1829. APM.

13 *"the highest worldly"*: JQA/W. B. Sprague, 10/31/1830. APM.

14 *"The absence of information stimulates"*: DJQA, 10/3/1823. APM.

15 *"In the ordinary"*: JQA, unfinished biography of JA, summer 1829. APM.

15 *"paternal mansion"*: JQA, unfinished biography of JA, summer 1829. APM.

16 *"amorous disposition"*: D&A3, 260.

20 *"Almighty God"*: AA/JA, 6/18/1775. AFC1, 222.

23 *"i have made"*: JQA/ESS, c. 1773. AFC1, 91.

24 *"used to walk"*: JQA/TBA, 4/3/1813. APM.

24 *"she had been"*: DJQA, 7/24/1831. APM.

24 *"Here is Solitude"*: JA, 8/13/1769. D&A1, 340.

26 *"the impression made"*: JQA/Joseph Sturge, 4/1846. APM. For "hooks of steel," see *Hamlet* 1, 3, 63.

27 *"Those were times"*: JQA/TBA, 3 April 1813. APM.

27 *"In my early childhood"*: JQA/Joseph Sturge, 4/1846. APM.

30 *"I wish to turn"*: JA/JQA, 8/11/1777. AFC2, 307.

31 *"Johnny sends"*: JA/AA, 2/13/1778. AFC2, 388.

31 *"Master Johnny"*: AA/John Thaxter, 2/15/1778. AFC2, 390.

31 *"Poor Johnny"*: John Thaxter/AA, 3/6/1778. AFC2, 401.

32 *"Injoin it upon him"*: AA/JA, 3/8/1778. AFC2, 403.

32 *"To describe"*: JA, 2/21/1778. D&A2, 275.

33 *"I often regretted"*: JA, 2/25/1778. D&A2, 276–277.

33 *"an excellent officer"*: JA, 3/8/1778. D&A2, 284.

34 *"The gun burst"*: JA, 3/14/1778. D&A2, 286.

34 *"I have a distinct"*: JQA/John W. Murdaugh, 5/20/1832. APM.

35 *"But here We are"*: JA, 3/19/1778. D&A2, 387.

35 *"It gives me"*: JA, 3/30/1778. D&A2, 292.

CHAPTER 2: A EUROPEAN EDUCATION, 1778–1783

37 *"a Scene of"*: JA, 5/27/1778. D&A4, 118.

37 *"a great genius"*: JA, *The Works of John Adams*, ed. Charles Francis Adams (Boston, 1851), vol. 3, p. 139.

38 *"extensive correspondence"*: JA/AA, 12/3/1778. AFC3, 129.

39 *"Life was new"*: DJQA, 4/26/1837. M9, 353–354.

39 *"many fine public"*: JQA/TBA, 4/11/1778. AFC3, 8.

39 *"very fine music"*: JQA/TBA, 4/11/1778. AFC3, 8.

39 *"not let slip one"*: JQA/AA, 4/12/1778. AFC3, 11.

40 *"I hope I shall"*: JQA/AA, 4/12/1778. AFC3, 11.

41 *"I answer that"*: JQA/CA, 10/3/1778. AFC3, 105.

41 *"You have entered"*: AA/JQA, 6/10/1778. AFC3, 38.

41 *"Papa laments"*: JQA/CA, 10/2/1778. AFC3, 102.

42 *"to read and speak"*: JA/AA, 7/26/1778. AFC3, 66.

42 *"packed up his things"*: JA, 4/14/1778. D&A2, 301.

43 *"very Well"*: JQA/AA, 4/10/1778. AFC3, 16.

43 *"Johnny . . . reads"*: JA/AA, 9/9/1778. AFC3, 88.

43 *"If I do not make"*: JQA/TBA, 10/1/1778. AFC3, 100.

43 *"We are sent"*: JQA/CA, 6/6/1778. AFC3, 34.

44 *"yet I have not"*: JQA/AA, 9/27/1778. AFC3, 92.

44 *"which if I had written"*: JQA/AA, 9/27/1778. AFC3, 93.

45 *"scenes of Magnificence"*: JQA/AA, 9/27/1778. AFC3, 93.

45 *"much esteemed here"*: JA/AA, 7/26/1778. AFC3, 66.

46 *"He lets me go"*: JQA/AA2, 9/27/1778. AFC3, 93.

46 *"The first woman"*: JQA/LCA, 8/28/1822. APM.

49 *"wearied to death"*: JA, 5/20/1778. D&A2, 314.

49 *"the Joy of my Heart"*: JA/AA, 12/2/1778. AFC3, 125.

49 *"much rather be"*: JQA/AA, 6/5/1778. AFC3, 33.

50 *"a man of whom nobody"*: JA, 2/11/1779. D&A2, 352.

51 *"On Dr. F."*: JA, 2/9/1779. D&A2, 347.

51 *"My evils here"*: JA/AA, 2/9/1779. AFC3, 160.

51 *"one of the most wicked"*: JA, 2/8/1779. D&A2, 345.

54 "My dear fellow traveller": JA/AA, 7/14/1779. AFC3, 205.

54 "enjoyed perfect health": JA/AA, 5/14/1779. AFC3, 196.

55 "in raptures with": JA, 6/20/1779. D&A2, 385.

56 "the young gentleman": François de Barbé-Marbois/ JA, 9/29/1779. D&A4, 174.

56 "My little son": JA/François de Barbé-Marbois, 10/17/1779. D&A4. 176.

57 "My dear sons": AA/JA, 11/14/1779. AFC3, 234.

60 "Let me entreat you": JA/AA, 2/21/1779. AFC3, 177.

61 "of my Mamma": AP-DJQA1, 1. 11/12/1779.

61 "The passengers are all": AP-DJQA1, 7. 11/29/1779.

62 "looks like a man's face": JA, 2/5/1780. D&A2, 434.

62 "I thank Almighty God": AP-DJQA1, 25. 1/3/1780.

62 "often undergone": JA, 1/11/1780. D&A2, 426.

63 "very content": JQA/TBA, 2/22/1780. AFC3, 279.

64 "My Work for a day": JQA/JA, 3/16/1780. AFC3, 307.

65 "I hope soon to hear": JA/JQA, 3/17/1780. AFC3, 309.

65 "after his own Image": AA/JQA, 3/20/1780. AFC3, 310–311.

69 "that my sons may": JA/AA, 5/12/1780. AFC3, 342.

70 "got the better of": JA/AA, 6/17/1780. AFC3, 366.

70 "Your son": Rector Verheyk/JA, 11/10/1780. AFC4, 11.

70 "I should not wish": JA/AA, 12/18/1780. AFC4, 35.

71 "If the gentlemen": Benjamin Waterhouse/JA, 12/13/1789. AFC4, 32.

72 *"Everything in life"*: JA/JQA, 12/28/1780. AFC4, 56.
74 *"distracted with more"*: JA/AA, 7/11/1781. AFC4, 170.
75 *"I was at first"*: AFC4, 171.
75 *"satiated with travel"*: JA/AA, 7/11/1781. AFC4, 170.
76 *"to favor and support"*: Francis Dana/JA, 8/28/1781. www.masshist.org/database/1701.
77 *"a village, inhabited"*: AP-DJQA1, 94. 7/12/1781.
77 *"There are 600"*: AP-DJQA1, 96. 7/16/1781.
78 *"the handsomest"*: AP-DJQA1, 99. 7/23/1781.
78 *"have a great reason"*: JQA/JA, 9/1/1781. AFC4, 206.
78 *"all the farmers are"*: JQA/JA, 9/1/1781. AFC4, 207.
83 *"I want you"*: JA/JQA, 5/13/1782. AFC4, 323.
83 *"I might very possibly"*: JQA/JA, 9/6/1782. AFC4, 378.
84 *"strangers are treated"*: AP-DJQA1, 161. 11/23/1782.
84 *"Sweden is the country"*: JQA/AA, 7/23/1783. AFC5, 216.
84 *"You cannot imagine"*: JA/JQA, 2/18/1783. AFC5, 97.
85 *"will I dare say"*: JQA/AA, 7/23/1783. AFC5, 216.
85 *"He is grown up"*: JA/AA, 8/14/1783. AFC5, 221.
86 *"the beauties of"*: JQA/Alexander Everett, 8/19/1811. Everett-Peabody Papers. MHS.

CHAPTER 3: SLOW VOYAGE HOME, 1783–1787

90 *"Awe and Veneration"*: JQA/Elizabeth Cranch, 4/18/1784. AFC5, 322.

90 *"made his most"*: AP-DJQA1, 202. 11/11/1783.

90 *"in the room with"*: D&A3, 150.

91 *"There is a cliff"*: JQA/PJM, 10/26/1783. APM.

92 *"formerly the Residence"*: JQA/PJM, 12/2/1783. APM.

92 *"A young lady"*: AP-DJQA2, 198. 10/31/1783.

92 *"I am told that"*: JQA/PJM, 11/2/1783. APM.

93 *"I am not a severe"*: JQA/PJM, 2/17/1784. APM.

93 *"a great number of"*: JQA/PJM, 12/19/1783. APM.

94 *"Now sacred to"*: JQA/PJM, 12/2/1783. APM.

95 *"I go pretty often"*: JQA/PJM, 12/7/1783. APM.

95 *"Mr. Joshua Johnson"*: JQA/PJM, 12/23/1783. APM.

95 *"Alas! Alas! I have"*: JQA/PJM, 2/13/1784. APM.

96 *"entirely a joke"*: JQA/PJM, 1/27/1784. APM.

96 *"Oh love, thou tyrant"*: enclosed in JQA/PJM, 4/16/1784. APM.

97 *"He is but a prodigal"*: JA/AA, 2/4/1783. AFC5, 88.

98 *"Will you come"*: JA/AA, 9/7/1783. AFC5, 236.

98 *"I have been in constant"*: JA/AA, 7/3/1784. AFC5, 354.

98 *"I am still here"*: JQA/PJM, 5/3/1784. APM.

99 *"is one of the prettiest"*: JQA/PJM, 1/27/1784. APM.

99 *"a real living person"*: JQA/PJM, 5/3/1784. APM.

99 *"differed in opinion upon"*: JQA/PJM, 5/3/1784. APM.

100 *"I will give you"*: JQA/PJM, 2/29/1784. APM.

101 *"not to let my"*: JQA/PJM, 12/19/1783. APM.

102 *"we shall not have"*: JQA/AA, 6/6/1784. AFC5, 340.

102 *"I send you a son"*: JA/AA, 7/26/1784. AFC5, 399.

102 *"Young Mr. Adams"*: AA/ESS, 7/30/1784. AFC5, 382.

103 *"I am twenty years"*: JA/AA, 7/26/1784. AFC5, 399.

104 *"a horrid dirty city"*: AA/ESS, 12/14/1784. AFC6, 30.

104 *"almost universally"*: AP-DJQA1, 214. 1/4/1785.

104 *"I heartily wish"*: JQA/PJM, 11/10/1784. APM.

104 *"began to mutter"*: AP-DJQA1, 235. 3/15/1785.

105 *"Thus does this"*: AP-DJQA1, 220. 2/7/1785.

105 *"I shall never forget"*: JQA/James H. Hackett, 11/4/1845. APM.

106 *"a man of great"*: AP-DJQA1, 262. 5/4/1785.

106 *"very well treated"*: AP-DJQA1, 262. 5/4/1785.

107 *"the greatest Traveller"*: JA/AA, 7/26/1784. AFC5, 399.

108 *"my return home"*: AP-DJQA, 4/26/1837. APM.

108 *"form connections"*: AA/Mercy Otis Warren, 5/10/1775. AFC6, 138.

109 *"we had nothing but"*: JQA/AA2, 5/25/1785. AFC6, 155.

109 *"every nation seems"*: JQA/AA2, 5/25/1785. AFC6, 158.

109 *"prefer being mistaken"*: JQA/AA2, 5/25/1785. AFC6, 160.

110 *"the greatest day"*: AP-DJQA1, 283. 7/4/1785.

111 *"I have been introduced"*: JQA/AA2, 8/10/1785. AFC6, 251.

112 *"The politicians here"*: JQA/JA, 8/3/1785. AFC6, 248.

112 *"You will perhaps think"*: JQA/JA, 8/3/1785. AFC6, 250.

112 *"an uncommon instance"*: AP-DJQA1, 307–308. 8/19/1785.

113 *"I am very impatient"*: JQA/AA2, 8/9/1785. AFC6, 255.

113 *"the mistress of the tavern"*: AP-DJQA1, 311. 8/23/1785.

113 *"I shall not attempt"*: AP-DJQA1, 313. 8/26/1785.

113 *"No person who"*: AP-DJQA1, 312. 8/26/1785.

114 *"quite a stranger"*: MSC/AA, 8/14/1785. AFC6, 269.

114 *"There is a character"*: DJQA, 4/26/1837. APM.

114 *"It reminded me"*: AP-DJQA1, 314–315. 8/28/1785.

114 *"went to the library"*: JQA/AA2, 8/20/1785. AFC6, 291.

115 *"When will they return"*: JQA/AA2, 8/20/1785. AFC6, 291.

115 *"had heard him"*: JQA/AA2, 9/25/1785. AFC6, 373.

115 *"to wait till next"*: AP-DJQA1, 317. 8/31/1785.

116 *"with natural Wit"*: ESS/AA, 10/15/1784. AFC5, 473.

118 *"rather a dangerous"*: Elizabeth Cranch/AA, 10/9/1785. AFC6, 421.

118 *"She either treats"*: AP-DJQA1, 335. 10/5/1785.

118 *"I have heretofore"*: AP-DJQA1, 351. 11/3/1785.

119 *"the passions are high"*: AP-DJQA1, 356. 11/12/1785.

119 *"in much better spirits"*: AP-DJQA1, 367. 12/6/1785.

120 *"Let poets boast"*: "To Delia," 12/12/1785. Miscellanies. APM.

120 *"His candle goeth out"*: ESS/AA2, 2/14/1786. AFC7, 58.

121 *"I am very much"*: AP-DJQA1, 395. 1/27/1786.

121 *"for what they could not"*: AP-DJQA1, 412. 3/5/1786.

122 *"whatever a man's"*: AP-DJQA1, 384. 1/7/1786.

122 *"with vice"*: AP-DJQA1, 381. 12/31/1785.

122 *"Her going away"*: AP-DJQA1, 400. 2/9/1786.

123 *"marching as the heroes"*: JQA/AA2, 3/15/1786. AFC7, 90.

123 *"the most extraordinary"*: JQA/AA2, 3/15/1786. AFC7, 92.

123 *"there are no misters"*: JQA/AA2, 3/15/1786. AFC7, 91.

124 *"Your brother is exceedingly"*: ESS/AA2, 2/14/1786. AFC7, 58.

124 *"esteemed and respected"*: JQA/AA2, 3/15/1786. AFC7, 91.

124 *"I have already come"*: JQA/AA2, 3/15/1786. AFC7, 90.

125 *"My short discipline"*: AP-DJQA, 4/26/1837. APM.

125 *"He had imbibed"*: ESS/AA, 5/18/1786. AFC7, 93.

126 *"I see the wise politician"*: ESS/AA, 5/18/1786. AFC7, 94.

126 *"I wish Mr JQA"*: ESS/AA, 5/18/1786. AFC7, 93.

127 *"Mathematics and natural"*: AP-DJQA2, 30. 5/9/1786.

128 *"It seems almost"*: AP-DJQA2, 29. 5/8/1786.

128 *"some of them got drunk"*: AP-DJQA2, 2. 3/15/1786.

128 *"Drunkenness is the mother"*: AP-DJQA2, 11. 3/28/1786.

129 *"make such a noise"*: MSC/AA, 10/22/1786. AFC7, 380.

129 *"I behold those qualities"*: ESS/AA, 3/18/1786. AFC7, 94.

130 *"as studious as a hermit"*: Charles Storer/AA, 2/12/1786. AFC7, 53.

130 *"to see them all together"*: MSC/AA, 3/22/1786. AFC7, 103.

130 *"upon a much better plan"*: JQA/JA, 4/2/1786. AFC7, 131.

132 *"the question of whether"*: AP-DJQA2, 14. 4/4/1786.

132 *"This is one of the"*: JQA/AA2, 4/1/1786. AFC7, 119.

132 *"But it so happens"*: JQA/AA, 5/15/1786. AFC7, 163.

132 *"the enslaved African"*: AP-DJQA2, 34. 5/16/1786.

133 *"Near as we are to Boston"*: JQA/JA, 5/21/1786. AFC7, 183.

134 *"It is against the law"*: JQA/AA2, 5/18/1786. AFC7, 169.

135 *"ideas of happiness"*: AP-DJQA2, 55. 6/26/1786.

135 *"whether civil discord"*: AP-DJQA2, 60. 7/6/1786.

135 *"Which so ever of the Party"*: AP-DJQA2, 61. 7/6/1786.

135 *"whether inequality among"*: AP-DJQA2, 9. 9/26/1786.

135 *"they did not either"*: MSC/AA, 9/28/1786. AFC7, 349.

136 *"whether love or fortune"*: AP-DJQA2, 169. 3/5/1787.

137 *"mutual esteem"*: AP-DJQA2, 171–172. 3/5/1787.

138 *"William Cranch"*: AP-DJQA2, 178. 3/19/1787.

138 *"Solomon Vose"*: AP-DJQA2, 231. 6/2/1787.

139 *"was 19 the 27th"*: AP-DJQA2, 211. 4/26/1787.

140 *"the inscriptions which"*: AP-DJQA2, 248. 7/2/1787.

141 *"Where this will end"*: AP-DJQA2, 92. 9/7/1786.

143 *"without the avocations"*: AP-DJQA2, 237. 6/8/1787.

143 *"saddens very much"*: AP-DJQA2, 168. 3/2/1787.

143 *"these disagreeable"*: AP-DJQA2, 237. 6/6/1787.

144 *"To me, he has been"*: AP-DJQA2, 174. 3/10/1787.

144 *"I was shocked"*: AP-DJQA2, 183. 3/22/1787.

144 *"I should be very glad to study"*: JQA/JA, 8/30/1786. AFC7, 326.

145 *"a man of great wit"*: AP-DJQA2, 238. 6/10/1787.

CHAPTER 4: MOST BEAUTIFUL AND BELOVED, 1787–1794

147 *"The importance and necessity"*: AP-DJQA2, 222. 5/18/1787.

147 *"I am led unawares"*: JQA/JA, 6/30/1787. AFC8, 97.

148 *"spirit of patriotism"*: AP-DJQA2, 261. 7/18/1787.

148 *"own insufficiency"*: AP-DJQA2, 261. 7/18/1787.

150 *"with humiliation"*: AP-DJQA2, 266. 7/18/1787.

150 *"a fine, handsome fellow"*: DJQA, 4/17/1839. APM.

151 *"But when with a"*: Cicero, *De Officiis*, I, 17. Loeb Classical Library, *On Duties* (Cambridge, Eng., 2005), pp.59–60.

151 *"He knew no one"*: MSC/AA, 9/2/1787. AFC8, 146.

152 *"the comparative utility"*: AP-DJQA2, 175. 3/12/1787.

152 *"the intimate connection between"*: AP-DJQA2, 202. 4/10/1787.

153 *"one of the best novels in the language"*: AP-DJQA2, 276. 8/17/1787.

153 *"the most extraordinary book"*: AP-DJQA2, 292. 9/19/1787.

153 *"books of entertainment"*: AP-DJQA2, 298. 10/2/1787.

153 *"In Thomson's* Castle of Indolence*"*: JQA/GWA, 12/19/1827. APM.

154 *"I nearly destroyed myself"*: JQA/CFA, 2/9/1827. APM.

154 *"My eyes and health begin"*: AP-DJQA2, 317. 11/14/1787.

154 *"My prospects appear darker"*: AP-DJQA2, 330. 12/18/1787.

155 *"When I look back"*: AP-DJQA2, 337. 1/1/1788.

157 *"I am still of opinion"*: JQA/WC, 12/8/1787. APM.

157 *"I must freely acknowledge"*: JQA/WC, 2/16/1788. APM.

158 *"Reading makes a full man"*: JQA/CFA, 11/7/1827. APM.

159 *"common-place: his ideas"*: AP-DJQA2, 376. 3/16/1788.

160 *"somewhat in the dumps"*: JQA/WC, 2/16/1788. APM.

160 *"I trust my Reason will"*: JQA/AA, 12/30/1787. AFC7, 420.

161 *"which you know is as good"*: JQA/WC, 2/16/1788. APM.

161 *"When my sullens leave me"*: JQA/WC, 2/16/1788. APM.

161 *"A Vision"*: Poems of Religion and Society (Buffalo, 1848), pp. 109–116.

163 *"the most beautiful and beloved"*: DJQA, 11/17/1838. APM.

163 *"The young Misses have assumed"*: AP-DJQA2, 442. 8/13/1788.

163 *"there was an exception"*: Recollections of Samuel Breck: With Passages from His Notebooks (1771– 1862), ed. H. E. Scudder (Philadelphia, 1877), pp. 120–121.

164 *"In the 15th year of her age"*: DJQA, 11/17/1838. APM.

165 *"You may know"*: James Bridge/JQA, 9/28/1790. AFC9, 43, and APM.

168 *"I must be pinched"*: JA/JQA, 7/9/1789. AFC8, 386.

199 *"in the evening with M.F"*: DJQA, 7/14/1790. APM.

169 *"troubles of the heart"*: JQA/CFA, 3/26/1828. APM.

170 *"It was a consuming flame"*: Recollections of Samuel Breck, pp. 120–121.

170 *"Dearly!—how dearly did the sacrifice"*: DJQA, 11/17/1838. APM.

170 *"I was too much agitated"*: JQA/AA, 10/17/1790. AFC9, 132–133.

171 *"Your agitation, your confusion"*: JA/JQA, 10/23/1790. AFC9, 139.

173 *"most magnificent and elegant"*: DJQA, 9/9/1789. APM.

173 *"they were doing nothing"*: DJQA, 9/9/1789. APM.

173 *"exhibited a melancholy picture"*: DJQA, 9/14/1789. APM.

174 *"the debates . . . upon the judiciary bill"*: DJQA, 9/17/1789. APM.

175 *"the difficulty of adjusting"*: DJQA, 9/17/1789. APM.

176 *"do it, my dear son"*: JA/JQA, 7/9/1789. AFC8, 386.

176 *"not conscious of an unworthy"*: DJQA, 5/6/1792. APM.

176 *"The consideration is equally"*: DJQA, 5/6/1792. APM.

177 *"Hurt my eyes much"*: DJQA, 4/3/1791. APM.

177 *"these sandy deserts"*: DJQA, 5/4/1792. APM.

180 *"has hunted for epigrams . . . established order of nature"*: Letters of Publicola, *Columbian Centinel* (Boston), 6–7/1791.

185 *"seems to have some talents"*: Brown's *Federal Gazette* (Philadelphia), 7/1791.

187 *"the whole body"*: JQA/TBA, 2/1/1792. APM.

187 *"Boston is highly boring"*: Raymond Walters Jr., *Albert Gallatin: Jeffersonian Financier and Diplomat* (New York, 1957; Pittsburgh, 1969), p. 11.

188 *"If your legislators have"*: Samuel Eliot Morrison, *The Life and Letters of Harrison Gray Otis, Federalist, 1765–1847* (Boston, 1913), pp. 37–38.

188 *"this evening will be delivered"*: *Independent Chronicle* (Boston), 11/30/1792.

189 *"In a free government"*: Menander [JQA], *Columbian Centinel* (Boston), 12/19/1792.

191 *"I am really astonished at"*: JA/AA, 12/28/1792. AFC9, 360.

192 *"No stipulation contained"*: Marcellus [JQA], *Columbian Centinel* (Boston), 5/4/1793, 5/11/1793.

193 *"Delivered the Oration"*: DJQA, 7/4/1793. APM.

194 *"when your coasts were infested"*: *An Oration Pronounced July 4th, 1793, at the Request of the Inhab-*

*itants of the Town of Boston; in Commemoration of
the Anniversary of American Independence* (Boston,
1793).

194 *"with a warmth and animation"*: JQA/CFA, 7/9/1828.
APM.

195 *"posted on the mast"*: DJQA, 8/10/1793. APM.

196 *"But for what purpose"*: DJQA, 11/22/1793. APM.

197 *"hare-brained Hotspur"*: WJQA1, 152, 155, 158,
164. 11–12/1793.

198 *"I have so long been"*: JA/AA, 12/26/1793. AFC9,
484.

199 *"Life in this State"*: DJQA, 3/26/1794. APM.

199 *"to go to The Hague"*: DJQA, 6/3/1794. APM.

199 *"I wrote those papers without"*: JQA/CFA,
3/26/1828. APM.

200 *"The pain of separation"*: DJQA, 9/17/1794. APM.

201 *"a beautiful rainbow"*: DJQA, 9/17/1794. APM.

CHAPTER 5: REMEMBER YOUR CHARACTERS, 1794–1797

202 *"a sound as of"*: DJQA, 10/15/1794. M1, 41.

203 *"a hair-breadth escape"*: DJQA, 10/15/1794. M1, 42.

203 *"I once more wish you"*: JA/JQA and TBA,
9/14/1794. AFC10, 230.

204 *"old, crazy and leaky"*: DJQA, 9/18/1794. APM.

205 *"I repeated with"*: DJQA, 10/16/1794. APM.

205 *"reserved and distant"*: DJQA, 7/7/1794. M1, 33.

205 *"thought of the new"*: JA/JQA, 12/24/1804. APM.

206 *"The President began"*: DJQA, 7/11/1794. M1,
34–35.

207 *"I told him that"*: JQA/JA, 10/23/1794. WJQA1, 201–209.

207 *"Since the peace with America"*: JQA/AA, 10/25/1794. AFC10, 240.

208 *"Mrs. Siddons appeared"*: DJQA, 10/18/1794. M1, 46.

208 *"thunder-clap of loud applause"*: DJQA, 10/20/1794. M1, 48.

208 *"Disaffection is silent"*: JQA/Thomas Welsh, 10/27/1794. APM.

208 *"from the sovereign"*: DJQA, 10/25/1794. APM.

209 *"extremely delicate"*: DJQA, 10/28/1794. APM.

210 *"Be always upon"*: JA/JQA, 3/26/1795. AFC10, 401.

211 *"frightful foggs"*: TBA/JA, 12/20/1794. AFC10, 315.

211 *"is well disposed"*: DJQA, 2/2/1795. APM.

211 *"broke out in transports"*: DJQA, 2/2/1795. APM.

211 *"everything so perfectly quiet"*: JQA/JA, 11/9/1794. APM.

212 *"has the appearance of profound"*: JQA/JA, 11/9/1794. APM.

212 *"live in perfect harmony"*: JQA/WC, 5/18/1795. APM.

213 *"the rights of Man"*: DJQA, 3/12/1795. APM.

213 *"it is in vain"*: DJQA, 2/23/1795. APM.

214 *"the course of events"*: DJQA, 4/13/1795. APM.

214 *"it is remarkable that"*: JQA/Edmund Randolph, 2/15/1795. WJQA1, 297–298.

215 *"An American Minister at the Hague"*: JQA/Edmund Randolph, 5/19/1795. WJQA 1, 348–362.

215 *"something like drudgery"*: DJQA, 7/22/1795. APM.

215 *"Mr. J Q Adams Your Son"*: AA/JQA, 9/15/1795. AFC11, 23.

216 *"I want your society"*: JA/JQA, 8/25/1795. AFC11, 21.

216 *"We are indeed once more"*: JQA/AA2, 4/15/1795. AFC10, 410.

216 *"the adventures of myself"*: JA/AA, 2/10/1795. AFC10, 376.

218 *"I shall therefore always"*: JQA/JA, 10/31/1795. AFC11, 49.

219 *"perhaps no one knew"*: ESS/JQA, 6/9/1794. AFC10, 203.

219 *"He is at last safe landed"*: AA2/JQA, 10/26/1795. AFC11, 43.

219 *"Charles has got the start"*: JQA/TBA, 11/2/1795. AFC11, 53.

220 *"five and twenty years ago"*: TBA/AA, 7/12/1795. AFC11, 8.

220 *"by voluntary violence"*: JQA/AA, 11/7/1795. AFC11, 61–62.

221 *"is very well written"*: DJQA, 10/25/1795. APM.

222 *"The woman, who is"*: DJQA, 10/25/1795. APM.

223 *"I most sincerely hope"*: JQA/JA, 12/29/1795. APM.

224 *"The internal evidence"*: JQA/AA, 11/24/1795. AFC11, 69–70.

225 *"I have not yet"*: JQA/AA, 11/24/1795. AFC11, 70.

225 *"It is a Mortifying Consideration"*: JA/AA, 5/3/1796. AFC11, 282.

226 *"handsome, if not beautiful"*: DJQA, 10/27/1794. APM.

227 *"He has made a good picture of it"*: DJQA, 4/4/1796.
   APM.

229 *"I wish he would come home"*: JA/AA, 3/29/1796.
   AFC11, 234.

229 *"the result of necessity"*: JQA/AA, 7/25/1796.
   AFC11, 339.

229 *"was that I never would marry"*: JQA/AA, 7/25/1796.
   AFC11, 339.

230 *"a very amiable young lady"*: JQA/AA, 2/28/1796.
   AFC11, 190.

230 *"Ring from Louisa's finger"*: DJQA, 3/2/1796. APM.

230 *"Perhaps you will hear"*: JQA/JA, 3/20/1796. AFC11,
   224.

231 *"Whom you call yours"*: AA/JQA, 5/25/1796. AFC11,
   299.

231 *"attachment had been made"*: JQA/AA, 7/25/1796.
   AFC11, 339.

231 *"the destiny which is said"*: JQA/AA, 7/25/1796.
   AFC11, 339.

231 *"always had a prejudice"*: D&AW1, 37.

233 *"She was his pride"*: D&AW1, 20.

233 *"The greatest fault"*: D&AW1, 8.

234 *"All the scenes of my infancy"*: D&AW1, 3.

234 *"Much, much depends"*: D&AW1, 17.

235 *"was the handsomest man"*: D&AW1, 7.

235 *"If I ever had any admirers"*: D&AW1, 33.

236 *"the matrimonial union"*: JQA/AA, 8/16/1796.
   AFC11, 362.

237 *"conversation with Louisa"*: DJQA, 4/18/17. APM.

237 *"The right and the reason"*: DJQA, 4/18/17. APM.

238 *"I shall return in all"*: JQA/CA, 6/9/1796. AFC11, 310.

238 *"upon the whole I have passed"*: JQA/AA, 5/5/1796. AFC11, 286.

239 *"evening of delight"*: DJQA, 5/27/1796. APM.

239 *"until you return to your"*: AA/TBA, 6/10/1796. AFC11, 316.

240 *"You mention the pain"*: LCA/JQA, 11/29/1796. AFC11, 426.

241 *"The time for original composition"*: DJQA, 7/27/1796. APM.

241 *"In the intercourse of friends"*: JQA/LCJ, 10/12/1796. AFC11, 388.

242 *"reason tells me I shall"*: LCJ/JQA, 12/6/1796. AFC11, 436.

242 *"You are the delight and pride"*: JQA/LCJ, 10/12/1796. AFC11, 389–390.

242 *"I will not say that I envy"*: JQA/CA, 10/25/1796. AFC11, 391.

242 *"the admiration of which increases"*: DJQA, 9/4/1796. APM.

243 *"been a time of as steady"*: DJQA, 12/31/1796. APM.

243 *"He said nothing to me upon"*: JQA/Rufus King, 2/9/1797. WJQA2, 111–113.

243 *"that your Mother has"*: AA/JQA, 11/28/1796. AFC11, 422.

245 *"with firmness & a good grace"*: JA/AA, 12/7/1796. AFC11, 439.

245 *"drive out of it"*: JA/AA, 3/13/1796. APM.

246 *"I trust his conduct will be"*: AA/Elbridge Gerry, 12/31/1796. AFC11, 475–476.

246 *"Jefferson and I"*: JA/AA, 1/1/1797. AFC11, 481.

247 *"As Events will not accommodate"*: JQA/LCJ, 3/6/1797. APM.

247 *"Go to Lisbon"*: JA/JQA, 3/31/1797. APM.

247 *"I advise you to marry"*: AA/JQA, 11/11/1796. AFC11, 399.

248 *"My last disappointment has taught"*: LCJ/JQA, 4/24/1797. APM.

CHAPTER 6: BEGIN ANEW THE WORLD, 1797–1801

249 *"this month as one"*: DJQA, 11/30/1797. APM.

249 *"The mind at least"*: DJQA, 11/17/1797. APM.

250 *"I know not a human being"*: JQA/JA, 7/22/1797. APM.

250 *"seems to me too refined"*: JA/JQA, 10/25/1797. APM.

251 *"directions concerning his affairs"*: DJQA, 7/21/1797. APM.

252 *"I have now the happiness"*: JQA and LCA/AA and JA, 7/28/1797. APM.

252 *"a long conversation"*: DJQA, 8/8/1797. APM.

252 *"without either yielding"*: JQA/JA, 8/10/1797. APM.

253 *"When I arose and found"*: D&AW1, 52.

253 *"She is indeed a most"*: TBA/JA, 9/10/1797. APM.

253 *"that no event of my life"*: Catherine Nuth Johnson/JQA, 9/18/1797. APM.

253 *"I need not attempt"*: Joshua Johnson/JQA, 9/12/1797. APM.

254 *"Find the affairs of Mr. J.":* DJQA, 10/9/1797. APM.

255 *"My connection Sir":* JQA/Frederick Delius, 10/9/1797. APM.

255 *"had forfeited all that":* D&AW1, 51.

256 *"dirty, noisy, and uncomfortable":* D&AW1, 67.

256 *"Three fourths of my time":* DJQA, 11/30/1797. APM.

256 *"This I believe saved":* D&AW1, 55.

257 *"determined to give me":* DJQA, 12/3/1797. APM.

258 *"the beau ideal":* D&AW1, 70.

259 *"These parties all have":* DJQA, 1/9/1798. APM.

260 *"Buonaparte affects great splendor":* JQA/AA, 12/28/1797. APM.

260 *"the military power":* JQA/JA, 2/15/1798. APM.

261 *"very much dissatisfied":* DJQA, 1/12/1798. APM.

261 *"presents difficulties which":* DJQA, 1/27/1798. APM.

261 *"Find myself in a state":* DJQA, 4/5/1798. APM.

261 *"Three quarters of my life":* DJQA, 4/28/1798. APM.

262 *"I find I can yet understand":* DJQA, 5/29/1798. APM.

262 *"I find some encouragement":* DJQA, 8/31/1798. APM.

262 *"tolerably easy":* DJQA, 11/13/1798. APM.

263 *"which is one of the best":* DJQA, 11/3/1799. APM.

263 *"I wish not to judge":* JQA/JA, 1/3/1798. APM.

264 *"At length Talleyrand":* DJQA, 6/19/1798. APM.

265 *"one half of the House":* JQA/William Vans Murray, 6/7/1798. WJQA2, 298–303.

265 *"A long and terrible War is":* JQA/AA, 7/25/1798. APM.

266 *"can hurt us little by sea"*: JQA/William Vans Murray, 3/6/1798. WJQA2, 265–267.

266 *"But if the negro keepers"*: JQA/William Vans Murray, 8/14/1798. WJQA2, 349–350.

267 *"In all my letters"*: JQA/CA, 2/14/1798. APM.

268 *"Happy to be Childless"*: JA/AA, 12/31/1798. APM.

268 *"It is hard upon you"*: AA/JQA, 11/15/1798. APM.

268 *"My confidence in him"*: JQA/AA, 7/25/1798. APM.

268 *"He is not at peace"*: AA/JQA, 11/15/1798. APM.

269 *"much more deeply afflicted"*: JQA/TBA, 7/1/1799. APM.

269 *"facility in writing"*: TBA/AA, 2/12/1798. APM.

269 *"it is something more than"*: DJQA, 9/30/1798. APM.

270 *"He seldom mentions you"*: LCA/TBA, 10/6/1798. APM.

270 *"and assured him"*: D&AW1, 52.

270 *"a solace in my moments"*: D&AW1, 88.

270 *"my wife yet very ill"*: DJQA, 1/18/1798. APM.

270 *"again better, to flatter us again"*: DJQA, 1/22/1798. APM.

271 *"My prophetic heart"*: DJQA, 3/21/1798. APM.

271 *"the blessings of wedded love"*: JQA/WC, 6/6/1798. APM.

271 *"some faint and feeble"*: DJQA, 7/14/1798. APM.

271 *"excessively painful"*: DJQA, 7/15/1798. APM.

271 *"the case appears"*: DJQA, 7/17/1798. APM.

272 *"The anticipation of evils"*: DJQA, 7/17/1798. APM.

272 *"from the loveliness of temper"*: DJQA, 7/26/1798. APM.

273 *"they mean to delay"*: DJQA, 10/30/1798. APM.

273 *"you judge and think"*: AA/JQA, 12/2/1798. APM.

273 *"she was blessed with a fine"*: D&AW1, 113.

274 *"a great proof how much"*: DJQA, 9/4/1799. APM.

274 *"sometimes wild and sublime"*: JQA/AA, 9/21/1799. APM.

275 *"as much improved"*: JQA/TBA, 10/22/1799. APM.

275 *"I brought her home"*: DJQA, 12/4/1799. APM.

276 *"suffering no pain"*: DJQA, 1/8/1800. APM.

276 *"bore the operation well"*: DJQA, 1/2/1800. APM.

276 *"I can only pray to God"*: DJQA, 1/9/1800. APM.

276 *"These violent, long continued"*: DJQA, 1/24/1800. APM.

276 *"sick from anxiety"*: D&AW1, 130.

277 *"although the President's nomination"*: TBA/JQA, 5/11/1800. APM.

277 *"You know I am very proud"*: LCA/Nancy Johnson Hellen, 9/27/1798. APM.

278 *"Five centuries had"*: JQA/TBA, 8/1/1800. APM.

279 *"it would astonish you"*: JQA/TBA, 8/3/1800. APM.

279 *"The spectator has but to"*: JQA/TBA, 8/7/1800. APM.

279 *"the supreme creator"*: JQA/TBA, 8/7/1800. APM.

280 *"From lands, beyond"*: JQA/TBA, 8/7/1800. APM.

281 *"Therefore make up your account"*: JQA/TBA, 7/20/1800. APM.

281 *"Even making every allowance"*: JQA/TBA, 8/3/1800. APM.

282 *"was to obtain information"*: JQA/TBA, 8/5/1800. APM.

282 *"There is perhaps no part"*: JQA/TBA, 7/26/1800. APM.

283 *"I see not what Moses Mendelssohn"*: DJQA, 10/17/1798. APM.

283 *"apparently at the point"*: DJQA, 1/24/1795. APM.

284 *"Heard their Devotions"*: DJQA, 6/3/1797. APM.

284 *"this ridiculous and barbarous"*: JQA/TBA, 8/21/1800. APM.

285 *"The word filth conveys"*: JQA/TBA, 7/20/1800. APM.

285 *"Mr. and Mrs. Cohen"*: D&AW1, 197.

286 *"new pamphlet"*: DJQA, 6/23/1799. APM.

286 *"has produced some sensation"*: DJQA, 6/24/1799. APM.

286 *"atheism and revolution"*: JQA/JA, 1/3/1798. APM.

287 *"wanders about Germany"*: JQA/TBA, 10/22/1799. APM.

288 *"at the gallery of pictures"*: DJQA, 9/12/1799. APM.

288 *"very unwell and glad to rest"*: LCA/Joshua Johnson, 9/5/1800. APM.

288 *"We have a dismal month"*: DJQA, 9/20/1800. APM.

288 *"more threatening than it has been"*: DJQA, 10/21/1800. APM.

288 *"I was sick almost"*: D&AW1, 136.

289 *"as if she had been my own mother"*: D&AW1, 138.

289 *"my health was so weak"*: D&AW1, 151.

290 *"I have this day to offer"*: DJQA, 4/12/1801. APM.

290 *"I was a Mother"*: D&AW1, 154.

291 *"I deplore"*: JQA/JA, 6/19/1800. APM.

291 *"painted . . . the misery"*: AA/WSS, 9/6/1800. APM.

291 *"Such was the infatuation"*: AA/JQA, 5/30/1801. APM.

292 *"I implore the favor"*: DJQA, 5/2/1801. APM.

292 *"in a crucifying state"*: DJQA, 5/18/1801. APM.

292 *"I am sure your brother"*: AA/TBA, 7/12/1801. APM.

294 *"The conduct of men"*: JA/William Vans Murray, 12/16/1800. WJQA2, 486.

294 *"You speak of it as"*: JQA/TBA, 12/3/1800. WJQA2, 484.

295 *"Those absurd principles"*: JQA/TBA, 12/3/1800. WJQA2, 485.

295 *"some of my late letters"*: JQA/JA, 5/1/1801. APM.

298 *"could not bear the idea"*: D&AW1, 157.

299 *"To begin the world anew"*: DJQA, 12/31/1801. APM.

299 *"I feel very anxious for him"*: AA/TBA, 7/5/1801. APM.

300 *"with a faint and"*: DJQA, 9/3/1801. APM.

CHAPTER 7: THE WHITE WORM, 1801–1804

302 *"a sort of fatherly look"*: TBA/AA, 9/20/1801. APM.

303 *"He has been unfortunate"*: DJQA, 10/26/1801. APM.

303 *"So you see what good"*: JQA/TBA, 9/16/1801. APM.

304 *"amongst a people where"*: AA/JQA, 9/13/1801. APM.

305 *"Threats of insurrection"*: JQA/TBA, 9/16/1801. APM.

305 *"to have no concern whatsoever"*: JQA/TBA, 9/27/1801. APM.

306 *"Every creature that has seen"*: LCA/JQA, 9/22/1801. APM.

306 *"as for those of his mother"*: JQA/LCA, 10/8/1801. APM.

306 *"better than she has for years"*: JQA/TBA, 10/24/1801. APM.

307 *"The views along the Potomac"*: DJQA, 10/27/1801. APM.

308 *"excessive violence"*: DJQA, 11/3/1801. APM.

308 *"more depressed in her spirits"*: DJQA, 11/14/1801. APM.

308 *"had the pleasure of introducing"*: DJQA, 11/25/1801. APM.

309 *"I know very well"*: AA/TBA, 12/27/1801. APM.

309 *"Her frame is so slender"*: AA/TBA, 12/27/1801. APM.

309 *"I thank God I can yet struggle"*: DJQA, 1/8/1802. APM.

310 *"I feel strong temptation"*: DJQA, 1/28/1802. APM.

310 *"A politician in this"*: DJQA, 1/28/1802. APM.

311 *"burdened with the minutest"*: JQA/TBA, 1/9/1802. APM.

311 *"experimental and natural philosophy"*: JQA/JA, 3/28/1801. APM.

311 *"It is late in the progress"*: DJQA, 1/29/1802. APM.

312 *"I hope if any attempt"*: AA/TBA, 2/7/1802. APM.

312 *"I have little desire"*: DJQA, 4/1/1802. APM.

313 *"You will see my name"*: JQA/TBA, 4/11/1802. APM.

313 *"The fed's will be glad"*: TBA/JQA, 4/19/1802. APM.

315 *"the largest portion"*: Speech, Charitable Fire Department, 5/28/1802. MHS.

316 *"excessively warm and sultry"*: DJQA, 8/17/1802. APM.

317 *"I see no chance for quiet"*: AA/TBA, 10/10/1802. APM.

317 *"rests upon the support"*: JQA/Rufus King, 10/8/1802. APM.

317 *"I must consider the issue"*: DJQA, 11/3/1802. APM.

318 *"Liberty, Religion and Philosophy"*: JQA/TBA, 12/12/1802. APM.

318 *"the rock of the first landing"*: DJQA, 12/22/1802. APM.

318 *"was not made for himself alone . . . noblest empire of time"*: *An Oration at Plymouth, December 22, 1802, at the Anniversary Commemoration of the First Landing of Our Ancestors at That Place* (Boston, 1802), pp. 6, 31.

324 *"residence in Boston"*: DJQA, 9/29/1803. APM.

325 *"Mr. Emerson told me"*: DJQA, 1/26/1803. APM.

325 *"supported by the principal"*: Fisher Ames/Christopher Gore, 2/24/1803. WJQA3, 11.

326 *"were upon the nomination list"*: DJQA, 2/2/1803. APM.

326 *"After we adjourned"*: DJQA, 2/4/1803. APM.

327 *"I feel myself in a great degree"*: DJQA, 4/2/1803. APM.

328 *"A Catastrophe so"*: AA/TBA, 4/26/1803. APM.

328 *"as anxious as though"*: AA/TBA, 6/20/1803. APM.

329 *"For this new blessing"*: DJQA, 7/4/1803. APM.

329 *"much bruised, and inwardly"*: AA/TBA, 6/20/1803. APM.

329 *"large and highly inflamed"*: DJQA, 8/14/1803. APM.

329 *"The heaviness of heart"*: DJQA, 8/19/1803. APM.

330 *"She has little or no"*: JQA/TBA, 8/21/1803. APM.

330 *"God be praised"*: DJQA, 9/18/1803. APM.

333 *"I asked him whether"*: DJQA, 10/28/1803. APM.

334 *"consistently with the Constitution"*: DJQA, 10/29/1803. APM.

334 *"Congress to admit"*: DJQA, 11/25/1803. WJQA3, 20–21.

335 *"The Hon. John Quincy Adams"*: Worcester Aegis, 12/4/1803. WJQA3, 22.

335 *"already seen enough"*: DJQA, 11/1/1803. APM.

335 *"between two rows"*: JQA/AA, 12/22/1803. APM.

336 *"my warmth of opposition"*: DJQA, 1/13/1804. APM.

336 *"If any gentleman"*: DJQA, 1/10/1804. WJQA3, 25–30.

337 *"But their suspicion of me"*: JQA/TBA, 2/18/1804. APM.

337 *"The workings of this question"*: DJQA, 1/26/1804. APM.

338 *"like drinking the Millennium"*: JQA/TBA, 1/30/1804. APM.

340 *"shall hold their offices during"*: U.S. Constitution, Art. III, sec. 1; Art. II, sec. 4.

341 *"Motions were made to"*: DJQA, 3/3/1804. APM.

341 *"Mr. Adams is so much engaged"*: LCA/AA, 2/11/1804. APM.

342 *"Impeachment is nothing more"*: DJQA, 12/20/1804. APM.

342 *"need not imply any criminality"*: DJQA, 12/21/1804. APM.

342 *"began a speech of about"*: DJQA, 2/27/1805. APM.

343 *"Sir Gravity himself"*: JQA/JA, 3/8/1805. APM.

343 *"political opponents of Mr. Chase"*: JQA/JA, 3/8/1805. APM.

343 *"furnace of affliction"*: JQA/Harrison Gray Otis, 3/31/1808. WJQA3, 189–232.

346 *"If there is any one thing"*: JQA/William Smith Shaw, 1/20/1804. APM.

346 *"to be silent upon anything"*: JQA/JA, 3/1/1804. APM.

346 *"I wish you had seen"*: JQA/TBA, 1/22/1804. APM.

347 *"we have but one week more"*: JQA/JA, 3/17/1804. APM.

347 *"I feel already"*: JQA/LCA, 4/15/1804. APM.

347 *"passing the summer months"*: LCA/JQA, 4/17/1804. APM.

348 *"The first wish of my heart"*: JQA/LCA, 4/9/1804. APM.

348 *"May you never feel a pang"*: JQA/LCA, 4/9/1804. APM.

348 *"I think of scarce anything"*: JQA/LCA, 5/2/1804. APM.

349 *"Good Night my best beloved"*: JQA/LCA, 5/20/1804. APM.

349 *"learn to read my heart"*: LCA/JQA, 5/29/1804. APM.

349 *"I shall know no peace"*: LCA/JQA, 8/12/1804. APM.

350 *"Sargent has had the misfortune"*: JQA/LCA, 8/3/1804. APM.

350 *"inexpressibly shocked"*: LCA/JQA, 8/14/1804. APM.

350 *"still retain'd"*: LCA/JQA, 8/14/1804. APM.

351 *"But without intending to affect"*: JQA/LCA, 9/2/1804. APM.

351 *"was very anxious to know"*: LCA/JQA, 7/4/1804. APM.

352 *"I take it for granted that"*: JA/JQA, 12/6/1804. APM.

353 *"a man of very insinuating"*: DJQA, 1/9/1804. APM.

354 *"I cannot conceive any"*: JQA/LCA, 7/19/1804. APM.

354 *"The conduct of Mr. Burr"*: JQA/LCA, 7/22/1804. APM.

354 *"was a man of considerable"*: JQA/LCA, 9/2/1804. APM.

355 *"As to Madame de Stael's"*: JQA/LCA, 5/25/1804. APM.

355 *"I pay so much attention"*: JQA/LCA, 5/31/1804. APM.

356 *"A most pernicious insect"*: JQA/LCA, 7/4/1804. APM.

357 *"It will not answer"*: DJQA, 10/3/1804. APM.

358 *"I miss you, beyond expression"*: JA/JQA, 11/6/1804. APM.

358 *"clasped each other together":* JA/JQA, 11/9/1804. APM.

CHAPTER 8: FIERY ORDEAL, 1805–1808

360 *"his itch for telling":* DJQA, 1/11/1805. APM.

361 *"the whole revenues of the US":* "Proposed Amendment to the Constitution on Representation," 12/1804. WJQA3, 87–100.

362 *"The President of the United States belongs":* Publius Valerius [JQA], *Serious Reflections, Addressed to the Citizens of Massachusetts. The Repertory* (Boston), No. V, 11/6/1804.

362 *"change is the only unchangeable":* JQA/TBA, 11/26/1804. APM.

362 *"Man is a social animal":* JA/JQA, 11/16/1804. APM.

363 *"fear of giving offense":* Publius Valerius [JQA], *Serious Reflections,* No. V, 11/6/1804.

364 *"There is more merit":* JQA/TBA, 11/19/1804. APM.

365 *"wish to employ me":* DJQA, 11/25/1805. APM.

366 *"entering upon a new mode":* DJQA, 4/25/1805. APM.

366 *"In these measures":* DJQA, 4/25/1805. APM.

366 *"informed me that the Corporation":* DJQA, 6/23/1805. APM.

368 *"If you were induced":* JQA/WSS, 3/26/1806. APM.

369 *"a very sorry prospect":* JQA/LCA, 11/28/1806. APM.

369 *"keeps up his Spirits":* JQA/LCA, 5/1/1806. APM.

370 *"As the adventure was followed"*: JQA/TBA, 1/20/1806. APM.

376 *"violent head-ache"*: DJQA, 4/22/1806. APM.

376 *"The loss of Mr. Adams's society"*: LCA/AA, 5/11/1806. APM.

377 *"I have continued very ill"*: LCA/JQA, 6/9/1806. APM.

377 *"a message of Misfortune"*: DJQA, 6/30/1806. APM.

377 *"deeply by its tenderness"*: DJQA, 6/30/1806. APM.

378 *"too bitterly the pangs"*: DJQA, 6/30/1806. APM.

378 *"My anxiety to see you"*: LCA/JQA, 7/20/1806. APM.

379 *"very awkward"*: JQA/LCA, 6/29/1806. APM.

380 *"when eloquence produced . . . heard in vain"*: Lectures on Rhetoric and Oratory (Cambridge, Mass., 1810), pp. 1, 31.

382 *"the last act of his political life"*: JQA/TBA, 4/1/1805. APM.

382 *"dignity, order, and morality"*: JQA/TBA, 4/1/1805. APM.

382 *"when he was a private"*: JQA/TBA, 4/1/1805. APM.

385 *"The hurricane of animosity"*: JQA/AA, 1/25/1806. APM.

385 *"be equivalent to"*: JQA/AA, 1/25/1806. APM.

386 *"to have committed"*: JA/JQA, 1/29/1806. APM.

388 *"a distressing consciousness"*: DJQA, 1/31/1806. APM.

389 *"Mr. Clay the new Member"*: DJQA, 1/2/1807. APM.

390 *"For your sake and for that"*: JQA/LCA, 1/6/1807. APM.

391 *"in consequence of"*: JQA/LCA, 1/14/1897. APM.

391 *"the effects which their exhibitions"*: JQA/LCA, 2/6/1807. APM.

392 *"the sauciest lines I ever"*: LCA/JQA, 2/17/1807. APM.

392 *"When first, in Eden's"*: JQA, 1/1807. APM.

393 *"God bless you my best beloved"*: LCA/JQA, 2/10/1807. APM.

393 *"Friend of my bosom"*: JQA, 2/12/1807. APM.

396 *"to familiarize him"*: DJQA, 10/18/1806. APM.

396 *"He is born to be lucky"*: JQA and LCA/Catherine Nuth Johnson, 8/20/1807. APM.

396 *"In about five minutes"*: DJQA, 8/17/1807. APM.

397 *"in remembrance of"*: DJQA, 9/13/1807. APM.

397 *"Though systematically a man"*: DJQA, 7/20/1807. APM.

399 *"I should* have my head*"*: DJQA, 7/11/1807. APM.

400 *"in the meantime your nation"*: DJQA, 11/25/1807. APM.

400 *"includes the whole compass"*: DJQA, 11/25/1807. APM.

400 *"very anxious to make"*: DJQA, 11/25/1807. APM.

401 *"that this country"*: DJQA, 11/14/1807. APM.

401 *"an obvious strong disposition"*: DJQA, 11/17/1807. APM.

402 *"The experiment has been"*: JQA/Joseph Hall, 12/11/1807. WJQA3, 164–166.

403 *"I do not believe"*: JQA/JA, 12/27/1807. APM.

404 *"You may as well drive"*: JA/JQA, 1/8/1808. APM.

404 *"for the appointment"*: DJQA, 1/11/1808. APM.

405 *"if our present rulers"*: JA/JQA, 1/8/1808. APM.

405 *"I have long since"*: JA/JQA, 1/17/1808. APM.

406 *"Some days after"*: JQA/TBA, 3/12/1808. APM.

407 *"mere spectator"*: JQA/TBA, 2/6/1808. APM.

407 *"Deeming it inconsistent"*: JQA/JA, 2/27/1808. APM.

408 *"making advances"*: JQA/TBA, 3/12/1808. APM.

408 *"principles were too pure"*: DJQA, 2/1/1808. APM.

408 *"On your return home"*: TBA/JQA, 3/24/1808. APM.

409 *"will bring upon me"*: DJQA, 3/31/1808. APM.

409 *"The embargo was the only"*: JQA/Harrison Gray Otis, 3/31/1808. WJQA3, 189–232.

410 *"Confidence is the only cement"*: JQA/Harrison Gray Otis, 3/31/1808. WJQA3, 189–232.

410 *"I found him as I expected"*: DJQA, 5/8/1808. APM.

411 *"a rude and indecent"*: DJQA, 5/18/1808. APM.

412 *"appears to be the political"*: James Sullivan/Thomas Jefferson, 6/3/1808. WJQA3, 236–237.

412 *"enforce the means"*: JQA/The Honorable Senate and House of Representatives of the Commonwealth of Massachusetts, 6/8/1808. WJQA3, 237–238.

CHAPTER 9: PARADISE OF FOOLS, 1809–1814

413 *"I have lost many friends"*: JQA/TBA, 8/7/1809. APM.

414 *"it is not magnanimous"*: JQA/AA, 10/14/1810. APM.

415 *"the year of my life"*: DJQA, 7/11/1809. APM.

415 *"Your exile will be"*: JA/JQA, 1/17/1808. APM.

416 *"politics threaten me"*: DJQA, 11/30/1808. APM.

416 *"to represent this"*: DJQA, 9/26/1808. APM.

417 *"came to a determination"*: DJQA, 11/29/1808. APM.

417 *"Let me hope"*: DJQA, 12/12/1808. APM.

418 *"For among themselves"*: JQA/Ezekiel Bacon, 11/17/1808. WJQA3, 248–253.

419 *"almost entirely upon"*: DJQA, 1/26/1809. APM.

419 *"the rage of politics"*: JQA/LCA, 1/1/1809. APM.

419 *"I find the pressure"*: DJQA, 1/16/1809. APM.

421 *"the slightest foundation"*: JQA/LCA, 2/26/1809. APM.

421 *"found the President reading"*: JQA/LCA, 2/21/1809. APM.

421 *"the Court did"*: DJQA, 3/2/1809. APM.

422 *"The crowd was excessive"*: DJQA, 3/3/1809. APM.

423 *"convenience would admit"*: DJQA, 3/6/1809. APM.

423 *"a whisper circulating"*: DJQA, 3/7/1809. APM.

423 *"In respect to ourselves"*: JQA/LCA, 3/9/1809. APM.

424 *"cannot be a citizen"*: WC/JQA, 3/15/1809. APM.

424 *"delighted to learn"*: JQA/TBA, 10/27/1810. APM.

425 *"this is itself a sort"*: DJQA, 5/31/1809. APM.

426 *"informed that he was"*: DJQA, 5/31/1809. APM.

426 *"the nomination to St. Petersburg"*: William Eustis/JQA, 6/10/1809. WJQA3, 317.

427 *"on American ground"*: William Eustis/JQA, 7/16/1809. WJQA3, 332–333.

427 *"to a man of active talents"*: Ezekiel Bacon/JQA, 6/29/1809. WJQA3, 321.

427 *"My personal motives"*: DJQA, 7/5/1809. APM.

428 *"a firm conviction"*: JQA/Robert Smith, 7/5/1809. WJQA3, 330–332.

428 *"This embassy to Russia"*: AA/ESS, 7/18/1809. APM.

429 *"O it was too"*: D&AW1, 283–284. 8/4/1809.

430 *"Youth is generous"*: JQA/TBA, 8/8/1809. APM.

430 *"with every pulsation"*: Lectures on Rhetoric and Oratory, pp. 2, 400.

433 *"which would have melted"*: DJQA, 8/6/1809. APM.

433 *"My dear children"*: AA/JQA and LCA, 8/5/1809. APM.

433 *"At this Commencement"*: DJQA, 8/5/1809. APM.

434 *"We landed on the quay"*: DJQA, 10/22/1809. APM.

436 *"the baseness of the mere"*: DJQA, 9/3/1809. APM.

437 *"the sight of so many"*: DJQA, 9/20/1809. APM.

438 *"I had objects on board"*: JQA/TBA, 11/16/1809. APM.

438 *"so full of rats"*: D&AW1, 293. 10/27/1809.

438 *"Everything here as to price"*: LCA/AA, 10/28/1809. APM.

439 *"powdered with a white"*: DJQA, 2/18/1810. APM.

440 *"I can take no pleasure"*: DJQA, 3/24/1810. APM.

440 *"extravagance and dissipation"*: JQA/AA, 2/8/1810. APM.

441 *"You can form no idea"*: LCA/AA, 1/7/1810. APM.

441 *"through the continual"*: DJQA, 1/28/1810. APM.

441 *"This mode of life"*: D&AW1, 310. 2/3/1810.

442 *"I am just recovering"*: LCA/AA, 7/19/1810. APM.

443 *"this day gave me information"*: DJQA, 2/12/1810. APM.

444 *"was a great Belle"*: D&AW1, 316. 4/29/1810.

444 *"attached to a young Russian"*: LCA, c. 1835. APM.

446 *"in a style of magnificence"*: JQA/AA, 3/19/1811. WJQA4, 22–27.

448 *"his subjugation of soul"*: DJQA, 4/15/1810. APM.

449 *"of a broken heart"*: DJQA, 11/9/1812. APM.

449 *"which of all the movable"*: JQA/AA, 3/19/1811. APM.

450 *"encumbered with innumerable"*: JQA/AA, 3/25/1813. APM.

450 *"suitable to a people"*: DJQA, 4/15/1814. APM.

450 *"I wish the time"*: LCA/MSC, 8/27/1810. APM.

450 *"the stagnant political atmosphere"*: JQA/TBA, 4/10/1811. APM.

451 *"The best of all possible"*: DJQA, 11/5/1813. APM.

451 *"This is an exile"*: LCA/AA, 5/13/1810. APM.

451 *"I hope you may have"*: AA/James Madison, 8/1/1810. APM.

452 *"As no communication"*: James Madison/JQA, 10/16/1810. APM.

452 *"We end this year"*: D&AW1, 334. 12/31/1810.

453 *"he was sorry for it"*: DJQA, 1/25/1811. APM.

453 *"Nothing short of the extremist"*: JQA/TBA, 4/10/1811. APM.

453 *"There is great anxiety"*: DJQA, 7/22/1812. APM.

454 *"I always shall be too much"*: JQA/TBA, 4/10/1811. WJQA4, 43.

454 *"the servile drudgery of caucuses"*: JQA/TBA, 4/10/1811. WJQA4, 43.

456 *"debilitated by internal dissensions"*: JQA/AA, 6/6/1810. APM.

456 *"likely to be long":* JQA/TBA, 5/13/1811. WJQA4, 65–71.

457 *"I felt a strong attachment":* JQA/AA, 7/29/1811. WJQA4, 157–159.

457 *"immediately saw":* D&AW1, 344, 5/23/1811.

457 *"The shock was sudden":* DJQA, 5/23/1811. APM.

458 *"the ill health which has":* DJQA, 7/26/1811. APM.

459 *"May the Mercy of God":* DJQA, 7/23/1811. APM.

459 *"as healthy and lively":* JQA/AA, 8/20/1811. APM.

459 *"a malignant bilious fever":* AA/JQA, 11/17/1811. APM.

459 *"My Poor Mother":* LCA/AA, 11/26/1811. APM.

460 *"consigned to the tomb":* AA/LCA, 11/26/1811. APM.

460 *"the frail tenure":* DJQA, 1/29/1812. APM.

460 *"You have a sweet":* LCA/GWA, 6/14/1812. APM.

461 *"Every one who sees her":* D&AW1, 354. 2/11/1812.

461 *"Warm baths, and injections":* DJQA, 9/14/1812. APM.

461 *"hair to be cut off":* DJQA, 9/14/1812. APM.

461 *"I pray that the calamity":* DJQA, 9/20/1812. APM.

461 *"My babes image flits":* D&AW1, 359. 11/5/1812.

462 *"cold blank and dreadful":* D&AW1, 358. 10/24/1812.

462 *"I am a useless being":* D&AW1, 359. 11/6/1812.

462 *"Her last moments":* DJQA, 9/15/1812. APM.

462 *"for the ordinary occupations":* DJQA, 9/10/1812. APM.

462 *"how keen and severe":* DJQA, 9/17/1812. APM.

462 *"If there be a moral Government":* JQA/AA, 9/21/1812. APM.

464 *"it has become a sort"*: JQA/AA, 12/31/1812. WJQA4, 422.

464 *"literally and really sick"*: JQA/AA, 12/31/1812. WJQA4, 424.

464 *"contrary to the course"*: JQA/AA, 12/31/1812. WJQA4, 424.

465 *"If all the people"*: JQA/R. G. Beasley, 4/29/1813. WJQA4, 478–483.

466 *"It is now for the God"*: JQA/TBA, 4/3/1813. APM.

466 *"to obtain a stipulation"*: DJQA, 7/24/1813. APM.

467 *"which would at least"*: AA/James Monroe, 4/3/1813. APM.

468 *"This day thirty years ago"*: JQA/JA, 9/3/1813. WJQA4, 512–516.

468 *"But from my Heart"*: "Lord of Creation," 8/1813. Miscellanies. APM.

469 *"very serious conversation"*: DJQA, 1/18/1813. APM.

469 *"looks dark"*: AA/ESS, 6/26/1813. APM.

469 *"May no lesson"*: DJQA, 10/28/1813. APM.

469 *"A nervous agitation"*: DJQA, 11/21/1812. APM.

470 *"at the idea of what"*: LCA/AA, 4/4/1813. APM.

470 *"Mr A is even more"*: LCA/AA, 9/2/1813. APM.

471 *"The plan of the Poem"*: DJQA, 10/20/1813. APM.

471 *"The chapter on emigration"*: DJQA, 5/20/1813. APM.

471 *"a philosophical Romance"*: DJQA, 12/2/1813. APM.

471 *"the afflictions of the righteous"*: DJQA, 12/2/1813. APM.

472 *"among the most distinguished"*: DJQA, 3/6/1814. APM.

473 *"The Empire of Napoleon"*: JQA/AA, 3/30/1814. WJQA5, 26.

CHAPTER 10: MY WANDERING LIFE, 1814–1817

475 *"the same house where"*: DJQA, 6/18/1814. APM.

475 *"several of the most"*: DJQA, 6/22/1814. APM.

478 *"a fair Lady"*: DJQA, 5/4/1814. APM.

479 *"his extraordinary genius"*: JQA/AA, 5/12/1814. WJQA5, 42–44.

479 *"the darling of the human"*: JQA/LCA, 7/2/1814. APM.

479 *"the prisoner of the ice"*: JQA/LCA, 5/13/1814. APM.

479 *"quickness of understanding"*: JQA/AA, 3/30/1814. APM.

481 *"His school has been"*: DJQA, 3/9/1821. APM.

483 *"which approved the war"*: DJQA, 11/19/1813. APM.

483 *"Divided among ourselves"*: JQA/Levett Harris, 11/15/1814. WJQA5, 186–188.

487 *"The continuance of the American war"*: Liverpool/Castlereagh, 11/2/1814. WJQA5, 176.

488 *"I think we have determined"*: Liverpool/Castlereagh, 11/18/1814. WJQA5, 179–180.

489 *"I confess that I think"*: Wellington/Liverpool, n.d. WJQA5, 179–180.

489 *"I was earnestly desirous"*: JQA/William Harris Crawford, 11/17/1814. WJQA5, 192–195.

490 *"the peace on the basis"*: DJQA, 11/24/1814. APM.

490 *"of the fixed determination"*: Henry Goulburn/ Bathurst, 11/25/1814. WJQA5, 182–183.

491 *"Mr. Gallatin brought us"*: DJQA, 11/29/1814. APM.

492 *"hoped it would be"*: DJQA, 12/24/1814. APM.

493 *"the study of his life"*: JQA/LCA, 12/16/1814. WJQA5, 237–241.

493 *"would say that one"*: JQA/LCA, 12/16/1814. WJQA5, 237–241.

494 *"The hatred [here] of the English"*: JQA/LCA, 10/28/1814. WJQA5, 174–175.

494 *"Thus ends the most"*: DJQA, 12/31/1814. APM.

494 *"the endless night"*: DJQA, 12/21/1814. APM.

494 *"Whenever the kneeling"*: DJQA, 1/29/1815. APM.

495 *"After an interval"*: JQA/AA, 3/4/1816. WJQA5, 522–526.

495 *"The tendency to dissipation"*: DJQA, 2/12/1815. APM.

495 *"its streets, its public walks"*: JQA/AA, 2/21/1815. WJQA5, 277–280.

496 *"objects of great wretchedness"*: DJQA, 2/4/1815. APM.

497 *"was any relation to"*: DJQA, 2/7/1815. APM.

497 *"The taste for frequenting"*: DJQA, 3/5/1815. APM.

497 *"the most enchanting music"*: DJQA, 2/13/1815. APM.

498 *"It is the contempt"*: DJQA, 3/8/1815. APM.

498 *"crowded as if it had been"*: DJQA, 3/9/1815. APM.

498 *"in many respects"*: JQA/AA, 3/4/1816. WJQA5, 522–526.

500 *"Presently I heard"*: D&AW1, 401–402.

502 *"concurred in the opinion"*: DJQA, 3/14/1815. APM.

502 *"If the slightest reliance"*: JQA/AA, 3/19/1815. WJQA5, 290–294.

502 *"The cries of Vive l'Empereur"*: DJQA, 5/7/1815. APM.

503 *"the soldiers would say"*: DJQA, 3/19/1815. APM.

503 *"the same garrison"*: DJQA, 3/21/1815. APM.

503 *"She and Charles"*: DJQA, 3/23/1815. APM.

503 *"sympathy of sentiment"*: JQA/AA, 4/22/1815. WJQA5, 299–304.

503 *"If the people of Paris"*: JQA/AA, 4/22/1815. WJQA5, 299–304.

504 *"the government of Louis XVIII"*: JQA/AA, 4/22/1815. WJQA5, 299–304.

504 *"will be to save"*: DJQA, 5/6/1815. APM.

504 *"is no fit person"*: JQA/AA, 3/4/1816. WJQA5, 522–526.

505 *"Had the name of Napoleon"*: JQA/AA, 3/4/1816. WJQA5, 522–526.

505 *"stood about five minutes"*: DJQA, 4/9/1815. APM.

505 *"Never at any public theatre"*: JQA/AA, 3/4/1816. WJQA5, 522–526.

506 *"on the same side"*: DJQA, 4/21/1816. APM.

506 *"upon the interest"*: JQA/AA, 3/4/1816. WJQA5, 522–526.

506 *"Envoy Extraordinary"*: James Madison/JQA, 2/28/1815. WJQA5, 276–277.

507 *"so odious that if"*: DJQA, 5/11/1815. APM.

507 *"I determined to proceed"*: DJQA, 5/8/1815. APM.

508 *"He wants Napoleon"*: DJQA, 5/15/1815. APM.

508 *"for a sample of Champagne"*: DJQA, 5/21/1815. APM.

508 *"a small assortment"*: DJQA, 5/22/1815. APM.

509 *"Dover Castle was"*: DJQA, 5/24/1815. APM.

509 *"recollected my father"*: DJQA, 5/24/1815. APM.

509 *"I had travelled"*: DJQA, 5/25/1815. APM.

510 *"the Prince took the letter"*: DJQA, 6/8/1815. APM.

510 *"would beat them all"*: DJQA, 6/8/1815. APM.

511 *"played bat and ball"*: DJQA, 7/12/1815. APM.

512 *"It has in relation"*: DJQA, 7/11/1815. APM.

512 *"a paradise compared"*: JQA/AA, 3/25/1816. WJQA5, 542–545.

512 *"They have never before"*: DJQA, 8/10/1815. APM.

514 *"each with a different"*: DJQA, 11/8/1816. APM.

514 *"saw a man, decently dressed"*: DJQA, 11/8/1816. APM.

516 *"With this assurance"*: DJQA, 7/4/1815. APM.

516 *"who was extremely ill"*: DJQA, 6/22/1815. APM.

516 *"the pistol flew out"*: DJQA, 10/13/1815. APM.

517 *"My whole course of life"*: DJQA, 10/14/1815. APM.

517 *"bear the light"*: DJQA, 10/25/1815. APM.

517 *"I was nearly delirious"*: DJQA, 10/28/1815. APM.

517 *"Thick purulent matter"*: DJQA, 10/28/1815. APM.

517 *"It was now that I knew"*: DJQA, 10/30/1815. APM.

518 *"that the disorder was"*: DJQA, 11/4/1815. APM.

518 *"After eighteen days"*: DJQA, 11/10/1815. APM.

519 *"There is more of nature"*: DJQA, 12/7/1815. APM.

519 *"My eyes and hand"*: DJQA, 11/16/1815. APM.

520 *"Table-Cloth Oratory"*: DJQA, 6/5/1816. APM.

520 *"distinguished by his love"*: JQA/AA, 6/6/1816. WJQA6, 40–45.

520 *"an object of more willing"*: JQA/AA, 6/6/1816. WJQA6, 40–45.

520 *"this is one of the many"*: DJQA, 8/8/1816. APM.

521 *"which has already felt"*: JQA/AA, 12/27/1815. WJQA5, 454.

522 *"That there is nothing"*: JQA/JA, 10/9/1815. APM.

522 *"May it please God"*: DJQA, 2/20/1816. APM.

523 *"the bias of my mind"*: JQA/JA, 8/31/1815. WJQA5, 360–364.

523 *"I find myself growing"*: DJQA, 4/11/1816. APM.

523 *"You caution me"*: JQA/JA, 1/3/1817. APM.

524 *"with the sublime Platonic"*: JQA/AA, 6/6/1816. WJQA6, 40–45.

524 *"They are all"*: DJQA, 5/28/1817. APM.

524 *"I have given up"*: JQA/AA, 6/6/1816. WJQA6, 40–45.

525 *"But I am aware that"*: JQA/AA, 6/6/1816. WJQA6, 40–45.

525 *"They are remarkably brilliant"*: DJQA, 12/18/1815. APM.

526 *"the suppression of the slave trade"*: DJQA, 6/6/1817. APM.

527 *"with the manacles of feudal"*: JQA/Samuel Dexter, 4/14/1816. WJQA6, 12–16.

528 *"has much of the ardor"*: DJQA, 2/17/1816. APM.

529 *"he is certainly at fifty"*: JQA/AA, 10/5/1816. WJQA6, 101–102.

530 *"poetical paroxysm"*: DJQA, 10/31/1816. APM.

530 *"Could I have chosen":* DJQA, 10/16/1816. APM.

530 *"a wayward boy":* GWA, "To My Mother." Miscellanies. APM.

531 *"great ingenuity and benevolence":* DJQA, 4/29/1817. APM.

531 *"very notorious":* DJQA, 7/29/1816. APM.

531 *"The train of thought":* DJQA, 12/10/1816. APM.

532 *"union of harmony":* DJQA, 9/15/1816. APM.

533 *"All which was necessary":* DJQA, 1/29/1816. APM.

533 *"improved, and amended":* DJQA, 2/3/1817. APM.

533 *"There is too much of rant":* DJQA, 2/3/1817. APM.

533 *"still a favorite":* DJQA, 5/10/1817. APM.

534 *"Respect for your talents":* James Monroe/JQA, 3/6/1817. WJQA6, 165–166.

534 *"Every man knows":* DJQA, 4/7/1817. APM.

534 *"It is probable that":* DJQA, 4/26/1817. APM.

535 *"London. Farewell":* DJQA, 6/10/1817. APM.

CHAPTER 11: THE TERRIBLE SUBLIME, 1817–1821

537 *"in the whirlwind":* JQA/Richard Rush, 8/21/1817. WJQA6, 185–186.

537 *"There is a capacity":* DJQA, 7/23/1817. APM.

538 *"roving about the world":* DJQA, 6/18/1817. APM.

539 *"Oh! the voracious maw":* DJQA, 12/31/1817. APM.

539 *"be radically and forever cured":* JQA/William Plumer, 10/5/1815. APM.

541 *"Much as I must disapprove":* JQA/CA, 6/9/1796. AFC11, 311.

542 *"revealed our darkest side"*: JQA/Alexander Everett, 3/16/1816. WJQA5, 537–542.

542 *"an anxious forecast"*: DJQA, 8/6/1817. APM.

543 *"The seeds of the Declaration"*: DJQA, 12/27/1819. APM.

543 *"contributing equally"*: DJQA, 8/22/1817. APM.

543 *"the magnificent Steam Boat"*: DJQA, 8/10/1817. APM.

544 *"inexpressible happiness"*: DJQA, 8/18/1817. APM.

544 *"Ancient friends and acquaintance"*: DJQA, 8/19/1817. APM.

546 *"to ride in the whirlwind"*: JQA/Richard Rush, 8/21/1817. WJQA6, 185–186.

548 *"The path before me"*: DJQA, 9/20/1817. APM.

549 *"have always over-estimated"*: JQA/AA, 4/23/1817 and 5/16/1817. WJQA6, 178–182.

549 *"I have known few"*: JQA/AA, 4/23/1817 and 5/16/1817. WJQA6, 178–182.

549 *"how much more delicate"*: DJQA, 5/16/1817. APM.

550 *"had the good fortune"*: DJQA, 5/16/1817. APM.

550 *"with a deep sense"*: JQA/AA, 4/23/1817 and 5/16/1817. WJQA6, 178–182.

550 *"to support, and not to counteract"*: JQA/AA, 4/23/1817 and 5/16/1817. WJQA6, 178–182.

550 *"as if his own"*: JQA/AA, 11/2/1817. WJQA6, 227–229.

550 *"Smith and his wife"*: DJQA, 12/6/1817. APM.

551 *"that it was impossible"*: DJQA, 9/25/1827. APM.

551 *"I cannot ask of heaven"*: JQA/JA, 8/1/1816. WJQA6, 58–62.

551 *"This sacrifice of all"*: DJQA, 3/28/1816. APM.

551 *"Business crowds upon me"*: JQA/AA, 11/2/1817. WJQA6, 227–229.

552 *"Yet when I look at"*: JQA/Thomas Jefferson, 10/9/1817. WJQA6, 217–221.

553 *"The subject is great"*: DJQA, 8/19/1820. APM.

556 *"mounted his South American"*: DJQA, 12/6/1817. APM.

557 *"had not taken the possession"*: DJQA, 1/10/1818. APM.

557 *"upon which we could"*: DJQA, 1/10/1818. APM.

558 *"to repair immediately"*: DJQA, 12/26/1817. APM.

558 *"that the arms"*: Andrew Jackson/James Monroe, 1/6/1818. http://thepapersofandrewjackson.utk.edu/documents_monroe.htm.

559 *"how it would suit"*: DJQA, 4/8/1818. APM.

560 *"The President's letter"*: DJQA, 12/24/1817. APM.

563 *"would approve General Jackson's"*: JQA/James Monroe, 7/8/1818. WJQA6, 383–384.

563 *"Spain must immediately"*: JQA/George Erving, 11/28/1818. WJQA6, 474–502.

564 *"the ablest State papers"*: D&AW2, 407. 1/24/1819.

566 *"the vast majority"*: DJQA, 2/2/1819. APM.

566 *"we are now approaching"*: DJQA, 2/11/1819. APM.

566 *"I observed there was"*: DJQA, 2/20/1819. APM.

566 *"involuntary exultation"*: DJQA, 2/21/1819. APM.

567 *"our hero looked depressed"*: D&AW2, 430. 3/4/1819.

568 *"an intuitive and natural"*: George Watterston, *Letters from Washington on the Constitution and Laws; with Sketches of Some of the Prominent Public*

Characters of the United States (Washington, D.C., 1818), pp. 43–48.

569 *"harmony of nature"*: LCA, "On the Portrait of My Husband," c. 1817. APM.

569 *"Nigh twenty years of lengthened span"*: "To My Wife," c. 1817. Miscellanies. APM.

570 *"confidential correspondent"*: JQA/JA, 12/14/1818. APM.

570 *"a sort of coldness"*: D&AW2, 12. 2/12/1819.

570 *"the duty of a man"*: DJQA, 6/4/1819. APM.

572 *"it was caused by a nest"*: DJQA, 8/22/1818. APM.

573 *"Antoine killed a brownish snake"*: DJQA, 8/12/1819. APM.

573 *"Among the desires"*: DJQA, 9/5/1818. APM.

574 *"deeply concerned"*: DJQA, 9/10/1818. APM.

574 *"a great relief to my mind"*: JQA/TBA, 12/18/1818. APM.

574 *"firm, prudent"*: DJQA, 12/13/1818. APM.

574 *"I am waiting with"*: JQA/GWA, 11/17/1818. APM.

575 *"my dear son George"*: JQA/GWA, 12/1818. APM.

576 *"a perpetual course"*: DJQA, 9/24/1818. APM.

576 *"dangerously ill"*: DJQA, 10/26/1818. APM.

576 *"Oh, that she could have been"*: DJQA, 11/1/1818. APM.

577 *"fondly hover o'er thy lonely friend"*: LCA, 10/1818. APM.

577 *"She was always cheerful"*: DJQA, 11/3/1818. APM.

579 *"an infamous fraud"*: DJQA, 3/8/1819. APM.

579 *"had a care clouded"*: DJQA, 3/8/1819. APM.

580 *"he should have insisted"*: DJQA, 3/10/1819. APM.

580 *"of all the transactions"*: DJQA, 8/11/1819. APM.

580 *"a minister of peace"*: JQA/Richard Rush, 5/20/1819. WJQA6, 545–550.

581 *"the ground upon which"*: DJQA, 1/7/1820. APM.

582 *"I told him that"*: DJQA, 2/17/1820. APM.

582 *"This will undoubtedly"*: DJQA, 4/29/1820. APM.

583 *"so immersed in the Spanish"*: D&AW2, 504–505. 5/1/1820, 5/4/1820.

583 *"we should look to"*: DJQA, 5/22/1820. APM.

583 *"an express declaration"*: DJQA, 2/12/1821. APM.

584 *"The excessive curiosity"*: DJQA, 2/16/1819. APM.

584 *"There is a darkness"*: AA/TBA, 2/28/1802. APM.

588 *"the intentions of the Society"*: DJQA, 10/11/1817. APM.

590 *"are men of all sorts"*: DJQA, 3/12/1819. APM.

590 *"upon the great earnestness"*: DJQA, 3/12/1819. APM.

590 *"that the mass"*: DJQA, 4/30/1819. APM.

591 *"highest respect for the motives"*: DJQA, 4/30/1819. APM.

591 *"I apprehend the Society"*: DJQA, 4/29/1819. APM.

592 *"The Missouri Question"*: DJQA, 1/10/1820. APM.

592 *"it was a shocking thing"*: DJQA, 2/13/1820. APM.

593 *"I have within these"*: DJQA, 1/10/1820. APM.

593 *"By what fatality"*: DJQA, 2/11/1820. APM.

594 *"cool judgment and plain sense"*: DJQA, 2/11/1820. APM.

594 *"this question will be"*: DJQA, 1/10/1820. APM.

594 *"I consider slavery"*: JQA/Jonathan Jennings, 7/17/1820. WJQA7, 52–54.

595 *"for a nursery of rattle-snakes"*: JQA/Jonathan Jennings, 7/17/1820. WJQA7, 52–54.

595 *"great concern at the re-appearance"*: DJQA, 11/12/1820. APM.

596 *"not destined to survive"*: DJQA, 1/10/1820. APM.

CHAPTER 12: THE MACBETH POLICY, 1821–1825

599 *"It demolished at a stroke . . . of her example"*: An Address . . . Celebrating the Anniversary of Independence, at the City of Washington on the Fourth of July 1821 (Washington, D.C., 1821), pp. 31–32.

602 *"I have considered"*: JQA/Edward Everett, 1/31/1822. WJQA7, 197–207.

604 *"political fabric"*: DJQA, 2/16/1821. APM.

605 *"The President is often"*: DJQA, 8/9/1823. APM.

605 *"He rather turns"*: DJQA, 6/23/1820. APM.

605 *"more governed by"*: DJQA, 6/23/1820. APM.

606 *"The Virginian opposition"*: DJQA, 10/20/1821. APM.

607 *"under the banners"*: DJQA, 10/20/1821. APM.

608 *"Wirt: Not so"*: DJQA, 11/17/1819. APM.

610 *"had rendered such services"*: DJQA, 1/2/1822. APM.

610 *"I have lived"*: JQA/John D. Heath, 1/7/1822. WJQA7, 191–195.

610 *"I have been entirely"*: JQA/Robert Walsh, 11/22/1822. WJQA7, 330–333

612 *"I have no countermining"*: JQA/Robert Walsh, 6/21/1822. WJQA7, 270–280.

613 *"Whenever a man resorts"*: DJQA, 4/25/1820. APM.

613 *"If there has ever been"*: DJQA, 2/25/1821. APM.

613 *"About one half the members"*: DJQA, 1/18/1821. APM.

615 *"I have been too much"*: JQA/Robert Walsh, 11/22/1822. WJQA7, 330–333.

615 The Duplicate Letters: *The Duplicate Letters, the Fisheries and the Mississippi: Documents Relating to Transactions at the Negotiation at Ghent* (Washington, D.C., 1822).

616 *"I am weary and sick"*: JQA/LCA, 7/23/1822. WJQA7, 285.

617 *"was tortured, thumb screwed"*: DJQA, 10/22/1820. APM.

618 *"inexpressible. The world"*: DJQA, 9/1/1819. APM.

618 *"He is near eighty-four-years"*: DJQA, 9/1/1819. APM.

619 *"as eccentric as ever"*: D&AW2, 355–356. 12/21/1820.

619 *"My Children seem"*: D&AW2, 519. 11/15/1820.

619 *"magnifies his joys"*: D&AW2, 595. 8/16/1821.

620 *"a void which cannot"*: DJQA, 2/8/1821. APM.

620 *"foolish entanglement"*: D&AW2, 555. 2/8/1821.

620 *"There are many objections"*: D&AW2, 555. 2/8/1821.

621 *"Cares for the welfare"*: DJQA, 9/30/1821. APM.

621 *"very much improved"*: D&AW2, 656. 2/2/1823.

622 *"perhaps five or six years"*: DJQA, 7/30/1823. APM.

623 *"John was to have taken"*: DJQA, 8/27/1823. APM.

624 *"This affair involves"*: DJQA, 7/12/1823. APM.

624 *"a gracious offer"*: DJQA, 7/12/1823. APM.

624 *"I have made a disposal"*: DJQA, 7/31/1823. APM.

624 *"All the labor of the year"*: DJQA, 6/9/1824. APM.

625 *"I am so thoroughly unfit"*: D&AW2, 456. 1/15/1820.

625 *"Very sick all day"*: D&AW2, 616–617. 12/13/1821.

626 *"very desirous of finding"*: D&AW2, 616. 12/12/1821.

626 *"Mr. A—acts for the best"*: D&AW2, 615. 12/6/1821.

626 *"I am absolutely refused"*: D&AW2, 618. 12/17/1821.

626 *"the most dreadful day"*: D&AW2, 542. 1/10/1821.

627 *"With the dawn"*: JQA/LCA, 7/26/1822. APM.

627 *"once celebrated in a different way"*: D&AW2, 587. 7/26/1821.

627 *"Yet while the ecstasies"*: JQA, 7/11/1822. APM.

628 *"some permanent employment"*: DJQA, 10/3/1819. APM.

628 *"a great, if not an insuperable"*: DJQA, 10/3/1819. APM.

629 *"I know my husbands"*: D&AW2, 486. 3/12/1820.

629 *"is a dejection of Spirits"*: DJQA, 6/1/1820. APM.

631 *"overpowered by John's arguments"*: D&AW2, 680, 687. 12/20/1823, 1/8/1824.

631 *"I took mankind"*: JQA/Christopher Hughes, 9/26/1833. APM.

632 *"Whatever talents I possess"*: DJQA, 1/25/1819. APM.

632 *"We had a dinner today"*: D&AW2, 680. 12/19/1823.

633 *"one of those mixed"*: DJQA, 3/11/1821. APM.

634 *"not worth a straw"*: DJQA, 11/8/1821. APM.

634 *"a supercilious prejudice"*: DJQA, 11/8/1821. APM.

634 *"'But, I now only have to say'"*: DJQA, 1/27/1821. APM.

636 *"suspected this bursting forth"*: DJQA, 3/12/1824. APM.

638 *"that all the Spanish Colonies"*: DJQA, 10/9/1810. APM.

641 *"It would be more candid"*: DJQA, 11/7/1823. APM.

642 *"alarmed far beyond"*: DJQA, 11/13/1823. APM.

643 *"My purpose would be"*: DJQA, 11/21/1823. APM.

644 *"called me out and gave me"*: DJQA, 11/27/1823. APM.

644 *"We owe it, therefore"*: James Monroe, State of the Union Address, 12/2/1823. http://www.presidency.ucsb.edu/ws/?pid=29465#axzz2gkFXGxPI.

645 *"Every day brings forth"*: D&AW2, 678. 12/13/1823.

645 *"the bitterness and violence"*: DJQA, 8/31/1824. M6, 415.

645 *"last bath in the Potomac"*: DJQA, 8/28/1824. APM.

646 *"the rate of ten miles"*: DJQA, 9/3/1824. APM.

646 *"the market-man"*: DJQA, 9/3/1824. APM.

646 *"all be moldering"*: DJQA, 9/20/1824. M6, 417–418.

647 *"Mr. Crawford's friends"*: DJQA, 5/25/1824. APM.

648 *"I further said, that although"*: DJQA, 4/22/1824. APM.

649 *"is an article of purchase"*: DJQA, 4/27/1824. APM.

651 *"would have received"*: David Walker Howe, *What Hath God Wrought: The Transformation of America, 1815–1848* (New York, 2007), p. 208.

652 *"ought to wish"*: DJQA, 5/8/1824. APM.

652 *"Yet a man qualified"*: DJQA, 5/8/1824. APM.

655 *"The elevation of the Hero"*: DJQA, 1/17/1825. M6, 469.

656 *"would be a member"*: DJQA, 1/21/1825. M6, 474.

656 *"spoke to me"*: DJQA, 1/29/1825. M6, 483.

657 *"the excitement of electioneering"*: DJQA, 2/3/1825. M6, 491.

657 *"Duplicity pervades the conduct"*: DJQA, 2/7/1825. M6, 499.

657 *"Never did I feel"*: JA/JQA, 2/28/1825. APM.

657 *"bed of roses"*: JQA/CFA, 5/16/1828. APM.

CHAPTER 13: NO BED OF ROSES, 1825–1829

659 *"Fellow citizens"*: Inaugural Address, 3/4/1825. http:// avalon.law.yale.edu/19th_century/qadams.asp.

659 *"Whatever vices there are"*: JA/Benjamin Rush, 1/23/1809. APM.

660 *"founded on geographical divisions . . . practical public blessing"*: Inaugural Address, 3/4/1825.

666 *"a sight loathsome"*: D&AW2, 613. 12/4/1821.

666 *"This is the most civilized"*: DJQA, 1/8/1824. APM.

666 *"There is not upon this globe"*: JQA/Benjamin Waterhouse, 10/24/1813. WJQA4, 526–527.

668 *"has accompanied one party"*: DJQA, 7/2/1828. APM.

668 *"speak with the great father"*: DJQA, 3/13/1828. APM.

668 *"I told them I had read"*: DJQA, 3/19/1828. APM.

669 *"They said the white people"*: DJQA, 11/29/1828. APM.

669 *"told them they had better"*: DJQA, 11/29/1828. APM.

670 *"to deliberate upon objects"*: First Annual Message, 12/6/1825. http://www.presidency.ucsb.edu/ws/?pid=29467#axzz2hPYyUcsg.

671 *"a project for opening"*: DJQA, 1/28/1825. APM.

672 *"conversation about cutting"*: DJQA, 7/30/1825. APM.

672 *"Would it not be worthwhile"*: DJQA, 4/26/1819. APM.

672 *"The Board of Engineers . . . common good"*: First Annual Message, 12/6/1825. http://www.presidency.ucsb.edu/ws/?pid=29467#axzz2hPYyUcsg.

678 *"was altogether unexpected"*: JQA/Levi Lincoln, 4/7/1826. APM.

679 *"be prejudicial to"*: Thomas Hart Benton, *Thirty Years' View; or, A History of the Working of the American Government for Thirty Years* (New York, 1858), vol. 1, pp. 65–69.

680 *"utterly foreign to"*: Benton, *Thirty Years' View*, vol. 1, pp. 65–69.

681 *"It is unquestionable . . . sky high"*: John Randolph, 3/30/1825. *Register of Debates* (Washington, D.C., 1825), 19th Congress, 1st session, pp. 391–403.

682 *"the eloquence of Hogarth's"*: *Documents Relating to New England Federalism* (Boston, 1877), p. 232.

682 *"My resentments . . . black leg"*: John Randolph, 3/30/1825. *Register of Debates*, pp. 391–403.

685 *"No president could"*: Benton, *Thirty Years' View*, vol. 1, p. 55.

687 *"It is among the rarest"*: JQA/WC, 8/17/1826. APM.

687 *"short and simple annals"*: *Works of John Adams, Second President of the United States, with a Life*

of the Author . . . By His Grandson Charles Francis Adams (Boston, 1856), vol. 1, p. 13.

688 *"oppressively hot night"*: DJQA, 8/10/1825. M8, 43.

688 *"the time piece of life"*: JQA/GWA, 11/28/1827. APM.

689 *"I consider every day"*: DJQA, 9/20/1820. APM.

689 *"took a firm resolve"*: DJQA, 4/11/1819. APM.

689 *"Visitors at home"*: DJQA, 4/11/1819. APM.

689 *"Above all an instinctive"*: DJQA, 4/11/1819. APM.

689 *"Had I spent upon any work"*: DJQA, 3/20/1821. APM.

689 *"will not accomplish"*: DJQA, 11/30/1821. APM.

690 *"I find it absolutely impossible"*: DJQA, 11/30/1828. APM.

690 *"I cannot indulge myself"*: DJQA, 5/10/1819. APM.

690 *"All the facts"*: DJQA, 4/2/1820. APM.

690 *"The idea that the execution"*: DJQA, 6/18/1820. APM.

691 *"The resurrection of the Spirit"*: DJQA, 7/23/1820. APM.

691 *"If Christian preachers"*: DJQA, 6/18/1820. APM.

691 *"I think God is too good"*: Samuel Hanson Cox, *Interviews: Memorable and Useful; from Diary and Memory Reproduced* (New York, 1853), pp. 213–274.

692 *"to which I was bred"*: DJQA, 10/24/1819. APM.

693 *"I can frequent without scruple"*: DJQA, 10/24/1819. APM.

693 *"sends forth the goodness"*: DJQA, 6/5/1825. APM.

693 *"the two pillars upon"*: DJQA, 1/7/1821. APM.

694 *"Religious liberty for ourselves"*: JQA/George Sullivan, 1/20/1821. WJQA7, 88–91.

695 *"The cock is a bird"*: JQA/CFA, 12/31/1827. APM.

695 *"between the peep of dawn"*: JQA/CFA, 12/31/1827. APM.

696 *"depression of spirits"*: D&AW2, 466. 2/5/1820.

696 *"a disease from which I have"*: JQA/CFA, 4/20/1828. APM.

696 *"the minstrel of morn"*: DJQA, "Sonnet to Chanticleer," 3/20/1827. M7, 244.

696 *"is the disorder of meditative"*: JQA/CFA, 4/20/1828. APM.

697 *"Having neither lot nor part"*: JQA/CFA, 1/29/1828. APM.

698 *"Oh! in the hour"*: DJQA37, 9/11/1825–6/24/1828. Inside cover. MHS.

699 *"violently sick"*: DJQA, 3/4/1825. APM.

699 *"inflammatory rheumatism"*: DJQA, 1/2/1829. APM.

699 *"the exchange to a more"*: DJQA, 2/19/1823. APM.

700 *"Go flattered image"*: LCA, "To My Sons with My Portrait," 12/17/1825. APM.

701 *"being discharged from"*: JQA/CFA, 9/15/1826. APM.

702 *"I do not hesitate to say"*: LCA/CFA, 5/25/1828. APM.

703 *"I have not in seven years"*: JQA/CFA, 11/11/1827. APM.

703 *"Do not think me"*: JQA/CFA, 11/7/1827. APM.

703 *"Early rising is so indissolubly"*: JQA/CFA, 2/9/1827. APM.

703 *"the master mind who"*: JQA/CFA, 2/9/1827. APM.

704 *"genius is the child of toil"*: JQA/CFA, 2/9/1827. APM.

704 *"the sport of idle coquetry":* GWA, *Diary of a Life,* 8/1825. APM.

704 *"My mother is half inclined":* DCFA1, 315. 9/6/1824.

705 *"to the Methodist camp":* DJQA, 8/8/1824. APM.

705 *"Without self control":* JQA/CFA, 12/24/1827. APM.

705 *"took it for consideration":* JQA/CFA, 12/24/1827. APM.

705 *"After the ceremony":* DJQA, 2/13/1828. APM.

706 *"licentious life":* JQA/GWA, 11/18/1827. APM.

706 *"I have seen in a late":* JQA/GWA, 11/18/1827. APM.

707 *"I rejoice heartily":* JQA/GWA, 2/25/1827. APM.

707 *"Irregular accounts are the direct":* JQA/GWA, 6/3/1827. APM.

707 *"Believe me when I say":* JQA/GWA, 5/19/1827. APM.

708 *"the last year of my":* DJQA, 3/2/1828. APM.

713 *"to seduce the passions":* DJQA, 1/30/1828. APM.

715 *"Slander and assassination":* JQA/CFA, 4/20/1828. APM.

717 *"It has been my endeavor":* JQA/CFA, 11/11/1827. APM.

717 *"about thirty thousand dollars":* JQA/GWA, 5/6/1827. APM.

717 *"to remember that I am":* JQA/CFA, 1/13/1829. APM.

718 *"I have laid the foundation":* JQA/CFA, 6/6/1828. APM.

718 *"the cultivation of forest trees":* JQA/CFA, 5/16/1828. APM.

719 *"without much success":* JQA/CFA, 5/16/1828. APM.

719 *"as a memento for them"*: JQA/CFA, 5/16/1828. APM.

720 *"Dendrology has become"*: JQA/GWA, 7/10/1828. APM.

720 *"If I had begun to plant"*: JQA/CFA, 5/16/1828. APM.

CHAPTER 14: THE USES OF ADVERSITY, 1829–1833

721 *"self-extinguished"*: DJQA, 1/1/1829. M8, 89.

722 *"witnessed the cloudless"*: DJQA, 1/1/1829. APM.

723 *"In looking back"*: DJQA, 11/1/1828. M8, 88.

725 *"For near two years and a half"*: Henry Clay, *Mr. Clay's Speech at the Dinner at Noble's Inn, Near Lexington, July 12, 1827* (Lexington, Ky., 1827), pp. 5–6.

726 *"be brought under the control"*: JQA/CFA, 3/3/1829. APM.

726 *"President, Vice President"*: JQA/CFA, 3/3/1829. APM.

727 *"disposed to hope"*: JQA/A Committee of a Numerous Meeting of the Citizens of Essex and Middlesex in New Jersey, 3/10/1829. APM.

728 *"there had never been"*: DJQA, 3/18/1808. M1, 521.

728 *"urged that a continuance"*: *Documents Relating to New England Federalism* (Boston, 1877), p. 41.

729 *"was rendered indispensable"*: JQA/GWA, 11/20/1828. APM.

729 *"My troubles thicken"*: DJQA, 12/3/1828. APM.

730 *"They persecuted my father"*: JQA/CFA, 12/31/1828. APM.

730 *"retirement will not shield"*: JQA/CFA, 12/31/1828. APM.

730 *"an unprofitable contest"*: JQA/CFA, 3/3/1829. APM.

731 *"the right of petition"*: JQA/To the Citizens of the United States, 2–3/1829. APM.

731 *"Such is the spirit"*: JQA/GWA, 1/13/1829. APM.

732 *"death by drowning"*: JQA/LCA, 6/15/1829. APM.

733 *"a temporary shelter"*: DJQA, 2/6/1829. APM.

733 *"My retirement is as complete"*: JQA/James Barbour, 4/4/1829. APM.

734 *"my wife, my son's wife"*: DJQA, 4/27/1829. APM.

734 *"not been disappointed"*: JQA/GWA, 1/13/1829. APM.

734 *"I had not"*: DJQA, 5/2/1829. APM.

735 *"Have compassion upon"*: DJQA, 5/3/1829. APM.

735 *"the impression that"*: DJQA, 5/4/1829. APM.

736 *"Oh! My unhappy son"*: DJQA, 5/5/1829. APM.

736 *"I believe that special Providences"*: DJQA, 5/7/1829. APM.

737 *"It is inexpressibly painful"*: DJQA, 5/9/1829. APM.

737 *"a grateful heart"*: JQA/Commodore Isaac Chauncey, 5/15/1829. APM.

737 *"the burial service"*: JQA/LCA, 6/15/1829. APM.

738 *"So long in memory"*: LCA, 4/12/1831. APM.

738 *"some remarks on parties"*: DJQA, 5/14/1829. APM.

738 *"The remnants of the Tories . . . imposing upon himself"*: Parties in the United States (New York, 1941), p. 9.

741 *"redeem his fame"*: DCFA2, 66. 7/9/1826.

741 *"He would have lived"*: DCFA2, 382. 5/28/1829.

742 *"But with a large"*: JQA/Joseph Gales Jr., 9/22/1830. APM.

742 *"My anxieties for both"*: JQA/LCA, 6/20/1829. APM.

743 *"The conversation was literary"*: DCFA3, 29. 9/27/1829.

743 *"with everything that"*: JQA/CFA, 12/18/1829. APM.

743 *"fair damsel"*: enclosed in JQA/Abigail Brooks Adams, 12/23/1829. APM.

744 *"There is beating of hearts"*: JQA/GWA, 3/13/1829. APM.

745 *"an administration one half"*: JQA/unknown, 4/1830. APM.

745 *"is that the affairs"*: JQA/William Plumer Jr., 10/25/1831. APM.

745 *"a holy and ever burning"*: JQA/CFA, 4/16/1829. APM.

747 *"now rages in the South"*: JQA/William Plumer, 9/24/1830. APM.

747 *"strengthen the ties"*: JQA/Robert Walsh, 12/26/1830. APM.

748 *"It is the odious nature"*: JQA/Henry Clay, 9/7/1831. APM.

749 *"No one knows me"*: DJQA, 2/7/1830. M8, 186.

749 *"memory returns to us"*: JQA/GWA, 9/23/1828. APM.

749 *"I have lost an always"*: JQA/Ann Harrod Adams, 3/17/1832. APM.

750 *"Mr. Jefferson's infidelity"*: JQA/Alexander Everett, 5/24/1830. APM.

751 *"came purposely to enquire"*: DJQA, 9/18/1830. M8, 239.

752 *"And so I am launched"*: DJQA, 10/13/1830. M8, 243.

752 *"I saw no warrantable"*: JQA/Samuel Southard, 12/6/1830. APM.

753 *"derogatory descent"*: DJQA, 11/6/1830. M8, 246.

753 *"has hung upon me"*: JQA/CFA, 12/22/1830. APM.

754 *"deserted by all mankind"*: JQA/CFA, 2/20/1832. APM.

754 *"My election as President"*: DJQA, 11/6/1830. M8, 247.

755 *"I am suffering some"*: JQA/JA2, 9/9/1831. APM.

755 *"Eloquence herself . . . union forever"*: An Oration Addressed to the Citizens of the Town of Quincy, on the Fourth of July, 1831, the Fifty-Fifth Anniversary of the Independence of the United States of America (Boston, 1831), pp. 5, 39.

757 *"it was a tribute"*: JQA/Charles J. Ingersoll, 9/8/1831. APM.

758 *"Your beau ideal of life"*: JQA/Charles J. Ingersoll, 9/8/1831. APM.

759 *"Mr. Clay asked me"*: DJQA, 12/26/1831. M8, 443.

759 *"Courage and force"*: JQA/CFA, 5/18/1828. APM.

760 *"is not exactly congenial"*: JQA/CFA, 5/18/1828. APM.

760 *"If I could take [Byron's] confession"*: JQA/Robert Walsh Jr., 7/25/1821. APM.

762 *"But further now"*: Dermot MacMorrogh; or, *The Conquest of Ireland: An Historical Tale of the Twelfth Century in Four Cantos* (Washington, D.C., 1832), pp. 40–41.

763 *"My style is the mock-heroic"*: DJQA, 3/10/1831. M8, 340.

767 *"I have petitioned almost"*: JQA/CFA, 12/15/1831. APM.

768 *"I repeated to him"*: DJQA, 9/14/1831. M8, 412–413.

768 *"I had voluntarily"*: JQA/Richard Rush, 8/3/1832. APM.

768 *"I clutch like a drowning man"*: JQA/LCA, 4/5/1832. APM.

770 *"There are two facts"*: George Wilson, *Pierson, Tocqueville and Beaumont in America* (New York, 1938); www.tocqueville.org.

772 *"the best part . . . convention"*: Report of the Minority of the Committee on Manufactures Submitted to the House of Representatives of the United States, February 28, 1833 (Boston, 1833), pp. 4–5.

775 *"Dermot MacMorrough is all prose"*: JQA/CFA, 4/1/1833. APM.

CHAPTER 15: A SUITABLE SPHERE OF ACTION, 1834–1839

777 *"It was so strong"*: JQA/CFA, 11/10/1833. APM.

777 *"about sixty feet in a second"*: JQA/CFA, 11/10/1833. APM.

778 *"Men, women, and a child"*: DJQA, 11/8/1833. M9, 31.

778 *"felt a strange and mingled"*: JQA/CFA, 11/10/1833. APM.

778 *"if no accident had happened"*: JQA/CFA, 11/10/1833. APM.

779 *"I brought myself to the conclusion"*: JQA/Christopher Hughes, 9/26/1833. APM.

779 *"As those we love"*: James Thomson, "On the Death of a Particular Friend," 1798.

780 *"Where is that Queen"*: JQA/Richard Rush, 11/24/1831. APM.

781 *"imaginary gold mine"*: JQA/JA2, 10/29/1830. APM.

783 *"Why will he waste"*: D&AW2, 704–705.

783 *"without risking a total"*: D&AW2, 702.

783 *"the extreme uneasiness"*: D&AW2, 704–705. 4/1/1836.

784 *"through the narrow pass"*: JQA/LCA, 9/1833. APM.

784 *"This is a subject"*: DJQA, 8/3/1834. APM.

785 *"more of a caucus club"*: JQA/Ward Nicholas Boylston, 5/24/1819. WJQA6, 551–553.

785 *"political coterie"*: JQA/Benjamin Waterhouse, 12/21/1829. APM.

785 *"can never regain his popularity"*: DJQA, 10/15/1834. APM.

786 *"We followed it until"*: DJQA, 7/31/1834. APM.

786 *"I have lost my appetite"*: DJQA, 7/23/1834. APM.

786 *"Independence Day"*: DJQA, 7/4/1834. APM.

786 *"in a declining and drooping state"*: DJQA, 10/23/1834. APM.

787 *"moral ruin"*: DCFA3, 46. 12/31/1834.

787 *"The thoughts of the future"*: DJQA, 7/26/1834. APM.

787 *"It is my undoubting conviction"*: JQA/JA2, 7/23/1834. APM.

788 *"a slovenly performance"*: DJQA, 8/28/1834. APM.

788 *"The dinner was excellent"*: DJQA, 8/28/1834. APM.

788 *"What an age of hope"*: DJQA, 8/28/1834. APM.

789 *"the most violent and long"*: JQA/JA2, 9/25/1834. APM.

789 *"half blind"*: DJQA, 9/2/1834. APM.

789 *"a bilious fever"*: DJQA, 9/24/1834. APM.

789 *"was much distressed"*: DJQA, 10/6/1834. APM.

789 *"John was extremely ill"*: DJQA, 10/18/1834. APM.

789 *"excessive sickness, faintings"*: DJQA, 10/18/1834. APM.

790 *"difficulty of resignation"*: DJQA, 10/19/1834. APM.

790 *"the controversial character"*: DJQA, 10/20/1834. APM.

790 *"very low"*: DJQA, 10/22/1834. APM.

790 *"At half past four"*: DJQA, 10/23/1834. APM.

791 *"he did not know"*: DJQA, 10/28/1834. APM.

791 *"his death was tranquil"*: DJQA, 10/24/1834. APM.

791 *"all the previous services"*: DJQA, 10/24/1834. APM.

792 *"Let me believe"*: DJQA, 10/24/1834. APM.

792 *"The resources of wretchedness"*: DJQA, 11/10/1834. APM.

792 *"melancholy pilgrimage"*: DJQA, 10/30/1834. APM.

793 *"Yet, strange as it may . . . career upon earth"*: *Oration on the Life and Character of Gilbert Motier De Lafayette* (Washington, D.C., 1835), pp. 7, 84–85.

795 *"for the plunder"*: DJQA, 12/12/1834. APM.

795 *"a great catastrophe"*: JQA/CFA, 11/19/1833. APM.

796 "SUPPRESSED BY": *Daily Advertiser and Patriot* (Boston, 1834).

798 *"I breasted them all"*: JQA/CFA, 3/5/1835. APM.

798 *"the last trophy"*: JQA/CFA, 4/8/1835. APM.

799 *"There is more of elementary"*: JQA/Benjamin Waterhouse, 9/11/1835. APM.

799 *"Every other leading"*: JQA/CFA, 1/25/1838. APM.

799 *"the battle of New Orleans"*: JQA/Benjamin Waterhouse, 9/11/1835. APM.

799 *"If I were the Grand Elector"*: JQA/Benjamin Waterhouse, 9/11/1835. APM.

800 *"a sense of duty"*: JQA/CFA, 4/8/1835. APM.

800 *"I must stay"*: JQA/CFA, 4/8/1835. APM.

801 *"Being Monday, the States were"*: DJQA, 2/20/1832. M8, 475.

802 *"read Shakespeare as"*: JQA/George Parkman, 11/19/1835. APM.

803 *"sensual passions"*: DJQA, 11/5/1831. APM.

803 *"the pleasure that we take"*: "Misconceptions of Shakespeare upon the Stage," *New England Magazine* 9 (December 1835): 435–440.

803 *"human action and passion"*: JQA/James H. Hackett, 2/19/1839. APM.

803 *"She not only violates"*: "Misconceptions of Shakespeare," pp. 435–440.

804 *"If the color of Othello"*: JQA/George Parkman, 12/31/1835. APM.

804 *"the moral of the tragedy"*: JQA/George Parkman, 12/31/1835. APM.

805 *"managers of our theatres"*: JQA/CFA, 5/13/1830. APM.

807 *"would lead to ill will"*: DJQA, 1/10/1832. M8, 454.

807 *"should petition"*: DJQA, 1/10/1832. M8, 454–455.

807 *"I abhorred slavery"*: DJQA, 1/10/1832. M8, 454.

808 *"was whether a population"*: DJQA, 12/24/1832. APM.

808 *"the day-dream of some"*: DJQA, 10/13/1833. APM.

808 *"Slavery is, in all probability"*: DJQA, 10/13/1833. APM.

810 *"a monument of the intellectual"*: DJQA, 10/13/1833. APM.

812 *"Freedom stands"*: John Greenleaf Whittier, "Stanzas for the Times," *POEMS, Written During the Progress of the Abolition Question in the United States, Between the Years 1830 and 1837* (Boston, 1837), pp. 41–44.

812 *"The slave drivers"*: JQA/CFA, 3/26/1833. APM.

813 *"If I am now finally"*: JQA/CFA, 3/26/1833. APM.

815 *"to sleep the sleep"*: Congressional Globe, 2/9/1837.

816 *"Slavery can never be"*: Congressional Globe, 2/4/1836.

817 *"Who shall dare"*: Documents Relating to New England Federalism (Boston, 1877), pp. 46–62.

818 *"Men of proud"*: JQA to TBA, 4/4/1801. APM.

818 *"Jean Jacques Rousseau"*: JQA/CFA, 5/24/1836. APM.

819 *"Congress possesses"*: Congressional Globe, 2/6/1836.

820 *"under the necessity"*: JQA/CFA, 5/2/1836. APM.

820 *"Am I being gagged"*: Congressional Globe, 5/25/1836.

820 *"Throughout the South"*: JQA/CFA, 5/2/1836. APM.

821 *"giving freedom"*: Congressional Globe, 6/9/1836.

822 *"has the effect"*: James Madison/Henry Clay, 6/1833. http://www.familytales.org/dbDisplay.php?id=ltr_mad5957.

823 *"The right of interference"*: JQA/Robert Walsh, 6/3/1836. APM.

824 *"threatened me"*: Letters from John Quincy Adams to His Constituents (Boston, 1837), pp. 13, 56.

824 *"called to the bar"*: Congressional Globe, 5/16/1837.

825 *"Let that gentleman"*: Congressional Globe, 2/9/1837.

825 *"If it is the law"*: Congressional Globe, 2/9/1837.

826 *"Dark Terror"*: LCA, "Threat of Assassination 1839," 2/18/1839. APM.

826 *"abolishing hereditary slavery"*: Congressional Globe, 2/25/1839.

826 *"The difference between"*: DJQA, 1/29/1840. M10, 206.

827 *"the most beautiful"*: DJQA, 11/18/1838. APM.

828 *"to the formation of other"*: DJQA, 11/18/1838. APM.

828 *"the most distressing"*: JQA/CFA, 1/25/1838. APM.

829 *"the fiftieth year"*: DJQA, 7/12/1834. APM.

829 *"I must wait"*: DJQA, 8/10/1834. APM.

829 *"Sculptor! Thy hand"*: JQA, "To Hiram Powers," 3/25/1837. APM.

830 *"his craft and duplicity"*: DJQA, 8/29/1836. M9, 305.

830 *"moderated some of"*: DJQA, 8/29/1836. M9, 306.

831 *"Jefferson was the father"*: JQA/Edward Everett, 10/10/1836. APM.

832 *"accordingly bad"*: DJQA, 9/27/1836. M9, 308–309.

833 *"What then is our duty . . . our fathers"*: A Eulogy on the Life and Character of James Madison, Fourth

President of the United States; Delivered at the Request of the Mayor, Aldermen, and Common Council of the City of Boston, September 27, 1836, by John Quincy Adams (Boston, 1836), pp. 86–87.

CHAPTER 16: ADHERING TO THE WORLD, 1838–1843

840 *"more than sixty years"*: DJQA, 3/23/1841. APM.

840 *"nothing better or more"*: DJQA, 7/12/1840. M10, 332.

840 *"to retire from"*: DJQA, 3/23/1841. APM.

841 *"and if my creditors"*: JQA/CFA, 4/29/1837. APM.

842 *"A divorce of Bank"*: JQA/Alexander Everett, 11/7/1837. APM.

842 *"Multitudes of [state] banks"*: JQA/CFA, 4/26/1837. APM.

842 *"Biddle broods"*: DJQA, 11/22/1840. M10, 361.

843 *"We are now in the midst"*: JQA/William Foster, 7/1/1837. APM.

843 *"Oh! Your Boston Banks"*: JQA/CFA, 2/7/1838. APM.

844 *"must have sulphuric acid"*: Ralph Waldo Emerson, *Journals* (Cambridge, Mass., 1982), vol. 6, pp. 349–350.

845 *"How they have weighed"*: JQA/CFA, 3/3/1837. APM.

845 *"Censure would be . . . from my hand"*: Letters from John Quincy Adams to His Constituents, pp. 19, 61.

848 *"the pestilence of exotic . . . thralldom of man"*: An Oration Delivered before the Inhabitants of the Town

of Newburyport (Newburyport, Mass., 1837), pp. 50–52.

850 *"My intent . . . Texian slavery"*: Speech of John Quincy Adams . . . upon the Right of the People, Men and Women, to Petition; On the Freedom of Speech and of Debate in the House of Representatives of the United States on the Resolutions of Seven Legislatures, and the Petitions of More Than One Hundred Thousand Petitioners Relating to the Annexation of Texas to This Union (Boston, 1838), pp. 63, 113, 131.

851 *"heartily approved"*: DJQA, 11/5/1834. APM.

852 *"lukewarm to the cause"*: DJQA, 11/5/1834. APM.

853 *"It is to the wives"*: Speech of John Quincy Adams . . . upon the Right of the People, Men and Women, pp. 55–56.

854 *"the tyrannical"*: Speech of John Quincy Adams . . . upon the Right of the People, Men and Women, p. 79.

855 *"it was the verdict of"*: Speech of John Quincy Adams . . . upon the Right of the People, Men and Women, pp. 79–80.

856 *"the laws of Nature"*: Social Compact Exemplified in the Constitution of the Commonwealth of Massachusetts (Providence, R.I., 1842), pp. 18–19, 25.

857 *"My principal apprehension"*: JQA/CFA, 4/23/1839. APM.

859 *"And now the future"*: Jubilee of the Constitution: A Discourse Delivered at the Request of the New-York Historical Society (New York, 1839), p. 118; Milton, Paradise Lost, Book 12, pp. 646–647.

859 *"the only mode of relieving"*: DJQA, 5/9/1840. M10, 285.

860 *"Man wants but little"*: "The Wants of Man," *Poems of Religion and Society* (Buffalo, N.Y., 1854), pp. 15–23.

863 *"a responsible hereafter"*: JQA/Alexander Everett, 11/7/1837. APM.

863 *"how little is the fruit"*: DJQA, 7/4/1834. APM.

863 *"in the existence of"*: DJQA, 3/19/1843. M11, 341.

863 *"involuntary and agonizing"*: DJQA, 7/25/1840. M11, 341.

865 *"unfortunate Africans"*: JQA/William Jay, 9/17/1839. APM.

865 *"their own right of liberty"*: JQA/William Jay, 9/17/1839. APM.

867 *"the right of capture"*: JQA/Ellis Gray Loring, 11/19/1839. APM.

867 *"piracies committed"*: JQA/Ellis Gray Loring, 11/19/1839. APM.

869 *"I endeavored to excuse"*: DJQA, 10/27/1840. M10, 358.

871 *"As soon as I heard"*: DJQA, 2/18/1841. M10, 425.

871 *"with a heart melted"*: DJQA, 2/20/1841. M10, 427.

872 *"bewildered"*: DJQA, 2/22/1841. M10, 429.

872 *"follow some order"*: DJQA, 2/21/1841. M10, 429.

872 *"fervent prayer that presence"*: DJQA, 2/22/1841. M10, 429.

872 *"sound and eloquent"*: DJQA, 2/22/1841. M10, 429.

873 *"powerful and perhaps"*: DJQA, 2/23/1841. M10, 430.

873 *"little short of agony"*: DJQA, 2/23/1841. M10, 430.

873 *"the very skeleton"*: DJQA, 2/23/1841. M10, 430.

873 *"that the Court was ready"*: DJQA, 2/24/1841. M10, 431.

873 *"perfectly simple"*: DJQA, 2/24/1841. APM.

873 *"a Court of JUSTICE . . . Justice"*: Argument of John Quincy Adams Before the Supreme Court of the United States, Appellants, vs. Cinque, and Others, Africans, Captured in the Schooner Amistad, by Lieut. Gedney, Delivered on the 24th of February and 1st of Match, 1841, with a Review of the Case of the Antelope, Reported in the 10th, 11th and 12th Volumes of Wheaton's Reports (New York, 1841), p. 3.

875 *"encouraged and cheerful"*: DJQA, 2/25/1841. APM.

876 *"May it please"*: Argument of John Quincy Adams, pp. 134–135.

878 *"tenterhooks"*: DJQA, 3/9/1841. M10, 441.

878 *"upon the merits"*: Supreme Court of the United States, 40 U.S. 518, January 1841 Term.

880 *"When you speak of Democracy"*: JQA/George Bancroft, 10/25/1835. APM.

883 *"May I be permitted . . . of human life"*: A Discourse on Education, Delivered at Braintree, Thursday, Oct. 24, 1839 (Boston, 1840), pp. 9, 14, 18, 32.

884 *"the Smithsonian Institution . . . sphere of existence"*: The Great Design: Two Lectures on the Smithson Bequest by John Quincy Adams, ed. Wilcomb E. Washburn (Washington, D.C., 1965), pp. 47–48.

887 *"The private interests"*: DJQA, 10/26/1839. M10, 139.

889 *"can it be wondered"*: JQA, *The Great Design: Two Lectures*, p. 93.

890 *"an incident of the most"*: DJQA, 7/4/1843. M11, 389.

891 *"saw the smoke ascending"*: DJQA, 6/17/1843. M11, 384.

892 *"Oh, my mother"*: DJQA, 8/1/1843. M11, 400.

892 *"a rash promise"*: DJQA, 7/25/1843. M11, 394.

892 *"that irresistible current"*: DJQA, 7/28/1843. M11, 398.

894 *"Some made faces"*: DJQA, 11/2/1843. M11, 419.

894 *"I write amidst"*: DJQA, 11/3/1843. M11, 420.

894 *"the only time, he said"*: DJQA, 10/6/1843. M11, 423.

894 *"the whole plain"*: DJQA, 11/9/1843. M11, 426.

895 *"spoke of the laws . . . arts and sciences"*: An Oration Delivered Before the Cincinnati Historical Society, on the Occasion of Laying the Cornerstone of an Astronomical Observatory (Cincinnati, 1843), pp. 16, 62.

CHAPTER 17: THE SUMMIT OF MY AMBITION, 1844–1848

897 *"dangerously ill"*: DJQA, 5/27/1843. M11, 378.

897 *"Some of the earliest"*: DJQA, 5/27/1843. M11, 378.

898 *"the thundering cannon"*: DJQA, 6/17/1843. M11, 383.

899 *"What have these to do"*: DJQA, 6/17/1843. M11, 384.

901 *"I see the hand of God"*: JQA/LCA, 8/25/1841. APM.

901 *"The most extraordinary feature"*: JQA/Henry Clay, 9/20/1842. APM.

901 *"almost to suffocation"*: JQA/LCA, 9/23/1841. APM.

902 *"double dealing"*: Address of John Quincy Adams to His Constituents of the Twelfth Congressional District, September 17, 1842 (Boston, 1842), p. 129.

903 *"Twice in the space"*: Address of John Quincy Adams to His Constituents, pp. 16, 57.

905 *"gentlemen from the South"*: Speech of Mr. J. R. Underwood upon the Resolution Proposing to Censure JQA for Presenting to the House of Representatives a Petition for the Dissolution of the Union (Washington, D.C., 1842), p. 14.

906 *"the acutest, the astutest"*: Barton Haxall Wise, The Life of Henry A. Wise (New York, 1899), p. 61.

906 *"Your prediction"*: JQA/CFA, 12/21/1842. APM.

908 *"the Slave power"*: JQA/CFA, 1/15/1844. APM.

908 *"The Treaty for the annexation"*: JQA/CFA, 4/15/1844. APM.

908 *"the conflict is between"*: JQA/William H. Seward, 4/16/1841. APM.

908 *"a beautiful ivory cane"*: JQA/Julius Pratt and Co., 4/23/1844. APM.

909 *"My time on earth"*: JQA/Julius Pratt and Co., 4/23/1844. APM.

909 *"Blessed, forever blessed"*: DJQA, 12/3/1844. M12, 116.

909 *"a daily-deepening consciousness"*: DJQA, 3/19/1843. M10, 341.

910 *"It has been on many"*: JQA/Henry Clay, 1/4/1845. APM.

910 *"From the day"*: JQA/Henry Clay, 9/20/1842. APM.

912 *"All this shall not"*: JQA/LCA, 9/23/1841. APM.

913 *"a perpetual harrow"*: DJQA, 6/30/1841. APM.

913 *"Their success was their misfortune"*: DJQA, 6/30/1841. APM.

914 *"The whole territory . . . or Pizarro"*: The New England Confederacy of MDCXLIII, A Discourse Delivered Before the Massachusetts Historical Society in Celebration of the Second Centennial Anniversary of the Event (Boston, 1843), pp. 14–15.

915 *"polemical porcupine"*: DJQA, 5/19/1843. M11, 376.

916 *"It is not for us . . . of the sky"*: JQA, *The New England Confederacy*, p. 11.

917 *"I believe that moral"*: DJQA, 5/6/1843. M11, 372.

918 *"an aged, grey-haired man"*: Charles Dickens, *American Notes* (London, 1842), ch. 8.

919 *"ate, and drunk"*: Dickens/Charles A. Davis, 4/4/1842. *The Letters of Charles Dickens*, vol. 3, *1842–1843* (Oxford, 1974), pp. 185–186.

919 *"Receive Lady"*: JQA/Catherine Dickens, 3/1842. APM.

922 *"always unable to write"*: DJQA, 5/20/1840. APM.

922 *"unclouded reason"*: DJQA, 3/20/1843. APM.

922 *"my boy companion"*: DJQA, 3/21/1843. APM.

922 *"There has perhaps not been"*: DJQA, 10/31/1846. APM.

923 *"If my intellectual powers"*: DJQA, 10/31/1846. APM.

923 *"that moral principle"*: DJQA, 5/6/1843. M11, 372.

924 *"I would, by the irresistible"*: DJQA, 10/31/1846. APM.

926 *"Texas will be only"*: JQA/Abigail Brooks Adams, 4/2/1845. APM.

926 *"would become indispensably"*: DJQA, 4/13/1820. APM.

926 *"not in the lands"*: DJQA, 4/13/1820. APM.

928 *"The memory of violent men"*: DJQA, 11/5/1844. M12, 101.

928 *"Shall we respond"*: JQA/Anna Quincy Thaxter, 7/29/1844. APM.

929 *"is written in the Book"*: DJQA, 2/19/1845. M12, 171.

929 *"a mere menstruous rag"*: DJQA, 2/19/1845. M12, 171.

929 *"I regard it as"*: DJQA, 2/28/1845. M12, 174.

929 *"an hour about democracy"*: DJQA, 1/31/1845. M12, 159.

930 *"She is heaping conquest"*: JQA/Joseph Sturge, 4/1846. APM.

930 *"that it was the meeting"*: DJQA, 2/13/1846. M12, 246.

933 *"Let me be told"*: Mr. Adams's Speech, on War with Great Britain, with the Speeches of Messrs. Wise & Ingersoll, to Which It Is a Reply (Boston, 1842), p. 34.

934 *"The war has never"*: JQA/Albert Gallatin, 12/26/1848. APM.

934 *"It is now established"*: JQA/Albert Gallatin, 12/26/1848. APM.

936 *"drawn by an irresistible"*: DJQA, 7/17/1846. M12, 268–269.

937 *"the desolation of the season"*: DJQA, 10/25/1844. M12, 96.

938 *"The recollection of the past"*: DJQA, 10/25/1844. M12, 97.

938 *"Will prayer to God"*: DJQA, 10/25/1844. M12, 97.

938 *"The pleasure I take"*: DJQA, 6/26/1845. M12, 200.

939 *"errors and delinquincies"*: DJQA, 10/31/1846. M12, 277.

939 *"Oh Lord my God"*: "A Prayer Composed in the Sleepless Hours of Last Christmas Night to Close the Year by J. Q. Adams," 12/25/1845. APM.

940 *"Jefferson, a hollow"*: DJQA, 7/26/1845. M12, 206.

940 *"We have enjoyed much"*: DJQA, 7/26/1845. M12, 206.

940 *"Pardon! pardon"*: D&AW2, 763. 4/12/1847.

940 *"the only member"*: DJQA, 7/26/1845. M12, 206.

941 *"a mere necrology"*: DJQA, 12/1/1838. M10, 47.

941 *"But I am wandering"*: DJQA, 7/26/1845. M12, 206.

942 *"My faculties are now"*: DJQA, 8/16/1846. M12, 271.

942 *"outrage upon the laws"*: DJQA, 9/14/1846. M12, 273.

942 *"a redoubled agitation"*: DJQA, 9/14/1846. M12, 273.

942 *"But there is no Christian church"*: DJQA, 8/25/1846. M12, 275.

942 *"checking and controlling"*: DJQA, 4/26/1840. M10, 276.

943 *"with a suspension"*: DJQA, 3/14/1847. M12, 279.

944 *"more powerful"*: Congressional Globe, 2/13/1847.

944 *"a paralytic complaint"*: JQA/Daniel King, 2/26/1847. APM; DJQA, 11/14/1847. M12, 279.

944 *"that I should remember"*: D&AW2, 762. 3/14/1847.

945 *"I have been disabled"*: JQA/CFA, 5/11/1847. APM.

945 *"Spare O Lord"*: D&AW2, 762–763. 4/7/1847.

946 *"I am unable"*: DJQA, 5/4/1847. APM.

946 *"I have got beyond"*: JQA/George H. Colton, 4/17/1847. APM.

947 *"can hardly be called"*: DJQA, 6/9/1847. APM.

947 *"O God in thy great"*: D&AW2, 766. 7/11/1847.

948 *"I shudder to think"*: D&AW2, 767. 7/30/1847.

948 *"passed most quietly"*: D&AW2, 767. 7/30/1847.

948 *"I am almost totally"*: DJQA, 10/9/1847. APM.

948 *"The existence of intelligent"*: DJQA, 10/18/1847. APM.

949 *"cast into the shade"*: DJQA, 10/8/1847. APM.

949 *"on leaving home"*: DJQA, 11/1/1847. APM.

949 *"a bad fall"*: DJQA, 11/22/1847. APM.

952 *"Fair Lady"*: DJQA, 2/20/1848. APM.

953 *"And when thy summons"*: DJQA, 11/4/1837. APM.

953 *"gallantry and military skill"*: Congressional Globe, 2/21/1848.

954 *"A sudden cry was heard"*: Addresses in the Congress of the United States and Funeral Solemnities on the Death of John Quincy Adams Who Died in the Capitol at Washington on Wednesday Evening, February 23, 1848 (Washington, D.C., 1848), p. 3.

954 *"painful announcement"*: Addresses in the Congress, p. 3.

954 *"This is the end"*: Addresses in the Congress, p. 4.

955 *"But neither the skill"*: Addresses in the Congress, p. 5.

955 *"The whole ceremony"*: Benton, Thirty Years' View, vol. 2, p. 709.

955 *"Thy fiat has gone forth"*: D&AW2, 770. 3/18/1849.

The illustrations appear courtesy of the following institutions and collections, with my gratitude:

First section of color plates:
Page 1, top: National Park Service, Adams National Historical Park.
Page 1, bottom: Massachusetts Historical Society.
Page 2, top left and right: Massachusetts Historical Society.
Page 2, bottom: New-York Historical Society.
Page 3, top: Massachusetts Historical Society.
Page 3, bottom: Museum of Fine Arts, Boston.
Page 4, top: Massachusetts Historical Society.
Page 4: bottom: National Park Service, Adams National Historical Park.
Page 5: National Park Service, Adams National Historical Park.
Page 6: Library and Archives Canada.
Page 7: Maps.com.
Page 8, all: The Granger Collection, New York.

Second section of color plates:
Page 1, top: National Gallery of Art, Washington.
Page 1, bottom: National Park Service, Adams National Historical Park.
Page 2, all: National Park Service, Adams National Historical Park.
Page 3: Massachusetts Historical Society.
Page 4, top: White House Historical Association.
Page 4, bottom: National Park Service, Adams National Historical Park.
Page 5, top: National Park Service, Adams National Historical Park.
Page 5, bottom: Library of Congress.
Page 6: Massachusetts Historical Society.
Page 7: Metropolitan Museum of Art, New York.
Page 8, top left: Smithsonian American Art Museum, Washington.
Page 8, top right: Massachusetts Historical Society.
Page 8, bottom: Library of Congress.

About the Author

DATE DUE

PRINTED IN U.S.A.

F red [   ] meritus
   of E[   ] raduate
Center of[   ] e is the
author of[   ] ; which
was name[   ] ew *York
Times* an[   ] r publications. [   ] a final-
ist for the[   ] and the
Pulitzer [